Methods in Enzymology

Volume 308
ENZYME KINETICS AND
MECHANISM
Part E
Energetics of Enzyme Catalysis

METHODS IN ENZYMOLOGY

EDITORS-IN-CHIEF

John N. Abelson   Melvin I. Simon

DIVISION OF BIOLOGY
CALIFORNIA INSTITUTE OF TECHNOLOGY
PASADENA, CALIFORNIA

FOUNDING EDITORS

Sidney P. Colowick and Nathan O. Kaplan

Methods in Enzymology

Volume 308

Enzyme Kinetics and Mechanism

Part E
Energetics of Enzyme Catalysis

EDITED BY

Vern L. Schramm

ALBERT EINSTEIN COLLEGE OF MEDICINE
BRONX, NEW YORK

Daniel L. Purich

UNIVERSITY OF FLORIDA COLLEGE OF MEDICINE
GAINESVILLE, FLORIDA

QP601
C71
v.308
1999

ACADEMIC PRESS

San Diego   London   Boston   New York   Sydney   Tokyo   Toronto

This book is printed on acid-free paper. ∞

Copyright © 1999 by ACADEMIC PRESS

All Rights Reserved.
No part of this publication may be reproduced or transmitted in any form or by any means, electronic or mechanical, including photocopy, recording, or any information storage and retrieval system, without permission in writing from the Publisher.
The appearance of the code at the bottom of the first page of a chapter in this book indicates the Publisher's consent that copies of the chapter may be made for personal or internal use, or for the personal or internal use of specific clients. This consent is given on the condition, however, that the copier pay the stated per copy fee through the Copyright Clearance Center, Inc. (222 Rosewood Drive, Danvers, Massachusetts 01923) for copying beyond that permitted by Sections 107 or 108 of the U.S. Copyright Law. This consent does not extend to other kinds of copying, such as copying for general distribution, for advertising or promotional purposes, for creating new collective works, or for resale. Copy fees for pre-1999 chapters are as shown on the chapter title pages. If no fee code appears on the chapter title page, the copy fee is the same as for current chapters.
0076-6879/99 $30.00

Academic Press
A Harcourt Science and Technology Company
525 B Street, Suite 1900, San Diego, California 92101-4495, USA
http://www.academicpress.com

Academic Press Limited
24-28 Oval Road, London NW1 7DX, UK
http://www.hbuk.co.uk/ap/

International Standard Book Number: 0-12-182209-5

PRINTED IN THE UNITED STATES OF AMERICA
99 00 01 02 03 04 MM 9 8 7 6 5 4 3 2 1

Table of Contents

CONTRIBUTORS TO VOLUME 308   vii

PREFACE   ix

VOLUMES IN SERIES   xi

Section I. Energetic Coupling in Enzymatic Reactions

1. Energetics of Substrate Binding, Catalysis, and Product Release — W. WALLACE CLELAND AND DEXTER B. NORTHROP — 3

2. Energy Coupling Through Molecular Discrimination: Nicotinate Phosphoribosyltransferase — CHARLES TIMMIS GRUBMEYER, JEFFREY W. GROSS, AND MATHUMATHI RAJAVEL — 28

3. On the Advantages of Imperfect Energetic Linkage — THOMAS S. LEYH — 48

4. Reaction Dynamics of G-Protein Catalyzed Hydrolysis of GFP as Viewed by X-Ray Crystallographic Snapshots of $G_{i\alpha 1}$ — DAVID E. COLEMAN AND STEPHEN R. SPRANG — 70

5. Energetics of Nucleotide Hydrolysis in Polymer Assembly/Disassembly: The Cases of Actin and Tubulin — DANIEL L. PURICH AND FREDERICK S. SOUTHWICK — 93

6. Fundamental Mechanisms of Substrate Channeling — KAREN S. ANDERSON — 111

Section II. Intermediates and Complexes in Catalysis

7. Intermediates and Energetics in Pyruvate Phosphate Dikinase — DEBRA DUNAWAY-MARIANO — 149

8. Raman Spectroscopic Studies of the Structures, Energetics, and Bond Distortions of Substrates Bound to Enzymes — HUA DENG AND ROBERT CALLENDER — 176

9. Crystallographic Analysis of Solvent-Trapped Intermediates of Chymotrypsin — GREGORY K. FARBER — 201

v

Section III. Detection and Properties of Low-Barrier Hydrogen Bonds

10. Nuclear Magnetic Resonance Methods for the Detection and Study of Low-Barrier Hydrogen Bonds of Enzymes — ALBERT S. MILDVAN, THOMAS K. HARRIS, AND CHITRANANDA ABEYGUNAWARDANA   219

11. Hydrogen Bonding in Enzymatic Catalysis: Analysis of Energetic Contributions — SHU-OU SHAN AND DANIEL HERSCHLAG   246

12. Application of Marcus Rate Theory to Proton Transfer in Enzyme-Catalyzed Reactions — A. JERRY KRESGE AND DAVID N. SILVERMAN   276

Section IV. Transition State Determination and Inhibitors

13. Enzymatic Transition-State Analysis and Transition-State Analogues — VERN L. SCHRAMM   301

14. Determining Transition States from Kinetic Isotope Effects — PAUL J. BERTI   355

15. Computational Methods for Transition State and Inhibitor Recognition — BENJAMIN B. BRAUNHEIM AND STEVEN D. SCHWARTZ   398

AUTHOR INDEX 427

SUBJECT INDEX 447

Contributors to Volume 308

Article numbers are in parentheses following the names of contributors.
Affiliations listed are current.

CHITRANANDA ABEYGUNAWARDANA (10), *Department of Biological Chemistry, Johns Hopkins University School of Medicine, Baltimore, Maryland 21205*

KAREN S. ANDERSON (6), *Department of Pharmacology, Yale University School of Medicine, New Haven, Connecticut 06520-8066*

PAUL J. BERTI (14), *Department of Biochemistry, Albert Einstein College of Medicine, Bronx, New York 10461*

BENJAMIN B. BRAUNHEIM (15), *Department of Physiology and Biophysics, Albert Einstein College of Medicine, Bronx, New York 10461*

ROBERT CALLENDER (8), *Department of Biochemistry, Albert Einstein College of Medicine, Bronx, New York 10461*

W. WALLACE CLELAND (1), *Institute for Enzyme Research, University of Wisconsin, Madison, Wisconsin 53705-4098*

DAVID E. COLEMAN (4), *Department of Biochemistry, University of Texas Southwestern Medical Center, Dallas, Texas 75235-9050*

HUA DENG (8), *Department of Biochemistry, Albert Einstein College of Medicine, Bronx, New York 10461*

DEBRA DUNAWAY-MARIANO (7), *Department of Chemistry, University of New Mexico, Albuquerque, New Mexico 87131*

GREGORY K. FARBER (9), *Department of Biochemistry and Molecular Biology, Pennsylvania State University, University Park, Pennsylvania 16802*

JEFFREY W. GROSS (2), *Department of Biochemistry and Fels Research Institute, Temple University School of Medicine, Philadelphia, Pennsylvania 19140*

CHARLES TIMMIS GRUBMEYER (2), *Department of Biochemistry and Fels Research Institute, Temple University School of Medicine, Philadelphia, Pennsylvania 19140*

THOMAS K. HARRIS (10), *Department of Biological Chemistry, Johns Hopkins University School of Medicine, Baltimore, Maryland 21205*

DANIEL HERSCHLAG (11), *Department of Biochemistry, Stanford University, Stanford, California 94305-5307*

A. JERRY KRESGE (12), *Department of Chemistry, University of Toronto, Toronto, Ontario MSS 3H6, Canada*

THOMAS S. LEYH (3), *Department of Biochemistry, Albert Einstein College of Medicine, Bronx, New York 10461*

ALBERT S. MILDVAN (10), *Department of Biological Chemistry, Johns Hopkins University School of Medicine, Baltimore, Maryland 21205*

DEXTER B. NORTHROP (1), *School of Pharmacy, University of Wisconsin, Madison, Wisconsin 53706*

DANIEL L. PURICH (5), *Department of Biochemistry and Molecular Biology, University of Florida College of Medicine, Gainesville, Florida 32610-0245*

MATHUMATHI RAJAVEL (2), *Department of Biochemistry and Fels Research Institute, Temple University School of Medicine, Philadelphia, Pennsylvania 19140*

VERN L. SCHRAMM (13), *Department of Biochemistry, Albert Einstein College of Medicine, Bronx, New York 10461*

STEVEN D. SCHWARTZ (15), *Departments of Physiology and Biophysics, and Biochemistry, Albert Einstein College of Medicine, Bronx, New York 10461*

SHU-OU SHAN (11), *Department of Biochemistry, Stanford University, Stanford, California 94305-5307*

DAVID N. SILVERMAN (12), *Department of Pharmacology, University of Florida College of Medicine, Gainesville, Florida 32610-0267*

FREDERICK S. SOUTHWICK (5), *Division of Infectious Diseases, Department of Medicine, University of Florida College of Medicine, Gainesville, Florida 32610-0277*

STEPHEN R. SPRANG (4), *Department of Biochemistry, Howard Hughes Medical Institute, University of Texas Southwestern Medical Center, Dallas, Texas 75235-9050*

Preface

The catalytic action of enzymes invokes wonder among enzymologists, fear and loathing in students, and hope and anticipation in venture capitalists. We hope that this volume of *Methods in Enzymology* (Part E of Enzyme Kinetics and Mechanism) will provide some solace for all of these readers. Parts A through D have been widely adopted as primary sources for graduate education in enzymology. For experimentalists, they have served as a staple of reference material for techniques to ferret out the truth from obfuscating enzymatic reactions. In Part E, we have attempted to emphasize several aspects of enzymology that have developed substantially since 1995 when Part D was published.

Stoichiometric energy coupling in catalysis has provided the classical thermodynamic explanation for established equilibria, substrate, and product flow. Recently it became apparent that a group of enzymes has evolved that disregard the matched stoichiometry of energy coupling. These enzymes use nonstoichiometric coupling of energetic substrates to drive unfavorable reactions or to accomplish macromolecular assembly or motion. The Energetic Coupling in Enzymatic Reactions section in this volume provides an introduction to what is certain to be an expansionist area of enzymology. The rapid advance and application of crystallographic techniques have illuminated the nature of protein–protein contacts as well as those of enzyme substrates, products, and inhibitors. Combined with rapid reaction methods, the results provide novel information on the nature of enzyme intermediates, complexes which can be trapped in protein channels or on covalent carriers to translocate reactive intermediates. Substrate channeling is no longer hypothetical, and the several aspects of the basis for sequestering reactants from solvent are summarized in the Intermediates and Complexes in Catalysis section. The publication of Part D coincided with the debut of the role of short hydrogen bonds in enzymatic catalysis. Evidence for the presence of substrate–enzyme interactions, often called "low-barrier hydrogen bonds," has accumulated at a rapid pace, and a section on Detection and Properties of Low-Barrier Hydrogen Bonds covers only the first ground of what promises to provide both insight and controversy in understanding catalysis by enzymes. Earlier Enzyme Kinetics and Mechanism volumes have contributed to the development of kinetic isotope effects as our front-line tool to understand enzymatic transition states. These advances are updated in the section on Transition State Determination and Inhibitors. The chapters document advances in both theory and practice. The first generation of transition state inhibitors have now

ix

been obtained from this technology, promising growing interest in experimental approaches to transition states and the predictive use of transition state information to design powerful inhibitors. We anticipate that this volume will serve as a stepping stone to the next generation of advances that continue to delight and surprise those with an interest in mechanistic enzymology.

VERN L. SCHRAMM
DANIEL L. PURICH

METHODS IN ENZYMOLOGY

VOLUME I. Preparation and Assay of Enzymes
Edited by SIDNEY P. COLOWICK AND NATHAN O. KAPLAN

VOLUME II. Preparation and Assay of Enzymes
Edited by SIDNEY P. COLOWICK AND NATHAN O. KAPLAN

VOLUME III. Preparation and Assay of Substrates
Edited by SIDNEY P. COLOWICK AND NATHAN O. KAPLAN

VOLUME IV. Special Techniques for the Enzymologist
Edited by SIDNEY P. COLOWICK AND NATHAN O. KAPLAN

VOLUME V. Preparation and Assay of Enzymes
Edited by SIDNEY P. COLOWICK AND NATHAN O. KAPLAN

VOLUME VI. Preparation and Assay of Enzymes (*Continued*)
Preparation and Assay of Substrates
Special Techniques
Edited by SIDNEY P. COLOWICK AND NATHAN O. KAPLAN

VOLUME VII. Cumulative Subject Index
Edited by SIDNEY P. COLOWICK AND NATHAN O. KAPLAN

VOLUME VIII. Complex Carbohydrates
Edited by ELIZABETH F. NEUFELD AND VICTOR GINSBURG

VOLUME IX. Carbohydrate Metabolism
Edited by WILLIS A. WOOD

VOLUME X. Oxidation and Phosphorylation
Edited by RONALD W. ESTABROOK AND MAYNARD E. PULLMAN

VOLUME XI. Enzyme Structure
Edited by C. H. W. HIRS

VOLUME XII. Nucleic Acids (Parts A and B)
Edited by LAWRENCE GROSSMAN AND KIVIE MOLDAVE

VOLUME XIII. Citric Acid Cycle
Edited by J. M. LOWENSTEIN

VOLUME XIV. Lipids
Edited by J. M. LOWENSTEIN

VOLUME XV. Steroids and Terpenoids
Edited by RAYMOND B. CLAYTON

VOLUME XVI. Fast Reactions
Edited by KENNETH KUSTIN

VOLUME XVII. Metabolism of Amino Acids and Amines (Parts A and B)
Edited by HERBERT TABOR AND CELIA WHITE TABOR

VOLUME XVIII. Vitamins and Coenzymes (Parts A, B, and C)
Edited by DONALD B. MCCORMICK AND LEMUEL D. WRIGHT

VOLUME XIX. Proteolytic Enzymes
Edited by GERTRUDE E. PERLMANN AND LASZLO LORAND

VOLUME XX. Nucleic Acids and Protein Synthesis (Part C)
Edited by KIVIE MOLDAVE AND LAWRENCE GROSSMAN

VOLUME XXI. Nucleic Acids (Part D)
Edited by LAWRENCE GROSSMAN AND KIVIE MOLDAVE

VOLUME XXII. Enzyme Purification and Related Techniques
Edited by WILLIAM B. JAKOBY

VOLUME XXIII. Photosynthesis (Part A)
Edited by ANTHONY SAN PIETRO

VOLUME XXIV. Photosynthesis and Nitrogen Fixation (Part B)
Edited by ANTHONY SAN PIETRO

VOLUME XXV. Enzyme Structure (Part B)
Edited by C. H. W. HIRS AND SERGE N. TIMASHEFF

VOLUME XXVI. Enzyme Structure (Part C)
Edited by C. H. W. HIRS AND SERGE N. TIMASHEFF

VOLUME XXVII. Enzyme Structure (Part D)
Edited by C. H. W. HIRS AND SERGE N. TIMASHEFF

VOLUME XXVIII. Complex Carbohydrates (Part B)
Edited by VICTOR GINSBURG

VOLUME XXIX. Nucleic Acids and Protein Synthesis (Part E)
Edited by LAWRENCE GROSSMAN AND KIVIE MOLDAVE

VOLUME XXX. Nucleic Acids and Protein Synthesis (Part F)
Edited by KIVIE MOLDAVE AND LAWRENCE GROSSMAN

VOLUME XXXI. Biomembranes (Part A)
Edited by SIDNEY FLEISCHER AND LESTER PACKER

VOLUME XXXII. Biomembranes (Part B)
Edited by SIDNEY FLEISCHER AND LESTER PACKER

VOLUME XXXIII. Cumulative Subject Index Volumes I–XXX
Edited by MARTHA G. DENNIS AND EDWARD A. DENNIS

VOLUME XXXIV. Affinity Techniques (Enzyme Purification: Part B)
Edited by WILLIAM B. JAKOBY AND MEIR WILCHEK

VOLUME XXXV. Lipids (Part B)
Edited by JOHN M. LOWENSTEIN

VOLUME XXXVI. Hormone Action (Part A: Steroid Hormones)
Edited by BERT W. O'MALLEY AND JOEL G. HARDMAN

VOLUME XXXVII. Hormone Action (Part B: Peptide Hormones)
Edited by BERT W. O'MALLEY AND JOEL G. HARDMAN

VOLUME XXXVIII. Hormone Action (Part C: Cyclic Nucleotides)
Edited by JOEL G. HARDMAN AND BERT W. O'MALLEY

VOLUME XXXIX. Hormone Action (Part D: Isolated Cells, Tissues, and Organ Systems)
Edited by JOEL G. HARDMAN AND BERT W. O'MALLEY

VOLUME XL. Hormone Action (Part E: Nuclear Structure and Function)
Edited by BERT W. O'MALLEY AND JOEL G. HARDMAN

VOLUME XLI. Carbohydrate Metabolism (Part B)
Edited by W. A. WOOD

VOLUME XLII. Carbohydrate Metabolism (Part C)
Edited by W. A. WOOD

VOLUME XLIII. Antibiotics
Edited by JOHN H. HASH

VOLUME XLIV. Immobilized Enzymes
Edited by KLAUS MOSBACH

VOLUME XLV. Proteolytic Enzymes (Part B)
Edited by LASZLO LORAND

VOLUME XLVI. Affinity Labeling
Edited by WILLIAM B. JAKOBY AND MEIR WILCHEK

VOLUME XLVII. Enzyme Structure (Part E)
Edited by C. H. W. HIRS AND SERGE N. TIMASHEFF

VOLUME XLVIII. Enzyme Structure (Part F)
Edited by C. H. W. HIRS AND SERGE N. TIMASHEFF

VOLUME XLIX. Enzyme Structure (Part G)
Edited by C. H. W. HIRS AND SERGE N. TIMASHEFF

VOLUME L. Complex Carbohydrates (Part C)
Edited by VICTOR GINSBURG

VOLUME LI. Purine and Pyrimidine Nucleotide Metabolism
Edited by PATRICIA A. HOFFEE AND MARY ELLEN JONES

VOLUME LII. Biomembranes (Part C: Biological Oxidations)
Edited by SIDNEY FLEISCHER AND LESTER PACKER

VOLUME LIII. Biomembranes (Part D: Biological Oxidations)
Edited by SIDNEY FLEISCHER AND LESTER PACKER

VOLUME LIV. Biomembranes (Part E: Biological Oxidations)
Edited by SIDNEY FLEISCHER AND LESTER PACKER

VOLUME LV. Biomembranes (Part F: Bioenergetics)
Edited by SIDNEY FLEISCHER AND LESTER PACKER

VOLUME LVI. Biomembranes (Part G: Bioenergetics)
Edited by SIDNEY FLEISCHER AND LESTER PACKER

VOLUME LVII. Bioluminescence and Chemiluminescence
Edited by MARLENE A. DELUCA

VOLUME LVIII. Cell Culture
Edited by WILLIAM B. JAKOBY AND IRA PASTAN

VOLUME LIX. Nucleic Acids and Protein Synthesis (Part G)
Edited by KIVIE MOLDAVE AND LAWRENCE GROSSMAN

VOLUME LX. Nucleic Acids and Protein Synthesis (Part H)
Edited by KIVIE MOLDAVE AND LAWRENCE GROSSMAN

VOLUME 61. Enzyme Structure (Part H)
Edited by C. H. W. HIRS AND SERGE N. TIMASHEFF

VOLUME 62. Vitamins and Coenzymes (Part D)
Edited by DONALD B. MCCORMICK AND LEMUEL D. WRIGHT

VOLUME 63. Enzyme Kinetics and Mechanism (Part A: Initial Rate and Inhibitor
Methods)
Edited by DANIEL L. PURICH

VOLUME 64. Enzyme Kinetics and Mechanism (Part B: Isotopic Probes and Complex Enzyme Systems)
Edited by DANIEL L. PURICH

VOLUME 65. Nucleic Acids (Part I)
Edited by LAWRENCE GROSSMAN AND KIVIE MOLDAVE

VOLUME 66. Vitamins and Coenzymes (Part E)
Edited by DONALD B. MCCORMICK AND LEMUEL D. WRIGHT

VOLUME 67. Vitamins and Coenzymes (Part F)
Edited by DONALD B. MCCORMICK AND LEMUEL D. WRIGHT

VOLUME 68. Recombinant DNA
Edited by RAY WU

VOLUME 69. Photosynthesis and Nitrogen Fixation (Part C)
Edited by ANTHONY SAN PIETRO

VOLUME 70. Immunochemical Techniques (Part A)
Edited by HELEN VAN VUNAKIS AND JOHN J. LANGONE

VOLUME 71. Lipids (Part C)
Edited by JOHN M. LOWENSTEIN

VOLUME 72. Lipids (Part D)
Edited by JOHN M. LOWENSTEIN

VOLUME 73. Immunochemical Techniques (Part B)
Edited by JOHN J. LANGONE AND HELEN VAN VUNAKIS

VOLUME 74. Immunochemical Techniques (Part C)
Edited by JOHN J. LANGONE AND HELEN VAN VUNAKIS

VOLUME 75. Cumulative Subject Index Volumes XXXI, XXXII, XXXIV–LX
Edited by EDWARD A. DENNIS AND MARTHA G. DENNIS

VOLUME 76. Hemoglobins
Edited by ERALDO ANTONINI, LUIGI ROSSI-BERNARDI, AND EMILIA CHIANCONE

VOLUME 77. Detoxication and Drug Metabolism
Edited by WILLIAM B. JAKOBY

VOLUME 78. Interferons (Part A)
Edited by SIDNEY PESTKA

VOLUME 79. Interferons (Part B)
Edited by SIDNEY PESTKA

VOLUME 80. Proteolytic Enzymes (Part C)
Edited by LASZLO LORAND

VOLUME 81. Biomembranes (Part H: Visual Pigments and Purple Membranes, I)
Edited by LESTER PACKER

VOLUME 82. Structural and Contractile Proteins (Part A: Extracellular Matrix)
Edited by LEON W. CUNNINGHAM AND DIXIE W. FREDERIKSEN

VOLUME 83. Complex Carbohydrates (Part D)
Edited by VICTOR GINSBURG

VOLUME 84. Immunochemical Techniques (Part D: Selected Immunoassays)
Edited by JOHN J. LANGONE AND HELEN VAN VUNAKIS

VOLUME 85. Structural and Contractile Proteins (Part B: The Contractile Apparatus and the Cytoskeleton)
Edited by DIXIE W. FREDERIKSEN AND LEON W. CUNNINGHAM

VOLUME 86. Prostaglandins and Arachidonate Metabolites
Edited by WILLIAM E. M. LANDS AND WILLIAM L. SMITH

VOLUME 87. Enzyme Kinetics and Mechanism (Part C: Intermediates, Stereochemistry, and Rate Studies)
Edited by DANIEL L. PURICH

VOLUME 88. Biomembranes (Part I: Visual Pigments and Purple Membranes, II)
Edited by LESTER PACKER

VOLUME 89. Carbohydrate Metabolism (Part D)
Edited by WILLIS A. WOOD

VOLUME 90. Carbohydrate Metabolism (Part E)
Edited by WILLIS A. WOOD

VOLUME 91. Enzyme Structure (Part I)
Edited by C. H. W. HIRS AND SERGE N. TIMASHEFF

VOLUME 92. Immunochemical Techniques (Part E: Monoclonal Antibodies and General Immunoassay Methods)
Edited by JOHN J. LANGONE AND HELEN VAN VUNAKIS

VOLUME 93. Immunochemical Techniques (Part F: Conventional Antibodies, Fc Receptors, and Cytotoxicity)
Edited by JOHN J. LANGONE AND HELEN VAN VUNAKIS

VOLUME 94. Polyamines
Edited by HERBERT TABOR AND CELIA WHITE TABOR

VOLUME 95. Cumulative Subject Index Volumes 61–74, 76–80
Edited by EDWARD A. DENNIS AND MARTHA G. DENNIS

VOLUME 96. Biomembranes [Part J: Membrane Biogenesis: Assembly and Targeting (General Methods; Eukaryotes)]
Edited by SIDNEY FLEISCHER AND BECCA FLEISCHER

VOLUME 97. Biomembranes [Part K: Membrane Biogenesis: Assembly and Targeting (Prokaryotes, Mitochondria, and Chloroplasts)]
Edited by SIDNEY FLEISCHER AND BECCA FLEISCHER

VOLUME 98. Biomembranes (Part L: Membrane Biogenesis: Processing and Recycling)
Edited by SIDNEY FLEISCHER AND BECCA FLEISCHER

VOLUME 99. Hormone Action (Part F: Protein Kinases)
Edited by JACKIE D. CORBIN AND JOEL G. HARDMAN

VOLUME 100. Recombinant DNA (Part B)
Edited by RAY WU, LAWRENCE GROSSMAN, AND KIVIE MOLDAVE

VOLUME 101. Recombinant DNA (Part C)
Edited by RAY WU, LAWRENCE GROSSMAN, AND KIVIE MOLDAVE

VOLUME 102. Hormone Action (Part G: Calmodulin and Calcium-Binding Proteins)
Edited by ANTHONY R. MEANS AND BERT W. O'MALLEY

VOLUME 103. Hormone Action (Part H: Neuroendocrine Peptides)
Edited by P. MICHAEL CONN

VOLUME 104. Enzyme Purification and Related Techniques (Part C)
Edited by WILLIAM B. JAKOBY

VOLUME 105. Oxygen Radicals in Biological Systems
Edited by LESTER PACKER

VOLUME 106. Posttranslational Modifications (Part A)
Edited by FINN WOLD AND KIVIE MOLDAVE

VOLUME 107. Posttranslational Modifications (Part B)
Edited by FINN WOLD AND KIVIE MOLDAVE

VOLUME 108. Immunochemical Techniques (Part G: Separation and Characterization of Lymphoid Cells)
Edited by GIOVANNI DI SABATO, JOHN J. LANGONE, AND HELEN VAN VUNAKIS

VOLUME 109. Hormone Action (Part I: Peptide Hormones)
Edited by LUTZ BIRNBAUMER AND BERT W. O'MALLEY

VOLUME 110. Steroids and Isoprenoids (Part A)
Edited by JOHN H. LAW AND HANS C. RILLING

VOLUME 111. Steroids and Isoprenoids (Part B)
Edited by JOHN H. LAW AND HANS C. RILLING

VOLUME 112. Drug and Enzyme Targeting (Part A)
Edited by KENNETH J. WIDDER AND RALPH GREEN

VOLUME 113. Glutamate, Glutamine, Glutathione, and Related Compounds
Edited by ALTON MEISTER

VOLUME 114. Diffraction Methods for Biological Macromolecules (Part A)
Edited by HAROLD W. WYCKOFF, C. H. W. HIRS, AND SERGE N. TIMASHEFF

VOLUME 115. Diffraction Methods for Biological Macromolecules (Part B)
Edited by HAROLD W. WYCKOFF, C. H. W. HIRS, AND SERGE N. TIMASHEFF

VOLUME 116. Immunochemical Techniques (Part H: Effectors and Mediators of Lymphoid Cell Functions)
Edited by GIOVANNI DI SABATO, JOHN J. LANGONE, AND HELEN VAN VUNAKIS

VOLUME 117. Enzyme Structure (Part J)
Edited by C. H. W. HIRS AND SERGE N. TIMASHEFF

VOLUME 118. Plant Molecular Biology
Edited by ARTHUR WEISSBACH AND HERBERT WEISSBACH

VOLUME 119. Interferons (Part C)
Edited by SIDNEY PESTKA

VOLUME 120. Cumulative Subject Index Volumes 81–94, 96–101

VOLUME 121. Immunochemical Techniques (Part I: Hybridoma Technology and Monoclonal Antibodies)
Edited by JOHN J. LANGONE AND HELEN VAN VUNAKIS

VOLUME 122. Vitamins and Coenzymes (Part G)
Edited by FRANK CHYTIL AND DONALD B. McCORMICK

VOLUME 123. Vitamins and Coenzymes (Part H)
Edited by FRANK CHYTIL AND DONALD B. McCORMICK

VOLUME 124. Hormone Action (Part J: Neuroendocrine Peptides)
Edited by P. MICHAEL CONN

VOLUME 125. Biomembranes (Part M: Transport in Bacteria, Mitochondria, and Chloroplasts: General Approaches and Transport Systems)
Edited by SIDNEY FLEISCHER AND BECCA FLEISCHER

VOLUME 126. Biomembranes (Part N: Transport in Bacteria, Mitochondria, and Chloroplasts: Protonmotive Force)
Edited by SIDNEY FLEISCHER AND BECCA FLEISCHER

VOLUME 127. Biomembranes (Part O: Protons and Water: Structure and Translocation)
Edited by LESTER PACKER

VOLUME 128. Plasma Lipoproteins (Part A: Preparation, Structure, and Molecular Biology)
Edited by JERE P. SEGREST AND JOHN J. ALBERS

VOLUME 129. Plasma Lipoproteins (Part B: Characterization, Cell Biology, and Metabolism)
Edited by JOHN J. ALBERS AND JERE P. SEGREST

VOLUME 130. Enzyme Structure (Part K)
Edited by C. H. W. HIRS AND SERGE N. TIMASHEFF

VOLUME 131. Enzyme Structure (Part L)
Edited by C. H. W. HIRS AND SERGE N. TIMASHEFF

VOLUME 132. Immunochemical Techniques (Part J: Phagocytosis and Cell-Mediated Cytotoxicity)
Edited by GIOVANNI DI SABATO AND JOHANNES EVERSE

VOLUME 133. Bioluminescence and Chemiluminescence (Part B)
Edited by MARLENE DELUCA AND WILLIAM D. MCELROY

VOLUME 134. Structural and Contractile Proteins (Part C: The Contractile Apparatus and the Cytoskeleton)
Edited by RICHARD B. VALLEE

VOLUME 135. Immobilized Enzymes and Cells (Part B)
Edited by KLAUS MOSBACH

VOLUME 136. Immobilized Enzymes and Cells (Part C)
Edited by KLAUS MOSBACH

VOLUME 137. Immobilized Enzymes and Cells (Part D)
Edited by KLAUS MOSBACH

VOLUME 138. Complex Carbohydrates (Part E)
Edited by VICTOR GINSBURG

VOLUME 139. Cellular Regulators (Part A: Calcium- and Calmodulin-Binding Proteins)
Edited by ANTHONY R. MEANS AND P. MICHAEL CONN

VOLUME 140. Cumulative Subject Index Volumes 102–119, 121–134

VOLUME 141. Cellular Regulators (Part B: Calcium and Lipids)
Edited by P. MICHAEL CONN AND ANTHONY R. MEANS

VOLUME 142. Metabolism of Aromatic Amino Acids and Amines
Edited by SEYMOUR KAUFMAN

VOLUME 143. Sulfur and Sulfur Amino Acids
Edited by WILLIAM B. JAKOBY AND OWEN GRIFFITH

VOLUME 144. Structural and Contractile Proteins (Part D: Extracellular Matrix)
Edited by LEON W. CUNNINGHAM

VOLUME 145. Structural and Contractile Proteins (Part E: Extracellular Matrix)
Edited by LEON W. CUNNINGHAM

VOLUME 146. Peptide Growth Factors (Part A)
Edited by DAVID BARNES AND DAVID A. SIRBASKU

VOLUME 147. Peptide Growth Factors (Part B)
Edited by DAVID BARNES AND DAVID A. SIRBASKU

VOLUME 148. Plant Cell Membranes
Edited by LESTER PACKER AND ROLAND DOUCE

VOLUME 149. Drug and Enzyme Targeting (Part B)
Edited by RALPH GREEN AND KENNETH J. WIDDER

VOLUME 150. Immunochemical Techniques (Part K: *In Vitro* Models of B and T Cell Functions and Lymphoid Cell Receptors)
Edited by GIOVANNI DI SABATO

VOLUME 151. Molecular Genetics of Mammalian Cells
Edited by MICHAEL M. GOTTESMAN

VOLUME 152. Guide to Molecular Cloning Techniques
Edited by SHELBY L. BERGER AND ALAN R. KIMMEL

VOLUME 153. Recombinant DNA (Part D)
Edited by RAY WU AND LAWRENCE GROSSMAN

VOLUME 154. Recombinant DNA (Part E)
Edited by RAY WU AND LAWRENCE GROSSMAN

VOLUME 155. Recombinant DNA (Part F)
Edited by RAY WU

VOLUME 156. Biomembranes (Part P: ATP-Driven Pumps and Related Transport: The Na,K-Pump)
Edited by SIDNEY FLEISCHER AND BECCA FLEISCHER

VOLUME 157. Biomembranes (Part Q: ATP-Driven Pumps and Related Transport: Calcium, Proton, and Potassium Pumps)
Edited by SIDNEY FLEISCHER AND BECCA FLEISCHER

VOLUME 158. Metalloproteins (Part A)
Edited by JAMES F. RIORDAN AND BERT L. VALLEE

VOLUME 159. Initiation and Termination of Cyclic Nucleotide Action
Edited by JACKIE D. CORBIN AND ROGER A. JOHNSON

VOLUME 160. Biomass (Part A: Cellulose and Hemicellulose)
Edited by WILLIS A. WOOD AND SCOTT T. KELLOGG

VOLUME 161. Biomass (Part B: Lignin, Pectin, and Chitin)
Edited by WILLIS A. WOOD AND SCOTT T. KELLOGG

VOLUME 162. Immunochemical Techniques (Part L: Chemotaxis and Inflammation)
Edited by GIOVANNI DI SABATO

VOLUME 163. Immunochemical Techniques (Part M: Chemotaxis and Inflammation)
Edited by GIOVANNI DI SABATO

VOLUME 164. Ribosomes
Edited by HARRY F. NOLLER, JR., AND KIVIE MOLDAVE

VOLUME 165. Microbial Toxins: Tools for Enzymology
Edited by SIDNEY HARSHMAN

VOLUME 166. Branched-Chain Amino Acids
Edited by ROBERT HARRIS AND JOHN R. SOKATCH

VOLUME 167. Cyanobacteria
Edited by LESTER PACKER AND ALEXANDER N. GLAZER

VOLUME 168. Hormone Action (Part K: Neuroendocrine Peptides)
Edited by P. MICHAEL CONN

VOLUME 169. Platelets: Receptors, Adhesion, Secretion (Part A)
Edited by JACEK HAWIGER

VOLUME 170. Nucleosomes
Edited by PAUL M. WASSARMAN AND ROGER D. KORNBERG

VOLUME 171. Biomembranes (Part R: Transport Theory: Cells and Model Membranes)
Edited by SIDNEY FLEISCHER AND BECCA FLEISCHER

VOLUME 172. Biomembranes (Part S: Transport: Membrane Isolation and Characterization)
Edited by SIDNEY FLEISCHER AND BECCA FLEISCHER

VOLUME 173. Biomembranes [Part T: Cellular and Subcellular Transport: Eukaryotic (Nonepithelial) Cells]
Edited by SIDNEY FLEISCHER AND BECCA FLEISCHER

VOLUME 174. Biomembranes [Part U: Cellular and Subcellular Transport: Eukaryotic (Nonepithelial) Cells]
Edited by SIDNEY FLEISCHER AND BECCA FLEISCHER

VOLUME 175. Cumulative Subject Index Volumes 135–139, 141–167

VOLUME 176. Nuclear Magnetic Resonance (Part A: Spectral Techniques and Dynamics)
Edited by NORMAN J. OPPENHEIMER AND THOMAS L. JAMES

VOLUME 177. Nuclear Magnetic Resonance (Part B: Structure and Mechanism)
Edited by NORMAN J. OPPENHEIMER AND THOMAS L. JAMES

VOLUME 178. Antibodies, Antigens, and Molecular Mimicry
Edited by JOHN J. LANGONE

VOLUME 179. Complex Carbohydrates (Part F)
Edited by VICTOR GINSBURG

VOLUME 180. RNA Processing (Part A: General Methods)
Edited by JAMES E. DAHLBERG AND JOHN N. ABELSON

VOLUME 181. RNA Processing (Part B: Specific Methods)
Edited by JAMES E. DAHLBERG AND JOHN N. ABELSON

VOLUME 182. Guide to Protein Purification
Edited by MURRAY P. DEUTSCHER

VOLUME 183. Molecular Evolution: Computer Analysis of Protein and Nucleic Acid Sequences
Edited by RUSSELL F. DOOLITTLE

VOLUME 184. Avidin–Biotin Technology
Edited by MEIR WILCHEK AND EDWARD A. BAYER

VOLUME 185. Gene Expression Technology
Edited by DAVID V. GOEDDEL

VOLUME 186. Oxygen Radicals in Biological Systems (Part B: Oxygen Radicals and Antioxidants)
Edited by LESTER PACKER AND ALEXANDER N. GLAZER

VOLUME 187. Arachidonate Related Lipid Mediators
Edited by ROBERT C. MURPHY AND FRANK A. FITZPATRICK

VOLUME 188. Hydrocarbons and Methylotrophy
Edited by MARY E. LIDSTROM

VOLUME 189. Retinoids (Part A: Molecular and Metabolic Aspects)
Edited by LESTER PACKER

VOLUME 190. Retinoids (Part B: Cell Differentiation and Clinical Applications)
Edited by LESTER PACKER

VOLUME 191. Biomembranes (Part V: Cellular and Subcellular Transport: Epithelial Cells)
Edited by SIDNEY FLEISCHER AND BECCA FLEISCHER

VOLUME 192. Biomembranes (Part W: Cellular and Subcellular Transport: Epithelial Cells)
Edited by SIDNEY FLEISCHER AND BECCA FLEISCHER

VOLUME 193. Mass Spectrometry
Edited by JAMES A. MCCLOSKEY

VOLUME 194. Guide to Yeast Genetics and Molecular Biology
Edited by CHRISTINE GUTHRIE AND GERALD R. FINK

VOLUME 195. Adenylyl Cyclase, G Proteins, and Guanylyl Cyclase
Edited by ROGER A. JOHNSON AND JACKIE D. CORBIN

VOLUME 196. Molecular Motors and the Cytoskeleton
Edited by RICHARD B. VALLEE

VOLUME 197. Phospholipases
Edited by EDWARD A. DENNIS

VOLUME 198. Peptide Growth Factors (Part C)
Edited by DAVID BARNES, J. P. MATHER, AND GORDON H. SATO

VOLUME 199. Cumulative Subject Index Volumes 168–174, 176–194

VOLUME 200. Protein Phosphorylation (Part A: Protein Kinases: Assays, Purification, Antibodies, Functional Analysis, Cloning, and Expression)
Edited by TONY HUNTER AND BARTHOLOMEW M. SEFTON

VOLUME 201. Protein Phosphorylation (Part B: Analysis of Protein Phosphorylation, Protein Kinase Inhibitors, and Protein Phosphatases)
Edited by TONY HUNTER AND BARTHOLOMEW M. SEFTON

VOLUME 202. Molecular Design and Modeling: Concepts and Applications (Part A: Proteins, Peptides, and Enzymes)
Edited by JOHN J. LANGONE

VOLUME 203. Molecular Design and Modeling: Concepts and Applications (Part B: Antibodies and Antigens, Nucleic Acids, Polysaccharides, and Drugs)
Edited by JOHN J. LANGONE

VOLUME 204. Bacterial Genetic Systems
Edited by JEFFREY H. MILLER

VOLUME 205. Metallobiochemistry (Part B: Metallothionein and Related Molecules)
Edited by JAMES F. RIORDAN AND BERT L. VALLEE

VOLUME 206. Cytochrome P450
Edited by MICHAEL R. WATERMAN AND ERIC F. JOHNSON

VOLUME 207. Ion Channels
Edited by BERNARDO RUDY AND LINDA E. IVERSON

VOLUME 208. Protein–DNA Interactions
Edited by ROBERT T. SAUER

VOLUME 209. Phospholipid Biosynthesis
Edited by EDWARD A. DENNIS AND DENNIS E. VANCE

VOLUME 210. Numerical Computer Methods
Edited by LUDWIG BRAND AND MICHAEL L. JOHNSON

VOLUME 211. DNA Structures (Part A: Synthesis and Physical Analysis of DNA)
Edited by DAVID M. J. LILLEY AND JAMES E. DAHLBERG

VOLUME 212. DNA Structures (Part B: Chemical and Electrophoretic Analysis of DNA)
Edited by DAVID M. J. LILLEY AND JAMES E. DAHLBERG

VOLUME 213. Carotenoids (Part A: Chemistry, Separation, Quantitation, and Antioxidation)
Edited by LESTER PACKER

VOLUME 214. Carotenoids (Part B: Metabolism, Genetics, and Biosynthesis)
Edited by LESTER PACKER

VOLUME 215. Platelets: Receptors, Adhesion, Secretion (Part B)
Edited by JACEK J. HAWIGER

VOLUME 216. Recombinant DNA (Part G)
Edited by RAY WU

VOLUME 217. Recombinant DNA (Part H)
Edited by RAY WU

VOLUME 218. Recombinant DNA (Part I)
Edited by RAY WU

VOLUME 219. Reconstitution of Intracellular Transport
Edited by JAMES E. ROTHMAN

VOLUME 220. Membrane Fusion Techniques (Part A)
Edited by NEJAT DÜZGÜNEŞ

VOLUME 221. Membrane Fusion Techniques (Part B)
Edited by NEJAT DÜZGÜNEŞ

VOLUME 222. Proteolytic Enzymes in Coagulation, Fibrinolysis, and Complement Activation (Part A: Mammalian Blood Coagulation Factors and Inhibitors)
Edited by LASZLO LORAND AND KENNETH G. MANN

VOLUME 223. Proteolytic Enzymes in Coagulation, Fibrinolysis, and Complement Activation (Part B: Complement Activation, Fibrinolysis, and Nonmammalian Blood Coagulation Factors)
Edited by LASZLO LORAND AND KENNETH G. MANN

VOLUME 224. Molecular Evolution: Producing the Biochemical Data
Edited by ELIZABETH ANNE ZIMMER, THOMAS J. WHITE, REBECCA L. CANN, AND ALLAN C. WILSON

VOLUME 225. Guide to Techniques in Mouse Development
Edited by PAUL M. WASSARMAN AND MELVIN L. DEPAMPHILIS

VOLUME 226. Metallobiochemistry (Part C: Spectroscopic and Physical Methods for Probing Metal Ion Environments in Metalloenzymes and Metalloproteins)
Edited by JAMES F. RIORDAN AND BERT L. VALLEE

VOLUME 227. Metallobiochemistry (Part D: Physical and Spectroscopic Methods for Probing Metal Ion Environments in Metalloproteins)
Edited by JAMES F. RIORDAN AND BERT L. VALLEE

VOLUME 228. Aqueous Two-Phase Systems
Edited by HARRY WALTER AND GÖTE JOHANSSON

VOLUME 229. Cumulative Subject Index Volumes 195–198, 200–227

VOLUME 230. Guide to Techniques in Glycobiology
Edited by WILLIAM J. LENNARZ AND GERALD W. HART

VOLUME 231. Hemoglobins (Part B: Biochemical and Analytical Methods)
Edited by JOHANNES EVERSE, KIM D. VANDEGRIFF, AND ROBERT M. WINSLOW

VOLUME 232. Hemoglobins (Part C: Biophysical Methods)
Edited by JOHANNES EVERSE, KIM D. VANDEGRIFF, AND ROBERT M. WINSLOW

VOLUME 233. Oxygen Radicals in Biological Systems (Part C)
Edited by LESTER PACKER

VOLUME 234. Oxygen Radicals in Biological Systems (Part D)
Edited by LESTER PACKER

VOLUME 235. Bacterial Pathogenesis (Part A: Identification and Regulation of Virulence Factors)
Edited by VIRGINIA L. CLARK AND PATRIK M. BAVOIL

VOLUME 236. Bacterial Pathogenesis (Part B: Integration of Pathogenic Bacteria with Host Cells)
Edited by VIRGINIA L. CLARK AND PATRIK M. BAVOIL

VOLUME 237. Heterotrimeric G Proteins
Edited by RAVI IYENGAR

VOLUME 238. Heterotrimeric G-Protein Effectors
Edited by RAVI IYENGAR

VOLUME 239. Nuclear Magnetic Resonance (Part C)
Edited by THOMAS L. JAMES AND NORMAN J. OPPENHEIMER

VOLUME 240. Numerical Computer Methods (Part B)
Edited by MICHAEL L. JOHNSON AND LUDWIG BRAND

VOLUME 241. Retroviral Proteases
Edited by LAWRENCE C. KUO AND JULES A. SHAFER

VOLUME 242. Neoglycoconjugates (Part A)
Edited by Y. C. LEE AND REIKO T. LEE

VOLUME 243. Inorganic Microbial Sulfur Metabolism
Edited by HARRY D. PECK, JR., AND JEAN LEGALL

VOLUME 244. Proteolytic Enzymes: Serine and Cysteine Peptidases
Edited by ALAN J. BARRETT

VOLUME 245. Extracellular Matrix Components
Edited by E. RUOSLAHTI AND E. ENGVALL

VOLUME 246. Biochemical Spectroscopy
Edited by KENNETH SAUER

VOLUME 247. Neoglycoconjugates (Part B: Biomedical Applications)
Edited by Y. C. LEE AND REIKO T. LEE

VOLUME 248. Proteolytic Enzymes: Aspartic and Metallo Peptidases
Edited by ALAN J. BARRETT

VOLUME 249. Enzyme Kinetics and Mechanism (Part D: Developments in Enzyme Dynamics)
Edited by DANIEL L. PURICH

VOLUME 250. Lipid Modifications of Proteins
Edited by PATRICK J. CASEY AND JANICE E. BUSS

VOLUME 251. Biothiols (Part A: Monothiols and Dithiols, Protein Thiols, and Thiyl Radicals)
Edited by LESTER PACKER

VOLUME 252. Biothiols (Part B: Glutathione and Thioredoxin; Thiols in Signal Transduction and Gene Regulation)
Edited by LESTER PACKER

VOLUME 253. Adhesion of Microbial Pathogens
Edited by RON J. DOYLE AND ITZHAK OFEK

VOLUME 254. Oncogene Techniques
Edited by PETER K. VOGT AND INDER M. VERMA

VOLUME 255. Small GTPases and Their Regulators (Part A: Ras Family)
Edited by W. E. BALCH, CHANNING J. DER, AND ALAN HALL

VOLUME 256. Small GTPases and Their Regulators (Part B: Rho Family)
Edited by W. E. BALCH, CHANNING J. DER, AND ALAN HALL

VOLUME 257. Small GTPases and Their Regulators (Part C: Proteins Involved in Transport)
Edited by W. E. BALCH, CHANNING J. DER, AND ALAN HALL

VOLUME 258. Redox-Active Amino Acids in Biology
Edited by JUDITH P. KLINMAN

VOLUME 259. Energetics of Biological Macromolecules
Edited by MICHAEL L. JOHNSON AND GARY K. ACKERS

VOLUME 260. Mitochondrial Biogenesis and Genetics (Part A)
Edited by GIUSEPPE M. ATTARDI AND ANNE CHOMYN

VOLUME 261. Nuclear Magnetic Resonance and Nucleic Acids
Edited by THOMAS L. JAMES

VOLUME 262. DNA Replication
Edited by JUDITH L. CAMPBELL

VOLUME 263. Plasma Lipoproteins (Part C: Quantitation)
Edited by WILLIAM A. BRADLEY, SANDRA H. GIANTURCO, AND JERE P. SEGREST

VOLUME 264. Mitochondrial Biogenesis and Genetics (Part B)
Edited by GIUSEPPE M. ATTARDI AND ANNE CHOMYN

VOLUME 265. Cumulative Subject Index Volumes 228, 230–262

VOLUME 266. Computer Methods for Macromolecular Sequence Analysis
Edited by RUSSELL F. DOOLITTLE

VOLUME 267. Combinatorial Chemistry
Edited by JOHN N. ABELSON

VOLUME 268. Nitric Oxide (Part A: Sources and Detection of NO; NO Synthase)
Edited by LESTER PACKER

VOLUME 269. Nitric Oxide (Part B: Physiological and Pathological Processes)
Edited by LESTER PACKER

VOLUME 270. High Resolution Separation and Analysis of Biological Macromolecules (Part A: Fundamentals)
Edited by BARRY L. KARGER AND WILLIAM S. HANCOCK

VOLUME 271. High Resolution Separation and Analysis of Biological Macromolecules (Part B: Applications)
Edited by BARRY L. KARGER AND WILLIAM S. HANCOCK

VOLUME 272. Cytochrome P450 (Part B)
Edited by ERIC F. JOHNSON AND MICHAEL R. WATERMAN

VOLUME 273. RNA Polymerase and Associated Factors (Part A)
Edited by SANKAR ADHYA

VOLUME 274. RNA Polymerase and Associated Factors (Part B)
Edited by SANKAR ADHYA

VOLUME 275. Viral Polymerases and Related Proteins
Edited by LAWRENCE C. KUO, DAVID B. OLSEN, AND STEVEN S. CARROLL

VOLUME 276. Macromolecular Crystallography (Part A)
Edited by CHARLES W. CARTER, JR., AND ROBERT M. SWEET

VOLUME 277. Macromolecular Crystallography (Part B)
Edited by CHARLES W. CARTER, JR., AND ROBERT M. SWEET

VOLUME 278. Fluorescence Spectroscopy
Edited by LUDWIG BRAND AND MICHAEL L. JOHNSON

VOLUME 279. Vitamins and Coenzymes (Part I)
Edited by DONALD B. MCCORMICK, JOHN W. SUTTIE, AND CONRAD WAGNER

VOLUME 280. Vitamins and Coenzymes (Part J)
Edited by DONALD B. MCCORMICK, JOHN W. SUTTIE, AND CONRAD WAGNER

VOLUME 281. Vitamins and Coenzymes (Part K)
Edited by DONALD B. MCCORMICK, JOHN W. SUTTIE, AND CONRAD WAGNER

VOLUME 282. Vitamins and Coenzymes (Part L)
Edited by DONALD B. MCCORMICK, JOHN W. SUTTIE, AND CONRAD WAGNER

VOLUME 283. Cell Cycle Control
Edited by WILLIAM G. DUNPHY

VOLUME 284. Lipases (Part A: Biotechnology)
Edited by BYRON RUBIN AND EDWARD A. DENNIS

VOLUME 285. Cumulative Subject Index Volumes 263, 264, 266–284, 286–289

VOLUME 286. Lipases (Part B: Enzyme Characterization and Utilization)
Edited by BYRON RUBIN AND EDWARD A. DENNIS

VOLUME 287. Chemokines
Edited by RICHARD HORUK

VOLUME 288. Chemokine Receptors
Edited by RICHARD HORUK

VOLUME 289. Solid Phase Peptide Synthesis
Edited by GREGG B. FIELDS

VOLUME 290. Molecular Chaperones
Edited by GEORGE H. LORIMER AND THOMAS BALDWIN

VOLUME 291. Caged Compounds
Edited by GERARD MARRIOTT

VOLUME 292. ABC Transporters: Biochemical, Cellular, and Molecular Aspects
Edited by SURESH V. AMBUDKAR AND MICHAEL M. GOTTESMAN

VOLUME 293. Ion Channels (Part B)
Edited by P. MICHAEL CONN

VOLUME 294. Ion Channels (Part C)
Edited by P. MICHAEL CONN

VOLUME 295. Energetics of Biological Macromolecules (Part B)
Edited by GARY K. ACKERS AND MICHAEL L. JOHNSON

VOLUME 296. Neurotransmitter Transporters
Edited by SUSAN G. AMARA

VOLUME 297. Photosynthesis: Molecular Biology of Energy Capture
Edited by LEE MCINTOSH

VOLUME 298. Molecular Motors and the Cytoskeleton (Part B)
Edited by RICHARD B. VALLEE

VOLUME 299. Oxidants and Antioxidants (Part A)
Edited by LESTER PACKER

VOLUME 300. Oxidants and Antioxidants (Part B)
Edited by LESTER PACKER

VOLUME 301. Nitric Oxide: Biological and Antioxidant Activities (Part C)
Edited by LESTER PACKER

VOLUME 302. Green Fluorescent Protein
Edited by P. MICHAEL CONN

VOLUME 303. cDNA Preparation and Display
Edited by SHERMAN M. WEISSMAN

VOLUME 304. Chromatin
Edited by PAUL M. WASSARMAN AND ALAN P. WOLFFE

VOLUME 305. Bioluminescence and Chemiluminescence (Part C) (in preparation)
Edited by THOMAS O. BALDWIN AND MIRIAM M. ZIEGLER

VOLUME 306. Expression of Recombinant Genes in Eukaryotic Systems
(in preparation)
Edited by JOSEPH C. GLORIOSO AND MARTIN C. SCHMIDT

VOLUME 307. Confocal Microscopy
Edited by P. MICHAEL CONN

VOLUME 308. Enzyme Kinetics and Mechanism (Part E: Energetics of Enzyme
Catalysis)
Edited by VERN L. SCHRAMM AND DANIEL L. PURICH

VOLUME 309. Amyloids, Prions, and Other Protein Aggregates (in preparation)
Edited by RONALD WETZEL

VOLUME 310. Biofilms (in preparation)
Edited by RON J. DOYLE

VOLUME 311. Sphingolipid Metabolism and Cell Signaling (Part A)
(in preparation)
Edited by ALFRED H. MERRILL, JR., AND Y. A. HANNUN

VOLUME 312. Sphingolipid Metabolism and Cell Signaling (Part B)
(in preparation)
Edited by ALFRED H. MERRILL, JR., AND Y. A. HANNUN

VOLUME 313. Antisense Technology (Part A: General Methods, Methods of De-
livery and RNA Studies) (in preparation)
Edited by M. IAN PHILLIPS

VOLUME 314. Antisense Technology (Part B: Applications) (in preparation)
Edited by M. IAN PHILLIPS

VOLUME 315. Vertebrate Phototransduction and the Visual Cycle (Part A) (in
preparation)
Edited by KRZYSZTOF PALCZEWSKI

VOLUME 316. Vertebrate Phototransduction and the Visual Cycle (Part B) (in
preparation)
Edited by KRZYSZTOF PALCZEWSKI

VOLUME 317. RNA-Ligand Interactions (in preparation)
Edited by DANIEL W. CELANDER AND JOHN N. ABELSON

VOLUME 318. Singlet Oxygen, UV-A, and Ozone (in preparation)
Edited by LESTER PACKER AND HELMUT SIES

Section I

Energetic Coupling in Enzymatic Reactions

[1] Energetics of Substrate Binding, Catalysis, and Product Release

By W. WALLACE CLELAND and DEXTER B. NORTHROP

As a starting point for discussion of the energetics of an enzyme-catalyzed reaction, one has the overall equilibrium constant. This is independent of the presence of any catalyst and thus cannot be altered by an enzyme. The kinetic parameters of the enzymatic reaction are constrained by the equilibrium constant, however, and this constraint is normally expressed in the Haldane relationship. For a reaction with one substrate and product, this is the ratio of V/K values in forward and reverse directions:

$$K_{eq} = \frac{(V_1/K_a)}{(V_2/K_p)} = \frac{V_1 K_p}{V_2 K_a} \tag{1}$$

When there are more substrates or products, the dissociation constants of all reactants except the last to add to the enzyme are included. Thus with two substrates and products, one has

$$K_{eq} = \frac{(V_1/K_b)K_{iq}}{(V_2/K_p)K_{ia}} = \frac{V_1 K_p K_{iq}}{V_2 K_{ia} K_b} \tag{2}$$

Haldane relationships have a number of uses,[1] but an important one is to establish the upper limit of the turnover number for an enzymatic reaction. Thus to optimize V_1/E_t in Eq. (1), one raises V_1/K_a and V_2/K_p until one of them reaches the limit set by diffusion.[2] Because these V/K values are the apparent first-order rate constants at low substrate concentration, they cannot exceed the rate at which enzyme and substrates diffuse together and collide. This rate is $\approx 10^9\ M^{-1}\ sec^{-1}$ for small substrates[3] and up to an order of magnitude smaller for large ones such as nucleotides. Once the V/K values are optimized, one has to keep K_a at or below the physiological level of A or the enzyme will not be at least half saturated with substrate. This puts an upper limit on V_1/E_t, the turnover number, and enzymes that are involved in major biochemical pathways all seem to be optimized within this limit.

Haldane relationships are discussed in detail in Cleland.[1]

[1] W. W. Cleland, *Methods Enzymol.* **87,** 366 (1982).
[2] W. J. Albery and J. R. Knowles, *Biochemistry* **15,** 5631 (1976).
[3] R. A. Alberty and G. G. Hammes, *J. Phys. Chem.* **62,** 154 (1958).

Copyright © 1999 by Academic Press
All rights of reproduction in any form reserved.
0076-6879/99 $30.00

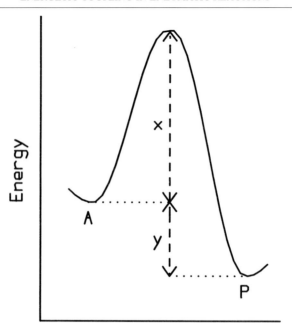

Reaction Coordinate

FIG. 1. Free energy profile for an uncatalyzed unimolecular reaction of A to give P. x is the activation energy and y is $-RT \ln K_{eq}$.

Free Energy Profile

The energetics of an enzyme-catalyzed reaction are usually discussed in terms of a free energy profile. This is a diagram showing the relative energy levels of all enzyme–reactant complexes and the transition states for conversion between them. Detailed instructions for the construction of free energy profiles for complex enzymatic reactions have been provided by Lumry.[4] For an uncatalyzed unimolecular reaction, such a profile is straightforward to draw (Fig. 1). In this profile, y is ΔG for the reaction, or $-RT \ln K_{eq}$. The parameter x is the activation energy for the forward reaction, and $x + y$ is the activation energy in the back reaction. The activation energies are related to rate constants in forward and reverse directions by equations such as Eq. (3):

$$k = ce^{-E/RT} \tag{3}$$

[4] R. Lumry, "Enzymes," 2nd ed., Vol. 1, p. 157. Academic Press, New York, 1959.

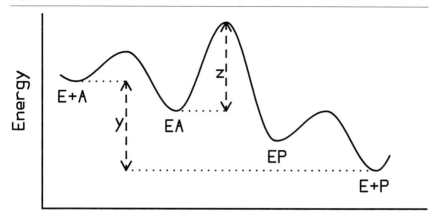

Reaction Coordinate

FIG. 2. Free energy profile for an enzyme-catalyzed reaction similar to the one shown in Fig. 1. z is the activation energy for the catalyzed reaction, but y is the same as in Fig. 1.

where E is the activation energy and c is the Arrhenius preexponential factor (usually written A, but written c in this article to avoid confusion with A as a substrate concentration). This equation can be rearranged to show the activation energy as a function of the rate constant:

$$E = RT\ln(c/k) = (RT \ln 10)\log_{10}(c/k) \tag{4}$$

$RT \ln 10$ is 1.36 kcal/mol at 25°, and with c about 6×10^{12} sec^{-1} for a unimolecular reaction ($\mathbf{k}T/h$ according to transition state theory, where $\mathbf{k}$ and h are Boltzmann's and Planck's constants), an activation energy of 3.8 kcal/mol corresponds to a rate constant of 10^{10} sec^{-1}, and an activation energy of 10.6 kcal/mol corresponds to a rate constant of 10^5 sec^{-1}.

The free energy profile for an enzyme-catalyzed reaction with the same K_{eq} as in Fig. 1 can be drawn as in Fig. 2 if A and P are at equilibrium. Here the enzyme and substrate combine to form an EA complex, which isomerizes to EP, followed by dissociation of P to give free enzyme again. The value of y ($= -RT \ln K_{eq}$) is the same as in Fig. 1, but z is now the activation energy for conversion of EA to EP, which will be less than x in Fig. 1 for the noncatalyzed reaction. The difference in the energy levels of EA and EP is not necessarily equal to y, but may differ considerably from the equilibrium constant of the reaction if one reactant is bound more tightly than the other. The energy levels of EA and EP are typically closer

[5] B. D. Nageswara Rao and M. Cohn, *J. Biol. Chem.* **254,** 2689 (1979).

to unity than those of E + A and E + P, i.e., the internal equilibrium constant on the enzyme is closer to unity than K_{eq} itself.[5]

Calculation of the rate acceleration resulting from enzymatic catalysis for a unimolecular reaction is simple; it corresponds to the antilog of the difference between x and z divided by $RT \ln 10$. Thus if z in Fig. 2 is 11 kcal/mol, x in Fig. 1 is 25.7 kcal/mol, and the rate acceleration is $10^{(25.7-11)/1.36} = 5 \times 10^{10}$.

The problem with profiles such as those in Fig. 2 is the choice of standard states. Because the combination of E and A to give EA is a bimolecular step, the relative energy levels of E + A and EA depend on the concentration of A. The same applies to the dissociation of EP to E + P, although the difference in energy levels of E + A and E + P must always equal $-RT \ln K_{eq}$, regardless of what standard state is chosen if one is representing an equilibrium free energy profile. If A and P are not at equilibrium, the levels of E + A and E + P will not differ by $-RT \ln K_{eq}$. Thus in Fig. 3, which shows a profile for a reaction with an equilibrium constant of 0.1 and with the concentration of A equal to its dissociation constant from EA, the level of E + P lies 1.36 kcal/mol higher than E + A when P and A are at equal concentrations, whereas if P = 0.01 A, the level of E + P lies 1.36 kcal/mol below E + A.

This situation can cause problems with extrapolated limits. Enzymatic reactions are analyzed in terms of effects on V (substrate concentration extrapolated to ∞) and V/K (substrate concentration extrapolated to near

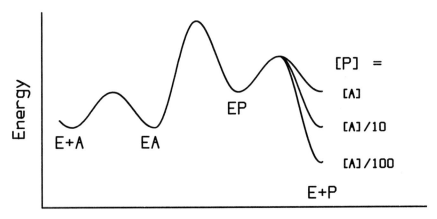

FIG. 3. Free energy profile showing the influence of product concentration on the energy level of E + P. The concentration of A is set equal to its dissociation constant from EA and $K_{eq} = 0.1$.

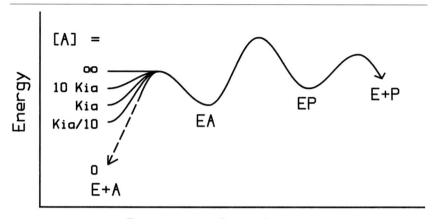

Fig. 4. Free energy profile showing the influence of substrate concentration on the energy level of E + A. K_{ia} is the dissociation constant of A from EA. The level of P is not specified in the profile.

zero). With A = ∞, the level of E + A is even with the top of the activation barrier for dissociation of EA, as the bimolecular combination of E and A goes at an infinite rate. With A → 0, however, the level of E + A lies at ∞, as shown in Fig. 4. If the level of E+A is to be plotted on the profile at all, a finite level of A should be picked and stated. Logical choices are A = K_{ia} (dissociation constant of A), A = K_a (Michaelis constant), or A = physiological level of A.

With two substrates and two products, the problem of bimolecular steps is even greater (Fig. 5). The levels of E + A + B, EA + B, and EAB depend on the concentrations of A and B, and the same is true of EPQ, EQ + P, and E + P + Q. The difference in the level of E + A + B and E + P + Q again corresponds to $-RT \ln K_{eq}$, regardless of the actual standard state chosen, if an equilibrium profile is to be plotted.

While it might seem appropriate in drawing equilibrium-free energy profiles to choose the concentration of all substrates as 1 M, with the concentrations of products then being calculated from K_{eq}, how does one choose the concentrations of the two products, since one can be high in concentration and the other lower, or vice versa?

As a result of these problems, it is really useful for most purposes only to plot the portion of the free energy profile that involves unimolecular steps. In Fig. 2 this is from EA to EP, and in Fig. 5, from EAB to EPQ. This avoids any problems with binding energies.[6]

[6] D. B. Northrop, *Adv. Enzymol.* **73,** 25 (1999).

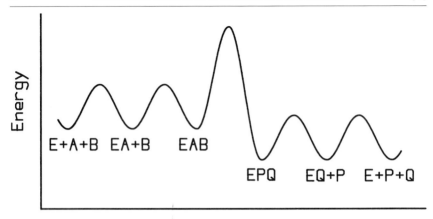

FIG. 5. Free energy profile for an enzyme-catalyzed reaction with two substrates and two products. Reactants are at levels equal to their dissociation constants.

The center part of the free energy profile usually does not consist of a single step, as in Figs. 2 and 5, but rather includes conformation changes before and after the chemical step and may involve chemical intermediates as well. Such a profile is shown in Fig. 6. In this case, the chemical reaction actually occurs between EA* and EP* via the relatively unstable intermedi-

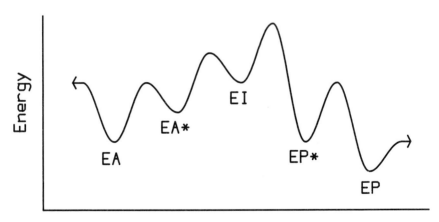

FIG. 6. Free energy profile showing conformation changes in enzyme–reactant complexes as well as an intermediate between the two activated complexes.

ate EI. This sort of free energy profile is needed for the triose-P isomerase reaction, e.g., where EI is an endiolate.

In a profile such as Fig. 6, the equilibrium constant between any two complexes is given by the vertical distance between them (corresponding to $-RT \ln K_{eq}$). Because the height of each barrier is related inversely to the log of the rate constant for the step according to Eq. (4), a high barrier corresponds to a slow rate constant and a low one to a fast rate constant. If one knows the partition ratio of an intermediate (i.e., $k_{forward}/k_{reverse}$), one can plot the relative heights of the barriers for a given complex, even if one does not know the absolute energy level of the complex. Figure 7 shows a typical situation, where a $= -RT \ln(k_{forward}/k_{reverse})$. To determine the level of EI, one must be able to quench the system at equilibrium and determine how much I (as opposed to A and P) is present on the enzyme or directly observe the species present by nuclear magnetic resonance (NMR) or a similar technique. Such an experiment must be carried out with a high enough enzyme level for all of the reactants to be bound or run at several enzyme levels and extrapolated to this point.

The presence of more than one substrate in a reaction causes problems in calculating rate accelerations for enzymatic reactions. Because the uncatalyzed reaction of A and B is a bimolecular reaction, we cannot compare the energy barrier here with that for the unimolecular reaction of EAB to give products. An appropriate way to make the comparison is to set B equal to its dissociation constant from EAB to give EA. Then the activation

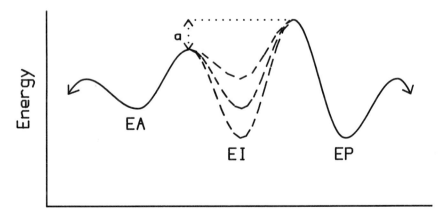

FIG. 7. Free energy profiles showing that the partition ratio of the EI intermediate is determined by the relative heights of the barriers and does not depend on the energy level of EI itself. The difference in barrier heights, a, is $-RT \ln(k_{forward}/k_{reverse})$.

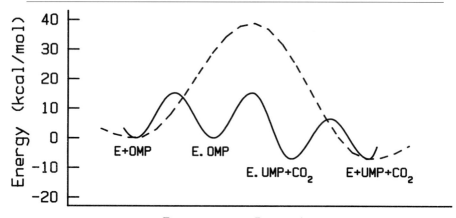

FIG. 8. Free energy profile for the reaction catalyzed by orotidylate decarboxylase. The dotted profile is for the uncatalyzed reaction. CO_2, 1 atmosphere; OMP and UMP levels equal to their dissociation constants. Calculated from data found in Radzicka and Wolfenden.[7]

energy for the uncatalyzed reaction of A and B at this level of B can be compared directly with that for the reaction of EAB. With three substrates, the levels of B and C should be picked to make the level of E + A + B + C equal to that of EABC. This permits comparison of the uncatalyzed reaction at these levels of B and C with the reaction of EABC, although if the kinetics of the uncatalyzed reaction are not cleanly termolecular, the comparison may not be completely valid.

So far we have assumed equal numbers of substrates and products so that K_{eq} does not have dimensions. However, a number of enzymatic reactions involve unequal numbers of substrates and products. For orotidylate decarboxylase, for example, two products (CO_2 and UMP) are formed from one substrate (OMP), and K_{eq} has dimensions of concentration. This enzyme has the greatest known rate acceleration (1.4×10^{17}).[7] The K_{eq} value is not known, but if we assume for purposes of example the same value as that for decarboxylation of oxalacetate (-7 kcal/mol for 1 atmosphere CO_2;[8] the value is a function of CO_2 concentration), we can draw the free energy profile shown in Fig. 8 by assuming the following concentrations: OMP and UMP equal to their dissociation constants, and CO_2 at 1 atmosphere over aqueous solution (the standard state that gives a ΔG value of

---

[7] A. Radzicka and R. Wolfenden, *Science* **267**, 90 (1995).
[8] M. J. Johnson, "Enzymes," 2nd ed., Vol. 3, p. 407. Academic Press, New York, 1960.

-7 kcal/mol for oxalacetate decarboxylation). A profile for the uncatalyzed reaction, employing the same concentrations of reactants, is also plotted for comparison.

Note that the barrier for release of OMP from E-OMP is the same height as that for the decarboxylation step (rate constant 39 sec^{-1}). This is deduced from the fact that the ^{13}C isotope effect at neutral pH is reduced from that seen at pH 4 by a commitment of 1.0.[9] The barrier for release of UMP is not known, but we have assumed a rate constant of 800 sec^{-1}, as the K_i for UMP is an order of magnitude higher than the K_m for OMP and the K_m should be twice the dissociation constant because of the equal rates for partitioning of E-OMP.

If we assume a different value for the CO_2 concentration, the vertical difference between E + OMP and E + UMP + CO_2 in this profile will differ from -7 kcal/mol. For example, the solubility of CO_2 in water at 1 atmosphere and 25° is about 33 mM; if we change the CO_2 concentration to 10 mM, the value of -7 becomes -7.7 kcal/mol. Lowering the value to 1 mM will change the energy difference to -9.06 kcal/mol. These changes affect the free energy profiles for both enzymatic and nonenzymatic reactions and simply reflect the fact that K_{eq} is not dimensionless.

An interesting application of free energy profiles involves so-called "Iso" mechanisms, where the isomerization of free enzyme is a kinetically detectable step.[10] Fisher and co-workers[11] have shown that proline racemase has such a mechanism, where one form of free enzyme reacts with L-proline and the other with D-proline. The free enzyme isomerization goes at 10^5 sec^{-1}, whereas the turnover numbers are 1600 sec^{-1}, but by raising the substrate concentrations to the point where most of the free enzyme is tied up in enzyme–reactant complexes, the free enzyme isomerization can be made rate limiting. This situation is diagrammed as an equilibrium free energy profile in Fig. 9. If c in Eq. (4) is taken as 6×12 sec^{-1}, the barrier heights are 12.7 kcal/mol between EA and E'P, 9.4 kcal/mol for dissociation of EA or E'P, and 10.6 kcal/mol for interconversion of E' and E. In the unsaturated region where substrate concentrations are below K_m values, the interconversion of EA and E'P is the slow step, which is still true at the point where half of the enzyme is tied up as EA or E'P. However, when the levels of substrates are raised still further, the barrier between

[9] J. A. Smiley, P. Paneth, M. H. O'Leary, J. B. Bell, and M. E. Jones, *Biochemistry* **30,** 6216 (1991).
[10] D. B. Northrop and K. L. Rebholz, *Anal. Biochem.* **216,** 285 (1994); K. L. Rebholz and D. B. Northrop, *Methods Enzymol.* **249,** 211 (1995).
[11] L. M. Fisher, W. J. Albery, and J. R. Knowles, *Biochemistry* **25,** 2529 (1986).

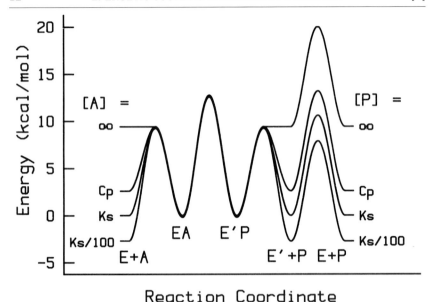

Fig. 9. Free energy profile for proline racemase at several reactant concentrations. See text for details.

E′ and E becomes as high as that between EA and E′P, and at still higher substrate levels (in a region Knowles called "oversaturating") the interconversion of E and E′ becomes rate limiting.

This type of mechanism is demonstrated readily by observing counter transport of label.[12] In this experiment, labeled D- and L-proline (16 mM each; about 6 K_m) were allowed to come to equilibrium in the presence of enzyme and then 183 mM unlabeled L-proline was added.[13] The fraction of label in L-proline rose to 65% and then gradually returned to 50% as chemical equilibrium was reached. The high level of proline present in this experiment makes the free enzyme interconversion rate limiting and thus forces the return of E′ to E to occur by reaction with labeled D-proline. This experiment allows calculation of the point where the barriers between EA and E′P and between E′ and E reach equal heights. Knowles called this point the "peak switch concentration" (C_P = 125 mM in Fig. 9).

Note that observation of oversaturation and countertransport will occur only if the sum of the barrier heights for release of P from E′P (or A from

[12] H. G. Britton, *Arch. Biochem. Biophys.* **117**, 167 (1966); *Biochem. J.* **133**, 255 (1973).
[13] L. M. Fisher, W. J. Albery, and J. R. Knowles, *Biochemistry* **25**, 2538 (1986).

EA) and for interconversion of E and E' is greater than the barrier height between EA and E'P. If not, the conversion of EA to E'P will remain rate limiting even at infinite substrate concentrations. In the proline racemase case, $9.4 + 10.6 = 20$ kcal/mol clearly exceeds 12.7 kcal/mol. Many other enzymes probably have Iso mechanisms, but if the E' to E rate is fast enough, the isomerization will not be detected. The isomerization rate for proline racemase is slow enough to detect because the general acid/base groups for the reaction are sulfhydryl groups, which dissociate protons slowly because of their low tendency to form hydrogen bonds.

There is another type of Iso mechanism which shows countertransport and is exemplified by carbonic anhydrase. An equilibrium-free energy profile at pH 8.2 is shown in Fig. 10.[14] In the unsaturated region, most of the enzyme is present as E (free enzyme with ZnOH), HE ($ZnOH_2$), or EH (ZnOH with protonated His-64) and the highest barrier on the profile is for binding or dissociation of CO_2. When the substrates are at K_m levels, the barrier heights for conversion of CO_2 to bicarbonate and for conversion of HE to E are equal and both portions of the mechanism limit the rate equally. In the saturated state, the interconversion of HE and E limits V_{max}. As the substrate level increases further, the levels of E-CO_2 and HE-bicarbonate will become even with the level of HE and E. At still higher substrate concentrations the height from HE-bicarbonate to the barrier between HE and EH will be greater than the barrier height between HE and EH; the enzyme is oversaturated.

In this type of mechanism, countertransport occurs above the K_m levels of substrates (the "peak switch concentration" according to Knowles; C_P in Fig. 10) because the barrier from HE back to E via HE-bicarbonate and E-CO_2 is lower than that for direct conversion. Simpson and Northrop[15] have been able to measure this countertransport with carbonic anhydrase by adding 50 mM bicarbonate (≈ 1.6 K_m) to an equilibrated mixture of ^{13}C-containing CO_2 and bicarbonate and by rapid sampling via a membrane of the CO_2 by isotope ratio mass spectrometry.

Effects of Temperature

The free energy of activation for a unimolecular chemical reaction is related to the enthalpy and entropy of activation by Eq. (5):

$$\Delta G^{\ddagger} = \Delta H^{\ddagger} - T\Delta S^{\ddagger} \tag{5}$$

[14] G. Behravan, B.-H. Jonsson and S. Lindskog, *Eur. J. Biochem.* **190,** 351 (1990).
[15] F. B. Simpson and D. B. Northrop, *Arch. Biochem. Biophys.* **352,** 288 (1998).

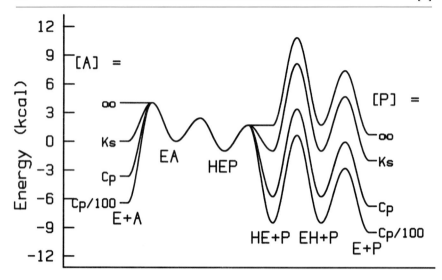

Reaction Coordinate

FIG. 10. Free energy profile for human carbonic anhydrase II at pH 8.2, with A being CO_2 and P being bicarbonate. E represents enzyme with Zn-OH, HE is enzyme with Zn-OH_2, and EH is enzyme with Zn-OH and protonated His-64. The profile was constructed by starting with the rate constants of Behravan et al.[14] who did not include a step for the chemical conversion of CO_2 to HCO_3^-. This step was added by setting $k_3/k_2 = 2$ (i.e., c_r for the dehydration of bicarbonate; derived from the ^{13}C isotope effect of 1.0101 of Paneth and O'Leary[29] and the equilibrium ^{13}C isotope effect of 1.009[30] and an assumed intrinsic ^{13}C isotope effect of 1.0147, which is the value reported for the uncatalyzed reaction[31]). The value of k_4 was adjusted to yield an equilibrium constant of 5.6 for the hydration step, based on the difference between pH 6.35, where the equilibrium constant equals unity and the pK of 7.1 for water bound to zinc on the enzyme. The rate constant for the transfer of the His-64 proton to buffer was adjusted to match the overall equilibrium constant of 70 at pH 8.2. Next, the rate constants for the dissociation of CO_2 and bicarbonate were raised in a constant ratio until the calculated rate of isotopic exchange between them reached the value of 1.6×10^6 sec^{-1} measured by NMR.[32] Finally, the rate constants for their association were raised until V/K for CO_2 was 1.2×10^8.[14] A value for C_P of CO_2 of 53 mM was calculated from Eqs. (21) and (22) of Northrop and Rebholz.[10] Final values of the rate constants used in the profile in M^{-1} sec^{-1} or in sec^{-1} are $k_1 = 1.99 \times 10^8$; $k_2 = 3.08 \times 10^6$; $k_3 = 6.16 \times 10^6$; $k_4 = 6.7 \times 10^6$; $k_5 = 2.91 \times 10^7$; $k_6 = 2.98 \times 10^8$; $k_7 = 1.2 \times 10^6$; $k_8 = 1.2 \times 10^6$; $k_9 = 9.2 \times 10^8$; and $k_{10} = 7.4 \times 10^7$.

Equating E in Eq. (3) to $\Delta G^{\ddagger}$ and c to $\mathbf{k}T/h$, we have

$$k = (\mathbf{k}T/h)e^{-\Delta G^{\ddagger}/RT} = (\mathbf{k}T/h)e^{-\Delta H^{\ddagger}/RT}e^{\Delta S^{\ddagger}/R} \tag{6}$$

When $\ln(k/T)$ is plotted vs l/T, one gets

$$\ln(k/T) = -\Delta H^{\ddagger}/RT + \Delta S^{\ddagger}/R + \ln(\mathbf{k}/h) \tag{7}$$

The slope of this plot is $-\Delta H^{\ddagger}/R$ and the intercept at $l/T = 0$ is $\Delta S^{\ddagger}/R +$ 23.76, where $R = 1.987 \times 10^{-3}$ kcal/mol · deg and k is given in sec^{-1}.

The estimation of $\Delta H^{\ddagger}$ is straightforward, and the value of $\Delta G^{\ddagger}$ from Eq. (4) is not likely to be in serious error, but the values of $\Delta S^{\ddagger}$ obtained in this way depend heavily on the validity of the theory behind Eq. (3) and thus are subject to greater uncertainty. We will not discuss the interpretation of these values further. The reader is referred to previous articles in this series that deal with the temperature variation of enzymic rates.[16]

Rate-Limiting Steps

There is considerable interest in determining the rate-limiting steps in enzymatic reactions, as study of chemical mechanisms (by isotope effects, for example) requires that the chemistry be rate limiting. Over the course of evolution, the turnover numbers of most enzymatic reactions have approached the theoretical limit set by the Haldane relationship and by diffusion rates in solution (see earlier) and the chemistry of the reaction is typically only partly rate limiting.

As noted earlier, enzymatic reactions are analyzed by separate consideration of V and V/K. The rate-limiting step for V/K is established easily and corresponds to the step with the highest activation barrier in the unimolecular portion of the free energy profile. When this barrier is at least several kcal/mol above all others, the step is essentially rate limiting, and isotope effects on this step will be fully expressed on V/K. If the chemical step does not have the highest barrier, isotope effects are suppressed by commitments to catalysis (usually simply called commitments). These are the ratio of the rate constant for the isotope-sensitive step to the net rate constant for the first irreversible step in the other direction (release of a substrate or product, or a chemically irreversible step). Commitments are made up of combinations of partition ratios of intermediates. Thus for the mechanism

$$\text{E} + \text{A} \underset{k_2}{\overset{k_1}{\rightleftharpoons}} \text{EA} \underset{k_4}{\overset{k_3}{\rightleftharpoons}} \text{EA*} \underset{k_6}{\overset{k_5}{\rightleftharpoons}} \text{EPQ*} \underset{k_8}{\overset{k_7}{\rightleftharpoons}} \text{EPQ} \overset{k_9}{\longrightarrow} \text{EQ} \overset{k_{11}}{\longrightarrow} \text{E} \quad (8)$$

where conversion of EA* to EPQ* is the chemical and isotope-sensitive step, the deuterium isotope effect on V/K is given by Eq. (9):

[16] K. L. Laidler and B. F. Peterman, *Methods Enzymol.* **63**, 234 (1979); R. Lumry, *Methods Enzymol.* **259**, 628 (1995).

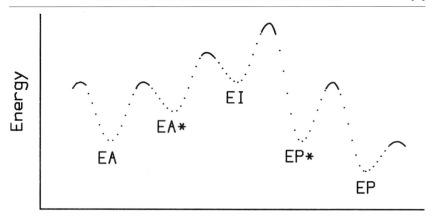

Reaction Coordinate

Fɪɢ. 11. At very low substrate levels (V/K conditions), only the heights of the barriers are important.

$$^D(V/K) = \frac{^Dk_5 + c_f + {}^DK_{eq}c_r}{1 + c_f + c_r} \qquad (9)$$

where $^Dk_5$ is k_{5H}/k_{5D}, $^D(V/K) = (V/K)_H/(V/K)_D$, and $^DK_{eq} = K_{eqH}/K_{eqD}$ Northrop[17] writes the commitments with capital letters (C_f, C_r). The forward commitment, c_f, is given by

$$c_f = (k_5/k_4)(1 + k_3/k_2) \qquad (10)$$

whereas the reverse commitment, c_r, is given by

$$c_r = (k_6/k_7)(1 + k_8/k_9) \qquad (11)$$

Equation (9) applies to any mechanism with a single isotope-sensitive step, but c_f and c_r depend on which step is isotope sensitive. Note that one only has to know the partition ratios k_5/k_4, k_3/k_2, and so on to calculate commitments, and one does not need to know the relative energy levels of EA*, EA*, EPQ*, and EPQ. Note also that once the first irreversible step has occurred (k_9), further steps do not contribute to V/K. In terms of a free energy profile, V/K calculations involve only the heights of the various barriers and are not involved with the troughs in the profile (Fig. 11). Also, the order of these other barriers is not important, so very different profiles can be associated with identical values of V/K. What is important to the magnitude of V/K is the relative vertical positioning of the first

[17] D. B. Northrop, *Biochemistry* **14**, 2644 (1975).

transition state compared to a collective function of all vertical positions of other transition states up to and including the first irreversible step because V/K is simply a fraction of k_1. This fraction is large for diffusion-controlled reactions in which other barriers are lower than the binding barrier and small under rapid equilibrium conditions in which other barriers are higher. Because no enzyme populates the troughs, V/K is not a measure of rates of reactive steps after the binding step, but rather is a measure of the rate of "capture"[6] of substrate by the enzyme into forms that are destined to produce successful catalytic turnovers.

The rate-limiting step for V, however, is more complicated and depends on the definition of rate limiting.[17] All unimolecular steps in the mechanism now become important, including ones that follow the first irreversible step. Common usage would suggest that the step traversing the highest point in an energy profile is the rate-limiting step. This is true for V/K (excluding steps after the first irreversible one), but not always true for V. Also not always true is the common reference to the "slowest step," which refers to the one with the largest individual barrier. This definition works within a series of irreversible steps, but is not always applicable when steps are reversible. In Fig. 12, for example, the second step has the largest individual barrier, but is the least rate determining.

A second definition of the rate-limiting step is the one involving the enzyme form present in the highest concentration in the steady state at substrate saturation. This is the "least conductive" step[18] and is the step in the free energy profile that sees the highest total barrier in the forward direction to reach an irreversible step. A third definition of rate limiting is the "most sensitive" step,[18] the step producing the largest percentage change in turnover number for a given percentage change in the forward rate constant. An isotope effect is expressed more fully on this step than on any other. This is usually the last step in a series of steps that have the greatest trough to peak distance, but may not be. In Fig. 12, the distances from EI and EP to the final barrier are the same, but the last step is the most sensitive. The least conductive step, which sees the highest cumulative barrier downstream, is the step from EA to EI.

Activation energy diagrams at equilibrium can misrepresent the distribution of intermediate forms of enzyme during catalytic turnover under saturating conditions. A better depiction is in Fig. 13, where net rate constants[19] at $A = 100\ K_s$ are plotted as irreversible steps.

Because all steps proceed at the same rate during steady state, the peaks are all shown at the same energy level; what differs are the levels for the troughs, with more enzyme piling up behind the steps with the lower net

[18] W. J. Ray, Jr., *Biochemistry* **22,** 4452 (1983).

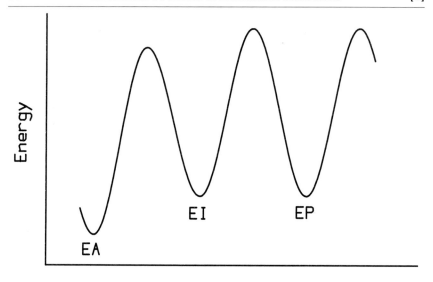

FIG. 12. Free energy profile at saturating substrate levels. See text for details.

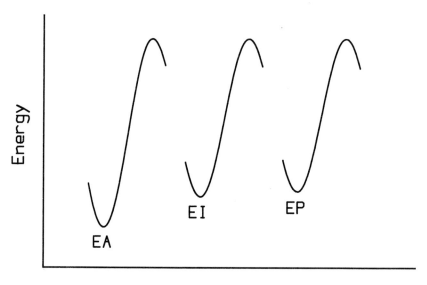

FIG. 13. Free energy profile for the reaction shown in Fig. 12 plotted in terms of net rate constants. All steps are thus by definition irreversible.

rate constants, depicted by larger barriers. These levels may differ considerably from the troughs in equilibrium diagrams. Figure 13 shows an example, because the level of EP is slightly higher than that of EI in Fig. 13, but at the same level as EI in Fig. 12. The slowest step, the least conductive step, and the most sensitive step may be the same, but need not be. The reason for this uncertainty is that where a step occurs within a series of steps it also contributes to its relative importance. Troughs are more important in the early steps and peaks more important in the later steps. In a series of steps described by identical rate constants, the first step will be the least conductive and will have the smallest net rate constant, while the last step will be the most sensitive and will express the largest isotope effect.

The isotope effect on V depends on whether the isotope-sensitive step is the most sensitive step and not on whether it is the slowest or least conductive one. In mechanism 8 the isotope effect on V (^DV) is given by

$$^DV = \frac{^Dk_5 + c_{Vf} + ^DK_{eq}c_r}{1 + c_{Vf} + c_r} \tag{12}$$

where c_r has the same definition as in Eq. (11), but c_{Vf} is given by

$$c_{Vf} = [k_5/(1 + k_4/k_3)][1/k_3 + (1/k_7)(1 + k_8/k_9) + 1/k_9 + 1/k_{11}] \tag{13}$$

Note that a low value of k_{11} (second product release) will produce a high value of c_{Vf} and thus suppress the isotope effect on V, while having no effect on V/K. This is a common situation with dehydrogenases, where nucleotide release is slow and isotope effects are seen only on V/K and not on V, at least at neutral pH.

Northrop[17] replaces c_{Vf} in Eq. (12) with R_f/E_f, where

$$R_f = k_5/k_3 + k_5/k_7' + k_5/k_9 + k_5/k_{11} \tag{14}$$

In Eq. (14), k_7' is a net rate constant[19] $[= k_7k_9/(k_8 + k_9)]$. If the first step of mechanism 8 were not substrate binding, but another unimolecular step, another ratio would be added to Eq. (14) $[k_5/k_1'',$ where $k_1'' = k_1k_3/(k_2 + k_3)]$. This sum of individual ratios comparing the isotopically sensitive step to each other step contributing to V is the origin of the "slowest" step being rate limiting, which would be the only term present in an equation describing a series of irreversible steps. E_f is the "forward equilibration preceding catalysis" and is given by

$$E_f = 1 + k_4/k_3 \tag{15}$$

[19] W. W. Cleland, *Biochemistry* **14**, 3220 (1975).

E_f is a thermodynamic function and $1/E_f$ equals the fraction of forms of enzyme preceding the isotopically sensitive step (in this case EA and EA*) that is poised to undergo this step. It is the reason that troughs are more important early in the mechanism.

Examination of Eqs. (9)–(13) shows that equations for V/K include the rate constant k_2 in mechanism 8 for substrate release and k_9 for first product release. Conversely, equations for V do not include k_2, but do include rate constants for release of all products and thus include k_{11} as well as k_9. When the first product to be released dissociates slowly, this step may limit both V/K and V, whereas release of later products affects only V.

When a substrate dissociates slowly relative to turnover, we say that it is "sticky" and we define the stickiness ratio S_r as the net rate constant for conversion of the initial EA complex through the first irreversible step $[(k_3k_5/k_4)/(1 + k_5/k_4 + (k_6/k_7)(1 + k_8/k_9))]$ divided by the off rate constant for the substrate (k_2 in mechanism 8).[20] In terms of the commitments defined by Eqs. (10) and (11), this equals

$$S_r = c_{f\text{-ex}}/(1 + c_{f\text{-in}} + c_r) \qquad (16)$$

where $c_{f\text{-ex}}[= k_3k_5/(k_2k_4)]$ is the external part of c_f and $c_{f\text{-in}}(= k_5/k_4)$ is the internal part. The sum of $c_{f\text{-ex}}$ and $c_{f\text{-in}}$ equals c_f.

When $c_{f\text{-ex}}$ is large, one will have a sticky substrate unless $c_{f\text{-in}}$ and/or c_r is even larger, and isotope effects on V/K will be suppressed. Those on V are not affected by $c_{f\text{-ex}}$ however. Even when a substrate is not sticky, isotope effects on V/K may be suppressed by large values of $c_{f\text{-in}}$ or c_r. The stickiness of substrates can distort the apparent pK values in V/K profiles, moving them outward and broadening the profile. Stickiness is detected by such alteration in pH profiles (compared to the true pK values seen in pK_i profiles of competitive inhibitors), by the isotope partition method, or by the effect of viscosity on V/K values.[20]

Energy for Catalysis

This is a subject about which many words have been written, but considerable controversy has erupted. The definition of catalysis is that the enzyme must bind the transition state more tightly than the initial EA complex and thus lower the activation energy for reaction of the EA complex. This statement does not, however, indicate how the enzyme does this or where the energy comes from to decrease the activation energy. Let us consider several ideas that have been put forth.

---

[20] W. W. Cleland, "Enzymes," 3rd ed., Vol. 19, p. 99, 1990; W. W. Cleland, *Methods Enzymol.* **87**, 390 (1982).

First, it has been proposed that the substrate is distorted or otherwise perturbed during initial binding to bring it closer to the transition state, with the inherent energy of binding used to accomplish this. This can be called ground state destabilization. The inherent energy of binding would come from favorable electrostatic charge interactions (such as a carboxyl group interacting with a lysine or arginine) and/or from the favorable entropy of shedding immobilized hydration shells of enzyme and substrate when they combine. London forces or hydrophobic effects are also invoked. These effects must cancel the entropy of bringing enzyme and substrate together.

While ground state destabilization may play a role in some enzymatic reactions, the important thing to remember is that the Michaelis constant of a substrate must be less than or equal to the physiological level for the enzymatic activity to be utilized, and thus whatever energetics are involved in binding must permit K_a to have an appropriate value. Too high a value does not permit at least half-saturation of the enzyme under physiological conditions, while too low a value limits the turnover number according to the limitations discussed earlier set by the Haldane relationship and the rate of diffusion of the substate in solution.

Let us focus, therefore, on the factors that permit a lowered activation energy for conversion of EA to the transition state. One is increased electrostatic interaction. This is probably responsible for most of the catalytic effect in the enolase reaction.[21] During this reaction a proton is removed by Lys-345 from C-2 of 2-phosphoglycerate to give an enolized "carbanion" intermediate that has an aci-carboxylate structure:

$$\text{HO—CH}_2\text{—}\underset{\underset{\text{H}}{|}}{\overset{\overset{\text{O—PO}_3^=}{|}}{\text{C}}}\text{—COO}^- \rightarrow \text{HO—CH}_2\text{—}\overset{\overset{\text{O—PO}_3^=}{|}}{\text{C}}\text{=C—O}^- \quad (17)$$

The transition states for forming such an unstable intermediate and for its further reaction will closely resemble it in structure. This intermediate is stabilized by electrostatic interaction with five positive charges supplied by two Mg^{2+} ions and a lysine:

[21] T. M. Larson, J. E. Wedekind, I. Rayment, and G. H. Reed, *Biochemistry* **35**, 4349 (1996).

It is important to remember that electrostatic interactions like these are more favorable in the low dielectric constant medium of an enzyme active site ($\approx$8–12) than in aqueous solution. Protonation of the hydroxyl at C-3 by Glu-211 permits its elimination from the intermediate to give water and phosphoenolpyruvate.

Another source of energy for enzyme catalysis is supplied by the formation of very strong hydrogen bonds in transition states or to intermediates where the transition states for formation of or further reaction of the intermediate will closely resemble them.[22,22a] Hydrogen bonds in aqueous solution are relatively weak, with values of ΔH for formation of $\leq$5 kcal/mol, and are $\geq$2.8 Å in length. In nonprotic solvents of lower dielectric constant, however, as well as in the gas phase and in crystals, hydrogen bonds can be shorter and stronger, with a few values of ΔH of formation reaching 25–30 kcal/mol in the gas phase.[23] What is essentially an electrostatic bond in aqueous solution becomes more and more covalent as the bond gets shorter and stronger, with the shortest bonds ($\approx$2.38 Å between oxygens) being largely covalent in nature.

In a weak hydrogen bond between groups of equal pK, there are two positions for the hydrogen, each $\approx$1 Å from one oxygen or nitrogen, with a longer bond ($\geq$1.8 Å) to the other heteroatom. As the bond shortens, the barrier between the two positions for the hydrogen decreases, and when it reaches the zero point energy level (at $\leq$2.5 Å for the overall bond length) there is no longer an effective barrier. The hydrogen is free to move anywhere between the heteroatoms, and we call such a bond a "low barrier hydrogen bond" or LBHB.

The strength of LBHBs depends critically on the pK match between the heteroatoms. Experimental studies suggest that the slope of the log of the formation constant of a hydrogen bond vs ΔpK is about 0.73 in nonprotic solvents.[24] Thus each pH unit mismatch in the pK value of the heteroatoms lowers the strength of the hydrogen bond by a factor of 5. Conversely, matching the pK value of atoms in a hydrogen bond when the pK values are initially quite different can strengthen the hydrogen bond greatly, thus providing the energy needed for catalysis.

A well-established example of this process is the reaction catalyzed by ketosteroid isomerase in which a nonconjugated ketone is converted into a conjugated one.[25] In the initial enzyme–steroid complex the keto group

[22] W. W. Cleland and M. M. Kreevoy, *Science* **264**, 1887 (1994).
[22a] J. A. Gerlt and P. G. Gassman, *Biochemistry* **32**, 11943 (1993).
[23] F. Hibbert and J. Emsley, *Adv. Phys. Org. Chem.* **26**, 255 (1990).
[24] S. Shan and D. Herschlag, *Proc. Natl. Acad. Sci. U.S.A.* **93**, 14474 (1996).
[25] C. Zhao, C. Abeygunawardana, P. Talalay, and A. S. Mildvan, *Proc. Natl. Acad. Sci. U.S.A.* **93**, 8220 (1996).

of the steroid ($pK < -5$) is hydrogen bonded to tyrosine-14 (measured pK 11.6). This will be a weak hydrogen bond because of the >15 pK mismatch. During the reaction, a proton is removed from C-4 by Asp-38 to give a dienolate whose pK in aqueous solution is 10 (the same as the pK of Tyr-14 under similar conditions). It was thus proposed that the matched pK values of the dienolate and of Tyr-14 permitted the hydrogen bond between them to become a very strong one, thus providing the energy to enolize the substrate and form the unstable intermediate.[22] The second half of the reaction involves the protonation of C-6 by Asp-38 to give a conjugated ketone and dissipates the strength of the hydrogen bond between the dienolate and Tyr-14.

The enzyme–dienolate complex has been mimicked by using an inhibitor aromatic in the ring containing a hydroxyl in the same position as the ketone of the substrate.[25] The pK of this phenolic hydroxyl of course matches that of tyrosine. Further, Asp-38 was changed to Asn to mimic the neutrality of protonated Asp-38 in the enzyme–intermediate complex. When the aromatic inhibitor was bound to this mutant, a proton NMR signal was seen at 18.15 ppm and a fractionation factor of 0.34 was measured for this proton; these are the characteristics of a LBHB. The increased strength of the LBHB relative to a hydrogen bond with the substrate was estimated to be 7.6 kcal/mol. This corresponds to a rate acceleration of 5.6 orders of magnitude, which is close to the effect caused by mutation of Tyr-14 to Phe. Later work has shown, however, that the proton at 18 ppm is actually between Tyr-14 and Asp-99 and that the proton between Tyr-14 and the dienolate resonates at 11.6 ppm.[26] Thus the proton of Tyr-14 is transferred largely to the dienolate in the intermediate, but at the transition state for formation of the intermediate the strongest hydrogen bond is probably between Tyr-14 and the dienolate.

Similar LBHBs are thought to be involved in the reactions of a number of enzymes that enolize their substrates and thus drastically alter the pK values of their reactants so that they match the pK values of groups that hydrogen bond to them.[22,22a] These include triose-P isomerase, aconitase, mandelate racemase, and citrate synthase.

Another well-characterized case of the involvement of LBHBs in enzymatic reactions is the mechanism of serine proteases.[27] The hydrogen bond between the Asp and the His of the catalytic triad is normally weak, as the aqueous pK of neutral imidazole is 14 and that of a carboxylic acid is ≈ 5 or higher in an enzyme active site. When the histidine becomes proton-

[26] C. Zhao, C. Abeygunawardana, A. G. Gittis, and A. S. Mildvan, *Biochemistry* **36,** 14616 (1997).

[27] P. Frey, S. Whitt, and J. Tobin, *Science* **264,** 1927 (1994).

ated as the result of serine adding to the substrate to form a tetrahedral intermediate, however, its inherent pK now matches that of Asp, and a LBHB forms between Asp and His (Fig. 14). The increased strength of the hydrogen bond lowers the activation energy for formation of the tetrahedral intermediate and thus facilitates catalysis of the reaction. This LBHB has been characterized by the low field proton NMR signal and low fractionation factor of its proton in complexes of serine proteases with fluorinated methyl ketone inhibitors that react to form stable tetrahedral adducts.[28] The pK of the proton in this LBHB is shifted upward as much as 5.5 pH units, suggesting an increase in hydrogen bond strength of a similar magnitude.

It is probable that strong hydrogen bonds play a role in another element of catalysis in enzymatic reactions: acid–base catalysis. When protons need to be removed from or added to an oxygen or nitrogen, enzymes employ various amino acid side chains as general acid–base catalysts. For example, in dehydrogenation of an alcohol the proton on the hydroxyl group must be removed either before or during the hydride transfer step. While primary alcohol dehydrogenases use Zn coordination to lower the pK of the hydroxyl group so that a Zn^{2+}-bound alkoxide forms, secondary alcohol dehydrogenases do not contain metal ions and the proton is transferred to a general base such as His, Asp, or Glu simultaneously with hydride transfer. In the initial enzyme–alcohol complex the pK of the alcohol will be ≈ 15, whereas that of the general base is usually 6–7. In the ketone product, the pK is ≤ -5. In neither case is the hydrogen bond to the catalytic base a strong one because of the pK mismatch.

There must be a point during the reaction, however, when the pK of the substrate crosses that of the general base, and it has been proposed that the transition state occurs close to this point and that a LBHB forms transiently.[22] (See Fig. 15.) That is, the strength of the hydrogen bond between substrate and catalytic base increases as the transition state is approached and decreases again once it is passed in concert with the pK changes. This increased strength at the transition state lowers the activation energy for the reaction and explains why acid–base catalysis in an enzymic reaction is often worth up to five orders of magnitude in rate acceleration.

Appendix

The free energy diagrams in this article were constructed with the use of the following computer program. PROFILE is written in IBM BASIC

[28] C. S. Cassidy, J. Lin, and P. A. Frey, *Biochemistry* **36,** 4576 (1997).
[29] P. Paneth and M. H. O'Leary, *Biochemistry* **24,** 5143 (1985); **26,** 1728 (1987).
[30] W. G. Mook, J. C. Bommerman, and W. H. Staverman, *Earth Planet. Sci. Lett.* **22,** 169 (1974).
[31] J. F. Marlier and M. H. O'Leary, *J. Am. Chem. Soc.* **106,** 5054 (1984).
[32] I. Simonsson, B.-H. Jansson, and S. Lindskog, *Eur. J. Biochem.* **93,** 409 (1979).

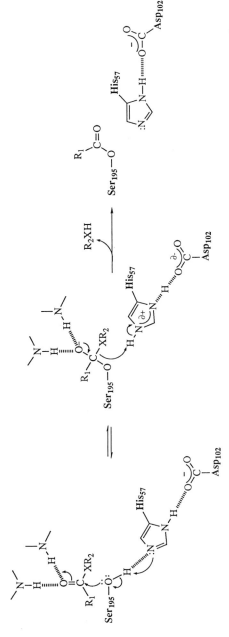

FIG. 14. Mechanism of serine proteases.[27] A low barrier hydrogen bond between His-57 and Asp-102 stabilizes the intermediate.

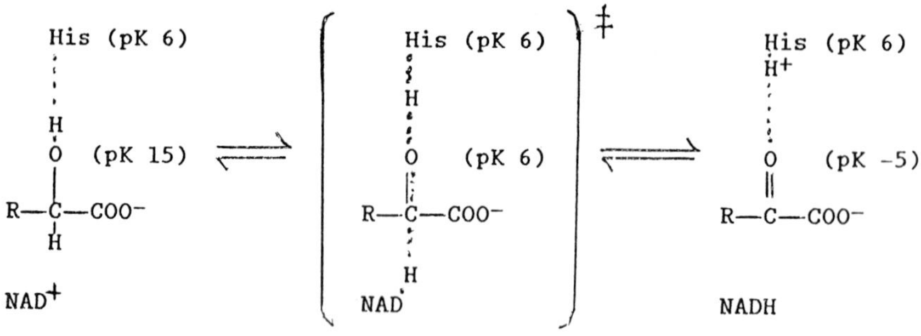

His (pK 6)

    ⋮

    H

    O (pK 15)

R—C—COO⁻

    H

NAD⁺

⇌

$$\left[\begin{array}{c} \text{His (pK 6)} \\ \vdots \\ \text{H} \\ \vdots \\ \text{O}\quad(\text{pK 6}) \\ | \\ \text{R—C—COO}^- \\ \vdots \\ \text{H} \\ \dot{\text{NAD}} \end{array} \right]^{\ddagger}$$

⇌

His (pK 6)
H⁺

    ⋮

    O (pK −5)
    ‖

R—C—COO⁻

NADH

Fig. 15. A low barrier hydrogen bond presumably forms at the transition state of the reaction catalyzed by lactate dehydrogenase as the pK of the substrate matches that of the histidine.

and should run in most other forms of BASIC as well (but the *natural* log function in statement 370 should be verified before use). For those not familiar with programming, the code can be prepared by word processing and saved as an ASCII file with an ASC extension. The user then executes IBM BASIC, types LOAD PROFILE.ASC, (enter key), and RUN. The program will call for the name of a data file to store the profile, the number of steps in the reaction mechanism, and the values for the rate constants. The rate constants are numbered consecutively up to twice the number of steps, with a maximum of 10 steps. PROFILE is intended to construct diagrams for reactions at equilibrium, but can accommodate portrayal of some nonequilibrium conditions, but not irreversibility. For a mechanism such as shown in Eq. (8), there are six steps but only 10 rate constants because the last two steps are irreversible. PROFILE will therefore ask for values for k_{10} and k_{12}, and small but finite values must be entered to avoid dividing by zero. To get an appropriate value for k_1 one must multiply the diffusion-controlled rate constant by a concentration for the substrate before entry, as illustrated in Fig. 4; similarly, the concentration of product must be multiplied times its associative rate constant as in Fig. 3. The calculations assume an Arrhenius preexponential of 6×10^{12} sec^{-1} for all steps. The temperature is set at 298° K, but can be changed in statement 240, if desired. The output is an array consisting of a series of 100 data points for each rate constant (plus 30-point trailers on each end for aesthetic balance), generated by modulating the amplitude of a cosine function in statement 370. The data points will be printed on the computer screen and in the named data file. In order to make use of the array in a graphics program, the user will have to leave IBM BASIC, load a graphics program, and import the point array from the data file according to procedures specific to the graphics program. A simple test of the program on the user's computer would be to enter "1" step followed by two rate constants of "10^5" each; the minimum value for the energy output in the data file should be "0" and the maximum value in the center of the array should be "10.6."

```
100    Rem    PROGRAM PROFILE   (JANUARY 1997, BY DBN)
110    Rem    X = REACTION COORDINATE
120    Rem    Y = ENERGY IN KCAL
130    Dim X(2000), Y(2000), K(20)
140    Print "ENTER THE NAME OF A FILE TO STORE PROFILE"
150    Input F$
160    Open F$ FOR OUTPUT AS #1
170    Print "ENTER THE NUMBER OF STEPS IN A SERIAL MECHANISM",
180    Input NUMBER
190    Print "ENTER THE VALUES FOR RATE CONSTANTS"
200    For I=1 to 2*NUMBER
210    Print "k",I,
220    Input K(I)
230    Next I
240    RT=1.987*298
250    POINTS=200*NUMBER+60
260    K(2*NUMBER+1)=K(2*NUMBER)
270    LAST=0
280    J=0
290    L=1
300    SIGN=-1
310    COUNT=-30
320    For I=1 to POINTS
330    COUNT=COUNT+1
340    J=J+1
350    X(J)=I
360    A0=COUNT*3.14159265/100
370    Y(J)= LAST-SIGN*RT*log(6*10^12/K(L))*(1-cos(A0))/2000
380    If COUNT<100 then 430
390    L=L+1
400    LAST=Y(J)
410    SIGN=-SIGN
420    COUNT=0
430    Print X(J),Y(J)
440    Print #1,X(J),Y(J)
450    Next I
460    Close #1
470    Stop
```

[2] Energy Coupling through Molecular Discrimination: Nicotinate Phosphoribosyltransferase

By CHARLES TIMMIS GRUBMEYER, JEFFREY W. GROSS, and MATHUMATHI RAJAVEL

Use of ATP Energy

ATP hydrolysis can be coupled energetically to other processes, either chemical or physical, forcing them toward completion. Mechanisms for energy use divide ATPases into two classes.[1] The enzymes of group 1, the synthetases, proceed via substrate-derived phosphorylated intermediates, such as the γ-glutamyl phosphate of glutamine synthetase, that chemically couple ATP hydrolysis to bond formation. Provided the γ-glutamyl phosphate intermediate undergoes productive nucleophilic attack rather than unproductive phosphohydrolytic decomposition, a high and unfavorable ratio of [Gln]/[Glu][NH$_3$], and a stoichiometry of one ATP cleaved per glutamine molecule formed will result. The classic determination of the free energy for ATP hydrolysis employing glutamine synthetase[2] requires this conservation of stoichiometry.

The ATPases of group 2 are fundamentally different in that no phosphorylated substrate-derived intermediates exist. Instead, the group 2 enzymes link the steps of ATP hydrolysis to changes in the properties of the enzyme, including discrimination toward free or bound ligands. The archetype for group 2 ATPases is the Ca-ATPase, in which the hydrolysis of ATP drives a sequence of protein-linked events in which Ca^{2+} is bound from low concentration in the sarcoplasm and released against a high concentration in the lumen of the sarcoplasmic reticulum, capturing the energy of ATP hydrolysis as an energetically equivalent Ca^{2+} concentration gradient.[3] The ability of the Ca-ATPase active site to differentially discriminate between two ligands (sarcoplasmic and cytoplasmic Ca^{2+}) at different reaction steps is the hallmark of the group 2 ATPases. Although the energetic linkage of group 2 ATPases has been referred to as "conformational coupling,"[1] this designation is not defining because it is not necessary to invoke profound or extensive protein conformational changes to produce the required discriminatory behavior. Because the essence of this coupling mecha-

---

[1] C. Tanford, *Annu. Rev. Biochem.* **52,** 379 (1983).
[2] J. Rosing and E. C. Slater, *Biochim. Biophys. Acta* **267,** 275 (1972).
[3] W. P. Jencks, *J. Biol. Chem.* **264,** 18855 (1989).

Copyright © 1999 by Academic Press
All rights of reproduction in any form reserved.
0076-6879/99 $30.00

TABLE I

ATP HYDROLYSIS COUPLED TO AFFINITY CHANGES[a]

Protein	Function	Transduction[b]
Rho	Transcription termination	ATP hydrolysis causes cycle of RNA affinity changes[1]
Lon, ClpA	Specific proteolysis	ATP hydrolysis coupled to binding and release of target protein[2,3]
Rec A	Strand exchange in recombination	ATP hydrolysis coupled to DNA strand exchange[4,5]
Gro EL	Prevents protein aggregation	ATP hydrolysis coupled to protein binding and release[6]
DNA	Unwinding DNA	Helicase II-promoted DNA unwinding reaction dependent on ATP hydrolysis[7,8]
Dna K	Protein folding	Binding and release of target protein coupled to ATP hydrolysis[9]

[a] Adapted from M. Rajavel et al., Biochemistry 35, 3909 (1996), with permission.

[b] Key to references: (1) B. L. Stitt, J. Biol. Chem. 263, 11130 (1988); (2) T. Armon, D. Ganoth, and A. Hershko, J. Biol. Chem. 265, 20723 (1990); (3) S. Gottesman and M. R. Maurizi, Microbiol. Rev. 56, 592 (1992); (4) T. T. Nguyen, K. A. Muench, and F. R. Bryant, J. Biol. Chem. 268, 3107 (1993); (5) S. C. West, Annu. Rev. Biochem. 61, 603 (1992); (6) J. Martin, T. Langer, R. Botera, A. Schramel, A. L. Horwich, and F. U. Hartl, Nature 352, 36 (1991); (7) J. G. Yodh and F. R. Bryant, Biochemistry 32, 7765 (1993); (8) P. M. Cullis, A. Maxwell, and D. P. Weiner, Biochemistry 31, 9642 (1992); and (9) A. Szabo, T. Langer, H. Schroder, J. Flanagan, B. Bukau, and F. U. Hartl, Proc. Natl. Acad. Sci. U.S.A. 91, 10345 (1994).

nism is the ability of the enzyme to discriminate ligands, it is appropriate to refer to the process as "ATP-driven molecular discrimination."

Jencks[3,4] has most clearly shown that the energetics of molecular discrimination follow the same logic as those of chemical coupling: enzyme-derived intermediates, to be energetically useful, must not be allowed to undergo futile decomposition, but only productive reaction. When the steps of a net ATPase reaction are intertwined with a second chemical process through such intermediates, energetic coupling must result.

The ion transporters and the ATPases of motility, both studied extensively, constitute two subgroups of enzymes using ATP-driven molecular discrimination, with the energy of hydrolysis transduced to physical work. A third subclass uses the energy of ATP hydrolysis to discriminate and process macromolecular targets. This subclass is large, mechanistically heterogeneous, and of very timely interest. Table I presents a number of examples.

[4] W. P. Jencks, Adv. Enzymol. 51, 75 (1980).

How can ATP-linked changes in ligand discrimination drive coupled reactions? While some superficially attractive mechanisms can be proposed, making them perform work in a kinetically reasonable fashion is difficult. A review of this subject[5] was subtitled "The importance of the details," emphasizing the need for complete elucidation of molecular mechanisms.

This article first explores how ATP hydrolysis might drive a hypothetical reaction through conformational linkage, explorations that predict three routes for intrinsic inefficiency. Second, nicotinate phosphoribosyltransferase is presented as a metabolic enzyme that provides an excellent model for conformationally coupled ATPases. NAPRTase successfully uses ATP to drive the formation of NAMN, but, as suggested by the theoretical analysis, the enzyme is a profligate consumer of ATP.

Some Theory

Many workers, including Jencks,[4] have presented thorough considerations of the energetics of molecular discrimination in ion transporters. For a generally applicable understanding of ATP use in molecular discrimination, ion transporters may not be ideal models, as the asymmetric barrier imposed by the membrane introduces a complicating factor not present in solution-phase reactions. Instead, a comparison of the well-studied S/P isomerase reaction with its ATP-coupled analog provides a groundwork for analysis. In the nonlinked version, Eq. (1), E can bind either S or P, and $K_{eq} = K_1 K_2 K_3$. As P builds up, its frequency of productive binding to E becomes the same as that for S, and equilibrium is reached.

$$E + S \rightleftharpoons E \cdot S \rightleftharpoons E \cdot P \rightleftharpoons E + P \tag{1}$$

For ATP to drive the reaction further toward P the enzyme would need to acquire a higher discrimination for S and against P than is allowed by K_{eq}. Well-documented cases of allostery show that it is possible to change the binding or catalytic properties of an enzyme readily, but the energy of allosteric binding is only "borrowed" and cannot drive the reaction. Equation (2) shows a typical case in which an allosteric activator remains bound to the enzyme throughout multiple reaction cycles, converting it to an enzyme · ligand complex, designated E', with properties different from E.

$$E' + S \rightleftharpoons E' \cdot S \rightleftharpoons E' \cdot P \rightleftharpoons E' + P \tag{2}$$

K_{eq}, a property of S and P, does not alter, and any attractive change in one K (e.g., an increase in K_1 to promote S binding) is offset by other deleterious changes (decreased K_2 or K_3).

[5] B. Alberts and R. Miake-Lye, *Cell* **68,** 415 (1992).

When ATP hydrolysis is added to the S/P conversion, even if it does not participate directly in the chemistry of S isomerization, the reaction can be driven far to the right. Perhaps the simplest intertwining of ATPase and P synthesis is presented in Eq. (3), most of whose features are maintained in the many other possible variants in which ATP use is inserted more directly into steps of the S/P conversion:

$$\overset{\displaystyle \text{ATP} \quad \text{ADP} + \text{P}_i}{\underset{\displaystyle \text{E} \rightleftharpoons \text{E}' + \text{S} \rightleftharpoons \text{E}' \cdot \text{S} \rightleftharpoons \text{E}' \cdot \text{P} \rightleftharpoons \text{E} \cdot \text{P} \rightleftharpoons \text{E}}{\displaystyle \qquad}} \qquad (3)$$

In Eq. (3), ATP hydrolysis converts E to an energized state, designated E', which later decomposes to E. In this scheme, $K'_{eq} = K_{eq}(K_{hyd\ ATP})$. For successful capture of ATP hydrolysis energy, the only essential feature of Eq. (3) is that no unproductive branches exist. For example, in Eq. (3), E' and E' · P cannot return to E without either net formation of P or resynthesis of ATP. Additionally, E cannot perform overall conversion of P to S unless E' is formed. Finally, E' is not allowed to bind free P, nor is E' · P allowed to release P. These "rules"[3,4] for molecular discrimination prevent uncoupled processes and are the required feature of energetically coupled reactions.

This logic explains the overall thermodynamics of coupling through molecular discrimination; it does not clarify its practical aspects. What strategies should the enzyme adopt to use ATP with maximal catalytic efficiency? How does the enzyme of Eq. (3) "know" that P has been produced and that it can now discharge the energized state? The difference between K_{eq} and K'_{eq} (i.e., the energy acquired from ATP hydrolysis) could be used to alter any or all of the five internal equilibrium constants for the scheme, allowing E' to assume ligand binding and catalytic properties substantially different from those of E. Which properties should be changed to maximize catalytic efficiency? Which *can* be changed?

The most obvious property to change is the affinity of E and E' for S and P. If the affinity of E' for S were high [i.e., an increased value for K_1 of Eq. (1)], the energized enzyme could bind S from low concentration. Similarly, if E were given a low affinity for P [i.e., an increased value for K_3 in Eq. (1)], the product could be released against a high concentration. Thus, it appears initially that a key useful discrimination is that of E' in favor of S (and thus against P) or of E against P (and thus in favor of S). However, this logic essentially states that both E and E' bind P poorly and that E and E' bind S tightly: no affinity changes have occurred between the two forms. Discriminatory ability could be assigned specifically to E', binding S tightly and P poorly, with E binding both S and P poorly. However, all schemes in which $E' \cdot P$ allows the escape of P contain the reversible path, Eq. (4):

$$E' + S \rightleftharpoons E' \cdot S \rightleftharpoons E' \cdot P \rightleftharpoons E' + P \qquad (4)$$

that is identical to Eq. (1) and must share its K_{eq}. The manipulation of variants of Eq. (4) shows that only one set of affinity changes is reasonable in Eq. (3): if the affinity of E' for S is high, then the affinity of E' for P is also high, whereas if that of E for P is low, then that of E for S must also be low. While the high affinity for S and the low affinity for P achieve the goals of efficiently binding substrate and expelling product, neither of the derived properties seems desirable. In particular, high affinity of E' for P would lead to competition between S and P for E', making P a potent inhibitor.

A route around the problem posed by tight binding of P by E' is to create an alternative route for its expulsion by allowing adventitiously bound P to cause the same discharge of the high energy state that results from nascent P. This branched scheme breaks a rule for energy coupling, but in the process becomes kinetically realistic:

$$
\begin{array}{c}
\overset{\displaystyle \text{ATP} \,\,\text{ADP} + P_i}{\overbrace{}} \\[2pt]
E \rightleftharpoons E' + S \rightleftharpoons E' \cdot S \rightleftharpoons E' \cdot P \rightleftharpoons E \cdot P \rightleftharpoons E \qquad (5) \\[6pt]
\updownarrow \qquad\qquad \updownarrow \\[4pt]
E \cdot P \rightleftharpoons E' \cdot P
\end{array}
$$

In the modified scheme of Eq. (5), E has low affinity for P, allowing product release from $E \cdot P$ complexes, and E' has high affinity for P, disallowing its escape from $E \cdot P$. The cost of this branch is the energy expended through the decomposition of dead end $E' \cdot P$ complexes. Through ATP use, the enzyme has become an expensive Maxwell's demon, binding a ligand and accepting it for processing if it is S or rejecting it (and in the process discharging the energized state) if it is P. This mechanism is formally identical to the kinetic proofreading proposed for various polymerases,[6,7] with discrimination between substrate and product rather than between alternative substrates. Generating a "smart" enzyme through ATP use is expensive—knowledge is energy, and discrimination requires work. In addition, a defining property of any enzyme, the reaction stoichiometry, has been set free to vary: at high P levels, the isomerase could require many rounds of ATP hydrolysis to reject bound P molecules before an S molecule is bound.

[6] J. J. Hopfield, *Proc. Natl. Acad. Sci. U.S.A.* **71**, 4135 (1974).
[7] S. M. Burgess and C. Guthrie, *Trends Biochem. Sci* **18**, 381 (1993).

Equation (5) can be reformulated in a way that allows easier quantitation of its inefficiency:

$$E' + S \rightleftharpoons E' \cdot S \rightleftharpoons E' \cdot P \rightleftharpoons E' + P$$
$$\Updownarrow \qquad \qquad (6)$$
$$E + S \rightleftharpoons E \cdot S \rightleftharpoons E \cdot P \rightleftharpoons E + P$$

In Eq. (6), the formation of E' from E and ATP has been omitted. The two horizontal pathways represent conversion of S to P in which ATP is not used. Each is drawn as reversible and must have the same K_{eq} as the chemical conversion of S and P. The pathway from upper left to lower right is the productive ATP-linked S isomerization. The upper right (binding of P to E') and lower left (dissociation of S from E $\cdot$ P) represent locations where coupling is lost. The pathway from upper left to lower left represents futile loss of E'. Product-induced ATP hydrolysis proceeds from upper right to lower right. The efficient use of ATP depends on the extent to which the nonproductive branches operate. Maximal efficiency requires the tight binding of ligands by E', loose binding by E, a favorable equilibrium for the conversion of E' $\cdot$ P to E $\cdot$ P, and a very poor rate for conversion of E $\cdot$ P to free S. If these affinity and catalytic differences between E and E' are maximized to those allowed by the energy of ATP hydrolysis, the lower left pathway becomes minimal and ATP use becomes optimal. This optimal ATP use is still very poor and is in direct proportion to the [P]/[S] ratio: if binding of P and S to E' occurs with equal rate constants, then at [P]/[S] ratios of 1:1, the stoichiometry of ATP hydrolyzed/P formed is 2, and at [P/S] ratios of 10:1, the stoichiometry increases to 11 ATP hydrolyzed/P formed. The maximal [P]/[S] gradient attainable is limited by the energy of ATP hydrolysis. However, any leaks in the pathway, such as a finite rate of conversion of E $\cdot$ P to E + S, add further inefficiency. In this case, the maximal [P]/[S] gradient attainable is the ratio of the partition coefficient for E $\cdot$ P toward free P over that toward free S, which reduces to $k_{off P}/K_{rev}$, where $k_{off P}$ is used to designate the off rate of P from E $\cdot$ P and k_{rev} represents the net rate constant[8] for conversion of E $\cdot$ P to free S.

So far, intuitively sound strategies to link ATP use to a solution phase reaction have led to a loss of reaction stoichiometry and efficiency. Two mechanisms are considered that appear to avoid these problems, but are themselves flawed. One mechanism that appears fundamentally different from Eq. (3) is that of making the collapse of the energetic conformer E' truly concerted with the S/P isomerization step:

[8] W. W. Cleland, *Biochemistry* **14**, 3220 (1975).

$$\text{E} \xrightleftharpoons{\overset{\text{ATP}\quad\text{ADP + P}_i}{\frown}} \text{E}' + \text{S} \rightleftharpoons \text{E}' \cdot \text{S} \rightleftharpoons \text{E} \cdot \text{P} \rightleftharpoons \text{E} \qquad (7)$$

In this scheme, no $\text{E}' \cdot \text{P}$ complexes occur, and $\text{E} \cdot \text{P}$ is not allowed to convert to $\text{E} \cdot \text{S}$. Therefore, E can be assigned a low affinity for both S and P and E' a high affinity for S, and a low affinity for P. In essence, Eq. (7) simply removes the connections between the productive pathway in Eq. (6) and the two paths for loss of coupling. However, this removal is arbitrary and instantaneous and the communication between substrate chemistry and protein conformation that could lead to a truly concerted conformational change is difficult to imagine.

Additional conformers of the enzyme can also be proposed:

$$\text{E} \xrightleftharpoons{\overset{\text{ATP}\quad\text{ADP + P}_i}{\frown}} \text{E}' \rightleftharpoons \text{E}' \cdot \text{S} \rightleftharpoons \text{E}'' \cdot \text{S} \rightleftharpoons \text{E}'' \cdot \text{P} \rightleftharpoons \text{E} \cdot \text{P} \rightleftharpoons \text{E} \qquad (8)$$

In this formulation, E has affinity for neither ligand, E' is a discriminatory form, binding S but not P, and E'' binds both S and P. In this scheme, as written, there is no problem of inhibitory P binding, as only E'' has any affinity for P, and E'' only results from $\text{E}' \cdot \text{S}$ complexes, not any form of apoenzyme. In essence, this scheme, and its more baroque variants, simply adds additional horizontal pathways to Eq. (6) and requires a succession of enzyme conformers with substantially different properties.

Thus, it is difficult to assign E and E' properties that result in a pathway that is unbranched, thermodynamically possible, and kinetically reasonable. In fact, in detail, Eq. (3) simply does not fit with enzymic reality. Only Eq. (5), which is energetically expensive, is functional. Some of the basic assumptions made so far could be altered, e.g., by changing the position of ATP use or introducing an E-P form of the enzyme, but neither of these appears to alter this fundamental conclusion.

Jencks' rules indicate that energetically coupled systems must link steps of macromolecular discrimination and ATP hydrolysis, but it is not immediately clear how this communication occurs. In the chemically coupled glutamine synthetase reaction, the electronic structure of the tetrahedral intermediate dictates productive Pi release or energy-conserving return to glutamyl phosphate, which is not the case when intermediates are coupled through molecular discrimination. One could envision "signal" and "clock" type mechanisms for the flow of information linking reaction steps. In a signaling mechanism, formation of product would indicate successful completion of the coupled reaction and cause discharge of the energized state. Equation (3) falls into this class: P formation triggers collapse of E' to E.

In a clock mechanism, rate constants for the collapse of the high energy state would allow time for individual steps of a second reaction to occur, without requiring direct feedback. Many examples of ATP-driven molecular clocks exist, most notably the $\alpha\beta\gamma$ GTPases, in which GTP binding allows dissociation and activation of α subunits, which react freely until subsequent GTP cleavage causes their reassociation with $\beta\gamma$.[9] Recent years have seen an abundance of accessory proteins that essentially serve to reset the timing mechanisms of GTPase clocks. In a sense these proteins serve as signals, but their signaling is extrinsic to the GTPase process. Superficially, clocks seem attractive as mechanisms to link reactions energetically. However, these mechanisms have flaws that obviate their ability to conserve ATP energy as a [P]/[S] gradient:

$$\text{ATP} \overbrace{\quad} \text{ADP} + P_i$$
$$E \rightleftharpoons E' + S \rightleftharpoons E' \cdot S \rightleftharpoons E' \cdot P \rightleftharpoons E' + P \rightleftharpoons E \qquad (9)$$

In the clock-type mechanism of Eq. (9), ATP hydrolysis generates the intermediate E'. The rate of decomposition of E' to E is intrinsic to the enzyme and can occur from all forms of E' (including E' and E' · S; these pathways are not shown). The desirable enzyme properties in this model would be high affinity of E' for S, allowing its binding from low concentration, low affinity of E for P, allowing its release, and a high $K_{internal}$.

Equation (9) has a fundamental violation of rules: E', a required intermediate in the pathway, is allowed to break down to E without obligatory P or ATP formation (as is E' · S) and will do so when low S concentrations limit the rate of E' · S formation. Second, because the free energy of E' + S equals that of E' + P, any gains made by driving $K_{internal}$ toward E' · P must result in tighter E' binding of P than S. In clock mechanisms, ATP is hydrolyzed, but K_{eq} is unchanged.

If generating thermodynamically perfect and kinetically reasonable linkages is so difficult, how is it possible that the ion transporters work at high efficiency and with very reasonable kinetic properties? Using Eq. (3), a scheme for a hypothetical inward-directed ion pump can be constructed:

$$\text{ATP} \overbrace{\quad} \text{ADP} + P_i$$
$$E \rightleftharpoons E' + I_{out} \rightleftharpoons E' \cdot I_{out} \rightleftharpoons E' \cdot I_{in} \rightleftharpoons E \cdot I_{in} \rightleftharpoons E + I_{in}$$
$$(10)$$

In this scheme, I_{out} and I_{in} designate the ion and are analogs of S and P, with the exception that the ion is chemically unchanged by its transport. The desirable property of E' is its high affinity for I_{out}, and the desirable

---

[9] E. J. Neer, *Cell* **80**, 249 (1995).

property of E is its low affinity for I_{in}. The step $E' \cdot I_{out} = E' \cdot I_{in}$ is analogous to $E' \cdot S = E' \cdot P$, and it is desirable, but not necessary, to have a high $K_{internal}$. As written, I_{in} represents a "signal" to promote the discharge of E' to E. Equation (10) suffers from the same flaws as Eq. (3). In particular, if E' is assigned high affinity for the ion, then inhibition by I_{out} will occur, and assigning E' a low affinity for I_{out} gives rise to a branch with no driving force. The solution to this problem is the asymmetry of the transporter with respect to the two separate compartments. E' can be assigned a high affinity for ions if it is given no access to internal ions, and E can be assigned low affinity for ions if it is not allowed access to the outside compartment. $E' \cdot I_{out}$ and $E' \cdot I_{in}$ are chemical isomers, but must be asymmetric with respect to the membrane. We have become aware of the many observations of ATP-linked systems with apparent variations in stoichiometry. In particular, the example of Ca-ATPase is analyzed convincingly by Yu and Inesi.[10]

In summary, for ATP to drive a reaction using enzyme-derived intermediates as energy carriers, linkage between ATP hydrolysis and that reaction is required. A good strategy for the enzyme is to use affinity changes toward S and P, although these changes are more limited than initially appreciated. We have also seen that linkage must involve signals rather than clocks. This raises a problem that is apparently fundamental: If the signal is the product, then it is possible for the adventitious product to cause discharge of the high energy state. This discharge of the energetic state by bound P can serve to expel adventitious P bound tightly to energized enzyme, but causes ATP hydrolysis superstoichiometric with product formation.

NAPRTase

The study of energy-linked ATPases has progressed somewhat slowly. One of the problems with many ATP-coupled systems is that intermediates are difficult to identify and quantitate, requiring functional assays or gel electrophoresis.[5]

NAPRTase (EC 2.4.2.11) presents a nearly unique metabolic enzyme in which ATP drives a solution phase chemical reaction through conformational intermediates, without substrate phosphorylation (note: an analogous reaction, ATP sulfurylase, or APS synthetase, is discussed in this volume[31]). NAPRTase was first identified from human erythrocytes in 1958 by Preiss and Handler[10a] and its phosphoribosyltransferase reaction characterized:

[10] X. Yu and G. Inesi, *J. Biol. Chem.* **270**, 4361 (1995).
[10a] J. Preiss and P. Handler, *J. Biol. Chem.* **233**, 492 (1958).

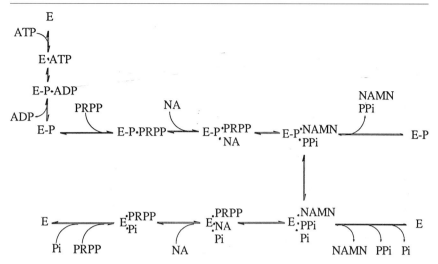

SCHEME I. Comprehensive NAPRTase mechanism. The ATP-coupled reaction starts at the upper left and proceeds to the lower right. NAPRTase binds to ATP, is covalently phosphorylated, and releases ADP. The high affinity E-P binds to PRPP and then NA, catalyzes phosphoribosyl transfer chemistry, and then hydrolyzes to the low affinity E conformer, from which products are released. The uncoupled reaction is represented by the lower horizontal pathway.

The formation of NAD proceeds through the NAMN product, which is generated by QAPRTase from the *de novo* substrate quinolinic acid or by NAPRTase through the recycling intermediate nicotinic acid (NA; Fig. 1).[11,12] NAPRTase is typical of the PRTase group in employing PRPP as the ribose 5-phosphate donor and generating PPi as one product.[13]

The unique feature of NAPRTase is its use of ATP. Imsande and Handler[14] first demonstrated that the reaction of the beef liver enzyme was stimulated by ATP. Honjo *et al.*[15] extended these results by showing that ATP is not simply an allosteric activator, but a substrate that is cleaved to ADP and P_i equimolar with NAMN synthesis. Surprisingly for a cosubstrate, ATP is not obligatory for NAPRTase; NAPRTases from mammalian sources and bacteria utilize ATP when present, but are able to catalyze

[11] G. J. Tritz, in "*Escherichia coli* and *Salmonella typhimurium*: Cellular and Molecular Biology" (F. C. Neidhardt, ed.), p. 557. ASM Press, Washington, DC, 1987.

[12] T. Penfound and J. Foster, in "*Escherichia coli* and *Salmonella typhimurium*: Cellular and Molecular Biology" (F. C. Neidhardt, ed.), p. 721. ASM Press, Washington, DC, 1996.

[13] W. D. L. Musick, *Crit. Rev. Biochem.* **11,** 1 (1981).

[14] J. Imsande and P. Handler, *J. Biol. Chem.* **236,** 525 (1961).

[15] T. Honjo, S. Nakamura, Y. Nishizuka, and O. Hayaishi, *Biochem. Biophys. Res. Commun.* **25,** 199 (1966).

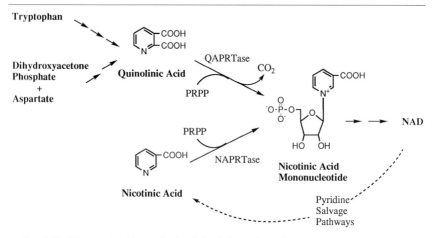

FIG. 1. Pyridine nucleotide synthesis. Quinolinic acid, the *de novo* substrate, is synthesized from tryptophan metabolism in eukaryotes and from dihydroxyacetone phosphate and aspartate in prokaryotes. Several enzymatic routes ("pyridine nucleotide cycles") exist for the formation of nicotinic acid from NAD.

ATP-independent NAMN synthesis,[14,16] and the enzyme from certain protozoa is unaffected by ATP.[17] In both mammalian and *Salmonella* NAPRTases, ATP reduces K_m values for substrates and increases V_{max}.[16,18]

ATP is not likely to react chemically with NAPRTase substrates to stimulate NAMN synthesis. The C1-pyrophosphate bond of PRPP is highly energetic[19] and activated toward nucleophilic attack, and phosphorylation of NA would not help chemistry. The fact that all the other PRTases catalyze analogous reactions without a requirement for ATP, and that NAPRTase is active in the absence of ATP, diminish the likelihood that ATP is required for the chemistry of N-glycosidic bond formation.

Is ATP solely a stimulator for NAPRTase or does its hydrolysis actually drive NAMN formation? In the presence of ATP, the NAPRTase equilibrium with the beef liver enzyme was too high to measure.[14] The catalysis of NAMN formation in the absence of ATP with the *Salmonella* enzyme allowed the determination of the equilibrium for that reaction as 0.67.[16] In the presence of ATP the product/substrate ratio [NAMN][PPi]/[PRPP][NA] was shifted far to the right, and a value of 1100 was determined. This result showed that ATP use was producing a 1640-fold increase in the

[16] A. Vinitsky and C. Grubmeyer, *J. Biol. Chem.* **268**, 26004 (1993).
[17] V. Kahn and J. J. Blum, *Biochim. Biophys. Acta* **146**, 305 (1967)
[18] L. D. Smith and R. K. Gholson, *J. Biol. Chem.* **244**, 68 (1969).
[19] P. A. Frey and A. Arabshahi, *Biochemistry* **34**, 11307 (1995).

product/substrate ratio, which is impressive, but far below the six orders of magnitude expected if ATP use was coupled perfectly to NAMN formation. As discussed later, even the value of 1640 exaggerates the efficiency of ATP use: in the steady state, ATP hydrolysis becomes superstoichiometric with NAMN formation, and many ATP molecules are hydrolyzed for each NAMN formed. Clearly there is a puzzle here: How is ATP use linked to NAMN formation and why isn't it linked better?

Cloning and Overproduction of NAPRTase

The *pncB* genes, which encode for NAPRTase, have been cloned and sequenced for both *Escherichia coli*[20] (accession No. J05568) and *Salmonella*[21] (accession No. M55986). The yeast sequence has also appeared in sequence banks (accession No. Z75117 and S51845). A newly available sequence from *Mycobacterium tuberculosis* (accession No. Z73902) shows similarity to NAPRTase, with a high degree of sequence identity in regions conserved between bacterial and yeast enzymes.

Methods have been published for the T7-based overexpression and purification of the *Salmonella* enzyme,[22] and the yeast NAPRTase has been overexpressed and purified by the same technique (M. Rajavel, unpublished). The *Salmonella* protein is composed of 399 amino acid residues with an M_r of 45,529 predicted by the DNA sequence and the known removal of the amino-terminal methionine.[21] Gel filtration studies demonstrated that the *Salmonella* protein is monomeric.

Protein measurements are made spectrophotometrically. An $E_{280\,nm}^{0.1\%} = 1.27$ for *Salmonella* NAPRTase[16] and a value of $E_{280\,nm}^{0.1\%} = 1.33$ for yeast NAPRTase were calculated using the method of Gill and von Hippel.[23] The *Salmonella* value was reexamined using quantitative amino acid analysis. The revised extinction coefficient, 1.65, is 1.3-fold higher and raises phosphorylation stoichiometry values published previously from 0.6–0.7 to 0.8–0.9 (J. W. Gross, unpublished). The high extinction coefficient for NAPRTase is the result of its eight tryptophan residues.

Assay of Enzymatic Activity

Spectrophotometric ATPase assays are performed using a pyruvate kinase/lactate dehydrogenase system.[16] The spectrophotometric assay of

[20] M. G. Wubbolts, P. Terpstra, J. van Beilen, J. Kingma, H. A. R. Meesters, and B. Witholt, *J. Biol. Chem.* **265,** 17665 (1990).

[21] A. Vinitsky, H. Teng, and C. Grubmeyer, *J. Bacteriol.* **173,** 536 (1991).

[22] M. Rajavel, J. Gross, E. Segura, W. T. Moore, and C. Grubmeyer, *Biochemistry* **35,** 3909 (1996).

[23] S. C. Gill and P. H. von Hippel, *Anal. Biochem.* **182,** 319 (1989).

PP_i-stimulated ATPase is conducted as for coupled ATPase, except PRPP and NA are omitted and PP_i is added to 1 mM. The [^{14}C]NA label transfer assay of Preiss and Handler[10a] is performed as described.[16]

[^{14}C]NA is also separated from [^{14}C]NAMN using high-pressure liquid chromatography. Aliquots (100 μl) of perchloric acid quenched and neutralized reactions are injected onto a μBondapack C_{18} column (3.9 × 300 mm) equilibrated in 100 mM sodium acetate, pH 4.5, and eluted at 1 ml/min under isocratic conditions. NAMN elutes at 4.5 ml and NA at 6.5 ml.

The Reaction Cycle

A kinetic scheme for NAPRTase is shown in Scheme I. Kosaka et al.[24] and Hanna et al.[25] used steady-state kinetics to delineate the major kinetic features of NAPRTase, including the ping-pong nature of the ATP-coupled productive pathway, the initial binding of ATP, and the order of product release. The kinetic mechanism of NAPRTase contains several nonproductive branches, discussed later.

The existence of a covalent phosphoenzyme (E-P) derived from ATP was inferred from the existence of relatively rapid ADP/ATP exchanges seen in the absence of NA and PRPP.[15,26] In addition, Kosaka et al.[24] showed that enzymes incubated with [γ-^{32}P]ATP and chromatographed by gel filtration retained radioactivity that could be discharged by the addition of NA and PRPP. The covalent nature of E-P was confirmed by Gross et al.[27] who demonstrated that radioactivity from [γ-^{32}P]ATP remained associated with denatured NAPRTase through gel filtration or SDS–PAGE. The $k_{\text{hydrolysis}}$ for denatured E-P at various pH values (0.0006 min^{-1} at pH 10.0 and 0.045 min^{-1} at pH values from 2 to 7, with a sharp increase at pH values below 2) and the rates of pyridine-, hydroxylamine-, and iodine-catalyzed E-P hydrolysis were indicative of a 1-phosphohistidine (π) linkage and ruled out phosphoester, acyl phosphate, and thiophosphate chemistries for E-P. Amino acid sequencing of tryptic and chymotryptic peptides showed that E-P is formed by the phosphorylation of His-219.[27] His-219 is conserved among known NAPRTase sequences. Comparison of sequences around His-219 with those of other phosphohistidines (Table II) revealed no convincing similarity. The 3-phosphohistidines of bacterial two-component regulatory systems show a consensus sequence HyAHEHy (in which Hy designates a hydrophobic amino acid) that is followed by

[24] A. Kosaka, H. O. Spivey, and R. K. Gholson, *Arch. Biochem. Biophys.* **179,** 334 (1977).
[25] L. S. Hanna, S. L. Hess, and D. L. Sloan, *J. Biol. Chem.* **258,** 9745 (1983).
[26] A. Kosaka, H. O. Spivey, and R. K. Gholson, *J. Biol. Chem.* **246,** 3277 (1971).
[27] J. Gross, M. Rajavel, E. Segura, and C. Grubmeyer, *Biochemistry* **35,** 3917 (1996).

TABLE II
AMINO ACID SEQUENCES SURROUNDING SOME KNOWN PHOSPHOHISTIDINES[a]

NAPRTase (*Salmonella typhimurium*)	A-L-T-P-M-G-T-Q-A-**H**-E-W-F-Q-A-**H**-Q-Q-I-S
VirA (*Agrobacterium tumefaciens*)	V-G-T-L-A-G-G-I-A-**H**-E-F-N-N-I-L-G-A-I-L
NtrB (*Escherichia coli*)	A-R-D-L-V-R-G-L-A-**H**-E-I-K-N-P-L-G-G-L-R
EnvZ (*E. coli*)	R-T-L-L-M-A-G-V-S-**H**-D-L-R-T-P-L-T-R-I-R
HPr (*S. typhimurium*)	V-T-I-T-A-P-N-G-L-**H**-T-R-P-A-A-Q-F-V-K-E
III^{LAC} (*Staphylococcus aureus*)	S-V-T-M-M-**H**-G-Q-D-**H**-L-M-T-T-I-L-L-K
EII^{mtl} (*E. coli*)	Y-L-G-E-S-I-A-V-P-**H**-G-T-V-E-A-K
CheA (*E. coli*)	Q-L-N-A-I-F-R-A-A-**H**-S-I-K-G-G-A-G-T-F-G
Histone H4 (calf thymus)	I-R-D-A-V-T-Y-T-E-**H**-A-K-R-K-T-V-T-A-M-D
Succinyl-CoA synthetase (*E. coli*)	T-A-P-K-G-K-R-M-G-**H**-A-G-A-I-I-A-G-G-K-G
Nucleoside diphosphate kinase (human RBC)	C-I-Q-V-G-R-N-I-I-**H**-G-S-D-S-V-E-S-A-E-K
Bisphosphoglycerate mutase (human RBC)	S-K-Y-K-L-I-M-L-R-**H**-G-E-G-A-W-N-K-E-N-R
Phosphoglycerate mutase (yeast)	P-K-L-V-L-V-R-**H**-G-Q-S-E-W-N-E-K-N-L
Pyruvate, phosphate dikinase (*Bacteroides symbiosus*)	G-G-M-T-S-**H**-A-A-V-V-A-R

[a] Reprinted with permission from Gross *et al.*, *Biochemistry* **35**, 3917 (1996).[27] Copyright 1996 American Chemical Society.

NAPRTase, but those enzymes contain other motifs that are not seen in NAPRTase.[27]

At pH 8.3, native E-P is unstable (0.18 min^{-1} with 5 mM Mg^{2+}; 0.068 min^{-1} with 1 mM EDTA) compared to denatured E-P (0.0042 min^{-1}). The fold increase in E-P stability on enzyme denaturation suggests that the phosphoramidate bond in native E-P is destabilized by about 2 kcal/mol.

H219N and H219E mutant NAPRTases have been prepared and purified. These enzymes do not become phosphorylated and fail to catalyze ADP/ATP exchanges (M. Rajavel, unpublished). The mutants have helped to clarify the relationship between binding of ATP and substrates. Like wild-type enzyme, His-219 mutants catalyze uncoupled NAMN synthesis (later), with similar k_{cat} values. In these mutants, ATP no longer stimulates NAMN formation, but instead acts as an inhibitor. In H219N, inhibition is competitive versus PRPP, $K_I = 0.4$ mM. A similar phenomenon can be seen with the wild-type enzyme in that the nonhydrolyzable AMP-PCP inhibits uncoupled NAMN formation weakly. The implication of these results is that ATP and PRPP binding are mutually exclusive.

The E-P binds PRPP and NA to produce a complete substrate complex. PRPP binding is with high affinity, $K_m = 22$ μM, as is that of NA, $K_m = 1.5$ μM.[16] Phosphoribosyl transfer occurs within this complex to produce bound products, E-P undergoes phosphohydrolysis, and all products are released. The overall k_{cat} is 2.9 sec^{-1} at 30° and 2.3 sec^{-1} at 23°.

To understand the order and relationship of the various catalytic events, rapid quench experiments at 23° were carried out (J. W. Gross, unpublished). An Update Instruments Model 1010 precision syringe ram was employed, with aging hoses and ram speed varied to produce aging times between 2.5 and 1000 msec. NAPRTase apoenzyme, E-P, or E-P · PRPP complexes can be formed in one syringe using appropriate conditions. These complexes are mixed with additional substrates in the second syringe and then quenched (NaOH for E-^{32}P measurements, perchloric acid for subsequent analysis of [^{14}C]NAMN). Rapid quench experiments have shown that enzyme phosphorylation from bound ATP occurs at about 29 sec^{-1} and is not significantly rate limiting for k_{cat}. Although a value for the rate of ADP release has not yet been determined, its dissociation appears not to participate in rate limitation. By measuring the concentration dependence of NAMN formation rates, it has been found that PRPP binding occurs with an on rate of at least $0.1 \times 10^6 \ M^{-1} \ sec^{-1}$. The minimal on rate for NA is $7.5 \times 10^6 \ M^{-1} \ sec^{-1}$. These values are very similar to those for PRPP ($0.2 \times 10^6 \ M^{-1} \ sec^{-1}$) and hypoxanthine ($20 \times 10^6 \ M^{-1} \ sec^{-1}$) binding to human HGPRTase.[28] On binding of NA to the E-P · PRPP complex, phosphoribosyl group transfer occurs at a rate of about 500 sec^{-1}, very rapid compared to k_{cat}. Finally, on addition of NA and PRPP to E-^{32}P, overall E-P hydrolysis occurs at about 6.3 sec^{-1}.

The major rate limitation seen in pre-steady-state reactions appears to come from E-P cleavage. However, that rate of 6.3 sec^{-1} is still too fast to explain the overall k_{cat} of 2.3 sec^{-1}. To explore this discrepancy further, quenching from the steady state was undertaken. Results show that in the steady state under V_{max} conditions, about half the enzyme is in covalently phosphorylated forms. This suggests the existence of a second rate-limiting process, occurring after E-P cleavage, but before rephosphorylation, that leads to the buildup of nonphosphorylated enzyme forms. This process may be product release, which has not yet been studied in detail.

The kinetics make it clear that E-P formation precedes phosphoribosyl group transfer, which is fast, and that group transfer is followed by E-P hydrolysis and product release. The kinetic statements that result from these experiments only compare rates and do not describe or quantitate the rules whose observance produces coupling. Interestingly, the kinetics demonstrate that the mode of coupling by NAPRTase is not chemical. Comparing the rates of E-P hydrolysis and group transfer initiated from identical complexes clearly shows that NAMN formation precedes E-P phosphohydrolysis, ruling out any possibility of substrate phosphorylation in the reaction.

[28] Y. Xu, J. Eads, J. C. Sacchettini, and C. Grubmeyer, *Biochemistry* **36**, 3700 (1997).

Side Reactions

NAPRTase is unusual among ATPases in that its use of ATP is "faculta-tive," with a weak NAMN synthesis activity displayed in the absence of ATP. The properties of this reaction with the *Salmonella* enzyme at 30° are tabulated below.[16]

Property	+ATP	−ATP
k_{cat} (sec^{-1})	2.9 ± 0.07	0.28 ± 0.05
K_m PRPP	22 ± 3.3 μM	4.5 ± 2.2 mM
K_m NA	1.5 ± 0.3 μM	0.29 ± 0.1 mM

Two major changes are immediately obvious when ATP is omitted: (1) a 10-fold lower k_{cat} and (2) 200-fold increases in K_m values for both PRPP and NA. Uncoupled NAMN synthesis is important for explorations of energetics because it allows the study of the properties of the nonphos-phorylated form of the enzyme, particularly quantitative comparison of substrate and product affinities. Previous work has suggested that yeast NAPRTase does not catalyze detectable uncoupled NAMN formation.[26] The yeast observations led to concern that the uncoupled NAMN synthesis seen with the bacterial enzyme might come from a small portion of the enzyme that had been damaged artifactually during isolation. We have found that the recombinant yeast enzyme in fact shows uncoupled NAMN synthesis at rates similar to the *Salmonella* NAPRTase (M. Rajavel, unpub-lished).

A second important side reaction is an ATPase stimulated by PP$_i$, which at 30° occurs at a rate of 4.3 sec^{-1}, slightly faster than k_{cat} for the coupled reaction (2.9 sec^{-1}). The existence of this reaction suggested that nascent PP$_i$ might be the signal that causes E-P cleavage in the normal reaction. The kinetic study of the PP$_i$-stimulated ATPase reaction has shown that PP$_i$ gives substrate inhibition, which is relieved at high ATP. The K_m for PP$_i$ (100 μM) represents its binding to E-P, whereas the K_I (330 μM) represents binding to E. An NAMN-stimulated ATPase may also occur, although it is apparently slow (unpublished).

Reaction Stoichiometry

Several groups have demonstrated convincingly that under initial veloc-ity conditions (usually with low levels of [^{14}C]NA and high levels of ATP and PRPP), the reaction stoichiometry of ATP cleaved to NAMN formed is 1:1.[15,16] Because of the existence of side reactions, we reinvestigated this value. When stoichiometry was measured at 100 μM PRPP and 120 μM

[^{14}C]NA with 1 mM ATP and increasing concentrations of PP$_i$, it rose in a linear fashion from a value of 1 NAMN formed/ATP cleaved at zero PP$_i$ to 10 ATP cleaved/NAMN formed at 3 mM PP$_i$. Clearly, the PP$_i$-induced ATPase represents a very significant use of ATP under conditions of high product concentrations.

Conformational Changes

The term conformational coupling implies protein structural differences between energized and unenergized states. NAPRTase has not yet been crystallized, and detailed structural statements cannot be made. Most physical work has been negative: no substantial differences between E and E-P were noted with tryptophan fluorescence or high-resolution gel filtration.[22] The thermostability of the *Salmonella*[22] and yeast[26] enzymes is increased by ATP and PRPP, but this might represent only effects of ligand binding on protein stability. However, it was found that the sensitivity of NAPRTase to inactivation by trypsin was affected directly by enzyme phosphorylation and ligand binding.[22] The enzyme rapidly lost activity in a first-order fashion ($T_{1/2}$ = 5 min) at 1 : 200 (w, w) trypsin, and ATP increased $T_{1/2}$ to 39 min, with a further increase to 158 min when both ATP and PRPP were present. Matrix-assisted laser desorption line of flight (MALDI-TOF) mass spectrometry showed that the initial inactivating cleavage occurs at Arg-384, followed by a secondary cleavage at Lys-374. The inactivation of NAPRTase with chymotrypsin also followed first-order kinetics, with cleavage occurring at Phe-382 and either ATP or ATP and PRPP offering similar levels of protection. In NAPRTase, protease susceptibility probes only the carboxy terminal of the protein; it is not yet clear if the inferred conformational change is similarly local or extends more globally. In analogous findings, phosphorylation of His-8 of phosphoglycerate mutase[29] results in the immobilization of a previously flexible solvent carboxy terminal.

Energetics

The chemical and kinetic questions that remain to be answered with NAPRTase are fundamental, but enough is known to allow a rational description of a mechanism for how the enzyme uses ATP hydrolysis to drive NAMN formation. The mechanism is that of Eq. (6), with the additional complications of being a two substrate reaction, with steps of E-P breakdown interleaved with NAMN formation. The mechanism is inherently flawed, but functionally successful.

[29] S. I. Winn, H. C. Watson, R. N. Harkins, and L. A. Fothergill, *Phil. Trans. R. Soc. Lond. B* **293**, 121 (1981).

One strategy used by NAPRTase is an ATP-driven cycle of affinity changes toward its substrates. For PRPP binding, unpublished equilibrium binding measurements have shown that K_D decreases from 1 mM to 0.5 μM when E becomes phosphorylated, a change in binding energy of about 4.5 kcal/mol. For NA, kinetic data show that K_m decreases 200-fold and k_{cat}/K_m increases 2000-fold with the E-P form. This strategy helps ensure that NAPRTase can bind substrates even at low concentrations.

Although substrate binding strategy is straightforward, product binding is not. Product NAMN binding has not yet been quantitated, but for PP$_i$, binding to E is with moderate affinity, K_D = 130 μM, and binding to E-P may be of similar affinity as judged by K_m for the PP$_i$-stimulated ATPase, 100 μM. Using this value to think about energetics neglects the possibility that dissociation of PP$_i$ from the complete products complex may follow that of NAMN or P$_i$ in compulsory order. As discussed earlier, the tight binding of product to E-P is desirable in that no product PP$_i$ escapes until E-P cleavage occurs, but undesirable in that E-P is a required intermediate, whose binding of PP$_i$ would result in a dead-end complex.

The linkage between ATP use (E-P hydrolysis) and NAMN synthesis is clearly of the signal type, with PP$_i$ the major signal (it is not yet clear if NAMN can also cause E-P cleavage at kinetically reasonable rates). The rate of E-P cleavage in native E-P $\cdot$ PRPP is 0.003 sec^{-1}, whereas that in the E-P $\cdot$ NAMN $\cdot$ PP$_i$ complex is about 6 sec^{-1}, a 2000-fold increase.

The second energy-coupling strategy employed by NAPRTase is to minimize the unproductive formation of free substrates from the E-P $\cdot$ NAMN $\cdot$ PP$_i$ complex. The complex is converted rapidly to E $\cdot$ P$_i$ $\cdot$ NAMN $\cdot$ PP$_i$ through E-P hydrolysis; it is not yet known if this reaction step is reversible. The E $\cdot$ P$_i$ $\cdot$ NAMN $\cdot$ PP$_i$ complex must not be allowed to revert to a E $\cdot$ P$_i$ $\cdot$ NA $\cdot$ PRPP complex and dissociate PRPP or NA. As discussed earlier, the ratio $k_{off\ P}/k_{rev}$ governs the maximal [P]/[S] value. One route for achieving this goal is the 10-fold decreased V_{max} for the phosphoribosyl transfer reaction in E compared to E-P. Although this difference is small, it may be enhanced in P$_i$-containing complexes. The competing productive pathway is product release, and if either PP$_i$ or NAMN is released rapidly, trapping products, the differential between the two pathways will be enhanced.

It is clear that NAPRTase violates the rules for energy coupling in Eq. (3) in several ways. First, the enzyme allows the reverse NAMN pyrophosphorolysis reaction to occur without ATP formation. Second, the product PP$_i$ triggers a wasteful ATPase reaction. These two leaks in NAPRTase energy coupling are demonstrated most dramatically in an experiment in which a molar excess of ATP is allowed to drive the forward reaction. As the ratio [NAMN][PP$_i$]/[PRPP][NA] increases and reaches a plateau,

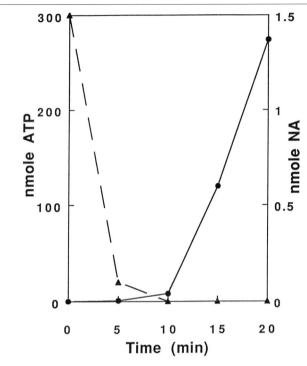

FIG. 2. Reversal of NAMN synthesis following hydrolysis of ATP. The reaction was started by the addition of 120 μg NAPRTase. The zero time point was taken 1 min after enzyme addition, at which time greater than 95% of the NA had been converted to NAMN, and 50% of the ATP was converted to ADP. At each time point, ATP and NA were determined. Reprinted with permission from A. Vinitsky and C. Grubmeyer, *J. Biol. Chem.* **268,** 26004 (1993).[16]

ATP hydrolysis continues. When ATP concentrations reach zero, the [NAMN][PP$_i$]/[PRPP][NA] ratio crashes (Fig. 2).

NAPRTase Physiology

Is there any value to ATP use by NAPRTase? Conventionally, biochemistry textbooks assume that PP$_i$-generating reactions like the phosphoribosyltransferases are "pulled" by pyrophosphatases. This logic would seem to make ATP use by NAPRTase unnecessary. However, it is clear that PP$_i$ does exist in cells, about 0.5 mM in *E. coli* under most conditions,[30] so that a thermodynamic "push" may be useful. It may also be important to render

[30] B. L. Wanner, *in* "*Escherichia coli* and *Salmonella typhimurium*: Cellular and Molecular Biology" (F. C. Neidhardt, ed.), p. 1357. ASM Press, Washington, DC, 1996.

the NAPRTase reaction irreversible because the irreversible QAPRTase reaction, which has NAMN as a product, would otherwise combine with NAPRTase to make a futile cycle. A second possibility is that ATP use controls NAPRTase as an allosteric activator. The concept that inefficient energy coupling is particularly valuable in this respect is considered in article [3] in this volume.[31]

There are alternatives to the use of ATP by NAPRTase being of direct metabolic value. One attractive possibility is that NAPRTase may be part of a multiprotein control system, with the NAPRTase E-P serving to phosphorylate other enzymes of NAD metabolism. Attempts to use NAPRTase to phosphorylate other proteins *in vitro* have provided negative results to date.

Relevance for Macromolecular Discrimination

What does NAPRTase tell us about how an enzyme can use ATP hydrolysis to impart directionality to a coupled process? One protein that has been well studied is the RecA enzyme of recombination, for which recent reviews provided the information discussed here.[32,33] In a standard *in vitro* assay, RecA protein interacts with two homologous pieces of DNA—one single stranded and one double stranded—and interchanges the two homologous strands between the two molecules.

The overall reaction (DNA binding, strand exchange, and regeneration of free RecA) requires ATP hydrolysis. Early literature suggested that ATPase drives the on-enzyme strand exchange process. However, results with slowly hydrolyzed substrates such as (γ-thio)ATP suggest that all events up to and including strand exchange can be catalyzed with ATP binding but without ATP hydrolysis. The enzyme also carried out an on-enzyme strand exchange with $ADP \cdot AlF_4$ as a nonhydrolyzable ATP analog.[34]

A scheme can be constructed for RecA using the formalism of Eq. (6):

$$E' + AB_1 + B_2 \rightleftharpoons E' \cdot AB_1 \cdot B_2 \rightleftharpoons E' \cdot AB_2 \cdot B_1 \rightleftharpoons E' + AB_2 + B_1$$
$$\updownarrow$$
$$E + AB_1 + B_2 \rightleftharpoons E \cdot AB_1 \cdot B_2 \rightleftharpoons E \cdot AB_2 \cdot B_1 \rightleftharpoons E + AB_2 + B_1 \qquad (11)$$

In this scheme, E' represents the $RecA \cdot ATP$ complex, which decomposes to E, representing the $RecA \cdot ADP \cdot P_i$ complex and RecA apoenzyme.

[31] T. S. Leyh, *Methods Enzymol.* **308,** Chap. 3, 1999 (this volume).
[32] S. C. Kowalczykowski, *Annu. Rev. Biophys. Biophys. Chem.* **20,** 539 (1991).
[33] S. C. Kowalczykowski, D. A. Dixon, A. K. Eggleston, S. D. Lauder, and W. M. Rehrauer, *Microbiol. Rev.* **58,** 401 (1994).
[34] S. C. Kowalczykowski and R. A. Krupp, *Proc. Natl. Acad. Sci. U.S.A.* **92,** 3478 (1995).

AB_1 designates a dsDNA, and B_2 a homologous ssDNA. Formation of E' causes a dramatic increase in the affinity for DNA. For ssDNA, lifetimes for release are about 24 hr with the $RecA \cdot ATP\gamma S \cdot DNA$ complex and seconds or minutes with the $RecA \cdot ADP \cdot DNA$ or $RecA \cdot DNA$ complexes, respectively. Equation (11) leads to useful questions about RecA. First, as drawn, it is clear that strand exchange is the signal for the cleavage of bound ATP, leading to questions about the structural or chemical nature of the signal. Second, it is clear that two alternative paths exist. In the upper right, release of DNA from E' is highly unlikely given the very low off-rates quoted earlier. However, to the extent that E can perform strand exchange, the path in the lower left will exist in which release of DNA with no net strand exchange has occurred. Finally, the very slow off-rates for DNA lead to a third wasteful pathway: rebinding of ATP before release of DNA. This failure is thought to account for the fact that only 1 in 50 ATP hydrolysis events leads to productive net strand exchange.

Much remains to be learned about NAPRTase energetics, but the enzyme does serve to provide a paradigm for energy use in other conformationally coupled systems. The paradigm portrays a system that is successful in driving the product formation, but is intrinsically wasteful in its use of ATP. It remains to be seen whether each of these cases will provide unique solutions to the energetic problems discussed here or if the NAPRTase paradigm is in fact general.

Acknowledgments

We thank the National Science Foundation for a grant (DMB-9103029) in support of this work. We gratefully acknowledge valuable conversations with lab group members, including Alexander Vinitsky and Yiming Xu.

[3] On the Advantages of Imperfect Energetic Linkage

By THOMAS S. LEYH

Metabolic flow is often controlled by the transfer of chemical potential from cellular reservoirs of potential, such as ATP and GTP, to specific points in metabolism. The connections between reservoirs and pathways are established by linking enzymes that act as conduits for the exchange of potential by providing a catalytic path for the transfer. This article treats the class of linking enzymes that transfer potential energy not through

Copyright © 1999 by Academic Press
All rights of reproduction in any form reserved.
0076-6879/99 $30.00

direct chemical reaction between the energy donor and acceptor, but, rather, through structural changes that occur in the catalyst itself during the reaction cycle. This class of enzymes, which has grown substantially during the last decade, includes molecular motors,[1,2] polymerases,[3,4] topoisomerases,[5,6] K^+ and Ca^{2+} pumps,[7–9] metabolite transporters,[10,11] GTPases,[12] ATP synthetases,[13] nitrogenase,[14] nicotinate phosphoribosyltransferase,[15] and ATP sulfurylase.[16] The mechanistic basis for energetic coupling in these so-called conformationally coupled systems is rooted in the allosteric interactions that occur in the reaction coordinate of the coupled reaction.

If the donor and acceptor reactions were coupled perfectly by the linking enzyme, i.e., if the potential were transferred with no energetic loss, the potential transferred to the acceptor would be precisely that available in the donor reaction prior to coupling. If the donor reaction were the hydrolysis of the β-γ bond of ATP or GTP, the potential transferred would be of the order of -7.7 kcal/mol, which would produce a 5×10^5-fold shift in the mass ratio of the acceptor reaction. It is sometimes beneficial, if not essential, that the cell transfer less than the maximum potential available from the donor. The cell must "titrate" the available chemical potential into the coupled pathways. One way to achieve this end is to regulate the coupling efficiency of the linking enzymes by the reactants that participate in the coupled reaction. This is the cornerstone of this article. The mechanistic bases for energetic linkage in three conformationally coupled systems are scrutinized in this review with a focus on how, and why, allosteric regulation by the coupling reactants is embedded in their mechanisms.

[1] D. D. Hackney, *Annu. Rev. Physiol.* **58,** 731 (1996).
[2] T. Hasson and M. S. Mooseker, *Curr. Opin. Cell Biol.* **7,** 587 (1995).
[3] P. H. von Hipple, F. R. Farifield, and M. K. Dolejsi, *Ann. N.Y. Acad. Sci.* **726,** 118 (1994).
[4] R. Guajardo and R. Sousa, *J. Mol. Biol.* **265,** 8 (1997).
[5] J. C. Wang, *Annu. Rev. Biochem.* **65,** 635 (1996).
[6] R. Menzel and M. Gellert, *Adv. Pharmacol.* **29A,** 39 (1994).
[7] W. P. Jencks, *Methods in Enzymol.* **171,** 145 (1989).
[8] W. P. Jencks, *J. Biol. Chem.* **264,** 18855 (1989).
[9] J. W. Keillor and W. P. Jencks, *Biochemistry* **35,** 2750 (1996).
[10] R. M. Krupka, *J. Membr. Biol.* **109,** 151 (1989).
[11] R. M. Krupka, *Biochem. Biophys. Acta* **1183,** 105 (1993).
[12] M. V. Rodnina, A. Savelsbergh, V. I. Katunin, and W. Wintermeyer, *Nature (London)* **385,** 18 (1997).
[13] H. Noji, R. Yasuda, M. Yoshida, and K. Kinosita, *Nature (London)* **386,** 299 (1997).
[14] H. Schindelin, C. Kisker, J. L. Schlessman, J. B. Howard, and D. C. Rees, *Nature (London)* **387,** 370 (1997).
[15] A. Vinitsky, and C. Grubmeyer, *J. Biol. Chem.* **268,** 26004 (1993).
[16] C. Liu, Y. Suo, and T. S. Leyh, *Biochemistry* **33,** 7309 (1994).

Background

Conformational Coupling

The conformational coupling of energetics is the linking of the chemical potentials of two, or more, molecular events through conformational changes in the catalyst that occur during the reaction cycle. Linking is achieved through allosteric interactions that require the events to occur in a concerted fashion, either simultaneously or sequentially. To satisfy this requirement, the events must be linked stoichiometrically. The stoichiometry, which can be any fixed number, determines the weighting of the potential of each reaction to that of the coupled reaction. Given an $n:m$ stoichiometry, the potential of the overall coupled reaction is calculated by adding n times the potential of its associated reaction to m times that of its reaction.

A Theoretical Coupling Model

To transfer potential, the coupling enzyme must catalyze each of the linked events in a defined stoichiometry, once per catalytic cycle. This requires that at least two steps in the mechanism, one from each reaction, be obliged, through allosteric interactions, to occur in a concerted fashion. A theoretical version of such a mechanism is presented in Fig. 1. The sets of shaded rectangles represent energetic barriers in the reaction coordinate that can restrict the progress of a given reaction. A barrier might represent an energetically unfavorable transition-state or enzyme-bound, ground-state equilibrium. Barriers can be modulated through allosteric interactions that develop as a consequence of an event such as the binding of substrate or the appearance of product. These manipulations can erode and/or construct barriers at various points in the reaction coordinate. A single protein conformational change might well erode one barrier while constructing another. The net effect of such a change is a shifting of the barrier among different positions in the reaction coordinate. Barrier shifting is represented in Fig. 1 by a repositioning of the sets of rectangles.

The reactants bind randomly to the enzyme in this example. For the sake of simplicity, only one binding sequence is shown. The nonligand bound, or free enzyme, has barriers that prevent it from catalyzing either reaction independently. The enzyme binds to A, but does not catalyze its conversion to B. The binding of C shifts the barrier for the A to B reaction such that the formation of B is encouraged energetically and the release of B is prevented. If the release of B occurred, A would be converted to B without the conversion of C to D, and the reactions would not be coupled energetically. The formation of B, in turn, shifts the barrier in the C to D reaction to foster the formation of D and prevent the release of C. The

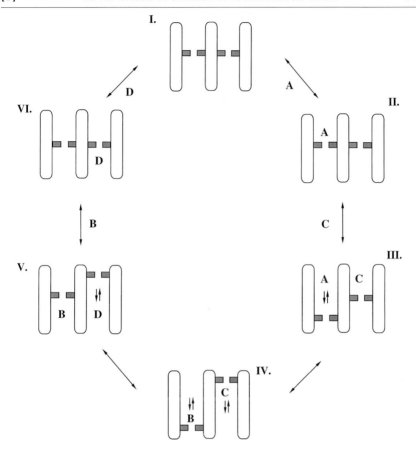

FIG. 1. A general model for the conformational coupling of the energetics of two reactions. Diagram I represents a free enzyme with two active sites indicated by the left and right "channels." The sets of shaded rectangles represent energetic barriers that inhibit specific chemical events. Barriers represent either an unfavorable transition-state or ground-state equilibrium. Double arrows represent the equilibrium that is established following removal of the barrier for that step. Moving clockwise through the reaction cycle, the barriers shift in response to the binding or appearance of ligands in such a way that the two reactions are linked stoichiometrically and energetically.

formation of D shifts the barrier for the A to B reaction back to that of the free enzyme, preventing the conversion of B back to A. The release of B reestablishes the central barrier that prevents the conversion of D back to C. The release of D regenerates the free enzyme. This cycle works because the changing allosteric interactions in the catalytic cycle establish an interdependence of the steps of the two reactions such that neither reaction can reach completion unless the other also reaches completion.

Origin of Leaks

The extent to which a mechanism succeeds in coupling the energetics of two reactions depends on how successfully its barriers prevent the linked steps from occurring independently. For example, if the barrier preventing the uncoupled conversion of A to B in Fig. 1 were not sufficiently "high," A would form B without the conversion of C to D. This would uncouple the chemistries in a barrier-height-dependent fashion and, under most circumstances, result in an ill-defined stoichiometry. In fact, most if not all enzymes that couple chemistry demonstrate a low, basal rate of catalysis of each independent reaction. The barriers, then, determine the tightness of the coupling; alternatively, they establish a "hole" or alternative catalytic path through which the chemical potential of the system can leak, with the rate of the leak being determined by the height of the barrier.

Necessary Imperfection

While an enzyme might achieve a coupling efficiency that is sufficiently high to preclude detectable leakage, it is not possible for it to perfectly couple potentials because this requires infinitely large barrier heights. As discussed earlier, barrier heights are determined by the equilibria, or potential differences, between two ground states or a ground and transition state. An infinite barrier between any two points in a reaction coordinate is the equivalent of no path between the two points; however, the existence of the path is proven by the fact that the step does occur when, say, a ligand from the complementary reaction binds to the enzyme. Hence, some fraction of the binding potential of the ligand–enzyme interaction has been used to alter the existing path such that the prohibited step can occur at a measurable rate. Regardless of how the binding of ligand achieves the enzyme form that accelerates the reaction, it is possible to argue that some, perhaps tiny, fraction of the enzyme can achieve that form in the absence of the bound ligand. Thus, the argument reduces to one of the accessibility of the path, not its existence. Infinite barriers, and therefore perfect coupling, are not possible.

Examples

ATP Sulfurylase

ATP sulfurylase (ATP:sulfate adenylyltransferase, EC 2.7.7.4), isolate from *Escherichia coli* K-12, catalyzes and energetically links the hydrolysis of GTP and the synthesis of activated sulfate, APS or adenosine-5'-phosphosulfate (reactions 1 and 2).[16] The enzyme is a tetramer of heterodimers.[17] The CysN (53 kDa) subunit

$$ATP + SO_4 \rightleftharpoons APS + PP_i \tag{1}$$
$$GTP + H_2O \rightleftharpoons GDP + P_i + H^+ \tag{2}$$

of the heterodimer catalyzes GTP hydrolysis; the other subunit, CysD (35 kDa), catalyzes the APS reaction. Because of its poor reactivity, sulfate must be activated to participate in its subsequent metabolism. Activation is accomplished by transferring the adenylyl moiety from the pyrophosphoryl group of ATP to sulfate. This transfer reaction produces the phosphoric–sulfuric acid anhydride bond of APS. The $\Delta G^{0'}$ associated with hydrolysis of the anhydride bond of APS, ~ -19.5 kcal/mol,[18,19] is much higher than that for the α-β bond of ATP, -10.9 kcal/mol.[18,20] Thus, the equilibrium constant for the formation of APS is extremely unfavorable, $\sim 1 \times 10^{-9}$ at pH 8.0.[18] The cell overcomes this formidible energetic barrier by coupling the chemical potential of GTP hydrolysis to the synthesis of APS. At a saturating concentration of GTP, the equilibrium constant for APS synthesis is shifted seven orders of magnitude, from ~ 1 e-9 to 0.059.[16]

After being transported actively into the bacterial cell, sulfate is activated chemically and then reduced, in a series of enzymatic reactions, to sulfide, which is incorporated into cysteine and, from there, into other reduced sulfur metabolites.[21,22] The fact that certain ancient anaerobic bacteria use the chemical potential of the APS bond to drive electron transport suggests that APS served as a primary metabolic driving force at an early stage of evolution.[23] The metabolic role of activated sulfate in mammals is very different from that in bacteria. In mammals, activated sulfate, in the form of PAPS (3'-phosphoadenosine-5'-phosphosulfate), serves as the sole sulfuryl group donor. Sulfuryl transfer, much like phosphoryl transfer, is used to regulate the activities of numerous metabolites.[18] Sulfotransferases, which transfer the sulfuryl group, act in conjunction with sulfatases, which

[17] T. S. Leyh, J. T. Taylor, and G. H. Markham, *J. Biol. Chem.* **263**, 2409 (1987).

[18] T. S. Leyh, *CRC Crit. Rev. Biochem.* **28**, 515 (1993).

[19] P. W. Robbins, and F. Lipmann, *J. Biol. Chem.* **233**, 681 (1958).

[20] P. A. Frey, and A. Arabshahi, *Biochemisry* **34**, 11308 (1995).

[21] L. M. Siegel, *in* "Metabolic Pathways" (D. M. Greenberg, ed.), p. 217. Academic Press, New York, 1975.

[22] R. M. De Meio, *in* "Metabolic Pathways" (D. M. Greenberg, ed.), p. 278. Academic Press, New York, 1975.

[23] R. J. Singleton, *in* "The Sulfate Reducing Bacteria: Contemporary Perspectives" (J. M. Odom and R. J. Singleton, eds.), p. 1. Springer-Verlag, New York, 1993.

remove this group, to regulate sulfation of a given metabolite.[24–35] An interesting example of regulation through sulfuryl transfer is the activation of the estrogen receptor.[32] The receptor is activated when estrogen binds to it and it does not bind sulfated estrogen. Hence, the receptor is regulated by estrogen sulfation/desulfation.

Mechanism. ATP sulfurylase catalyzes GTP hydrolysis and APS synthesis through either coupled or uncoupled catalytic paths. The maximum rates of the uncoupled reactions are one to two orders of magnitude less than that of the coupled reaction.[36,37] Comparing these mechanisms demonstrates how the uncoupled reaction coordinates are altered to achieve the rate enhancements and linking observed in the coupled reaction. The coupled mechanism is outlined in Fig. 2 and is discussed in detail later.

Forming the Central Complex. ATP and GTP bind randomly to the enzyme, and the assembly and disassembly of the binary and ternary nucleotide complexes are rapid compared to the turnover of the coupled reaction.[38] Experiments in this laboratory have identified a rate-limiting conformational step immediately preceding the hydrolysis of GTP. This conformational intermediate, indicated by the prime in Fig. 2, is present at low levels during turnover because the rate at which it disappears, the intrinsic rate of hydrolysis, is much faster than the rate at which it forms. Stopped-flow fluorescence experiments indicate that the enzyme also undergoes a conformational change in the GTP-binding reaction. These experiments have not resolved whether the isomerization occurs prior or subsequent to

[24] D. H. Atha, J. C. Lormeau, M. Petitou, R. D. Rosenberg, and J. Choay, *Biochemistry* **24,** 6723 (1985).
[25] S. J. Brand, B. N. Andersen, and J. F. Rehfeld, *Nature (London)* **309,** 456 (1984).
[26] P. R. Brauer, K. A. Keller, and J. M. Keller, *Development* **110,** 805 (1990).
[27] R. P. Garay, J. P. Labaune, D. Mesangeau, C. Nazaret, T. Imbert, and G. Moinet, *J. Pharmacol Exp. Ther.* **255,** 415 (1990).
[28] R. T. Jensen, G. F. Lemp, and J. D. Gardner, *J. Biol. Chem.* **257,** 5554 (1982).
[29] A. Leyete, H. B. van Schijndel, C. Niehrs, W. B. Huttner, M. P. Verbeet, K. Mertens, and J. A. van Mourik, *J. Biol. Chem.* **266,** 740 (1991).
[30] R. M. Maarouei, J. Tapon-Bretaudiere, J. Mardiguian, C. Sternberg, M. D. Dartzenberg, and A. M. Fischer, *Thromb. Res.* **59,** 749 (1990).
[31] K. D. Meisheri, J. J. Oleynek, and L. Puddington, *J. Pharmacol. Exp. Ther.* **258,** 1091 (1991).
[32] J. R. Pasqualini, B. Schatz, C. Varin, and B. L. Nguyen, *J. Steroid Biochem. Mol. Biol.* **41,** 323 (1992).
[33] J. R. Roth, and A. J. Rivett, *Biochem. Pharm.* **31,** 3017 (1982).
[34] S. R. Stone, and J. Hofsteenge, *Biochemistry* **25,** 4622 (1986).
[35] C. D. Unsworth, J. Hughes, and J. S. Morley, *Nature (London)* **295,** 519 (1982).
[36] T. S. Leyh and Y. Suo, *J. Biol. Chem.* **267,** 542 (1992).
[37] R. Wang, C. Liu, and T. S. Leyh, *Biochemistry* **34,** 490 (1995).
[38] C. Liu, E. Martin, and T. S. Leyh, *Biochemistry* **33,** 2042 (1994).

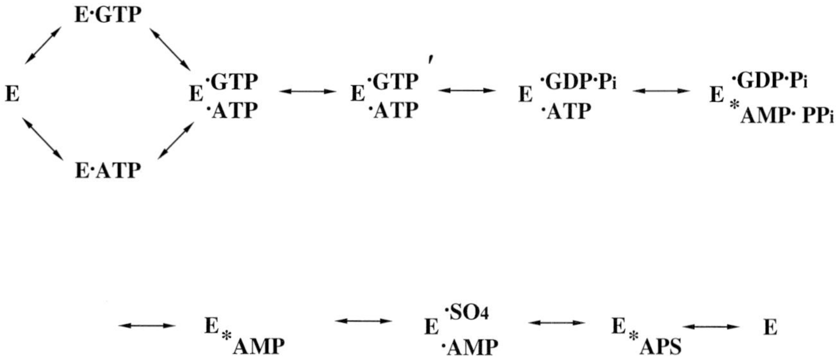

FIG. 2. The coupled mechanism of ATP sulfurylase. ATP and GTP bind randomly to the enzyme. Following the addition of nucleotides, the ternary complex isomerizes to a reactive form, indicated by the prime. GTP hydrolysis precedes or is concomitant with breaking the α-, β-bond of ATP, which produces a putative reactive AMP intermediate, indicated by the asterisk. The points of P_i and GDP release are not yet known and therefore are not included. Pyrophosphate departs from the enzyme prior to the addition of sulfate, which reacts with the E*AMP intermediate to form APS. The dissociation of APS completes the catalytic cycle.

nucleotide binding.[39] The substrate-binding region of the reaction coordinate is complex. It contains at least three distinguishable enzyme forms connected through two isomerizations and it offers numerous possibilities for allosteric interactions between the active sites in the coupled mechanism.

Chemistries of the Central Complexes. Two reactions are catalyzed by ATP sulfurylase in the absence of sulfate: hydrolysis of the β-γ bond of GTP and hydrolysis of the α-β bond of ATP. Each of these reactions is stimulated by the presence of the substrate for the other reaction. The formation of $[\alpha$-$^{32}P]$AMP is not detected under catalytic or single turnover conditions in the absence of GTP. The k_{cat} for ATP hydrolysis increases to 0.46 min^{-1} at a saturating concentration of GTP.[38] The enzyme is less stringent in restricting the independent hydrolysis of GTP. This reaction, albeit slow, is detected easily. The k_{cat} for GTP hydrolysis, 0.01 sec^{-1}, is stimulated 180-fold at saturating ATP.[37] The GTP dependence of AMP formation is an important linking aspect of the coupled mechanism. It means that some step(s) in the GTP hydrolysis reaction is coupled to transfer of the adenylyl group of ATP to H_2O. The chemical steps of this transfer are not yet known; however, stereochemical experiments underway in this laboratory will distinguish between transfer through a reactive E*AMP intermediate, as is indicated in Fig. 2, and nucleophilic attack of H_2O at the α-phosphorous of ATP.

[39] M. Yang and T. S. Leyh, *Biochemistry* **36**, 2370 (1997).

Interdependence of Bond Breaking. The nucleotide bond-breaking sequence was dissected using the hydrolysis resistant analogs of GTP and ATP, GMPPNP and AMPCPP.[38] AMPCPP and ATP behave similarly in activating GTP hydrolysis, suggesting that adenine nucleotide binding is sufficient to activate GTP hydrolysis. GMPPNP inhibits APS synthesis competitively versus GTP. Single turnover experiments were performed to assess whether GMPPNP binding drives the synthesis of enzyme-bound APS. APS synthesis was not observed in experiments in which ~0.4% of an enzyme active site equivalent could have been detected.[16] Thus, the binding of GMPPNP, and presumably GTP, establishes an energetic barrier that prevents APS synthesis prior to GTP hydrolysis. To determine at which step subsequent to GTP binding APS formation occurred, GDP and P_i were tested alone and in combination as activators of APS synthesis.[16] GDP is a modest activator of APS synthesis, whereas P_i has no detectable effect, even at saturating GDP. Guanine nucleotide product complexes have little effect on the APS chemistry. Thus, it appears that some event(s) associated with the transition from $E \cdot GTP$ to $E \cdot GDP \cdot P_i$ potentiates cleavage of the α-β bond of ATP and that this event(s) links these two reactions energetically.

A Partial Structural Model for the Coupling Step. The structure of ATP sulfurylase is not yet available; however, the sequence of CysN positions it solidly within the GTPase superfamily.[40] This family demonstrates a remarkably high degree of structural conservation, particularly in the vicinity of the nucleotide-binding pocket.[41] ATP sulfurylase is most similar to the elongation factor subgroup of the GTPase family. The primary sequence of ATP sulfurylase is 29% identical and 49% similar to elongation factor-Tu (EF-Tu).[40] Structures of the GMPPNP- and GDP-bound complexes of the *Thermus thermophilus* EF-Tu are very different.[42] Residues move by as much as 42 Å in the transition from the GDP to GMPPNP structure. Conformational changes appear to nucleate at the GTP hydrolysis site. In the GDP structure, Gly-84, located at the N terminus of a nine-residue α-helical element in the structure, "hovers" over the vacant γ-phosphate position. At the C terminus of the helix is a four- to five-residue stretch of random coil. When P_i adds at the γ-phosphate site and/or bond formation occurs, the helix "slides" through the random coil, leaving random coil behind it. The repositioning of the helix allows subdoman 2 of EF-Tu to

[40] T. S. Leyh, T. F. Vogt, and Y. Suo, *J. Biol. Chem.* **267**, 10405 (1992).
[41] A. Valencia, M. Kjeldgaard, E. F. Pai, and C. A. Sander, *Proc. Natl. Acad. Sci. U.S.A.* **88**, 5443 (1991).
[42] H. Berchtold, L. Reshetnikova, C. O. Reiser, N. K. Schirmer, and R. Hilgenfeld, *Nature (London)* **365**, 126 (1993).

clamp down onto it, causing the large structural changes mentioned earlier. The section of sequence that undergoes the helical slide is among the most conserved regions between EF-Tu and CysN (35% identity, 82% similarity). Thus, it appears likely that such a conformational change will attend GTP hydrolysis in the CysN subunit and that it will form part of the chain of structural events that ultimately reorganize the CysD active site to cause cleavage of the α-β bond of ATP.

Events Subsequent to α-β Bond Cleavage. To demonstrate that the events that lead to the hydrolysis of ATP form part of the catalytic path for APS synthesis, the reaction products were monitored as a function of sulfate concentration at saturating ATP and GTP. As the enzyme is titrated with sulfate, AMP formation is suppressed and is supplanted by APS synthesis. AMP formation is almost completely inhibited at near-saturating concentrations of sulfate. Thus, the putative E*AMP intermediate reacts with sulfate to produce APS. The fact that PP_i is competitive versus SO_4 for APS synthesis suggests that it dissociates prior to the addition of sulfate.[38]

Stoichiometry and Energy Transfer. Leaks, or uncoupled chemistry, prevent coupled systems from reaching equilibrium. For this reason, the stoichiometry of GTP hydrolysis to APS synthesis was determined kinetically. At saturating concentrations of GTP, ATP, and SO_4, the ratio of the initial rates of APS synthesis and GTP hydrolysis is 1.00 (± 0.05).[16] Thus, the efficiency of the transfer of chemical potential is quite high when the enzyme is saturated with substrates and very little product has formed. High coupling efficiency is acheived when experimental conditions minimize the enzyme complexes that catalyze the uncoupled reactions.

If the GDP produced by the coupled reaction is converted back into GTP by a regenerating system, the APS-forming reaction reaches a steady state, or quasi-equilibrium, in which the rates of the forward and reverse chemistries are identical.[16] The mass-action ratio ($[APS][PP_i]/[ATP][SO_4]$) in this steady state is 0.059, which is seven to eight orders of magnitude greater than that predicted by the energetics of the uncoupled reaction.

Transient Product Synthesis. In the early stage of an APS-forming reaction initiated at saturating concentrations of ATP, GTP, and SO_4, the synthesis of APS is driven by GTP hydrolysis with a coupling efficiency that approaches 100%. Over time, the accumulation of product results in enzyme forms that catalyze the uncoupled reaction, eroding the coupling efficiency. As flux through the uncoupled pathway increases, the chemical potential that had been transferred to the APS reaction "leaks" out of it into the chemical mileu. This transient transfer of potential results in a pulse-shaped progress curve for the reaction that is driven against its potential gradient. The first or earlier pulse shown in Fig. 3 is a typical progress curve for the APS reaction when GTP is not regenerated. The second pulse

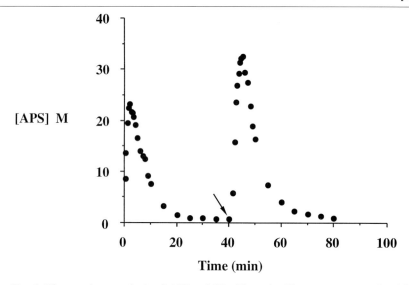

FIG. 3. The transient synthesis of APS and PP_i. The pulse-like progress curve for APS synthesis is shown. At $t = 0$, the enzyme was saturated with substrates. The experimental conditions were ATP sulfurylase (4.0 μM), ATP (0.50 mM), GTP (0.25 mM), [^{35}S]SO$_4$ (0.10 mM), MgCl$_2$ (1.75 mM), pyruvate kinase (6.0 mU/ml), HEPES (50 mM, pH/K$^+$ 8.0), and T = 25 ± 2°. At 41 min (indicated by the arrow), GTP was regenerated by the addition of PEP to 0.29 mM. One unit (U) converts 1.0 μmol of substrate to product at V_{max}.

occurs when the system is "recharged" by converting the GDP produced during the first pulse to GTP.

Underlying the time-dependent changes in the APS concentration that define the pulse are changes in the rate of its formation and decay. To determine how these rates change, the initial velocity of the forward and reverse direction of each reaction was determined at numerous, short time intervals throughout the pulse. The results of these measurements are compiled in the acceleration/deceleration profiles in Fig. 4. It is important to realize that the fact that GTP synthesis was not observed during the pulse (Fig. 4D) means that the synthesis of ATP occurs through an uncoupled catalytic path. Moreover, the energetics of the uncoupled synthesis of APS are so unfavorable that any detectable APS must have been produced through the coupled pathway. Hence, Figs. 4A and 4B represent flux through the coupled and uncoupled mechanisms, respectively. The uncoupled forward flux decreases smoothly through the pulse as GTP is consumed (Fig. 4A). The uncoupled reverse reaction accelerates until the velocities of the coupled and uncoupled reactions are equivalent (Fig. 4B). This equivalence point corresponds to the apex of the pulse. Subsequent to this

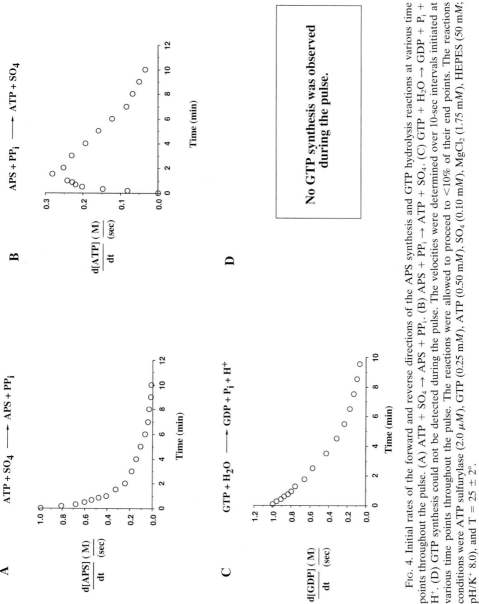

FIG. 4. Initial rates of the forward and reverse directions of the APS synthesis and GTP hydrolysis reactions at various time points throughout the pulse. (A) ATP + SO$_4$ → APS + PP$_i$. (B) APS + PP$_i$ → ATP + SO$_4$. (C) GTP + H$_2$O → GDP + P$_i$ + H$^+$. (D) GTP synthesis could not be detected during the pulse. The velocities were determined over 10-sec intervals initiated at various time points throughout the pulse. The reactions were allowed to proceed to <10% of their end points. The reactions conditions were ATP sulfurylase (2.0 μM), GTP (0.25 mM), ATP (0.10 mM), SO$_4$ (0.50 mM), MgCl$_2$ (1.75 mM), HEPES (50 mM; pH/K$^+$ 8.0), and T = 25 ± 2°.

point, the concentrations of APS and PP_i decrease, slowing the reverse reaction to the point where the system reaches what is essentially, but not quite, the non-GTPase-mediated equilibrium position.

The concomitant uncoupling of the APS reaction and hydrolysis of GTP suggests that GDP and/or P_i regulates the coupling efficiency. GDP is a modest activator of the APS reaction. (The K_m values of APS and PP_i are decreased 2.9- and 1.9-fold, respectively, k_{cat} increases 1.7-fold, and the K_m of GDP for the $E \cdot APS \cdot PP_i$ complex is 0.54 μM.) GDP competes with GTP for the guanine nucleotide-binding site.[38] Thus, GDP causes the system to uncouple in two ways: by providing an uncoupled catalytic path for the APS reaction and by preventing the coupled pathway by blocking GTP binding. The increasing [GDP]/[GTP] ratio during the pulse shifts the enzyme from the central complexes, $E \cdot GTP \cdot ATP \cdot SO_4$ and $E \cdot GDP \cdot P_i \cdot APS \cdot PP_i$, which catalyze the linked chemistry, to the analogous $\cdot GDP$ complexes, $E \cdot GDP \cdot ATP \cdot SO_4$ and $E \cdot GDP \cdot APS \cdot PP_i$, which catalyze the APS reaction but cannot transfer chemical potential.

Phosphate Plugs the Leak. Because the only complexes capable of catalyzing the coupled chemistries are the central $E \cdot GTP \cdot ATP \cdot SO_4$ and $E \cdot GDP \cdot P_i \cdot APS \cdot PP_i$ complexes, one expects that saturating the leaky $E \cdot GDP \cdot APS \cdot PP_i$ with phosphate would prevent leaking and enforce coupling. Data presented in Fig. 5 support this model strongly. The reaction profiles in Fig. 5 were obtained under identical conditions except that phosphate was present at 0.20 M in the reaction whose profile achieves the long-lived plateau. The binding of P_i to the leaky $E \cdot GDP \cdot APS \cdot PP_i$

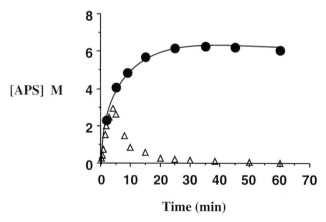

FIG. 5. The effect of a high concentration of phosphate on the progress curve for APS synthesis. The conditions of this experiment were ATP sulfurylase (1.0 μM), ATP (0.20 mM), [^{35}S]SO$_4$ (0.20 mM), GTP (0.10 mM), P_i [0.20 M (●) or 0.0 M (△)], HEPES (50 mM, pH/K$^+$ 8.0), and T = 25 ± 2°.

complex drives the system into the coupled catalytic path. Thus, P_i acts as a molecular plug by stopping the leak. The apparent equilibrium constant, calculated from the concentration of APS in the plateau, is 1.1×10^{-3}. This 1×10^5-fold shift in the equilibrium position of the APS reaction corresponds to a transfer of -6.8 kcal/mol of chemical potential from the GTP hydrolysis reaction to the APS-forming reaction.

The regulation of coupling efficiency by GTP, GDP, and P_i means that fixing the mass balance of the GTP reaction, $[GDP] \cdot [P_i]/[GTP]$, as is believed to be the case in the cell, will fix the position of the APS reaction, $[APS] \cdot [PP_i]/[ATP] \cdot [SO_4]$. That this is indeed the case is demonstrated by the experiments represented in Fig. 6 in which GMPPNP, which uncouples the system, was used to adjust the steady-state mass balance of the APS reaction. GMPPNP, like GDP, is a modest activator of the uncoupled pathway and its binding is competitive versus GTP. The GTP/GDP ratio was held fixed in these experiments at >20/1 using a GTP-regenerating system and the P_i concentration was small, <2 mM, compared with its K_m for the $E \cdot GDP \cdot APS \cdot PP_i$ complex, 44 ($\pm$13) mM. Data clearly indicate that as the flux through the uncoupled pathway is increased, by increasing the concentration of uncoupler, the mass ratio for the APS reaction de-

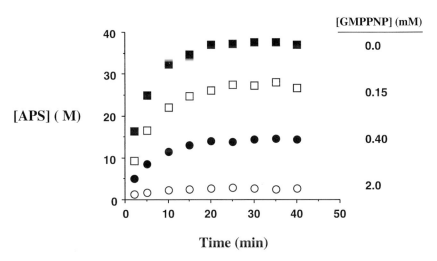

Time (min)

Fig. 6. GMPPNP influences the mass-action ratio of the APS reaction. The binding of GMPPNP, like GDP, prevents GTP binding and allows uncoupled chemistry. The reaction conditions were as follows: ATP sulfurylase (0.50 μM), ATP (0.20 mM), [^{35}S]SO$_4$ (0.20 mM), GTP (1.0 mM), GMPPNP (as indicated), MgCl$_2$ ([total nucleotide] + 1.0 mM), pyruvate kinase (8.0 U/ml), PEP (5.0 mM), HEPES (50 mM, pH/K$^+$ 8.0), and T = 25 $\pm$ 2°. One unit (U) converts 1.0 μmol of substrate to product at V_{max}.

creases to near that of the uncoupled system. Thus, uncoupling can be used to "titrate" the transfer of chemical potential between these two reactions.

Product formation in the APS reaction is regulated, within the energetic boundaries of the coupled and uncoupled reactions, by the concentrations of GTP, GDP, and P_i. Our ability to assess the extent to which the cell uses this allosteric regulation of energy transfer to control flux through the sulfate activation pathway is limited by an incomplete understanding of the *in vivo* mileu. The cellular concentrations of GTP, GDP, and P_i have been been determined in extracts of *E. coli* in asynchronous, logarithmic growth in rich media; they are 920 μM, 130 μM, and 10 mM, respectively.[43,44] How these metabolite concentrations vary as a function of media nutrients, cell cycle, or within microenvironments of the cell are unknown. The quoted values represent the weighted average of the metabolite concentrations over all of the states of the cell. Given these concentrations, one can calculate a mass ratio for the APS reaction of $\sim 2 \times 10^{-3}$, which is 30-fold below that associated with optimal coupling. Thus, the ATP sulfurylase reaction might well be substantially uncoupled in the cell.

The submaximal mass balance predicted for the APS reaction *in vivo* begs the question: Of what advantage is it to the cell to downregulate APS levels? On this point, genetic studies have demonstrated convincingly that the accumulation of PAPS, or a derivative of it, is toxic and can be lethal.[45-47] When the cysteine biosynthetic pathway of a cell is blocked at points of PAPS consumption, the cell becomes sick. To survive, it must acquire second-site mutations in genes upstream of PAPS synthesis (see Fig. 7). The inactivation of APS kinase, which synthesizes PAPS, protects the cell against the lethal consequences of inactivating the enzymes that consume PAPS.

At some point in evolution the synthesis of APS in *E. coli* was linked energetically to the hydrolysis of GTP. In a fixed chemical environment, the potential associated with the hydrolysis of the β-γ bond of GTP is immutable, $\Delta G^{0\prime} \sim -7.7$ kcal/mol.[49] If the system were coupled perfectly,

[43] J. Neuhard and P. Nygaard, *in* "*Escherichia coli* and *Salmonella typhimurium:* Cellular and Molecular Biology" (F. C. Neidhardt, ed.), p. 445. Am. Soc. for Microbiology, Washington, DC, 1987.

[44] N. N. Rao, M. F. Roberts, A. Torriani, and J. Yashphe, *J. Bact.* **175**, 74 (1993).

[45] D. Gillespie, M. Demerec, and H. Itikawa, *Genetics* **59**, 433 (1968).

[46] A. F. Newald, B. R. Krishnan, I. Brikun, S. Kulakauskas, K. Suziedelis, T. Tomcsanyi, T. S. Leyh, and D. E. Berg, *J. Bact.* **174**, 415 (1992).

[47] M. Russel, P. Model, and A. Holmgren, *J. Bact.* **172**, 1923 (1990).

[48] N. M. Kreditch, *in* "*Escherichia coli* and *Salmonella typhimurium:* Cellular and Molecular Biology" (F. C. Neidhardt, ed.), p. 419. Am. Soc. for Microbiology, Washington, DC, 1987.

[49] R. A. Alberty, *Pure Appl. Chem.* **66,** 1641 (1994).

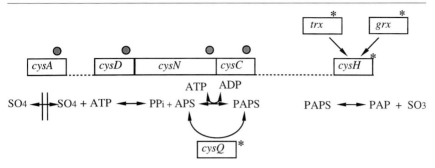

FIG. 7. The genetics of PAPS toxicity. Inactivating mutations in any of the genes labeled with an asterisk blocks the metabolic consumption of PAPS and causes cells to sicken and die. Cells that survive acquire second site mutations in the genes upstream of PAPS synthesis (shaded circles). The vertical, parallel lines beneath *cysA* represent the bacterial cell wall. The GTP hydrolysis reaction associated with APS synthesis has been omitted for simplicity. Genes not connected or connected by dotted lines are located in separate operons in the *cys* regulon. The gene pneumonics are defined as follows: *cysA*, sulfate permease[48]; *cysD* and *cysN*, subunits of ATP sulfurylase[40]; *cysC*, APS kinase[40]; *cysH*, PAPS reductase[48]; *cysQ*, putative PAPS phosphatase[46]; *trx*, thioredoxin[45–47]; and *grx*, glutaredoxin.[45–47] Thioredoxin and glutaredoxin provide reducing equivalents for the reduction of the sulfuryl group of PAPS.

the potential transferred is neither more nor less than that available from the hydrolysis reaction. This corresponds to an enormous (5×10^5-fold) shift in the mass balance of the coupled reaction. Changes of this magnitude can disrupt metabolic balance and can be detrimental to the organism. The challenge then is for the cell to evolve a mechanism that allows it to titrate the chemical potential, as needed, into the coupled metabolic pathway. Imperfect coupling provides such a mechanism.

DNA Gyrase

The conversion of the chemical potential of ATP hydrolysis into torsional stress in DNA is a fascinating example of mechano-chemical coupling. This transfer of potential is catalyzed by the enzyme DNA gyrase, a type II topoisomerase (type I and II topoisomerases are classified according to whether, respectively, one or both strands of DNA are broken and resealed in a complete isomerization cycle).[50,51] Virtually all DNA isolated from natural sources contains torsional stress, which manifests itself in a variety of ways, including supercoils, cruciforms, Z-DNA, and

[50] N. R. Cozzarelli, *Science* **207,** 953 (1980).
[51] L. M. Fisher, C. A. Austin, R. Hopewell, E. E. C. Margerrison, M. Oram, S. Patel, K. Plummer, J. Sng, and S. Sreedharan, *Phil. Trans. R. Soc. Lond. B* **336,** 83 (1992).

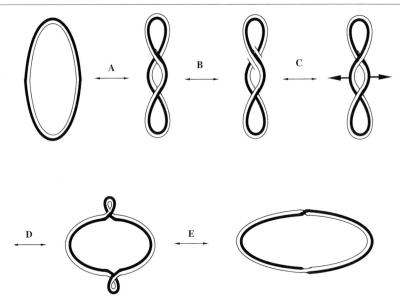

FIG. 8. The supercoiling of DNA by DNA gyrase. After being folded on itself (step A), each strand of the DNA is opened by the nucleophilic attack of a tyrosyl residue located in an A subunit of DNA gyrase (step B). The strand lying above the break is passed throughout it in an ATP hydrolysis-dependent reaction and the break is then resealed (step C). At this point the isomerization reaction is complete and one supercoil has been added to the DNA. The effect of the isomerization on the helicity of the DNA is depicted in steps C and D in which the coiled DNA is "stretched" into a circle. This procedure reveals that two helical turns with the same helical sense have been added to the DNA. Thus, a single isomerization, or supercoil, changes the linking number of the DNA by two.

promoter "melting."[52] These strain-induced structures are essential for cellular and viral DNA replication, as well as gene expression.[50,53,54] Type II topoisomerases appear to be the only class of enzymes that catalyze these isomerizations.

An overview of the reaction catalyzed by type II topoisomerases is outlined schematically in Fig. 8. In Fig. 8, the two strands of a DNA duplex are represented by white and black "strings." The helicity of the duplex has been omitted for simplicity. A typical DNA duplex forms a right-handed helix with a pitch of one helical turn per 10 bp. By convention, the right-handed twists of the duplex define the positive (+) helical sense. In

[52] R. W. Bauer, *Annu. Rev. Biophys. Bioeng.* **7,** 287 (1978).

[53] W. G. Miller and R. W. Simons, *Mol. Microbiol.* **10,** 675 (1993).

[54] M. van Workum, S. J. M. van Dooren, N. Oldenburg, D. Molenaar, P. R. Jensen, J. L. Snoep, and H. V. Westerhoff, *Mol. Microbiol.* **20,** 351 (1996).

the first step of the scheme (step A), one of the duplex strands of a closed, relaxed circle of DNA is wrapped over the other. The bottom strand is then opened at a single point (step B). The top strand is passed through the opening in an ATP-dependent reaction and the opening is then resealed (step C). The resulting structure is supercoiled DNA. The effect of a single DNA gyrase supercoiling event on the helicity of the DNA is demonstrated by converting the supercoiled DNA into circular DNA (steps D and E). As the central loop of the supercoil is stretched toward the circular form, the upper and lower loops tighten (step D). Further stretching reveals two helical turns with the same sense (step E). The result of the isomerization is that the number of helical turns in the DNA (the linking number) increases or decreases by two, depending on whether the sense of the added helical turns is $(+)$ or $(-)$. Positive helices "tighten" the duplex; negative helices unwind it.

Native DNA gyrase is an A_2B_2 tetramer that can be reconstituted from purified A and B subunits. The current models for the mechanism of DNA gyrase include a transesterification reaction[55,56] that opens and closes the DNA duplex, the gating reaction, and a set of conformational changes driven by the binding and hydrolysis of ATP that transport the unbroken DNA strand through the opened gate, the clamping reaction.[57,58] DNA scission is accomplished by the nucleophilic attack of two conserved tyrosine residues, located in the A subunits, at the $5'$-phosphate of the two complementary DNA strands. The strands are resealed by transferring the DNA strand back from the tyrosyl-hydroxyls to the conjugate $3'$-ribosyl-hydroxyl of the base, reforming the bond that had been broken.

While many details of the mechanism of this enzyme have been elucidated, certain issues, central to the energetic linkage, remain unresolved. The enzyme binds to and cleaves DNA in the absence of nucleotide. This nucleotide-independent gating reaction allows the enzyme to relax $(-)$ supercoils.[50] The intrinsic ATPase activity of DNA gyrase is stimulated 20-fold at a saturating concentration of double-stranded DNA.[59,60] Stimulation of basal hydrolytic activity by the coupling reactant(s) is typical of energetically linked systems. ATPase activation requires the binding of DNA to at least two sites on the protein, which are most likely the gating

[55] Y.-C. Tse, K. Kirkegaard, and J. C. Wang, *J. Biol. Chem.* **255,** 5560 (1980).
[56] T. C. Rowe, K. M. Tewey, and L. F. Liu, *J. Biol. Chem.* **259,** 9177 (1984).
[57] J. Roca and J. C. Wang, *Cell* **71,** 833 (1992).
[58] J. Roca, and J. C. Wang, *Cell* **77,** 609 (1994).
[59] A. Sugino and N. R. Cozzarelli, *J. Biol. Chem.* **255,** 6299 (1980).
[60] A. D. Bates, M. H. O'Dea, and M. Gellert, *Biochemistry* **35,** 1408 (1996).

and translocation sites.[61] The gating site is located in the A subunits, each of which contains one of the nucleophilic tyrosines. The translocation site is formed at the interface of the B subunits.[62] The binding of AMPPNP causes the translocation site to "clamp" shut.[57,63] If the clamp contained DNA on closing, it is trapped; otherwise, it is locked out. Thus, nucleotide binding drives a conformational change that links the nucleotide reaction to DNA translocation. How the hydrolysis of ATP and the subsequent release of products are linked to DNA translocation and closure remains unknown.

Kinetic measurements have shown that the ratio of the initial rates of ATP hydrolysis and negative supercoiling is between one and two. A stoichiometry of 2 : 1 is supported by energetic analyses that predict that the hydrolysis of two ATP is required to achieve the levels of ($-$) supercoiling observed in the coupled reaction,[64] and the fact that DNA gyrase binds to two molecules of ATP.[65,66] This stoichiometry only applies in the initial stage of the coupled reaction. As the reactions proceed, the net rate of supercoiling decreases to near zero, while the hydrolysis reaction continues unabated at essentially its initial rate. If the reactions were linked perfectly, the rates of both reactions would decrease in parallel as they approached equilibrium. Thus, the reactions progressively uncouple as the DNA becomes more supercoiled. This means that embedded in the system is a feedback mechanism that "senses" the extent of supercoiling and regulates the coupling efficiency accordingly. This behavior has strong parallels with the coupled ATP sulfurylase reaction in which the rate of the GTPase chemistry is essentially unaffected by the position of the APS-forming reaction. The uncoupling mechanism is also reminiscent of the ATP sulfurylase reaction in which GDP, which competes with GTP, uncouples the reactions. As discussed previously, DNA gyrase catalyzes a nucleotide-independent gating reaction that allows it to relax negatively supercoiled DNA. Relaxation, which represents a leak in the system, is not affected significantly by ADP, which is a competitive inhibitor versus ATP (M. Gellert, personal communication). Thus, at saturating nucleotide concentrations the distribution of DNA gyrase among its ATP and ADP complexes

[61] A. Maxwell and M. Gellert, *J. Biol. Chem.* **259**, 14472 (1984).
[62] D. B. Wigley, G. J. Davies, E. J. Dodson, A. Maxwell, and G. Dodson, *Nature (London)* **351**, 624 (1991).
[63] A. Sugino, N. P. Higgins, P. O. Brown, C. L. Peebles, and N. R. Cozzarelli, *Proc. Natl. Acad. Sci. U.S.A.* **75**, 4838 (1978).
[64] P. M. Cullis, A. Maxwell, and D. P. Weiner, *Biochemistry* **31**, 9642 (1992).
[65] J. K. Tamura and M. Gellert, *J. Biol. Chem.* **265**, 21342 (1991).
[66] J. K. Tamamura, A. D. Bates, and M. Gellert, *J. Biol. Chem.* **267**, 9214 (1992).

will determine the coupling efficiency and ultimately the degree of supercoiling of the DNA.

The *E. coli* chromosome contains approximately two to five negative superhelical twists per kilobase,[67] roughly one-half of the maximum supercoiling density attainable *in vitro*. Chromosomal supercoiling is regulated, in part, by the activities of DNA gyrase, which introduces supercoils, and topoisomerase I, which removes them.[53] Expression of the genes that encode DNA gyrase, *gyrA* and *gyrB*, and topoisomerase I, *topA*, is regulated by the density of (−) supercoils. Decreasing density upregulates *gyrA* and *gyrB* and downregulates *topA*. Supercoiling depends on the growth conditions of the cell. Changes in oxygenation,[68,69] pH,[70] nutrient composition,[71] temperature,[72] and osmolarity[68,69] influence supercoiling. Superhelical density has also been shown to be regulated, both *in vivo* and *in vitro*, by the ATP/ADP ratio.[54,73] These data support the position that the cell uses the ATP/ADP ratio to regulate gyrase leakage and "tune" its DNA tension to the environmental conditions.

Kinetic Proofreading

Kinetic proofreading is a well-documented example of the fruitful uncoupling of a conformationally linked system.[74,75] In this case, the hydrolysis of GTP by elongation factor-Tu (EF-Tu) is involved intimately in determining the fidelity of protein synthesis. Perfectly matched (cognate) aminoacylated-tRNAs (aa-tRNA) are committed, by GTP hydrolysis, to participate in polypeptide elongation. The same editing mechanism is used to expel near-cognate aa-tRNA from the synthetic machinery before its amino acid can be incorporated in the growing polypeptide chain. The mechanism is based on the relative rates of desorption of the aa-tRNA to what are believed to be irreversible steps in the hydrolysis reaction. The near-cognate species desorb quickly compared with the irreversible steps, while the cognate-tRNA species desorb slowly. Because the irreversible steps commit the aa-tRNA to the elongation cycle, these rates kinetically select cognate over near-cognate aa-tRNA for elongation. Once a near-cognate species

[67] R. R. Sinden and D. E. Pettijohn, *Proc. Natl. Acad. Sci. U.S.A.* **78**, 224 (1981).
[68] N. N. Bhriain, C. J. Dorman, and C. F. Higgins, *Mol. Microbiol.* **7**, 351 (1993).
[69] C. P. O'Byrne, N. N. Bhriain, and C. J. Dorman, *Mol. Microbiol.* **6**, 2467 (1993).
[70] K. Karem and J. W. Foster, *Mol. Microbiol.* **10**, 75 (1993).
[71] V. L. Balke, and J. D. Gralla, *J. Bacteriol.* **4499** (1987).
[72] E. K. Goldstein, *Proc. Natl. Acad. Sci. U.S.A* **81**, 4046 (1984).
[73] H. V. Westerhoff, M. H. O'Dea, A. Maxwell, and M. Gellert, *Cell Biophys.* **12**, 157 (1988).
[74] J. J., Hopfield, *Proc. Natl. Acad. Sci. U.S.A.* **71**, 4135 (1974).
[75] R. C. Thompson, *Trends Biochem. Sci.* **13**, 91 (1988).

departs, the system continues through its cycle of GTP hydrolysis and release. From a strictly thermodynamic point of view, this is a leak, uncoupled energy is released. Of course the uncoupling is integral to the proper metabolic functioning of this system; perfect linkage would have been detrimental.

The basic steps of the kinetic proofreading mechanism are summarized in Fig. 9, which includes typical values for the rate constants of the scheme. A single value is given for constants that are independent of the aa-tRNA. Constants that depend on the nature of the aa-tRNA are assigned two values, Cog and N-Cog, which are associated with the cognate- and near-cognate aa-tRNA, respectively. In the initial recognition step, the ternary complex (aa-tRNA · EF-Tu · GTP) binds to the ribosome, R, which is programmed with poly(U) RNA and AcPhe-tRNA at its P site to carry out peptide elongation. As indicated in Fig. 9, the rate constant governing the

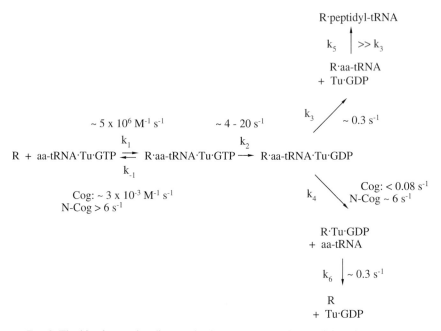

FIG. 9. The kinetic proofreading mechanism. The mechanism is divided into two phases: initial recognition, which includes the rate constants k_1 and k_{-1}, and proofreading, which includes all of the other rate constants. Values associated with constants are typical values. Rate constants having only one value are independent of aa-tRNA. Rate constants with two values depend on whether the aa-tRNA is cognate (Cog) or near-cognate (N-Cog), R represents a ribosome complex that is programmed with poly(U) and AcPhe-tRNA at its P site to catalyze polypeptide elongation. Rate constants were determined at 5° and were obtained from the following references: k_1 and k_{-1} (79) and k_2–k_6 (77 and references therein).

association of the ternary complex and ribosome is independent of whether the complex harbors cognate or near-cognate tRNA. The dissociation of the near-cognate species is roughly 2000-fold greater than that of the cognate species. This rate difference is the major discriminator in this system and is due to the differences in binding affinity between perfect and imperfect codon : anticodon interactions.

The remaining steps of the mechanism are referred to as the proofreading steps. Notice that GTP hydrolysis commits the quaternary complex to the next step in the elongation cycle and that the rate constant for hydrolysis, k_2, is similar to that for the dissociation of the near-cognate complex. This means that the discrimination available in the binding reaction is not fully realized. It has been suggested that this incomplete discrimination might have biological merit.[75,76] Once hydrolysis occurs, the aa-tRNA is "accepted" by the system. Release of the EF-Tu · GDP complex is considered irreversible and precedes peptide bond formation.[77] The rate constant for this release step, 0.3 sec^{-1}, is independent of the presence or absence of the aa-tRNA.[77] The rates at which cognate and near-cognate aa-tRNA depart from the R · aa-tRNA · EF-Tu · GDP complex are very different. The constant governing dissociation of the noncognate species, 6 sec^{-1}, is 20-fold greater than that for EF-Tu · GDP dissociation, 0.3 sec^{-1}, which in turn is more than fourfold faster than that for dissociation of the cognate species, <0.08 sec^{-1}. Thus, the cognate species are again kinetically selected to donate their amino acids to the growing peptide chain.

Kinetic proofreading offers an excellent example of how energetic barriers that control leaking are manipulated allosterically by "recognition" elements inherent in the physicochemical architecture of the interacting components of the system. The escape barrier of aa-tRNA is modulated by the very nature of the molecule itself. The cognate barriers are sufficiently high that GTP hydrolysis and polypeptide elongation are essentially perfectly linked, with a stoichiometry of 1 : 1. The near-cognate barriers are lowered to the point that the stoichiometry reduces to a very leaky 20 : 1.[78] It is worth mentioning that the coupling of the hydrolysis reaction is not solely at the point of β, γ-bond cleavage. Linking can occur at any of the steps, from the binding of GTP and EF-Tu to the release of P$_i$ and GDP, and that it is the product of the linking at each of these steps that determines the overall coupling efficiency of the system.

[76] T. Jacks and H. Varmus, *Science* **230,** 1237 (1985).
[77] R. C. Thompson, D. B. Dix, and A. M. Karim, *J. Biol. Chem.* **261,** 4868 (1986).
[78] R. C. Thompson and D. B. Dix, *J. Biol. Chem.* **257,** 6677 (1982).
[79] J. F. Eccleston, D. B. Dix, and R. C. Thompson, *J. Biol. Chem.* **260,** 16237 (1985).

Conclusions

The mechanistic basis of coupling and uncoupling in three conformationally coupled systems has been evaluated. In each case, the reactants that participate in the coupled reaction regulate the efficiency with which chemical potential is transferred between donor and acceptor reactions. The key to whether a system will link or leak energy lies in the distribution of the forms of the coupling enzyme. Certain complexes couple the energetics of two reactions while other leaky complexes catalyze the reactions independently. Each complex has its own set of kinetic and thermodynamic parameters. The extent to which a given complex is represented in the ensemble of complexes available to the system is determined by these parameters and the concentrations of coupling reactants. Thus, the distribution of complexes, and therefore the coupling efficiency, is set by the steady-state concentration of the cellular metabolites involved in the linking.

Imperfect linking provides the opportunity to regulate energy transfer within the energetic limits of the linked and unlinked chemistries. The examples presented indicate that the cell uses imperfect linking to avoid toxicity, regulate gene expression, and control the fidelity of protein synthesis. These systems leak with purpose. Regulating the efficiency of energy transfer has considerable metabolic utility and, at least for certain systems, it appears to have been optimized to satisfy the complex environmental and metabolic constraints under which the organism must operate.

Acknowledgment

This work was supported by National Institutes of Health Grant GM 54469.

[4] Reaction Dynamics of G-Protein Catalyzed Hydrolysis of GTP as Viewed by X-Ray Crystallographic Snapshots of $G_{i\alpha 1}$

By David E. Coleman and Stephen R. Sprang

Introduction

The α subunits (G_α) of heterotrimeric G proteins ($G_{\alpha\beta\gamma}$) are members of the Ras superfamily of GTP hydrolases, enzymes that use the free energy of GTP hydrolysis to regulate cellular processes (e.g., Ras, EF-

Copyright © 1999 by Academic Press
All rights of reproduction in any form reserved.
0076-6879/99 $30.00

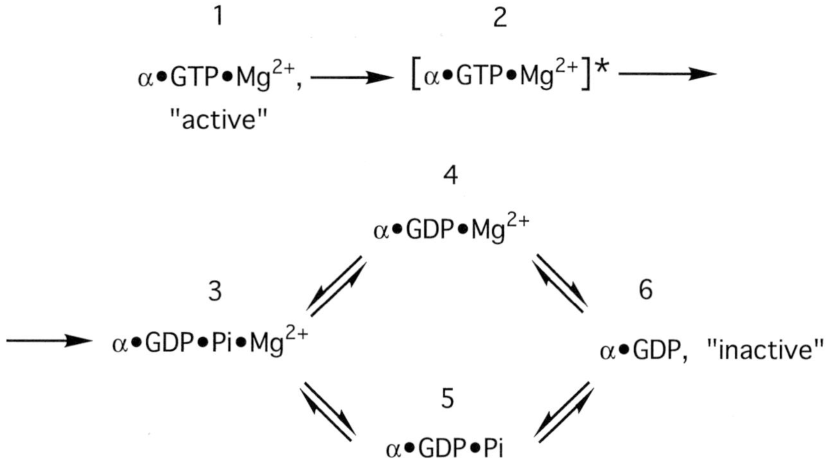

1   2

$\alpha \bullet GTP \bullet Mg^{2+}$, $\longrightarrow$ $[\alpha \bullet GTP \bullet Mg^{2+}]^* \longrightarrow$

"active"

4

$\alpha \bullet GDP \bullet Mg^{2+}$

3   6

$\longrightarrow$ $\alpha \bullet GDP \bullet Pi \bullet Mg^{2+}$ $\alpha \bullet GDP$, "inactive"

5

$\alpha \bullet GDP \bullet Pi$

SCHEME I

Tu, EF-G).[1,2] G_α subunits, in concert with $G_{\beta\gamma}$ heterodimers, transduce intracellular signals by coupling ligand-activated hepta-helical transmembrane receptors to downstream effectors such as adenylyl cyclase, phospholipase C, ion channels, and GMP phosphodiesterase. A key feature of every G_α subunit is that its ability to interact with other proteins in the signal transduction pathway is determined by the ligand bound to its active site. Thus, the stable enzyme–product complex $G_\alpha \cdot GDP$ assumes an inactive signaling conformation in which it cannot bind to effectors but, rather, binds to $G_{\beta\gamma}$ to form $G_{\alpha\beta\gamma}$ heterotrimers. Conversely, the enzyme–substrate complex $G_\alpha \cdot GTP \cdot Mg^{2+}$ assumes an active signaling conformation that causes it to dissociate from $G_{\beta\gamma}$ and to bind to and regulate downstream effectors. The active GTP-bound conformation is achieved by the receptor-mediated exchange of GDP for GTP, whereas the inactive GDP-bound conformation is generated by hydrolysis of GTP to GDP catalyzed by G_α itself. Thus, the rate of hydrolysis of GTP by G_α subunits serves as a timer for regulating the duration of downstream signaling events.

Scheme I describes the hydrolytic pathway and expected intermediates of G_α-catalyzed hydrolysis of GTP. Compared with other nucleotide hydrolases, the rate of hydrolysis leading from the ground state (species 1) to the binary product complex (species 6) is slow (typically $k_{cat} = 2–5$

[1] H. R. Bourne, D. A. Sanders, and F. McCormick, *Nature* **348,** 125 (1990).
[2] H. R. Bourne, D. A. Sanders, and F. McCormick, *Nature* **349,** 117 (1991).

min^{-1}).[3] This low rate is in keeping with the role of G_α in signal transduction and is indicative of a substantial kinetic barrier to the formation or breakdown of the transition state. Factors contributing to the low rate of hydrolysis might include insufficient activation of the hydrolytic water nucleophile, weak stabilization of the transition state, obligate conformational changes, or slow release of product.

Isotopic labeling experiments conducted with the Ras superfamily proteins Ras,[4] EF-G,[5] and EF-Tu[6] demonstrate that GTP hydrolysis proceeds by a direct in-line transfer of the γ-phosphate group from GTP to an attacking water molecule, with inversion of configuration at the γ-phosphate group. This mechanism implies that the reaction transition state (species 2, Scheme I) has bipyramidal geometry about the phosphate atom, with the three oxygen substituents in the equatorial plane and with the water nucleophile and the β/γ-phosphate bridging oxygen-leaving group in the axial positions (Fig. 4). It is reasonable to assume that other members of the Ras GTPase family, including G_α subunits, follow a similar reaction pathway.

A bipyramidal transition state may be achieved either by an associative or a dissociative mechanism.[7] In an associative mechanism, bond formation and bond cleavage occur simultaneously. This transition state exhibits substantial bond formation between the attacking water nucleophile and the γ-phosphate atom. Bond formation is accompanied by an increase of negative charge on both the β-phosphate oxygen-leaving group and the equatorial oxygen substituents. Collapse of this transition state inverts the configuration of the γ-phosphate group. In a dissociative mechanism there is extensive bond cleavage between the γ-phosphate and the β-phosphate oxygen-leaving group, which leads to a metaphosphate-like transition state prior to attack by water. In this mechanism, negative charge on the equatorial oxygen atoms decreases in the transition state concomitant with the increase of negative charge on the oxygen-leaving group. A dissociative mechanism may also exhibit strict inversion of configuration if the metaphosphate-like intermediate is attacked by the nucleophile at a rate that is faster than the diffusion rate[8] or if the intermediate is anchored by the enzyme to form a bipyramidal complex with the GDP-leaving group and the attacking nucleophile.

[3] A. G. Gilman, *Annu. Rev. Biochem.* **56,** 615 (1987).
[4] J. Feuerstein, R. S. Goody, and M. R. Webb, *J. Biol. Chem.* **264,** 6188 (1989).
[5] M. R. Webb and J. F. Eccleston, *J. Biol. Chem.* **256,** 7734 (1981).
[6] J. F. Eccleston and M. R. Webb, *J. Biol. Chem.* **257,** 5046 (1982).
[7] S. J. Admiraal and D. Herschlag, *Chem. Biol.* **2,** 729 (1995).
[8] W. P. Jencks, *Acc. Chem. Res.* **13,** 161 (1980).

Associative mechanisms have been proposed for Ras family GTPases based on X-ray crystal structures[9,10,36] and theoretical studies.[12] Alternatively, a dissociative mechanism has been presented based on the dissociative nature of the hydrolysis of nucleotides and model compounds in aqueous solution.[13] The actual transition state may lie on a spectrum between the two mechanisms. In either case the active site of G_α subunits would be expected to provide some means for stabilizing a bipyramidal transition state and to sense the formation of product by undergoing a transition from the active to the inactive signaling conformation.

Structural Studies of the Mechanism of $G_{i\alpha1}$: Snapshots of the Active Site during GTP Hydrolysis

In order to further elucidate the hydrolytic and regulatory mechanisms of G_α proteins, we have determined, by X-ray crystallography, structures of wild-type and mutant $G_{i\alpha1}$ complexed with products or substrate analogs (Table I). $G_{i\alpha1}$ is activated by α_2-adrenergic[14] and m2 muscarinic receptors,[15] among others, and inhibits certain isoforms of adenylyl cyclase.[16,17] These structures correspond to several of the species indicated in Scheme I and can be arrayed as a series of snapshots of the active site during GTP hydrolysis. The following account describes this series of structures and emphasizes conformational changes associated with the binding of Mg^{2+} and the γ-phosphate of GTP within the active site. As with any analysis based on X-ray crystallographic studies of proteins and substrate analogs, the following account may at best be an approximation of the actual events that occur within the enzyme.

Methods

The following is a summary of the methods employed in published crystallographic studies of several $G_{i\alpha1}$ complexes (see Table I for refer-

[9] G. G. Privé, M. V. Milburn, L. Tong, A. M. deVos, Z. Yamaizumi, S. Nishimura and S.-H. Kim, *Proc. Natl. Acad. Sci. U.S.A.* **89**, 3649 (1992).
[10] D. E. Coleman, A. M. Berghuis, E. Lee, M. E. Linder, A. G. Gilman, and S. R. Sprang, *Science* **265**, 1405 (1994).
[11] D. G. Lambright, J. P. Noel, H. E. Hamm, and P. B. Sigler, *Nature* **369**, 621 (1994).
[12] T. Schweins and A. Warshel, *Biochemistry* **35**, 14232 (1996).
[13] K. A. Maegley, S. J. Admiraal, and D. Herschlag, *Proc. Natl. Acad. Sci. U.S.A.* **93**, 8160 (1996).
[14] S. Cotecchia, B. K. Kobilka, K. W. Daniel, R. D. Nolan, E. Y. Lapetina, M. G. Caron, R. J. Lefkowitz, and J. W. Regan, *J. Biol. Chem.* **265**, 63 (1990).
[15] J. C. Migeon, S. L. Thomas, and N. M. Nathanson, *J. Biol. Chem.* **270**, 16070 (1995).
[16] R. Taussig, J. A. Iñiguez-Lluhi, and A. G. Gilman, *Science* **261**, 218 (1993).
[17] R. Taussig, W.-J. Tang, J. R. Hepler, and A. G. Gilman, *J. Biol. Chem.* **269**, 6093 (1994).

TABLE I
X-RAY DATA SETS AND REFINEMENT STATISTICS FOR $G_{i\alpha 1}$ · LIGAND COMPLEXES

Ligand(s)	Data set	Temperature	Space group	Resolution range (Å)	R factor[a] (R free[b]) (%)	Reference
GTPγS · Mg^{2+}						
	Wild type	7°	$P3_221$	8.0–2.0	17.5(22.8)	10
	Q204L	100 K	$P3_221$	8.0–2.3	23.0(31.0)	10
	R178C	100 K	$P3_221$	8.0–2.3	21.7(28.0)	10
	G42V	100 K	$P3_221$	8.0–2.0	20.6(25.9)	37
GDP · AlF_4^-						
	Wild type	100 K	$P3_221$	8.0–2.2	22.2(28.2)	10
GDP · P_i						
	G203A	20°	$P4_32_12$	8.0–2.6	19.2(28.3)	39
	G42V	20°	$P4_32_12$	8.0–2.8	18.9(25.0)	37
GDP						
	Wild type	100 K	$I4$	6.0–2.2	21.8(29.1)	22
	G203A	100 K	$I4$	8.0–2.6	21.4(28.0)	22
	G42V	100 K	$I4$	8.0–2.4	21.2(27.4)	37
GDP · Mg^{2+}						
	Wild type (5 mM Mg^{2+})	100 K	$I4$	15.0–2.6	21.4(25.9)	47
	Wild type (10 mM Mg^{2+})	100 K	$I4$	15.0–2.2	21.5(26.6)	47
	Wild type (100 mM Mg^{2+})	100 K	$I4$	15.0–2.6	21.5(25.5)	47
	Wild type (200 mM Mg^{2+})	100 K	$I4$	15.0–2.2	23.0(26.7)	47

[a] R factor = $\Sigma |F_o - F_c|/\Sigma |F_o|$, where F_o and F_c are the observed and calculated structure factor amplitude of the reflections used for refinement.
[b] R free = R factor for a set of reflections (approximately 10% of the observed reflections) removed prior to refinement and used to monitor the reliability of the refinement.

ences). All crystals were grown using intact, nonmyristoylated, nonpalmi-toylated recombinant rat $G_{i\alpha 1}$ protein (354 residues, 40.5 kDa) purified and crystallized as described in Table II and references therein. To facilitate comparisons between the structures, all crystals (with the exception of those of the GDP · P_i complex) were transferred to a standard stabilization solution containing the appropriate nucleotide or nucleotide analog and 100 mM N,N-bis[2-hydroxyethyl]-2-aminoethanesulfonic acid (BES) (pH 7.0), 5 mM dithiothreitol (DTT), and either 88% saturated Li_2SO_4 for crystals irradiated at 7° or 20° or 75% saturated Li_2SO_4 and 15% (v/v) glycerol (added as a cyroprotectant) for crystals irradiated at 100 K. Crystals of the GDP · P_i complex of $G_{i\alpha 1}$ could neither be transferred into the stan-dard stabilization solution nor cyroprotected and frozen. Therefore, data

TABLE II

CRYSTALLIZATION CONDITIONS FOR $G_{i\alpha1}$ · LIGAND COMPLEXES[a]

Ligand(s)	Protein	Reference	Space group	Protein solution	Reservoir
GTPγS · Mg²⁺[b]	Wild type	10	P3₂2₁1	3–6 µl	1000 µl
	Q204L	10		8 mg/ml protein (complexed with GTPγS · Mg²⁺), 0.5 mM GTPγS, 1.0 mM MgSO₄, 10 µM EDTA, 25 mM NaEPPS (pH 8.0), 25 mM DTT, 1.4 M K-phosphate (pH 7.0)	100 mM NaCl, 1.45–1.65 K-phosphate (pH 7.0)
	R178C	10			
	G42V	37			
GDP · AlF₄	Wild type	10	P3₂2₁1	10 µl	800 µl
				25–30 mg/ml protein[c] (complexed with GDP, Al³⁺, and F⁻), 300 µM AlCl₃, 5 mM NaF, 5 mM MgCl₂, 10 mM DTT mixed with 10 µl of reservoir solution	100 mM Na-acetate (pH 6.0), 1.8–1.9 M [d]ammonium sulfite (pH 8.0)
GDP · Pᵢ	G203A	39	P4₃2₁2	10 µl	1000 µl
	G42V	37		10–20 mg/ml protein (complexed with GDP), 5 mM GDP, 0.5 mM EDTA, 25 mM NaEPPS (pH 8.0), 5 mM DTT mixed with 10 µl of reservoir solution	2.2–2.5 M ammonium phosphate (pH 5.5–5.8)
GDP	Wild type	22	I₄	3 µl	1000 µl
	G203A	22		10 mg/ml protein (complexed with GDP), 5 mM GDP, 5 mM EDTA, 50 mM NaEPPS (pH 8.0), 5 mM DTT, mixed with 3 µl of reservoir solution	100 mM Na-acetate (pH 6.0), 1.9–2.0 M [d]ammonium sulfite (pH 8.0)
	G42V	37			
GDP · Mg²⁺	Wild type	47			

[a] All crystals were grown by the vapor equilibrium method at 20° using either hanging drops (for 3- to 6-µl drops) or sitting drops (for 20-µl drops).
[b] Crystals containing $G_{i\alpha1}$ complexed with GTPγS · Mg²⁺ can also be grown using conditions identical to those used to grow crystals of the GDP complex. These crystals are isomorphous to those grown in K-phosphate; X-ray structural analysis indicates that the structures are equivalent.
[c] A higher concentration of protein is necessary to grow these crystals relative to the GTPγS-bound form.
[d] Ammonium sulfite solution must be freshly made (no older than 1 week.)

obtained from these crystals were collected at 20° from crystals mounted directly from their mother liquor [5 mM GTPγS or GDP, 5 mM MgSO$_4$, 0.5 mM EDTA, 5 mM DTT, 25 mM N-(2-hydroxyethyl)pipera-zine-N'-(3-propanesulfonic acid) (EPPS), and ammonium phosphate (2.2–2.5 M, pH 5.5–5.8)]. In order to mimic the $G_{i\alpha 1} \cdot GDP \cdot Mg^{2+} \cdot P_i$ complex, which could not be crystallized directly, crystals of the $G_{i\alpha 1} \cdot GDP$ complex were soaked in standard stabilization solution (which contains SO_4^{2-}, a phosphate analog) and MgSO$_4$ at concentrations of 5, 10, 100, and 200 mM. Structures were determined and refined by the use of standard crystallo-graphic procedures as described (Table I and references therein.)

Molecular Architecture of $G_{i\alpha 1}$

The overall structure of $G_{i\alpha 1}$ as determined in the GTPγS $\cdot$ Mg^{2+} crystal form is shown in Fig. 1.[10] The enzyme is composed of two domains: a 244 residue α/β guanine nucleotide binding and hydrolytic domain that is similar in overall structure to p21^{ras} [18,19] and corresponding domains in other members of the Ras superfamily of GTPases[20] (the Ras-like domain) and, in-serted into it, a 110 residue α-helical domain (Fig. 1). The helical domain is unique to trimeric G protein α subunits and was first identified in the structure of the homologous G protein α transducin ($G_{t\alpha}$).[21] The nucleotide binds within a pocket between the Ras-like and the helical domains. Within this pocket the nucleotide is shielded from solvent by the helical domain, but makes direct contacts with the Ras-like domain only. Three segments of the Ras-like domain have been identified as switch regions because they adopt different conformations in the GDP and GTP analog-bound complexes.[11,22] Switch I (residues 177–187) and switch II (residues 199–219, including the α2 helix) are involved in binding Mg^{2+} and the γ-phosphate group, respectively, and are present in all Ras superfamily members. Switch III (residues 231–242) is anchored by ion pair interactions with switch II and is unique to heterotrimeric G protein α subunits.

The guanine nucleotide is bound to $G_{i\alpha 1}$ by direct interactions with the Ras-like domain (Fig. 2). The purine ring is sandwiched between the side chains of Lys-270 and Thr-327. Specificity for the guanine ring is provided primarily by the side chain of Asp-272, which forms hydrogen bonds with

[18] M. V. Milburn, L. Tong, A. M. deVos, A. Brünger, Z. Yamaizumi, S. Nishimura, and S.-H. Kim, *Science* **247**, 939 (1990).

[19] E. F. Pai, U. Krengel, G. A. Petsko, R. S. Goody, W. Kabsch, and A. Wittinghofer, *EMBO J.* **9**, 2351 (1990).

[20] M. Kjeldgaard, J. Nyborg, and B. F. C. Clark, *FASEB J.* **10**, 1347 (1996).

[21] J. P. Noel, H. E. Hamm and P. B. Sigler, *Nature* **366**, 654 (1993).

[22] M. B. Mixon, E. Lee, D. E. Coleman, A. M. Berghuis, A. G. Gilman, and S. R. Sprang, *Science* **270**, 954 (1995).

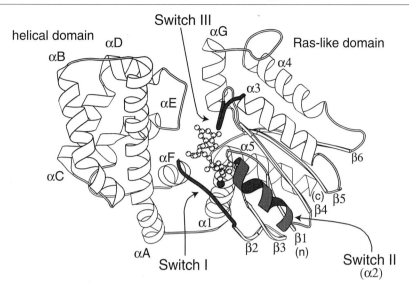

FIG. 1. Overall view of the structure of $G_{i\alpha1} \cdot GTP\gamma S \cdot Mg^{2+}$. The protein is displayed as a backbone trace with secondary structure, and the switch regions are labeled according to the nomenclature used originally for $G_{t\alpha}$.[11,21] The bound GTPγS molecule is rendered as a ball and stick model and the Mg^{2+} ion appears as a dark sphere. Residues 1–33 and 344–354 are disordered in this structure and are not shown. This figure was generated with MOLSCRIPT [P. J. Kraulis, *J. Appl. Cryst.* **24**, 946 (1991)] and rendered with Raster3D [D. Bacon and W. F. Anderson, *J. Mol. Graph*, **6**, 219 (1988)].

the purine ring nitrogen at position 1 and the exocyclic NH_2 at position 2. Mutation of the equivalent Asp in Ras[23,24] or EF-Tu[25] switches the nucleotide specificity to xanthosine. Hydrogen bonds are also formed between the side chain amido group of Asn-269 and purine ring N7. Asparagine-269, Lys-270, and Asp-272 are part of a conserved sequence motif, NKXD, which is found in members of the Ras GTPase superfamily.[2] Hydrogen bonds are also formed between the exocyclic oxygen and the main chain amide of Ala-326 and between N3 and a tightly bound water. The ribose ring is bound via hydrogen bonds between the exocyclic hydroxyl groups and main chain carbonyl groups of the switch I region. The 3' OH group is also bound to a water molecule that is anchored by Ser-151 of the helical domain. This water-mediated interaction and that between Ser-151 and the water bound to the guanine ring are the sole nucleotide/helical domain

[23] J. M. Zhong, M. C. Chen-Hwang, and Y. W. Hwang, *J. Biol. Chem.* **270**, 10002 (1995).
[24] G. Schmidt, C. Lenzen, I. Simon, R. Deuter, R. H. Cool, R. S. Goody, and A. Wittinghofer, *Oncogene* **12**, 87 (1996).
[25] Y.-W. Hwang and D. L. Miller, *J. Biol. Chem.* **262**, 13081 (1987).

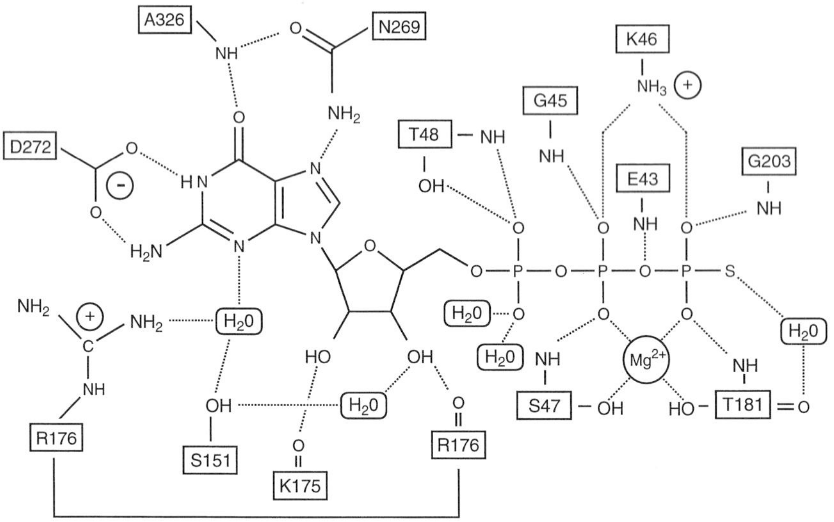

FIG. 2. Schematic diagram showing the major hydrogen-bonding interactions between the nucleotide and $G_{i\alpha 1}$ in the $G_{i\alpha 1} \cdot GTP\gamma S \cdot Mg^{2+}$ complex. (Hydrogen bonds are indicated by dashed lines.)

interactions. The α- and β-phosphate groups are bound by hydrogen bonds to side and main chain hydrogen donor groups within the conserved "P-loop" region (residues 40–46), which contains the conserved sequence motif GXXXXGK found in many nucleotide-binding proteins.[2] These interactions anchor the nucleoside diphosphate moiety of GTPγS rigidly within the active site and are observed, with minor variations, in all of the $G_{i\alpha 1}$ structures discussed later. The protein/nucleotide interactions observed in $G_{i\alpha 1}$ are similar to those found in the Ras-like domains of other Ras superfamily members.[20]

Ground State Enzyme–Substrate Complex

The crystal structure of the $G_{i\alpha 1} \cdot GTP\gamma S \cdot Mg^{2+}$ complex (guanosine 5'-O-3-thiotriphosphate, GTPγS, is a slowly hydrolyzed analog of GTP) provides a model of the ground state enzyme–substrate complex in the active signaling conformation (corresponding to species 1, Scheme I).[10] Key features of the binding site of the γ-thiophosphate moiety of GTPγS are shown in Fig. 3a. These are, with some exceptions, similar to those observed in α transducin.[21] The γ-thiophosphate group is bound and oriented by Lys-46 of the P loop, the bound Mg^{2+}, and by the main chain amide of Gly-203 of switch II. The interaction of Gly-203 with the γ-phosphate

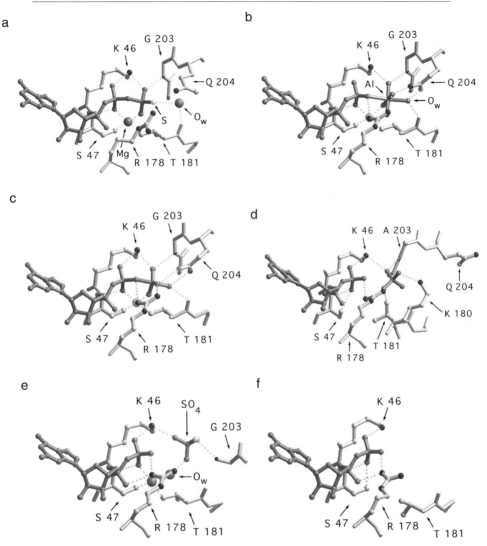

FIG. 3. Snapshots of the active site along the catalytic pathway (see text). Key nitrogen atoms are darkened. (a) $G_{i\alpha1} \cdot GTP\gamma S \cdot Mg^{2+}$. (b) $G_{i\alpha1} \cdot GDP \cdot AlF_4^- \cdot H_2O$. (c) Structure of the pentacoordinate transition state modeled on the $G_{i\alpha1} \cdot GDP \cdot AlF_4^- \cdot H_2O$ structure shown in (b). (d) $G_{i\alpha1} \cdot GDP \cdot P_i$ complex modeled on the structure of $G203A \cdot GDP \cdot P_i$. (e) $G_{i\alpha1} \cdot GDP \cdot SO_4^{2-} \cdot Mg^{2+}$. (f) $G_{i\alpha1} \cdot GDP$. Mg^{2+} and water molecules (labeled as O_W) are shown as large, gray spheres. This figure was rendered using SETOR [S. V. Evans, *J. Mol. Graph.* **11**, 134 (1993)].

plays a critical role in switching the enzyme from the GDP-bound, inactive signaling conformation to the GTP-bound, active signaling conformation. The Mg^{2+} coordination sphere contains six ligands and exhibits octahedral geometry. The four equatorial ligands are provided by two oxygens from the β- and γ-phosphate groups, and the side chain hydroxyl groups of conserved residues Thr-181 (part of switch I) and Ser-47. The two axial ligands are water molecules, one of which forms a hydrogen bond with Asp-200. Aspartate-200 and Gly-203 are part of a conserved sequence motif, ^{200}DXXG, which, like the P loop, is found in all regulatory GTP-binding proteins.[2]

The active site contains a well-ordered (B = 22.2 Å^2) water molecule that appears to be poised for an in-line nucleophilic attack on the γ-phosphate atom (Fig. 3a). This water molecule is located 3.85 Å from the γ-phosphate and is anchored by hydrogen bonds with the γ-thiophosphate group sulfur atom and the carbonyl oxygen of Thr-181 in switch I. A water molecule in the corresponding position has been observed in the enzyme : substrate–analog complexes of transducin ($G_{t\alpha}$),[21] Ras,[9,19] and EF-Tu.[26,27] As has been observed in these proteins, there is no residue within the active site that is positioned such that it could act as a general base to activate this putative hydrolytic water molecule. (Reorientation of the side chain of Glu-207 could permit it to serve as a catalytic base, but mutagenesis studies indicate that this residue serves no catalytic role.[28]) Studies of Ras provide evidence that a general base with a pK_a near 3 is involved in catalysis[29] and that several mutations that reduce catalytic activity are correlated with decreased basicity of the bound nucleotide.[30] These observations, along with theoretical studies,[31] have led to the proposal that the γ-phosphate of GTP, rather than any protein group, may correspond to this catalytic base.[31] It has been suggested that this hypothesis may partially explain the stability of GTPγS toward enzymatic hydrolysis because the γ-thiophosphate group has a lower pK_a than that of a GTP γ-phosphate (2.5 vs 2.9 as measured in Ras).[30]

The impairment of GTP hydrolytic activity that results from mutation

[26] H. Berchtold, L. Reshetnikova, C. O. A. Reiser, N. K. Schirmer, M. Sprinzl, and R. Hilgenfeld, *Nature* **365**, 126 (1993).
[27] M. Kjeldgaard, P. Nissen, S. Thirup, and J. Nyborg, *Structure* **1**, 35 (1993).
[28] C. Kleuss, A. S. Raw, E. Lee, S. R. Sprang, and A. G. Gilman, *Proc. Natl. Acad. Sci. U.S.A.* **91**, 9828 (1994).
[29] T. Schweins, M. Geyer, K. Scheffzek, A. Warshel, H. R. Kalbitzer, and A. Wittinghofer, *Nature Struct. Biol.* **2**, 36 (1995).
[30] T. Schweins, M. Geyer, H. R. Kalbitzer, A. Wittinghofer, and A. Warshel, *Biochemistry* **35**, 14225 (1996).
[31] T. Schweins, R. Langen, and A. Warshel, *Nature Struct. Biol.* **1**, 476 (1994).

of the highly conserved residues Gln-204 or Arg-178 indicates that these residues play important roles in the catalytic mechanism.[10] It is therefore surprising that, in the structure of the $G_{i\alpha 1} \cdot GTP\gamma S \cdot Mg^{2+}$ ground state complex, these residues do not interact with either the nucleotide or the hydrolytic water molecule (Fig. 3a). The side chain of Gln-204 is positioned near the γ-thiophosphate group but is poorly ordered and makes no contacts with the nucleotide, the hydrolytic water, or other protein residues. The side chain of Arg-178, although well ordered, does not interact with the nucleotide; the guanidinium group projects away from the active site and is exposed to solvent. This is in contrast to the conformation of the analogous arginine residue in the transducin–GTPγS complex ($G_{t\alpha} \cdot GTP\gamma S \cdot Mg^{2+}$), Arg 174, where the side chain makes hydrogen bonds to the γ-thiophosphate sulfur atom and the β/γ-phosphate bridging oxygen.[21] The different conformations of Arg-178 in $G_{i\alpha 1}$ and Arg-174 in $G_{t\alpha}$ might be attributed to the different solvent systems used for stabilization of the crystals prior to data collection: PEG in the case of $G_{t\alpha} \cdot GTP\gamma S \cdot Mg^{2+}$ [21] and Li_2SO_4 for $G_{i\alpha 1} \cdot GTP\gamma S \cdot Mg^{2+}$.[10]

Except for the side chain substitutions, the structures of GTPγS · Mg^{2+} complexes of the GTPase-deficient mutants Gln-204 → Leu and Arg-178 → Cys are identical to that of the wild-type $G_{i\alpha 1} \cdot GTP\gamma S \cdot Mg^{2+}$ complex and do not reveal any perturbations of the active site that could explain the loss of catalytic activity.[10] Significantly, all of the contacts among $G_{i\alpha 1}$, GTPγS, and the putative water nucleophile are retained in the GTPγS · Mg^{2+} complexes with the two $G_{i\alpha 1}$ mutants. Thus, neither Gln-204 nor Arg-178 is necessary for binding of the substrates (GTP and the hydrolytic water molecule) to the α subunit.

In summary, the structures of the native and mutant $G_{i\alpha 1} \cdot GTP\gamma S \cdot Mg^{2+}$ complexes reveal an active site that is complementary to the substrate GTP and in which a water molecule is positioned for an in-line nucleophilic attack on the γ-phosphate of GTP. However, these structures otherwise reveal little about the catalytic mechanism, the transition state, or the role of residues Gln-204 and Arg-178.

Transition State as Modeled by $GDP \cdot AlF_4^- \cdot Mg^{2+}$

A curious feature of the G_α class of regulatory GTPases is their ability to be activated by aluminum fluoride in concert with GDP and Mg^{2+}.[32] It has been pointed out that this behavior implies that a GDP/aluminum fluoride complex mimics the activated GTP-bound state, with aluminum

---

[32] P. C. Sternweis and A. G. Gilman, *Proc. Natl. Acad. Sci. U.S.A.* **79**, 4888 (1982).

fluoride serving as an analog for the γ-phosphate.[32–35] However, mutation of the catalytically important residues Gln-204 and Arg-178 has no effect on GTPγS binding but reduces the affinity of the enzyme for GDP/aluminum fluoride.[10] Hence, the mode in which GDP/aluminum fluoride is bound to $G_{i\alpha1}$ appears to differ from that in which GTP binds.

The crystal structure of the $G_{i\alpha1} \cdot GDP \cdot AlF_4^- \cdot Mg^{2+}$ complex[10] is essentially identical overall to that of the GTPγS $\cdot Mg^{2+}$ complex, but differs dramatically with respect to the γ-phosphate-binding site, indicating, as discussed later, that the $G_{i\alpha1} \cdot GDP \cdot AlF_4^- \cdot Mg^{2+}$ structure mimics the transition state (species 2, Scheme I). A single aluminum atom exhibiting octahedral-binding geometry is bound at the γ-phosphate site (Fig. 3b). The aluminum ligation sphere consists of four fluoride atoms in the equatorial plane and a GDP β-phosphate oxygen and the hydrolytic water molecule in the axial positions. Relative to its position in the GTPγS $\cdot Mg^{2+}$ complex, the putative hydrolytic water molecule has moved toward the γ-phosphate site and is 1.9 Å from the aluminum atom; in this position it becomes part of the aluminum ligation sphere. A key feature of this complex is the change in the side chain positions of the catalytically important residues Gln-204 and Arg-178. In contrast to the ground state GTPγS $\cdot Mg^{2+}$ structure, the side chains of both residues interact directly with the GDP $\cdot AlF_4^- \cdot H_2O$ complex. The side chain of Gln-204 becomes more ordered and rotates to make hydrogen bonds to one of the equatorial fluorine atoms and to the axial water molecule. As in the GTPγS structure, this water molecule is also hydrogen bonded to the carbonyl oxygen of Thr-181. In order to maintain this hydrogen bond as the axial water molecule moves toward the aluminum atom, the main-chain atoms of Thr-181 undergo a slight rotation. This rotation in turn causes the backbone of residues 182–186 to move by up to 1.4 Å. Glutamine-204 and Thr-181 support the water nucleophile at a position 1.9 Å away from the aluminum atom by forming hydrogen bonds with the water hydrogen atoms. The side chain of Arg-178 also changes conformation relative to $G_{i\alpha1} \cdot GTP\gamma S \cdot Mg^{2+}$, rotating such that the guanidinium imido groups interact with an equatorial fluorine atom and with the GDP oxygen atom that bridges the aluminum and β-phosphate atoms. Identical interactions are observed in the GDP $\cdot AlF_4^- \cdot H_2O$ complex of α transducin.[36]

The remarkable complementarity of the $G_{i\alpha1}$ active site to GDP $\cdot$

[33] J. Bigay, P. Deterre, C. Pfister, and M. Chabre, *FEBS Lett.* **191,** 181 (1985).
[34] B. Antonny and M. Chabre, *J. Biol. Chem.* **267,** 6710 (1992).
[35] R. B. Martin, *Coord. Chem. Rev.* **141,** 23 (1996).
[36] J. Sondek, D. G. Lambright, J. P. Noel, H. E. Hamm, and P. B. Sigler, *Nature* **372,** 276 (1994).

$AlF_4^- \cdot H_2O \cdot Mg^{2+}$ and the similarity of this complex to the proposed bipyramidal transition state suggest that $GDP \cdot AlF_4^- \cdot H_2O$ is a transition-state analog.[10,36] Accordingly, an approximate model of the transition state can be constructed by replacing the AlF_4^- moiety with PO_3 in the $G_{i\alpha1} \cdot GDP \cdot AlF_4^- \cdot H_2O \cdot Mg^{2+}$ structure. The plane of the trigonal PO_3 group is superimposed on that of the square-planar AlF_4^-. The PO_3 group is oriented (about an axis passing through the phosphorus and perpendicular to the plane joined by the equatorial oxygen atoms) such that respective PO_3 oxygen atoms are positioned within hydrogen-bonding distance of Lys-46, Arg-178, Gly-203, Gln-204, and Mg^{2+} (Fig. 3c). This model is similar to a transition-state model first proposed for Ras, in which a Gln residue analogous to Gln-204 is involved in direct stabilization of a bipyramidal transition state.[9] The transition-state model of $G_{i\alpha1} \cdot GTP$ may be used to discuss the roles of Gln-204, Arg-178, and Mg^{2+} in stabilizing either an associative or a dissociative transition state.

The orientation of the side chain of Gln-204 in the transition-state model may be identical to that observed in the $G_{i\alpha1} \cdot GDP \cdot AlF_4^- \cdot H_2O \cdot Mg^{2+}$ structure. In a dissociative mechanism the attacking nucleophile may be a neutral water molecule[13] (Fig. 4), in which case Gln 204 could serve to

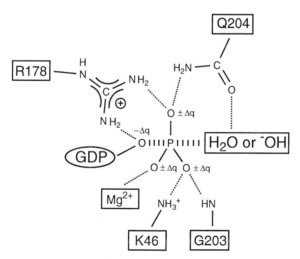

FIG. 4. Schematic diagram of the bipyramidal transition state of the hydrolytic reaction catalyzed by $G_{i\alpha1}$ as modeled from the $G_{i\alpha1} \cdot GDP \cdot AlF_4^- \cdot H_2O$ complex. $\pm\Delta q$ indicates atoms in which negative character, relative to the ground state, would increase in an associative mechanism or decrease in a dissociative mechanism. The leaving group oxygen atom acquires negative charge $(-\Delta q)$ in both cases. Key interactions are indicated.

position the nucleophile near the metaphosphate-like intermediate. For an associative transition state, Gln 204 could both position and polarize the hydrolytic water and stabilize developing negative charge on an equatorial oxygen. A theoretical analysis of GTP hydrolysis in Ras has shown that attack by a neutral water is energetically unfavorable in an associative mechanism, and that a hydroxyl ion is the likely nucleophile.[31] Furthermore, it was proposed that the γ-phosphate abstracts a proton from the hydrolytic water to generate this species. In this case, the side chain of Gln 204 makes a hydrogen bond between its amine group and a γ-phosphate oxygen, and between its carbonyl oxygen and the attacking hydroxyl ion, and thus positions and stabilizes the attacking hydroxyl ion.

In the transition-state model the guanidinium group of Arg-178 forms a hydrogen bond with an equatorial phosphoryl oxygen atom and with the leaving group oxygen of the GDP moiety and thus can stabilize negative character at both of these positions (Fig. 4). Therefore, Arg-178 could stabilize a dissociative transition state by neutralizing the developing negative charge on the leaving group or could stabilize an associative transition state by neutralizing the negative character of both the leaving group and an equatorial oxygen. For either mechanism, a negative charge on the leaving group oxygen can be stabilized further by interaction with the main chain amide group of Glu-43 as proposed for the analogous residue in Ras.[13] Negative charge on an equatorial oxygen of an associative transition state can also be stabilized by the side chain amino group of Lys-46 and the main chain amide of Gly-203 (Fig. 4).

Magnesium ion is required for the GTPase activity of G_α subunits.[3] The Mg^{2+} could provide a direct catalytic effect by shielding the attacking nucleophile from negative charges on the γ-phosphate and by stabilizing the development of the negative charge on the equatorial oxygens of an associative transition state.[19] The role of Mg^{2+} in a dissociative mechanism is less clear. Nucleotide hydrolysis in aqueous solution proceeds via a dissociative mechanism and is not affected greatly by Mg^{2+}.[7] Therefore, it has been proposed that Mg^{2+} would not be expected to directly stabilize a dissociative transition state during enzymatic hydrolysis,[7] although some degree of leaving-group stabilization could be affected by an interaction with a β-phosphate nonbridging oxygen.[13] However, the catalytic role of Mg^{2+} in the $G_{i\alpha1}$-catalyzed reaction may differ from that which proceeds in aqueous solution.[12] Magnesium ion may exert a more indirect effect on catalysis in a structural role by organizing the active site and orienting the β- and γ-phosphate groups. Also, as discussed later, Mg^{2+} may be required in order to correctly position Thr-181 to bind the hydrolytic water in the ground state.

The overall picture provided by the transition-state model is of an active site in which key residues must be reoriented in order to stabilize the transition state. The model also supports the hypothesis that GTPγS is refractory to hydrolysis because the thiophosphate sterically hinders the reorientation of Arg-178 and Gln-204 that is required in order to stabilize the transition state.[21]

Further evidence that the conformation of residues seen in the $G_{i\alpha 1} \cdot$ $GDP \cdot AlF_4^- \cdot H_2O \cdot Mg^{2+}$ structure represents the catalytically active state is provided by the structure of the GTPγS-bound form of the GTPase-deficient mutant Gly-42 → Val[37] (equivalent to the GTPase-deficient and oncogenic Gly-12 → Val mutation in Ras[38]). Comparison of the latter with the wild-type GTPγS and $GDP \cdot AlF_4^- \cdot H_2O$ complexes indicates that the side chain of Val-42 in the mutant prevents the side chain of Gln-204 from achieving the conformation observed in the transition state, thus partially explaining the reduction in the catalytic activity of the mutant. A similar interaction between Val-12 and Gln-61 in the transition state of the GTPase-deficient mutation Gly-12 → Val of Ras has been proposed.[9]

GDP · P_i Ternary Complex: Trapping a Transient Product Complex

Crystals grown at low pH (5.5–5.8) of the Gly-203 → Ala mutant of $G_{i\alpha 1}$ have trapped a species not observed using wild-type protein under similar conditions: the $G_{i\alpha 1} \cdot GDP \cdot P_i$ ternary complex (species 5, Scheme I).[39] The G203A $G_{i\alpha 1}$ mutant differs from wild-type $G_{i\alpha 1}$ primarily by its inability to dissociate from $G_{\beta\gamma}$ subunits after activation by $GTP \cdot Mg^{2+}$.[40] Significant differences between the ternary complex and the GTPγS-bound forms include the presence of GDP and P_i at the active site, absence of the Mg^{2+} at its binding site (despite inclusion of Mg^{2+} in the crystallization media), slight rearrangement of switch I, and, most significantly, a major rearrangement of the switch II helix. The latter conformational change is manifested by the formation of a 60° kink in the switch II helix and an extension of the helix in the N-terminal direction by half a turn. This conformation, in concert with changes in switch I, opens up the active site and generates a binding site for P_i (Fig. 3d). In this novel conformation

[37] A. Raw, D. E. Coleman, A. G. Gilman, and S. R. Sprang, *Biochemistry* **36**, 15660 (1997).
[38] M. Barbacid, *Annu. Rev. Biochem.* **56**, 779 (1987).
[39] A. M. Berghuis, E. Lee, A. S. Raw, A. G. Gilman, and S. R. Sprang, *Structure* **4**, 1277 (1996).
[40] N. Ueda, J. A. Iñiguez-Lluhi, E. Lee, A. V. Smrcka, J. D. Robishaw, and A. G. Gilman, *J. Biol. Chem.* **269**, 4388 (1994).

the amino terminus of the first helical segment of switch II is directed at the phosphate ion and a hydrogen bond is formed between the backbone amide nitrogen of residue Gly-203 and a phosphate oxygen, thereby retaining an interaction that is formed with the γ-phosphate in the $G_{i\alpha1} \cdot GTP\gamma S \cdot Mg^{2+}$ complex. The inorganic phosphate is also anchored by hydrogen bonds with the side chains of Lys-46, Lys-180, and Thr-181. Neither Lys-180 nor Thr-181 is a phosphate ligand in the $G_{i\alpha1} \cdot GTP\gamma S \cdot Mg^{2+}$ and $GDP \cdot AlF_4^- \cdot Mg^{2+}$ complexes; in those structures, Lys-180 is disordered and Thr-181 adopts a different side chain conformation in order to ligate Mg^{2+}. The switch III region becomes partially disordered as a result of the loss of stabilizing ionic contacts with switch II. The position of the catalytic residue Arg-178 is only slightly different from that occupied in the transition-state complex, whereas Gln-204 is directed completely away from the active site as a result of the rearrangement of switch II. Changes in the active site almost completely disassemble the catalytic arrangement observed in active and transition-state structures.

Attempts to crystallize the $GTP\gamma S \cdot Mg^{2+}$ complex of G203A $G_{i\alpha1}$ have been unsuccessful, suggesting that the mutation destabilizes the ground state in addition to stabilizing the $GDP \cdot P_i$ ternary complex. Indeed, the rate at which $GTP\gamma S$ dissociates from G203A $G_{i\alpha1}$ is substantially greater than the k_{off} for wild-type $G_{i\alpha1}$.[39] In contrast, both the $GTP\gamma S \cdot Mg^{2+}$ and the $GDP \cdot P_i$ complexes of G42V $\rightarrow G_{i\alpha1}$ can be crystallized (Tables I and II). Although the mutation perturbs the catalytic site in the $GTP\gamma S \cdot Mg^{2+}$ complex, the structure of the $GDP \cdot P_i$ complex is identical to that formed by G203A $G_{i\alpha1}$.[37] This observation of the same active site conformation in the $GDP \cdot P_i$ complexes supports the hypothesis that the $GDP \cdot P_i$ structures represent a conformation of the wild-type enzyme that occurs only transiently during catalysis, but is stabilized by the G203A and G42V mutations. The overall view provided by the $GDP \cdot P_i$ structures is of an enzyme that, after the collapse of the transition state, assumes an intermediate state prior to phosphate product release. In this intermediate state much of the catalytic machinery is disassembled.

Binary GDP Complex

A view of the binary $G_{i\alpha1} \cdot GDP$ complex (species 6, Scheme 1) has been obtained from crystals of wild-type $G_{i\alpha1}$ and the G203A[22] and G42V mutants.[37] Wild-type and mutant subunits (which are essentially identical) bind GDP through interactions that are identical to those observed in the GTPγS-, $GDP \cdot AlF_4 \cdot H_2O^-$, and $GDP \cdot P_i$-bound subunits, but differ significantly from these in the regions of the structure that involve γ-phosphate binding (Fig. 3f). Switch I moves away from the active site, which partially dismantles the Mg^{2+}-binding site. In contrast to the GDP-bound structures of Ras[41] and EF-Tu,[42] no Mg^{2+} is observed bound to the active site of $G_{i\alpha 1} \cdot$ GDP. The absence of bound Mg^{2+} reflects the weak affinity of $G_\alpha \cdot$ GDP complexes for Mg^{2+},[43] in contrast to the higher Mg^{2+}-binding affinity of Ras $\cdot$ GDP[44] and EF-Tu $\cdot$ GDP.[45] (The α transducin/GDP complex[11] contains a bound Mg^{2+}, but this may be a consequence of the presence of 200 mM Mg^{2+} in the crystallization solution.) Switch II (containing catalytic residue Gln-204) is completely disordered, a consequence, in part, of the loss of the hydrogen bond between the main chain amide group of Gly-203 and the γ-phosphate group. Switch III is also completely disordered due to the loss of stabilizing ionic contacts with the disordered switch II region. The analogous switch II and switch III regions are ordered in the structure of the α transducin GDP $\cdot Mg^{2+}$ complex, although they adopt different conformations from those exhibited in the GTPγS-bound complex. The different switch I and switch II conformations exhibited by $G_{i\alpha 1}$ and $G_{t\alpha} \cdot$ GDP may be attributed to (a) primary structure ($G_{i\alpha 1}$ and $G_{t\alpha}$ share 65% sequence identity), (b) the presence of Mg^{2+} bound at the active site of $G_{t\alpha}$ and its absence in $G_{i\alpha 1}$, (c) the use of N-terminally truncated $G_{t\alpha}$ but full-length $G_{i\alpha 1}$, or (d) differences in crystal lattice packing.

The catalytic residue Arg-178 is poorly ordered in the $G_{i\alpha 1} \cdot$ GDP complex and displays only weak electron density for the guanidinium group, which is associated with the α- and β-phosphate groups of GDP. Much of the catalytic apparatus is thus rearranged or disordered in the GDP complex. Because both GDP- and GTPγS-bound crystal forms can be grown and stabilized under identical conditions (see methods and Tables I and II), it is clear that the conformational changes observed in the crystals are solely the result of the presence or absence of the γ-phosphate group and hence are likely to reflect the differences between active and inactive signaling conformations. Also, in contrast to the $G_{i\alpha 1} \cdot$ GTPγS $\cdot Mg^{2+}$, GDP $\cdot AlF_4^- \cdot$ $H_2O \cdot Mg^{2+}$, or GDP $\cdot P_i$ complexes, the N terminus in the GDP-bound form is ordered and, in concert with the C-terminal residues, forms a small minidomain. The implications of this structure are discussed elsewhere.[22]

[41] L. Tong, A. M. De Vos, M. V. Milburn, and S.-H. Kim, *J. Mol. Biol.* **217,** 503 (1991).

[42] M. Kjeldgaard and J. Nyborg, *J. Mol. Biol.* **223,** 721 (1992).

[43] T. Higashijima, K. M. Ferguson, P. C. Sternweis, M. D. Smigel, and A. G. Gilman, *J. Biol. Chem.* **262,** 762 (1987).

[44] J. John, H. Rensland, I. Schlicting, I. Vetter, G. D. Borasio, R. S. Goody, and A. Wittinghofer, *J. Biol. Chem.* **268,** 923 (1993).

[45] A. Wittinghofer and R. Leberman, *Eur. J. Biochem.* **93,** 95 (1993).

The binary $G_{i\alpha1} \cdot$ GDP complex has also been observed bound to $G_{\beta\gamma}$ subunits in the heterotrimer.[46] Binding to the β/γ complex induces reordering of the switch II region and additional conformational changes in switch I, the amino and carboxyl-terminal residues, and other regions. These changes cause the GDP molecule to be sequestered more tightly in the active site, in part by the formation of an ion-pair hydrogen bond between Glu-43 and Arg-178. In the heterotrimer, switch II interacts directly with the G_β subunit and adopts a conformation that is distinctly different from that found in GTPγS-bound forms. Thus, when the heterotrimer undergoes receptor-mediated nucleotide exchange, binding GTP, rather than GDP, the presence of the γ-phosphate group of GTP disrupts the conformation of switch II that binds to G_β and stabilizes the conformation observed in the GTPγS $\cdot$ Mg^{2+}-bound complex; this then leads to the dissociation of the heterotrimer and release of the $G_{i\alpha1} \cdot$ GTP $\cdot$ Mg^{2+} complex in the active signaling conformation. The conformation of switch II observed in the GDP $\cdot$ P$_i$-bound form appears to be intermediate between that of GTP-bound and heterotrimeric forms.

Forcing Intermediates within Crystals

A quasi-dynamic view of the $G_{i\alpha1}$ product complex has been obtained by titrating $G_{i\alpha1} \cdot$ GDP crystals with MgSO$_4$.[47] The addition of 5, 10, 100, or 200 mM Mg^{2+} to the $G_{i\alpha1} \cdot$ GDP crystals successively transforms the structure into a conformation that has, in the region of the γ-phosphate-binding site, features of both the $G_{i\alpha1} \cdot$ GTPγS $\cdot$ Mg^{2+} and the $G_{i\alpha1} \cdot$ GDP $\cdot$ P$_i$ structures (Fig. 3e). As estimated from normalized electron density maps, the Mg^{2+}-binding site is partially (20–30%) occupied at 5 mM Mg^{2+}, is occupied more fully (30–60%) at 10 mM Mg^{2+}, and appears to be fully occupied at 100 and 200 mM Mg^{2+}. As the occupancy of this site increases, residues 180–182 of the switch I region move toward and reconstitute the Mg^{2+}-binding site and assume the conformation observed in the $G_{i\alpha1} \cdot$ GTPγS $\cdot$ Mg^{2+} structure. In this Mg^{2+}-induced conformation, the side chain of Thr-181 coordinates Mg^{2+} and the main chain carbonyl oxygen of this residue reorients to occupy a position identical to that observed in the GTPγS complex, where it binds the hydrolytic water molecule. This indicates that, in addition to other catalytic effects, the presence of Mg^{2+} to the active site may be necessary in order to bind the hydrolytic water. However, the presumptive water nucleophile is not present in crystals

[46] M. A. Wall, D. E. Coleman, E. Lee, J. A. Iñiguez-Lluhi, B. A. Posner, A. G. Gilman, and S. R. Sprang, *Cell* **80**, 1047 (1995).
[47] D. E. Coleman and S. R. Sprang, *Biochemistlry* **37**, 14376 (1998).

of the $G_{i\alpha 1} \cdot GDP$ complex that have been exposed to $MgSO_4$, presumably due to the absence of the γ-phosphate. In $G_{i\alpha 1} \cdot GDP$ crystals soaked in 200 mM Mg^{2+}, the Mg^{2+} coordination sphere is identical to that of the $G_{i\alpha 1} \cdot GTP\gamma S \cdot Mg^{2+}$ complex, except that a water molecule replaces the missing γ-phosphate oxygen. The Mg^{2+}-binding site also differs from that observed in the $GDP \cdot Mg^{2+}$ complexes of $G_{t\alpha}$, Ras, and EF-Tu.[20] In those structures, switch I moves away from, rather than toward, the binding site, and the ligand corresponding to the side chain hydroxyl of Thr-181 is replaced with a water molecule. In concert with increased occupancy of the Mg^{2+} site in the $G_{i\alpha 1} \cdot GDP$ complex, discrete electron density appears in the active site near the position occupied by P_i in the $GDP \cdot P_i$ complex. Based on its size, temperature factor, and protein contacts, this density has been identified as a sulfate ion derived from the crystal stabilization media. The occupancy of the SO_4^{2-} site increases as a function of the concentration of $MgSO_4$ in the soaking solution and parallels the occupancy of the Mg^{2+}-binding site. Accompanying the increase in fractional occupancy of the SO_4^{2-} site, continuous electron density appears for switch II residues 201–203, along with additional disconnected density throughout the switch II helix region (residues 204–217). The sulfate ion is bound by interactions with a water molecule (the same water ligand that replaces the γ-phosphate oxygen in the equatorial plane of the Mg^{2+} coordination sphere), Lys-46, and the backbone amide of Gly-203. The presence of this sulfate and its interaction with the Mg^{2+}-binding site may explain some of the differences between the $GDP \cdot Mg^{2+}$ structure of $G_{i\alpha 1}$ and those of Ras and transducin. With SO_4^{2-} serving as an analog of P_i, the $G_{i\alpha 1} \cdot GDP \cdot SO_4^{2-} \cdot Mg^{2+}$ complex may represent a snapshot of the active site with bound GDP, P_i, and Mg^{2+} just prior to P_i and Mg^{2+} release (species 3, Scheme I).

The structure of the $G_{i\alpha 1} \cdot GDP \cdot SO_4^{2-} \cdot Mg^{2+}$ complex differs from that of the G203A and G42V $G_{i\alpha 1}$ complexes with GDP and P_i. In contrast to the Mg^{2+}-free $GDP \cdot P_i$ structures, in crystals of $G_{i\alpha 1} \cdot GDP$ soaked in $MgSO_4$, switch I assumes the conformation observed in the GTPγS-bound ground state, and switch II assumes an altered and partially disordered state. However, the structure is similar to the $GDP \cdot P_i$ complex in that the phosphate product analog SO_4^{2-} is bound within the active site near the P_i site, and switch II, although partially disordered, exhibits increased order relative to the $G_{i\alpha 1} \cdot GDP$ binary complex and assumes a conformation that differs from that exhibited in the GTPγS ground state structure. The $G_{i\alpha 1} \cdot GDP \cdot SO_4^{2-} \cdot Mg^{2+}$ complex demonstrates that the wild-type enzyme can bind the $GDP \cdot P_i$ ternary complex at neutral pH provided that Mg^{2+} is present and that switch II undergoes a conformational change in response to formation of the ternary product complex. The simultaneous binding of Mg^{2+} and SO_4^{2-} in the $G_{i\alpha 1} \cdot GDP \cdot SO_4^{2-} \cdot Mg^{2+}$ structures also implies that

Mg^{2+} and P_i bind, and conversely are released, from the active site in a concerted manner. Binding of P_i and Mg^{2+} to their respective subsites is coupled to concerted conformational changes in the switch I and switch II regions.

Conclusions

The catalytic pathway described by the series of snapshots of the active site of $G_{i\alpha1}$ can be summarized in four stages: (1) binding and orientation of the substrates GTP and water (thus leading to the active signaling conformation), (2) generation and stabilization of the transition state, (3) breakdown of the pentacoordinate phosphoryl intermediate, and (4) release of P_i and Mg^{2+} (thus leading to the inactive signaling conformation).

In the ground state, GTP and a presumptive hydrolytic water molecule are bound to the active site by numerous contacts. The γ-phosphate group is positioned such that it forms part of the Mg^{2+}-binding site and tethers the switch II region in its active conformation. The hydrolytic water is positioned by the γ-phosphate and Thr-181 of switch I for an in-line nucleophilic attack on the γ-phosphate group. Binding of this water molecule may provide a substantial intramolecular effect that contributes to catalysis.[48] The highly conserved catalytic residues Arg-178 and Gln-204 do not interact with either GTP or the hydrolytic water. No residues can be identified that could serve as a general base catalyst for abstraction of a proton from the hydrolytic water. It has been proposed that, in Ras, the γ-phosphate of GTP may serve as the catalytic base rather than any protein group.[31] The ground state thus binds GTP and the hydrolytic water and achieves an active signaling conformation, but does not appear to be completely organized for catalysis.

The transition-state complex, as modeled by the $G_{i\alpha1} \cdot GDP \cdot AlF_4 \cdot H_2O \cdot Mg^{2+}$ structure, suggests possible catalytic roles for Gln-204 and Arg-178. Residue Arg-178 is reoriented such that it interacts with the negative charges on the transition state in a manner that is consistent with either a dissociative mechanism (in which case its primary role is to stabilize the developing negative charge on the leaving group) or an associative mechanism (in which case it stabilizes negative charge on the equatorial oxygens in addition to stabilizing the leaving group). Glutamine-204 is also reoriented such that it interacts directly with the attacking nucleophile and an equatorial group of the pentavalent phosphoryl group and can potentially stabilize the transition state at either or both of these positions. Reorganization of Gln-204 and Arg-178 prior to achieving the transition state may

---

[48] M. I. Page and W. P. Jencks, *Proc. Natl. Acad. Sci. U.S.A.* **68,** 1678 (1971).

contribute to the slow catalytic rate of $G_{i\alpha1}$, as has been proposed in general for conformational changes in enzymes.[49]

Magnesium ion may have either a direct catalytic role or an indirect structural role during catalysis. Mg^{2+} may shield the approaching water or hydroxyl ion nucleophile from the negative charges on the phosphate group. In an associative mechanism, Mg^{2+} can act as an electrophilic catalyst by stabilizing the developing negative charge on the equatorial oxygens of the transition state. In a dissociative mechanism, a negative charge on the equatorial oxygens decreases in the transition state and thus Mg^{2+} would have little role in its direct stabilization, as has been found in nonenzymatic nucleotide hydrolysis.[50] However, Mg^{2+} may be required in a structural role to form the active site and bind and orient the substrates. The set of $G_{i\alpha1} \cdot GDP \cdot Mg^{2+}$ structures indicates that Mg^{2+} binding stabilizes the active conformation of the switch I region in which Thr-181 is oriented to bind both Mg^{2+} and the hydrolytic water molecule.

The $G_{i\alpha1} \cdot GDP \cdot P_i$ complex stabilized by the G203A and G42V mutations at low pH and the $G_{i\alpha1} \cdot GDP \cdot Mg^{2+} \cdot SO_4^{2-}$ complex formed at neutral pH both exhibit conformations that are different from those of the ground state, the transition state, and the binary product complex. The structures of these enzyme–product complexes demonstrate that the enzyme can assume intermediate conformations that accompany cleavage of the β/γ phosphate bond and subsequent release of Mg^{2+} and P_i, thus generating the fully inactive signaling conformation. The overall rate of the $G_{i\alpha1}$-catalyzed GTPase reaction is potentially determined not only by chemical steps (vis deprotonation of the nucleophile) and the kinetic barrier to the transition state, but also by conformational and solvent rearrangement events required to affect cleavage and product release.

The principle of microscopic reversibility[51] implies that the conformational changes and intermediates observed during the forward hydrolytic pathway can also occur during the reverse or synthetic pathway (GDP + $P_i \rightarrow$ GTP + H_2O). Thus, the conformation of the switch II region observed in the $G_{i\alpha1} \cdot GDP \cdot P_i$ structure at low pH could be thought of as an intermediate state that captures a phosphate ion from the surrounding media and, via the conformational change leading from the $GDP \cdot P_i$ to the $GDP \cdot AlF_4^-$-bound structure, forces the phosphate through the active site on the pathway leading to GTP synthesis. For G_α subunits, this process is, of course, not energetically favorable, but may resemble the steps used by

[49] A. Warshel, "Computer Modeling of Chemical Reactions in Enzymes and Solutions." Wiley, New York, 1991.

[50] D. Herschlag and W. P. Jencks, J. Am. Chem. Soc. 109, 4665 (1987).

[51] A. Fersht, "Enzyme Structure and Mechanism," 2nd ed., p. 1. Freeman, New York 1985.

other enzymes that use mechanical energy to synthesize nucleotides, such as mitochondrial F1-ATPases.[52,53]

Regulators of G-protein signaling (RGS proteins) have been shown to stimulate the GTPase activity of the G_i family of α subunits[54] and to bind more tightly to the GDP $\cdot$ AlF$_4^-$-bound enzymes than to either GTPγS- or GDP-bound forms.[55] From the structure of RGS4 bound to the GDP $\cdot$ AlF$_4^-$ $\cdot$ Mg^{2+} complex of G$_{i\alpha1}$[56] it can be inferred that RGS molecules enhance the catalytic rate by stabilizing these enzymes in a state that is similar to the transition-state model based on the G$_{i\alpha1}$ $\cdot$ GDP $\cdot$ AlF$_4^-$ $\cdot$ H$_2$O structure. It is possible that other proteins that stimulate GTP hydrolysis in Ras superfamily members act in a similar way. In the case of Ras itself, the stimulatory factors (GAPs) may act as coenzymes by introducing an arginine residue into the active site that performs a function analogous to that of Arg-178 in G$_{i\alpha1}$. In contrast, RGS4 contributes an Asn residue to the active site of G$_{i\alpha1}$ that could interact with the water nucleophile in the ground state.[57]

Acknowledgments

We are grateful to all of our collaborators who have contributed to the crystallographic studies described here. We thank Alfred Gilman, Bruce A. Posner, André Raw, and John Tesmer for critical reviews of the manuscript. This research has been supported in part by NIH Grant DK 46371 and a grant from the Welch Foundation (I-1229) to S.R.S.

[52] E. J. Goldsmith, *FASEB J.* **10,** 702 (1996).
[53] H. Noji, T. Amano, and M. Yoshida, *J. Bioenerg. Biomemb.* **28,** 451 (1996).
[54] D. M. Berman, T. M. Wilke, and A. G. Gilman, *Cell* **86,** 1 (1996).
[55] D. M. Berman, T. Kozasa, and A. G. Gilman, *J. Biol. Chem.* **271,** 27209 (1996).
[56] J. J. G. Tesmer, D. M. Berman, A. G. Gilman, and S. R. Sprang, *Cell* (1997).
[57] K. Scheffzek, A. Lautwein, W. Kabsch, M. R. Ahmadian, and A. Wittinghofer, *Nature* **384,** 591 (1996).

[5] Energetics of Nucleotide Hydrolysis in Polymer Assembly/Disassembly: The Cases of Actin and Tubulin

By DANIEL L. PURICH and FREDERICK S. SOUTHWICK

The self-assembly of cytoskeletal structures involves a repetitive set of elongation reactions that were originally embodied in the condensation equilibrium model of Oosawa.[1] Although actin was then recognized to be an ATP-binding protein, his generalized model did not consider all of the ways in which monomer-bound nucleotides might modulate the self-assembly process. It is clear that the free energy of ligand binding plays a pivotal role in determining the strength of protein–protein interactions in allosteric regulation; so too can the energy of nucleoside-5′ tri- and diphosphate (hereafter pppN and ppN) binding modulate the protein–protein interactions attending the assembly of certain supramolecular structures. Moreover, if pppN hydrolysis is an essential step in the polymerization reaction, the Gibbs free energy of pppN hydrolysis can also be utilized to create additional conformationally energized states. The maturation of a protein monomer (M) is now recognized to include the following minimal set of steps:

$$P + M \cdot pppN \rightarrow P \bullet M \cdot pppN \tag{1}$$
$$P \bullet M \cdot pppN \rightarrow P \bullet M \cdot ppN \cdot P_i \tag{2}$$
$$P \bullet M \cdot ppN \cdot P_i \rightarrow P \bullet M \cdot ppN + P_i \tag{3}$$
$$P \bullet M \cdot ppN \rightarrow P \bullet M + ppN \tag{4}$$

Enzymatic transphosphorylation is not required to regenerate M · pppN, and spontaneous exchange often occurs between unbound *pppN* with actin-bound ppN:

$$pppN + M \cdot ppN \rightarrow M \cdot pppN + ppN \tag{5}$$

(*Note:* Italics are added to distinguish the added nucleotide from the monomerbound nucleotide). In certain cases, this exchange process can be facilitated by the action of an exchange-promoting factor, F:

$$F + M \cdot ppN \text{ complex} = F \bullet M \cdot ppN \tag{6}$$
$$F \bullet M \cdot ppN \rightarrow F \bullet M\_\_ + ppN \tag{7}$$
$$F \bullet M\_\_ + pppN \rightarrow F \bullet M \cdot pppN \tag{8}$$
$$F \bullet M \cdot pppN \rightarrow F + M \cdot pppN \tag{9}$$

---

[1] F. Oosawa, *J. Polymer Sci.* **26,** 29 (1957).

Copyright © 1999 by Academic Press
All rights of reproduction in any form reserved.
0076-6879/99 $30.00

Furthermore, because proteins often partake in head-to-tail polymerization, formation of the linear polymer actually involves two distinct cycles of reactions: one occurring at one polymer end (designated here as $_AP$) and another set at the opposite end (P_B).

Classical examples of pppN hydrolysis-associated polymerization reactions are the ATP-dependent polymerization of actin and the GTP-dependent polymerization of tubulin. Under a variety of conditions influencing the rate and extent of actin[2] or tubulin[3] self-assembly, ^{18}O-labeled ATP or ^{18}O-labeled GTP was hydrolyzed without any evidence of multiple reversals that are characteristic of reversible phosphoanhydride-bond cleavage. These data suggest that the reactions do not involve the formation of energy-rich intermediates in their respective hydrolytic reaction mechanisms. Nucleotide interactions are also thought to be of underlying importance in the rapid remodeling of the actin cytoskeleton during cell crawling.[4-7] Beyond the role of GTP in promoting microtubule self-assembly,[8] the stored energy in a microtubule appears to be harnessed in mitosis as microtubules depolymerize.[9] Another recently discovered cytoskeletal component, known as septin, appears to promote the septation of cells that have recently undergone mitosis.[10] The discovery of the GTPase activity of septin[11] serves to underscore the broader potential for regulating indefinite polymerization reactions through assembly-induced pppN hydrolysis. This article considers several schemes for linking the Gibbs free energy of pppN hydrolysis to the self-assembly process, and by using actin filaments and microtubules as examples, we illustrate how the thermodynamics and kinetics of the reaction cycle may be used to regulate the assembly and disassembly processes.

Models for Nucleotide Involvement in Self-Assembly Processes

Self-assembly involves a number of distinct phases, depending on whether the polymerization takes place by a condensation equilibrium

[2] M. F. Carlier, D. Pantaloni, J. A. Evans, P. K. Lambooy, E. D. Korn, and M. R. Webb, *FEBS Lett.* **235,** 211 (1988).
[3] J. M. Angelastro and D. L. Purich, *Eur. J. Biochem.* **191,** 507 (1990).
[4] T. P. Stossel, *Science* **260,** 1086 (1993).
[5] J. Condeelis, *Annu. Rev. Cell Biol.* **9,** 411 (1993).
[6] M. F. Carlier, *J. Biol Chem.* **266,** 1 (1991).
[7] F. S. Southwick and D. L. Purich, *N. Engl. J. Med.* **334,** 770 (1996).
[8] D. L. Purich and J. M. Angelastro, *Adv. Enzymol.* **69,** 121 (1994).
[9] D. E. Koshland, T. J. Mitchison, and M. W. Kirschner, *Nature* **331,** 499 (1988).
[10] M. Yagi, B. Zieger, G. J. Roth, and J. Ware, *Gene* **212,** 229 (1998).
[11] C. M. Field, O. Al-Awar, J. Rosenblatt, M. L. Wong, B. Alberts, and T. J. Mitchison, *J. Cell Biol.* **133,** 605 (1996).

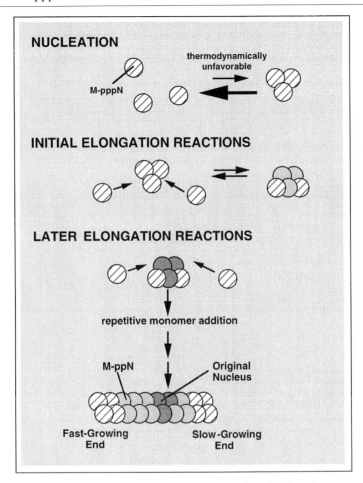

FIG. 1. Condensation equilibrium model for the formation of polymeric structures from protein subunits. Nucleation is highly cooperative and thermodynamically disfavored. The initial elongation reactions allow monomers to add to the unstable nucleus, and each of the later elongation reactions takes place with identical rates of monomer association and dissociation.

process or by a template-directed process.[12] For systems obeying the condensation equilibrium model (Fig. 1), one assumes an initial thermodynamically unfavorable step, known as nucleation, that typically requires three or more subunits to combine to form an unstable nucleus. Nucleation is followed by elongation, which can be treated as a repetitive set of identical

[12] D. L. Purich and D. Kristofferson, *Adv. Protein Chem.* **36,** 133 (1984).

monomer addition reactions. Elongation persists until the monomer concentration drops to a value corresponding to the critical concentration. When the monomer concentration equals the critical concentration, there can be no net change in the amount of protein present as monomer and polymer. Below this threshold concentration, the assembled state cannot persist and there is net disassembly until the critical concentration of monomers is restored. In template-directed assembly, there is no nucleation phase, although enzyme-catalyzed conversion from an inactive template to an active template can still introduce a lag phase in the polymerization process Because elongation is a common feature of condensation equilibrium and template models, and for our purposes, we will only consider the roles of nucleotide hydrolysis in the elongation process.

Polymerization in the Absence of pppN Hydrolysis

In this first case, the head-to-tail polymerization of a polymer of protein subunits occurs without any nucleotide hydrolysis. Hence, the only change in Gibbs free energy is that associated with the stability of the initial and final states of assembly. Elongation can occur by either of two reversible reactions:

$$M + {}_A(P_n)_B = [M \cdot {}_A(P_n)_B]^{\ddagger} = {}_A(P_{n+1})_B \tag{10}$$
$$_A(P_n)_B + M = [_A(P_n)_B \cdot M]^{\ddagger} = {}_A(P_{n+1})_B \tag{11}$$

where M is the polymerization monomer and $_A(P_n)_B$ is the self-assembled polymer that contains binding sites for monomer on both A and B ends. Note also that $_A(P_n)_B$ is merely the indefinite polymer $_A(M_1M_2M_3 \cdots M_i \cdots M_{n-2}(M_{n-1}M_n)_B$, such that any polymeric form $_A(P_{n+1})_B$ can arise from monomer addition to the A or B ends of $_A(P_n)_B$. The reader should note that the transition states $[M \cdot {}_A(P_n)_B]^{\ddagger}$ and $[_A(P_n)_B \cdot M]^{\ddagger}$ need not be identical. For this reason, the respective on- and off-rate constants for monomer addition to the A and B growth sites likewise need not be identical (i.e., $k_{A\text{-}on} >$, $=$, or $< k_{B\text{-}on}$, and $k_{A\text{-}off} >$, $=$, or $< k_{B\text{-}off}$). However, because the initial and final states are identical for both of these reactions, then the free energy change for monomer addition must be identical (i.e., $k_{A\text{-}on}/k_{A\text{-}off} = k_{B\text{-}on}/k_{B\text{-}off}$), irrespective of the ways in which the monomers originally added. The conclusion is that *both* ends grow or *both* ends shorten, depending on the instantaneous monomer concentration, but the growth or loss is inevitably *bidirectional*.

Another intrinsic property of this model is that the amount of protein in the polymer and monomer states is strictly determined by the overall equilibrium constant for each successive and identical elongation step.

Oosawa and Asakura[13] treated these steps in terms of their condensation equilibrium model, which is characterized by critical concentration behavior. Note that the polymer weight concentration is defined as the amount of total monomers present in polymer forms. We may consider a single representative step:

$$M + (P_n) = (P_{n+1}) \tag{12}$$

where the polarity of addition is unspecified. The kinetics of this elongation may be written as

$$d[M]/dt = -k_{on}[M][P_n] + k_{off}[P_{n+1}] \tag{13}$$

The first summation is made over all polymer lengths from l to $(m - 1)$ where m is the maximal length and represents the polymer weight concentration; similarly, the second summation is made over all polymer lengths from $(l - 1)$ to $(m - 1)$ and again represents the polymer weight concentration. At equilibrium, $d[M]/dt = 0$, and we get

$$k_{on}[M][P_n] = k_{off}[P_{n+1}] \tag{14}$$

or

$$[M] = k_{off}/k_{on} \tag{15}$$

because $[P_n]$ and $[P_{n+1}]$ are alternative and equivalent representations of the polymer weight concentration. If $[M] = k_{off}/k_{on}$, there can be neither a net increase or a net decrease in the polymer weight concentration. When $[M] > k_{off}/k_{on}$, there can only be a net increase in the polymer weight concentration. When $[M] < k_{off}/k_{on}$, there can only be a net decrease in the polymer weight concentration.

Although actin and tubulin polymerization involve ATP and GTP hydrolysis, respectively, nonhydrolyzable analogs, such as App(CH$_2$)p, App(NH)p, Gpp(CH$_2$)p, and Gpp(NH)p, can also efficiently promote *in vitro* assembly of microfilaments and microtubules.[6,8,12] This fact clearly indicates (a) that pppN hydrolysis cannot be a *sine qua non* for polymerization and (b) that nucleotide hydrolysis is likely to be involved in modulating the polymer disassembly process.

Unidirectional Polymerization with pppN Hydrolysis at the Time of Monomer Addition

As shown in Fig. 2, the simplest case involving nucleotide hydrolysis in filament self-assembly is the unidirectional growth model in which pppN

---

[13] F. Oosawa and S. Asakura, "Thermodynamics of Protein Polymerization." Academic Press, New York, 1975.

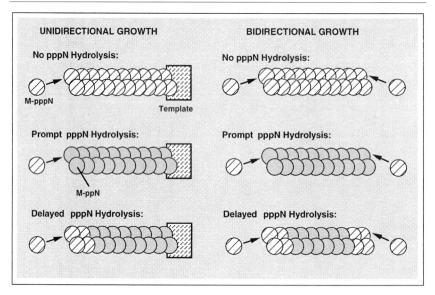

FIG. 2. Models for unidirectional and bidirectional polymerization of protein subunits in the absence or presence of nucleotide hydrolysis. The presence of a template prevents bidirectional growth.

hydrolysis is needed for reversible monomer incorporation exclusively at one end of a growing filament. Each elongation step is treated as a reversible process. No assembly/disassembly takes place at the opposite end, which in principle could be blocked by some other cellular component, such as a template structure that only promotes unidirectional growth. One step in unidirectional elongation can be represented as

$$M \cdot pppN + {}_A(P_n) = [M \cdot pppN \cdot P_i \bullet {}_A P_n]^{\ddagger}$$
$$= P_i + M \cdot pppN \bullet {}_A P_n \ (\text{or} \ {}_A P_{n+1}) \qquad (10)$$

where ${}_A(P_n)$ has only one growing end and where P_i is inorganic phosphate released during the hydrolytic reaction. Here, hydrolysis of the terminal P–O–P (or phosphoanhydride) bond of pppN occurs in the initial association step and is associated with a significant release of free energy ($\Delta G°$ -7 kcal/mol). Accordingly, the affinity of $M \cdot ppN$ for the single polymer end is determined solely by the extent to which the newly added monomeric unit retains some of the free energy of nucleotide hydrolysis. This retention of free energy would most likely be associated with a conformational change in the newly added monomer or the neighboring monomers with which the newly added monomer interacts. If there is significant retention of NTPase free energy, then the polymerization would be highly unfavorable because the $M \cdot pppN + {}_A(P_n)$ would be far more stable than $M \cdot ppN \bullet {}_A P_n$. If no free energy of hydrolysis is retained, then pppN hydrolysis cannot influence the magnitude of the overall critical concentration.

Bidirectional Polymerization with pppN Hydrolysis at the Time of Monomer Addition

In bidirectional growth, the two ends may have different transition states $[M \cdot pppN \bullet {}_A(P_n)_B]^{\ddagger}$ and $[{}_A(P_n)_B \bullet M \cdot pppN]^{\ddagger}$, and the associated on- and off-rate constants for monomer addition to the A and B growth sites again may not be identical (i.e., $k_{A\text{-on}} >, =,$ or $< k_{B\text{-on}}$, and $k_{A\text{-off}} >, =,$ or $k_{B\text{-off}}$). Because the initial and final states now depend on the extent to which each of the ends differentially retains a fraction of the free energy of pppN hydrolysis, the more stable end will be the one retaining the least amount of the G of pppN hydrolysis. Now, the equilibrium constants need not be identical, and $k_{A\text{-off}}/k_{A\text{-on}} <, =, > k_{B\text{-off}}/k_{B\text{-on}}$).

Wegner[14] was the first to recognize how this inherent property makes the macroscopic critical concentration $K_{\text{macroscopic}}$ a composite of two microscopic processes occurring on each end of the polymer:

$$K_{\text{macroscopic}} = (k_{A\text{-off}} + k_{B\text{-off}})/(k_{A\text{-on}} + k_{B\text{-on}}) \qquad (17)$$
$$K_A = k_{A\text{-off}}/k_{A\text{-on}} \qquad (18)$$
$$K_B = k_{B\text{-off}}/k_{B\text{-on}} \qquad (19)$$

As illustrated in Fig. 3, the monomer concentration equal to $K_{\text{macroscopic}}$ is called the macroscopic critical concentration (often designated [M]). At this concentration, the system is inherently metastable with respect to the binding interactions at each end of the polymer. The higher value of K_B relative to $K_{\text{macroscopic}}$ requires that the instantaneous monomer concentration is too low to stabilize the B end. The lower value of K_A relative to [M] means that the instantaneous monomer concentration is also more than sufficient to promote growth at the A end. This phenomenon, termed "treadmilling," results in the net loss of monomers from the B end and a net addition of monomers to the A end. Such behavior only occurs when $K_A < K_B$, a condition that can only occur as a consequence of pppN hydrolysis.

One may consider the *in vitro* assembly/disassembly dynamics of microtubules as an example of treadmilling. Margolis and Wilson[15] relied on the fact that microtubules trap the [³H]GDP formed during assembly-dependent [³H]GTP hydrolysis to follow the course of the treadmilling

[14] A. Wegner, *J. Mol. Biol.* **108**, 139 (1976).
[15] R. L. Margolis and L. Wilson, *Cell* **13**, 1 (1978).

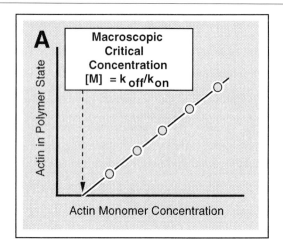

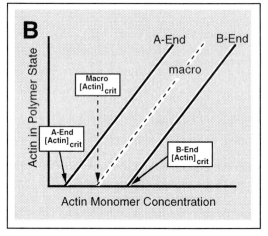

FIG. 3. Critical concentration behavior of polymeric structures whose assembly is independent or dependent on pppN hydrolysis. (A) The critical concentration behavior is that expected if a nonhydrolyzable analog is utilized in place of ATP. (B) The "A" end of actin is the fast-growing barbed end and the "B" end is the slow-growing pointed end.

reaction. They obtained a rate constant of 0.07 hr^{-1} for treadmilling during steady-state polymerization. Terry and Purich[16] confirmed that this steady-state rate behavior persists over the course of several hours in the presence of 1 mM GTP and demonstrated that no treadmilling occurred when the nonhydrolyzable analogs p(CH$_2$)ppG and p(NH)ppG were used in place of

[16] B. J. Terry and D. L. Purich, *J. Biol. Chem.* **255,** 10532 (1980).

GTP. Likewise, the microtubule-stabilizing drug taxol arrests microtubule disassembly,[17] and Wilson *et al.*[18] found that taxol completely suppressed microtubule treadmilling.

Before leaving the case of hydrolysis-dependent bidirectional polymerization, we should note that this model implicitly requires the existence of a structural or energetic discontinuity within the assembled filament. For example, suppose (i) that the monomers added to the A end release more free energy of pppN hydrolysis and (ii) that monomers incorporated at the A end (marked below as M*). Suppose further that the subunit addition at the B end retained more energy (designated M**). The assembled polymer will necessarily contain a discontinuity, shown here in bold type:

$$\cdots\cdot M^*M^*M^*M^*M^*\mathbf{M^*M^{**}}M^{**}M^{**}M^{**}M^{**}M^{**}\cdots\cdots \qquad (20)$$

As we shall see shortly, such a complication is eliminated for models involving the delayed hydrolysis of pppN.

Polymerization with Delayed pppN Hydrolysis (i.e., Hydrolysis after Monomer Addition)

There is no reason to assume that nucleotide hydrolysis must occur when a monomer is incorporated into the assembled filament. The more general case involves delayed hydrolysis, a condition that may occur if assembly and hydrolysis are kinetically decoupled. If the rate of elongation is then especially rapid, numerous $M \cdot pppN$ molecules could be incorporated initially, thereby forming a pppN-rich "cap" that stabilizes the polymer. This outcome would be expected when nucleotide hydrolysis proceeds at a relatively slow and constant rate. Delayed hydrolysis models can also overcome the difficulty described in the last paragraph of the previous section dealing with bidirectional polymerization with pppN hydrolysis occurring precisely at the time of monomer addition.

Another type of delayed pppN hydrolysis mechanism can be called a penultimate hydrolysis model wherein the previously terminal pppN-containing monomer is induced to undergo hydrolysis only after a new $M \cdot pppN$ adds. This mechanism would require that only those monomers at the polymer ends would contain pppN; all "interior" units would be either $M \cdot ppN \cdot P_i$ and $M \cdot ppN$ or the conformationally energized species $M^* \cdot ppN \cdot P_i$ and $M^* \cdot ppN$. The boundary stabilization model[19] for tubulin polymerization allows one to consider this point further: the

[17] P. B. Schiff and S. B. Horwitz, *Proc. Natl. Acad. Sci. U.S.A.* **77,** 1561 (1980).

[18] L. Wilson, H. P. Miller, K. W. Farrell, K. B. Snyder, W. C. Thompson, and D. L. Purich, *Biochemistry* **24,** 5254 (1985).

[19] T. L. Karr, A. E. Podrasky, and D. L. Purich, *Proc. Natl. Acad. Sci. U.S.A.* **76,** 5475 (1979).

tubulin·GTP complex is thought to be persist only at microtubule ends, whereas tubulin·GDP molecules are located throughout the interior lattice of the microtubule. In this model, microtubules remain stable as long as tubulin·GTP is bound at their termini; tubulin·GDP cannot maintain an intact microtubule end. Thus, extensive endwise depolymerization is postulated to ensue after the loss of the stabilizing boundary of tubulin·GTP molecules on the microtubule ends. Figure 4 illustrates how the terminally bound tubulin·GTP complex affects the polymerization and depolymerization properties of microtubules.

pppN Hydrolysis as a "Clock" Governing Intracellular Turnover of Self-Assembled Structures

Delayed pppN hydrolysis during polymerization also presents the opportunity to create a "molecular clock" that controls the timing of assembly and disassembly. A firmly established principle in cellular signal transduc-

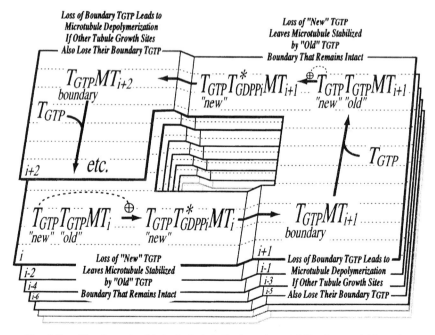

Fig. 4. Penultimate nucleotide hydrolysis in the boundary stabilization model for microtubules. The top and bottom of each of the spiraling layers schematically depicts a round of assembly-induced GTPase, incorporating a newly formed terminal GTP-containing tubulin dimer into the microtubule. The term "old" refers to the previous terminal GTP-containing dimer just before undergoing assembly-induced nucleotide hydrolysis.

tion is that members of the GTP regulatory protein (or G-protein) super-family can either activate or inhibit signaling processes, depending on the phosphorylation state of their bound guanine nucleotide.[20] G-proteins have two states, each characterized by the presence of bound GTP or GDP. Conversion of bound GTP to bound GDP is accomplished by an intrinsic GTPase activity, whereas the GDP-bound state is converted to the GTP-bound state by the release of GDP and binding of GTP. The latter reaction requires the participation of one or more nucleotide-exchange factors. The rate of interconversion between active and inactive states is limited by the rate constant for GTP hydrolysis, and GTPase-activating proteins (GAPs) can accelerate hydrolysis. For example, $G_{i\alpha 1}$ has a turnover number for GTP hydrolysis of around 3 min^{-1}, Ras has a k_{cat} of 0.3 min^{-1}, and EF-Tu has a value of 0.003 min^{-1}.

G-proteins are "clocks" controlling the duration of signaling events, and, by analogy, the delayed pppN hydrolysis during self-assembly reactions provides a mechanism for distinguishing recently created structures from those that have persisted for a longer period. The hydrolysis reaction can potentially form a variety species (e.g., polymer-bound $M \cdot pppN$, polymer-bound $M^* \cdot ppN \cdot P_i$, polymer-bound $M^* \cdot ppN$, polymer-bound $M \cdot ppN$). Therefore, an assembled filament can have many different hypothetical compositions, of which only a few are shown in Table I. Depending on the kinetics of hydrolysis, these chemical and/or conformational species could establish a compositional gradient within each assembled filament. Various capping, cross-linking, severing, and depolymerizing proteins could then selectively recognize the regions along a filament that contain monomers having the appropriate nucleotide form (i.e., pppN, ppN $\cdot P_i$, or ppN). Such selectivity would then control the timing of severing and depolymerization strictly on the basis of the kinetics of delayed pppN hydrolysis. This scheme also offers the added advantage that other regulatory factors might bind to assembled filaments and promote filament maturation by accelerating certain interconversions.

In terms of actin assembly and disassembly, members of the actin de-polymerizing factor (ADF)/cofilin family are believed to play a major role in controlling actin-based motility by promoting the polarized depolymeri-zation of actin (i.e., monomer loss from the slow-growing or pointed end). ADF may also enhance the turnover of actin filaments during the reorgani-zation of the actin cytoskeleton. Carlier et al.[21] reported that an actin depolymerizing factor from Xenopus is effective in depolymerizing normal

[20] A. G. Gilman, Annu. Rev. Biochem. **56,** 615 (1987).
[21] M. F. Carlier, V. Laurent, J. Santolini, R. Melki, D. Didry, G. X. Xia, Y. Hong, N. H. Chua, and D. Pantaloni, J. Cell Biol. **136,** 1307 (1997).

Table I. Hypothetical Compositions of Linear Polymers Based on Nucleotide Hydrolysis Kinetics

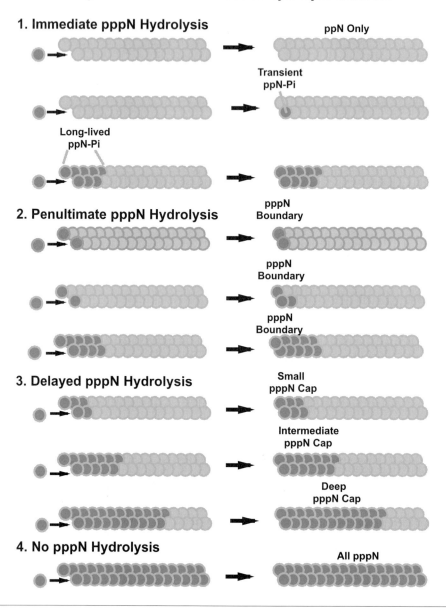

1. Immediate pppN Hydrolysis

ppN Only

Transient ppN-Pi

Long-lived ppN-Pi

2. Penultimate pppN Hydrolysis

pppN Boundary

pppN Boundary

pppN Boundary

3. Delayed pppN Hydrolysis

Small pppN Cap

Intermediate pppN Cap

Deep pppN Cap

4. No pppN Hydrolysis

All pppN

filaments containing ADP, but is unable to completely depolymerize actin filaments containing p(NH)ppA, a poorly hydrolyzed ATP analog. Their observations suggest that the substrate is actin–ADP in the filament and that the lifetime of the filament may be controlled by its nucleotide content. Similarly, Rosenblatt et al.[22] found that ADF from Arabidopsis binds to the ADP-bound forms of G- or F-actin with an affinity that is about two orders of magnitude higher than the ATP– or ADP–P_i-bound forms. This results in a 25 times increase in the rate of actin dissociation from the pointed ends, whereas the rate of actin dissociation from the barbed ends remains unchanged. The authors suggest that this speeding up of the rate-limiting step of the monomer–polymer cycle at steady serves to accelerate actin-based motile processes.

The biological implications of models for GTPase involvement became especially evident when Mitchison and Kirschner[23] presented a model for microtubule steady-state dynamics in terms of GTP hydrolysis and the stochastics of losing the stabilizing cap (or boundary) of tubulin molecules containing unhydrolyzed GTP at their E sites. In their "dynamic instability" model, microtubule length changes at steady state arise from an overall balance of two phases: the first involving slow growth of the majority of microtubule polymers and the second arising from the rapid disassembly of a smaller fraction of polymers. They proposed that microtubules at steady state contain Tb · GTP protomers at the polymer ends, forming a cap of stably bound protomers. In contrast, the microtubule interior lattice contains largely Tb · GDP protomers that are lost by endwise depolymerization whenever microtubules lose Tb · GTP protomers stabilizing their ends.

There is mounting evidence against any significant Tb · GTP cap; the dynamic instability model most probably arises from the presence or loss of the stabilizing boundary[17] formed by a few terminally bound Tb · GTP protomers. Tubules grow and remain stable as a consequence of Tb · GTP bound at the growth sites. They undergo stochastic disassembly whenever all (or a critical fraction of the) growth sites lose their Tb · GTP molecules. To explain microtubule stability, Carlier et al.[24] suggested that microtubule-bound tubulin-GDP-P_i is a stable intermediate and that the dynamic instability of microtubules could be governed by the loss of such "caps" or by P_i release into the medium. Two other laboratories subsequently concluded that such is not the case for either reassembled brain tubules[25] or avian

[22] J. Rosenblatt, B. J. Agnew, H. Abe, J. R. Bamburg, and T. J. Mitchison, J. Cell Biol. 136, 1323 (1997).
[23] T. J. Mitchison and M. W. Kirschner, Nature 312, 237 (1984).
[24] M. F. Carlier, D. Didry, R. Melki, M. Chabre, and D. Pantaloni, Biochemistry 27, 3555 (1988).
[25] M. Caplow, R. Ruhlen, J. Shanks, R. A. Walker, and E. D. Salmon, Biochemistry 28, 8136 (1989).

erythrocyte marginal band microtubules.[26] In both studies, the presence of elevated P_i concentrations had no significant effect on the observed dynamics of subunit addition during elongation, subunit release during the rapid shortening phase, or the frequency of transitions from shortening to regrowth phases.

Exchange-Limited Polymerization Zone Model for Intracellular Actin Filament Growth

One can now consider a model describing those nucleotide interactions that are likely to promote localized, polar growth of actin filaments during actin-based motility. Our model is based on actin-based locomotion of the intracellular pathogens *Listeria monocytogenes*[7,27–30] and *Shigella flexneri*[31,32] in infected host cells. The polymerization zone is characterized by the presence of docking sequences[33] that concentrate essential filament growth-promoting factors. Within the polymerization zone, the concentration of polymerizable actin–ATP far exceeds that present in the bulk phase of the cytoplasm. Ample evidence also suggests that an actin regulatory component, known as profilin, plays a critical role in promoting the explosive growth of filaments within the polymerization zone.

Actin filaments grow rapidly within cells, and the leading edge of the cell can advance at rates exceeding 0.5 to 1 μm/sec.[2,3] Likewise, the actin-based motility of *Listeria* and *Shigella* can attain rates of nearly 0.5 μm/sec.[5] Because microfilaments contain about 360 actin monomers/μm of length, a motility rate of 0.5 to 1 μm/sec corresponds to an apparent first-order rate constant (i.e., $k_{apparent} = k_{on}[\text{actin-ATP}]$) of about 180–360 sec^{-1}. The bimolecular rate constant k_{on} for actin–ATP addition to the barbed end has a nominal value of 2–3 $\times$ 10^6 M^{-1} sec^{-1}. Therefore, one can estimate that 60–180 μM actin–ATP (a value equal to $k_{apparent}/k_{on}$ or 180–360 sec^{-1}/ 2–3 $\times$ 10^6 M^{-1} sec^{-1}) would be needed to sustain a filament growth rate of 0.5 to 1 μm/sec. This local actin–ATP concentration greatly exceeds the

[26] B. Trinczek, A. Marx, E. M. Mandelkow, D. B. Murphy, and E. Mandelkow, *Mol. Cell Biol.* **4**, 323 (1993).
[27] G. A. Dabiri, J. M. Sanger, D. A. Portnoy, and F. S. Southwick, *Proc. Natl. Acad. Sci. U.S.A.* **87**, 6068 (1990).
[28] C. Kocks, E. Gouin, M. Tabouret, P. Berche, H. Ohayon, and P. Cossart, *Cell* **68**, 521 (1992).
[29] F. S. Southwick and D. L. Purich, *Proc. Natl. Acad. Sci. U.S.A.* **91**, 5168 (1994).
[30] T. Chakraborty, F. Ebel, E. Domann, K. Niebuhr, B. Gerstel, S. Pistor, C. J. Temm-Grove, B. M. Jockusch, M. Reinhard, U. Walter, and J. Wehland, *EMBO J.* **14**, 1314 (1995).
[31] W. L. Zeile, D. L. Purich, and F. S. Southwick, *J. Cell Biol.* **133**, 49 (1996).
[32] R. O. Laine, W. Zeile, F. Kang, D. L. Purich, and F. S. Southwick, *J. Cell Biol.* **138**, 1255 (1997).
[33] D. L. Purich and F. S. Southwick, *Biochem. Biophys. Res. Commun.* **231**, 686 (1997).

0.25–0.3 μM cytoplasmic G-actin concentration thought to coexist with pools of assembled actin filaments and sequestered monomeric actin.

How then can this reconcile this obvious discrepancy between the effective actin concentration in the polymerization zone and the much lower cytoplasmic actin concentration? First, we may quickly dispense with any trivial explanation suggesting that the value of the bimolecular rate constant k_{on} is unreasonably low. From numerous kinetic studies on protein subunit addition to filaments, the observed bimolecular rate constants have an upper bound of around 10^7 M^{-1} sec^{-1}. Second, although the total intracellular actin concentration often lies in the 200–300 μM concentration range, the uncomplexed actin–ATP monomer concentration appears to fall into the submicromolar range. Cells typically have about 10–30% of their total actin present as filaments, and the remainder is present as thymosin-β4–actin complex (60–70%), profilin–actin complex (10–20%), and substantially less than 1% as uncomplexed actin–ATP monomers. The most attractive idea is that regions of very active growth are apt to contain high local concentrations of actin–ATP that can be mobilized readily to fulfill the needs of rapidly elongating filaments.

We proposed a "polymerization zone" model[7,34] that relies on a hierarchy of oligoproline sequences. Actin-based motility involves a cascade of binding interactions designed to assemble actin regulatory proteins into functional locomotory units. Listeria ActA surface protein contains four nearly identical EFPPPPTDE-type oligoproline sequences that bind to vasodilator-stimulated phosphoprotein (VASP). The latter, a tetrameric protein with 20–24 GPPPPP docking sites, can in turn bind numerous molecules of profilin, a 15-kDa regulatory protein known to promote actin filament assembly. Actin-based motility (ABM) sequences are distinct oligoproline-containing recognition sequences that serve as docking sites[33] in actin-based motility. The ABM-1 consensus sequence (D/E)FPPPPX (D/E) [where X = P or T] was defined by the oligoproline modules in *L. monocytogenes* ActA surface protein and two host cell homologs known as vinculin and zyxin. The ABM-2 consensus sequence XPPPPP [where X = G, A, L, P, or S] was deduced from the analysis of human platelet VASP as well as the related proteins Mena and N-Mena. As shown in Fig. 5, this cascade of binding interactions can result in the concentration of profilin and/or profilin–actin complex within an activated cluster region. Within this polymerization zone, high local concentrations of profilin can facilitate ATP exchange with actin–ADP, ensuring that high levels of actin–ATP will be available for the barbed-end growth of nearby actin filaments.

[34] F. Kang, R. O. Laine, M. R. Bubb, F. S. Southwick, and D. L. Purich, *Biochemistry* **36**, 8384 (1997).

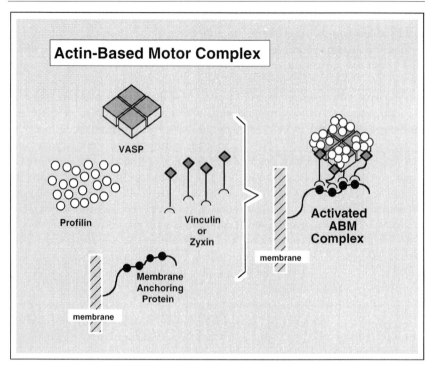

FIG. 5. Assembly of an activated actin-based motility complex using binding interactions at ABM-1 and ABM-2 docking sequences. The scheme presented here allows vasodilator-stimulated phosphoprotein (VASP) to bind at an ABM-1 sequence module, and this binding is thought to expose the numerous ABM-2 sites for concentrating profilin in the polymerization zone.

Because actin–ADP is much less efficient in actin assembly, conversion to the actin–ATP complex favors the rapid elongation of filaments.

Nucleation of actin filament formation is another important step in actin filament formation, and other components, such as actin-related proteins (or ARPs), may be recruited to the polymerization zone to promote nucleation. The Arp2/3 complex is a stable complex composed of seven protein subunits (two actin-related proteins Arp2 and Arp3 as well as five novel proteins) that is concentrated at the leading edges of motile nonmuscle cells.[34,35] The Arp2/3 complex appears to play a fundamental role in regulating actin dynamics during cell motility and is apparently highly conserved

[35] L. M. Machesky, *Curr. Biol.* **7,** R164 (1997).

among *Saccharomyces cerevisiae,*[36] amebas,[37,38] and mammals.[39] The basic idea is that the Arp2/3 complex may stabilize actin dimer and trimer formation, thereby improving the odds that the otherwise thermodynamically unfavorable nucleation will occur. As noted earlier, actin filament assembly at the cell surface of *L. monocytogenes* requires the bacterial surface protein ActA, and evidence suggests that the host cell Arp2/3 complex is likewise required.[40] The latter accelerates the nucleation of actin polymers *in vitro,* whereas ActA alone is without effect. Upon combination of the Arp2/3 complex and ActA, nucleation of actin filaments is stimulated synergistically. Because the Arp2/3 complex caps actin filament pointed ends, it should not be surprising that the filament-bound complex also inhibits both monomer addition/dissociation at the pointed ends.[41] It has been suggested that nearly all actin filament pointed ends are capped by the Arp2/3 complex *in vivo,*[42] suggesting that the Arp2/3 complex mediates most, if not all, nucleation. In this case, actin filaments can only elongate by monomer addition to the barbed ends.

As illustrated in Fig. 6, cells must also contain a depolymerization zone where filament disassembly occurs by loss of actin–ADP from the pointed ends. This depolymerization zone is also likely to contain docking sites for depolymerizing proteins such as ADF or cofilin.[21,22] The much lower concentration of untethered profilin in the depolymerization zone probably limits facilitated nucleotide exchange. The presence of the substantial actin–ADP complex in the polymerization zone may also suppress the spontaneous formation of polymerization nuclei, the uncontrolled formation of which could disturb the polarity of actin filaments formed in the peripheral cytoplasm. Actin–ADP is also likely to have a similar inhibitory effect on Arp2/3 complex-induced nucleation that might otherwise take place outside the polymerization zone.

Concluding Remarks

The universal coupling of pppN hydrolysis to the synthesis and degradation of key metabolites provides a series of checkpoints that ensure that

[36] S. H. Zigmond, *Curr. Biol.* **8,** R654 (1998).
[37] D. Winter, A. V. Podtelejnikov, M. Mann, and R. Li, *Curr. Biol.* **7,** 519 (1997).
[38] L. M. Machesky, S. J. Atkinson, C. Ampe, J. Vandekerckhove, and T. D. Pollard, *J. Cell Biol.* **127,** 107 (1994).
[39] R. D. Mullins, W. F. Stafford, and T. D. Pollard, *J. Cell Biol.* **136,** 331 (1997).
[40] M. D. Welch, A. Iwamatsu, and T. J. Mitchison, *Nature* **385,** 265 (1997).
[41] M. D. Welch, J. Rosenblatt, J. Skoble, D. A. Portnoy, and T. J. Mitchison, *Science* **281,** 105 (1998).
[42] R. D. Mullins, J. A. Heuser, and T. D. Pollard, *Proc. Natl. Acad. Sci. U.S.A.* **95,** 6181 (1998).

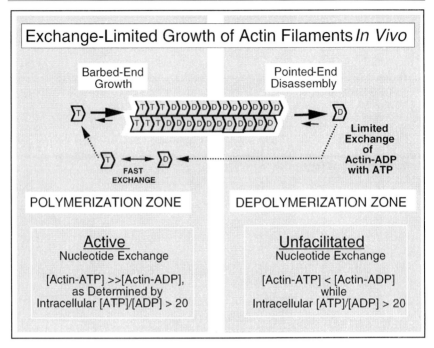

FIG. 6. An exchange-limited model that defines actin polymerization and depolymerization zones within cells. Actin–ATP is concentrated within the polymerization zone as a consequence of its binding interaction with profilin, which is bound to GPPPPP sequences within VASP (see Fig. 4). Any actin–ADP within this polymerization region is also converted rapidly to actin–ADP by profilin-promoted nucleotide exchange. However, the depolymerization zone allows the actin–ADP complex to dissociate from the pointed ends of actin filaments.

effective regulation is enforced in all living cells. The assembly and disassembly of organized structural elements, such as actin filaments and microtubules, can be likened to the anabolic and catabolic reactions in intermediary metabolism. In this respect, ATP and GTP provide the thermodynamic driving forces that control the dynamics of assembly and disassembly of these cytoskeletal fibers. As described in this article, the various stages of pppN hydrolysis also create a molecular clock for controlling how actin filaments and microtubules interact with other proteins within the cytomatrix. The restricted occurrence of pppN hydrolysis in the polymeric states of actin and tubulin limits the use of mechanistic tools that are available to other areas of enzymology. Moreover, because the hydrolytic reaction takes place within subunits at or near the filament ends, the detailed characterization of reaction kinetics and other parameters is made all the more challenging. Nonetheless, one can also recall how daunting the task of unraveling the salient features of F_0F_1–ATPase action in oxidative phosphorylation seemed only a decade ago. There have been major strides in efforts to understand how the energy of ligand binding and pppN hydrolysis alters the strength of protein–protein interactions, and the prospect for applying those principles to biochemical self-assembly appears to be most promising.

[6] Fundamental Mechanisms of Substrate Channeling

By KAREN S. ANDERSON

Introduction

There are many interesting enzyme systems that may exhibit substrate channeling between active sites. This article is not intended to provide an exhaustive survey of this topic but rather to provide an in-depth discussion for several examples of enzyme systems for which there is compelling evidence for substrate channeling based on structural and kinetic studies. A general definition of substrate channeling will be presented as well the biological relevance and the various criteria that may be used for providing evidence for channeling. This will be followed by examples of enzyme systems that display channeling behavior. Finally, the question will be considered as to whether there may be general features that govern the channeling of substrates in multienzyme or multireaction complexes.

Definition and Relevance of Substrate Channeling

Physiological Relevance

Substrate channeling is a process by which two or more sequential enzymes in a pathway interact to transfer a metabolite (or intermediate) from one enzyme active site to another without allowing free diffusion of the metabolite into bulk solvent.[1–3] For many years, it has been suggested that substrate channeling is of fundamental importance in regulating meta-

[1] J. Ovadi, C. Salerno, T. Keleti, and P. Fasella, *Eur. J. Biochem.* **90,** 499 (1978).
[2] J. Ovadi, *Trends Biochem. Sci.* **13,** 486 (1988).
[3] P. A. Srere, *Annu. Rev. Biochem.* **56,** 89 (1987).

Copyright © 1999 by Academic Press
All rights of reproduction in any form reserved.
0076-6879/99 $30.00

bolic pathways in the cell.[3-9] For example, enzyme complexes could serve to shuttle intermediates along a particular pathway, especially in cases where substrates are present at concentrations lower than their cognate enzymes.[1,5,10] There are numerous examples of sequential enzyme pairs believed to exhibit channeling behavior, such as those found in glycolysis, nucleic acid biosynthesis, amino acid biosynthesis, and fatty acid synthesis.[7,10-13] The cited physiological advantages for channeling include control of metabolic flux, protection of reactive and/or toxic intermediates, increased catalytic efficiency, and decreased diffusion of intermediates away from the catalytic sites. Several excellent reviews have been written on these aspects of channeling.[1,3-5,9]

Despite the apparent wide acceptance of substrate substrate channeling, there is a surprising lack of detailed mechanistic and structural information available on the processes that may govern the direct transfer of a metabolite between any two enzyme active sites. Several examples exist in which the evidence for channeling is controversial.[14-21] The focal point of this discussion will be enzyme systems for which there is general agreement that substrate channeling occurs based on the criteria described in this article. An understanding of the functional and structural aspects of channeling in these enzyme systems is important in order to define the fundamental parameters governing efficient channeling of metabolities. In addition, an understanding of these processes may serve as a guide toward understanding protein–protein communication and signaling pathways.

[4] G. R. Welch, *Prog. Biophys. Mol. Biol.* **32,** 103 (1977).
[5] T. Keleti and J. Ovadi, *Curr. Top. Cell. Regul.* **29,** 1 (1988).
[6] B. Vertessy and J. Ovadi, *Eur. J. Biochem.* **164,** 655 (1987).
[7] S. J. Wakil, J. K. Stoops, and V. C. Joshi, *Annu. Rev. Biochem.* **52,** 556 (1983).
[8] J. S. Easterby, *Biochem. J.* **264,** 605 (1989).
[9] J. Ovadi, *J. Theor. Biol.* **152,** 1 (1991).
[10] J. Batke, *Trends Biochem. Sci.* **14,** 481 (1989).
[11] C. K. Mathews and N. K. Sinha, *Proc. Natl. Acad. Sci. U.S.A* **79,** 302 (1982).
[12] L. J. Reed, F. Pettit, and S. Yeaman, "Microenvironments and Metabolic Compartmentation," p. 305. Academic Press, New York, 1978.
[13] J. Ovadi and T. Keleti, *Eur. J. Biochem.* **85,** 157 (1978).
[14] D. K. Srivastava and S. A. Bernhard, *Science* **234,** 1081 (1986).
[15] D. K. Srivastava and S. A. Bernhard, *Biochemistry* **24,** 623 (1985).
[16] D. K. Srivastava and S. A. Bernhard, *Biochemistry* **26,** 1240 (1987).
[17] B. P. Chock and H. Gutfreund, *Proc. Natl. Acad. Sci. U.S.A* **85,** 8870 (1988).
[18] J. Kvassman and G. Petterson, *Eur. J. Biochem.* **186,** 261 (1989).
[19] J. Kvassman and G. Pettersson, *Eur. J. Biochem.* **186,** 265 (1989).
[20] D. K. Srivastava, P. Smolen, G. F. Betts, T. Fukushima, H. O. Spivey, and S. A. Bernhard, *Proc. Natl. Acad. Sci. U.S.A* **86,** 6464 (1989).
[21] J. P. Weber and S. A. Bernhard, *Biochemistry* **21,** 4189 (1982).

Criteria for Establishing Substrate Channeling

Consider the enzymatic conversion of substrate, A, to product, B, by enzyme 1 (E_1) and the subsequent conversion of B to C by enzyme 2 (E_2). The transfer of the metabolic intermediate, B, from one enzyme site E_1 to the next enzyme in the sequence E_2 may occur by a diffusion pathway (left) or by a channeling pathway (right). The diffusion pathway involves the dissociation of B at E_1 and rebinding at E_2 in order to form product, C. In this case, when B dissociates, it will be diluted into bulk solvent. However, the channeling pathway would involve the direct transfer of B from E_1 to E_2 without direct exposure to solution.

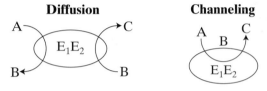

Steady-State Methods to Test for Channeling. Evidence for the channeling of metabolic intermediates between sequential pairs of enzymes has been based largely on steady-state measurements. A variety of techniques for demonstrating channeling behavior have been employed, including enzyme buffering, ligand exchange kinetics, isotope dilution, and estimation of the transient time. One of the most frequently used methods to establish that an intermediate is channeled from one enzyme to another is the transient time approximation.[22-24] This method involves the determination of a key phenomenological parameter known as the transient time, τ.[25,26] In this type of analysis (Fig. 1), the rate for the conversion of B > C by E_2 (Exp. 1) is compared to the rate of product formation, C, in the coupled enzyme system: A > B > C by E_1E_2 activities (Exp. 2). A theoretical estimate for the rate of formation of C for the coupled enzyme system is made based on the ratio of V_{max} and K_m (dashed line).[8,24,27] The transient time, τ, is then defined as the the lag time that precedes the attainment of a steady-state concentration of B. If the transient time for the formation of C is shorter (curve 2) than the theoretical estimate (curve 1), then it is assumed that B was transferred directly from E_1 to E_2. This analysis requires

[22] T. Keleti, J. Ovadi, and J. Batke, *Prog. Biophys. Mol. Biol.* **53,** 105 (1989).
[23] F. Orosz and J. Ovadi, *Biochim. Biophy. Acta* **915,** 53 (1987).
[24] J. S. Easterby, *Biochim Biophys Acta* **293,** 552 (1973).
[25] P. Tompa, J. Batke, and J. Ovadi, *FEBS Lett.* **214,** 244 (1987).
[26] J. Ovadi, P. Tompa, B. Vertessy, F. Orosz, T. Keleti, and G. R. Welch, *Biochem. J.* **257,** 187 (1989).
[27] F. B. Rudolph, B. W. Baugher, and R. S. Beissner, *Methods Enzymol.* **63,** 22 (1979).

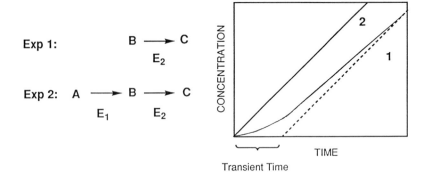

Analysis:

 Exp 1. Measure rate of conversion of B $\rightarrow$ C by E_2

 Exp 2. Measure lag in reaching steady state $\Rightarrow$ transient time

 • If transient time is shorter than predicted $\Rightarrow$ Channeling

Assumption:

 Both reactions are irreversible

  a. Binding of B to E_1 does not alter rate in Exp 1

  b. Reversal of reactions does not affect transient time

FIG. 1. Evidence for substrate channeling by the determination of transient time.

several assumptions. First, the reactions for A > B and B > C must be irreversible. Second, the transient time should not be affected by the reversal of reactions. Finally, the binding of B to E_1 should not affect the rate of B > C by E_2.

A second technique commonly employed for establishing channeling behavior is the isotope dilution method.[23] This method involves determining the rate of conversion of radiolabeled A* > C* in the presence of excess unlabeled B for individual and coupled enzyme reactions.

Diffusion   **Channeling**

A*⟶   ⟶C, C* A* C*

E_1E_2 B*

B*⟵ ⟵B, B* E_1E_2⟶C

            B

The relative specific activity may be predicted based on the kinetic parameters for the reaction of A* > B* and B* > C*. Based on the specific activity of the end product C* formed, a channeling mechanism may be implied. The limitations to this technique include the fact that recovery of C* may be affected by the reaction of B and B* from solution and errors that may arise in the estimation of kinetic parameters.

Transient Kinetic Analysis as a Test for Channeling. Metabolic channeling has been studied most thoroughly by the application of transient kinetic analysis (see references[28–30] for reviews on transient kinetic analysis). A rigorous kinetic analysis includes a measurement of the rates of substrate binding and dissociation from enzyme active site E_1 and rebinding to active site E_2. In order to rule out a free diffusion pathway and invoke channeling from one active site to another, one must not be able to explain the rates by a pathway that involves rapid dissociation and rebinding at the second active site. The use of rapid kinetic analysis, including rapid chemical quench and stopped-flow fluorescence methods, has been extremely useful in examining enzyme systems that may exhibit substrate channeling. This approach allows one to measure individual steps in a kinetic pathway, including both binding and dissociation rates for substrates and products as well as to define the reaction kinetics of intermediates that may be formed. This strategy has several advantages in examining substrate channeling since, in principle, this technique allows one to directly monitor chemical catalysis at each active site as well as the transit of the putative intermediate or metabolite from one active site to another. This is accomplished by using single turnover kinetic methods in which enzyme is in excess over limiting radiolabeled substrate. The rate of reaction of B* from solution at E_2 (conversion of B* > C*) in a single turnover (Exp. 1) is compared to the rate of A* > B* > C* in a single enzyme turnover (Exp. 2). If the formation of C* in Exp. 2 is faster, this indicates that B* was transferred directly from E_1 to E_2 without dissociation from E_1 and rebinding to E_2. Conceivably, depending on the relative rates of chemical catalysis for E_1 and E_2, one could also observe the transient formation of intermediate, B in Exp. 2.

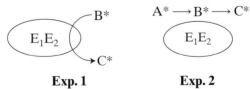

Exp. 1   **Exp. 2**

[28] K. A. Johnson, *in* "The Enzymes" (P. D. Boyer, ed.), 4th ed., Vol. 20, p. 1. Academic Press, New York, 1992.
[29] K. A. Johnson, *Methods Enzymol.* **249**, 38 (1995).
[30] C. A. Fierke and G. Hammes, *Methods Enzymol.* **249**, 3 (1995).

It is important to keep in mind that unlike steady-state methods in which the rate-limiting step is observed (often product release), transient kinetic methods under single turnover conditions measure directly the steps that limit *chemical catalysis*.

Structural Definition. Many enzymes have been suggested to form a transient association between two enzyme pairs to allow direct substrate transfer. We will focus on enzyme pairs where there is evidence for a physical association that is tight enough to allow the subunit complex to be isolated for the two active sites or for which the two active sites are located on a single polypeptide chain as a bifunctional protein.

In this article, four enzyme systems are presented for which structural information is available. In these examples, the crystal structures indicate that there are indeed two active sites that are too far apart to allow direct transfer of intermediate and therefore indicate that there may be a necessity for a channel to serve as a conduit to connect the two sites.

Tryptophan Synthase as a Classic Model

Perhaps the most well-defined enzyme to exhibit substrate channeling is tryptophan synthase, an $\alpha_2\beta_2$ tetrameric enzyme complex that catalyzes the final two steps in the biosynthesis of tryptophan (for reviews see references[31–33]). The α subunit catalyzes the cleavage of indole 3-glycerol phosphate (IGP) to indole and glyceraldehyde 3-phosphate (G3P) (α reaction) whereas the β subunit catalyzes the condensation of indole with serine in a reaction mediated by pyridoxal phosphate (β reaction). The physiologically important reaction is termed the $\alpha\beta$ reaction and involves the conversion of IGP and serine to tryptophan and water (Fig. 2).[34–36] The solution of the three-dimensional crystal structure of the enzyme from *Salmonella typhimurium* provides physical evidence for a hydrophobic tunnel 25 Å in length that connects the active sites of the α and β subunits.[37,38] In this case, the metabolic intermediate (indole) is transferred from the active site of the α subunit to the active site of the β subunit through the connecting tunnel as illustrated in Fig. 2.

[31] E. W. Miles, *Adv. Enzymol.* **49,** 127 (1979).
[32] E. W. Miles, *Adv. Enzymol. Rel. Areas Mol. Biol.* **64,** 93 (1991).
[33] E. W. Miles, "Subcellular Biochemistry," Vol. 24, p. 207. Plenum Press, New York, 1995.
[34] J. A. Demoss, *Biochim. Biophys. Acta* **62,** 279 (1962).
[35] T. E. Creighton, *Eur. J. Biochem.* **13,** 1 (1970).
[36] W. M. Matchett, *J. Biol. Chem.* **249,** 4041 (1974).
[37] C. C. Hyde, S. A. Ahmed, E. A. Padlan, E. W. Miles, and D. R. Davies, *J. Biol. Chem.* **263,** 17857 (1988).
[38] C. C. Hyde and E. W. Miles, *Biotechnology* **8,** 27 (1990).

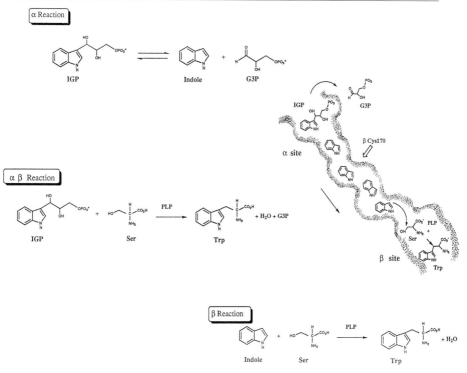

Fig. 2. Diagram illustrating the hydrophobic tunnel and catalysis at α and β active sites for tryptophan synthase.

Transient Kinetic Studies to Understand Substrate Channeling

Previous rapid transient kinetic analysis of tryptophan synthase has shown that the product of the α reaction, indole, is transferred to the β site where it reacts with an activated serine to form tryptophan.[39–41] A combination of transient kinetic experiments using rapid chemical quench and stopped-flow methods has provided definitive evidence for substrate channeling in tryptophan synthase. The rapid chemical quench analysis provides information about the chemical catalysis occurring at the active site whereas stopped-flow fluorescence analysis provides information about substrate binding and protein conformational changes, as well as catalysis

[39] A. Lane and K. Kirschner, *Biochemistry* **30**, 479 (1991).
[40] P. Brzovic, K. Ngo, and M. Dunn, *Biochemistry* **31**, 3831 (1992).
[41] K. S. Anderson, E. W. Miles, and K. A. Johnson, *J. Biol. Chem.* **266**, 8020 (1991).

in some cases. These techniques are complementary as the interpretation of optical signals by stopped-flow fluorescence can be corroborated by examining the conversion of radiolabeled substrate to product using rapid chemical quench techniques to monitor directly the chemistry occurring at the active site. This analysis has not only established that indole is indeed "channeled" but has also indentified three essential kinetic features that govern efficient channeling.

An α-β Intersubunit Model for Communication. Efficient channeling depends on effective communication between α and β sites to attenuate the formation of indole in the presence of serine.[41,42] Subunit communication was first observed in analysis of effects of α-site ligands on the interaction of serine with pyridoxal phosphate (PLP) at the β site. Although the relationship of this communication to substrate channeling is still not completely understood, these reactions have been well characterized by stopped-flow fluorescence, absorbance difference spectroscopy,[41,43,44] and emerging insights from structural studies.[37] The reactions occurring at the β subunit for the activation of serine via PLP are summarized in Fig. 3. In the absence of serine, the PLP is bound covalently to β Lys-87 (**I**). When serine binds, the amino group of serine replaces the ε amino group of the lysine, forming an external aldimine species (Ex-Ald, **II**). This is followed by a deprotonation to form a quininoid species (**III**) that subsequently loses H_2O, forming an aminoacrylate species (E-AA, **IV**). In the absence of indole, the aminoacrylate species slowly hydrolyzes, most likely through an iminopyruvate species to form pyruvate, ammonia, and H_2O. If instead indole is present, it will react with the E-AA to form similar types of PLP-mediated species, including a quininoid derivative that is ultimately converted to tryptophan. Transient kinetic studies using rapid scanning stopped-flow analysis (RSSF) indicate that formation of this quininoid species deactivates the α reaction.[45]

An in-depth transient kinetic analysis of wild-type tryptophan synthase using rapid chemical quench and stopped-flow methods has led to the kinetic model shown in Fig. 4 that illustrates the intersubunit communication between α and β subunits.[41]

The reactions in the left-hand column of Fig. 4 occur at the α site in the absence of serine. The substrate, IGP, binds to the enzyme to form IGP-E. Binding is followed by a rate-limiting conformational change to

[42] M. F. Dunn, V. Aguilar, P. Brzovic, W. F. Drewe, K. F. Houben, C. A. Leja, and M. Roy, *Biochemistry* **29**, 8598 (1990).
[43] W. F. Drewe and M. F. Dunn, *Biochemistry* **25**, 2494 (1986).
[44] A. Lane and K. Kirschner, *Eur. J. Biochem.* **129**, 561 (1983).
[45] C. A. Leja, E. U. Woehl, and M. F. Dunn, *Biochemistry* **34**, 6552 (1995).

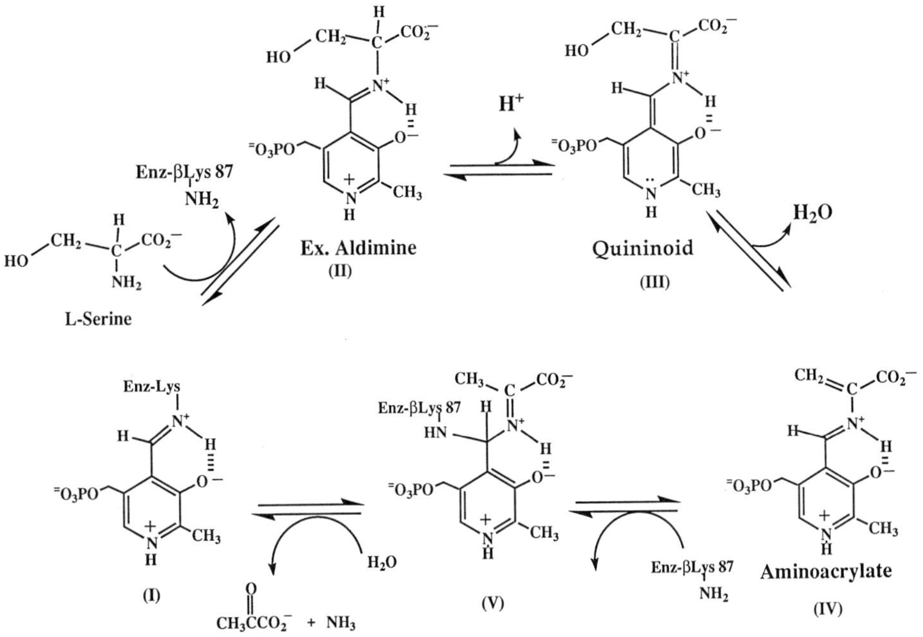

FIG. 3. PLP-mediated activation of serine at the β active site of tryptophan synthase.

IGP-E*. If serine is not present at the β site, the rate of this conformational change is 0.16 sec^{-1}. The reactions in the right-hand column of Fig. 4 occur at the β site. Serine binds to the enzyme and reacts with PLP to form a Schiff base external aldimine species (E-Ser or Ex-Ald).[43,44] This species is then deprotonated at the α carbon of serine to form a quininoid species, which is subsequently dehydrated to form an activated aminoacrylate derivative (E ~ AA). If indole is added (as in the β reaction), it condenses rapidly with the aminoacrylate to form tryptophan (Trp). In the physiological $\alpha\beta$ reaction, the presence of serine at the β site activates the α site to cleave IGP to indole. The formation of the E ~ AA species induces a protein conformational change to E* ~ AA. This activation enhances the rate of E-IGP to E*-IGP such that it is no longer rate limiting (24 sec^{-1}). Once an indole is formed, it is channeled rapidly to the β site where it reacts with the E* ~ AA to form Trp.

As summarized in Fig. 5, the salient features of this model involve the binding of the β subunit ligand serine and its activation via a PLP-dependent reaction to form a reactive enzyme-bound aminoacrylate (PLP ~ AA) species, which in turn triggers a conformational change that promotes the cleavage of IGP to indole at the α subunit.[41]

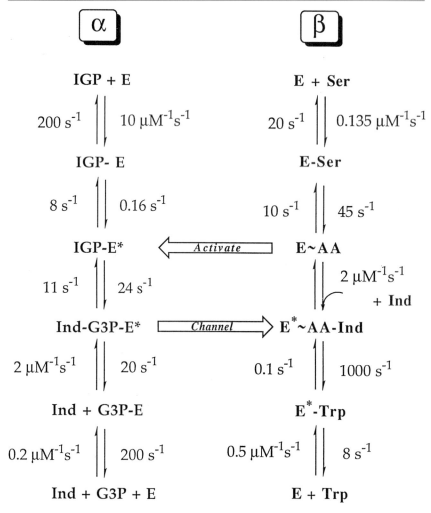

FIG. 4. Kinetic scheme for α-β intersubunit communication in tryptophan synthase. Reprinted with permission from K. S. Anderson *et al.*, *J. Biol. Chem.* **266**, 8020 (1991). Copyright (1991) The American Society for Biochemistry and Molecular Biology.

When a molecule of indole is formed it diffuses rapidly through the hydrophobic tunnel to the β subunit and reacts with the PLP $\sim$ AA, also very rapidly, to form the product tryptophan. This intersubunit communication keeps the α and β reactions in phase such that the intermediate indole does not accumulate.

Three features of the reaction kinetics are essential to ensure that indole

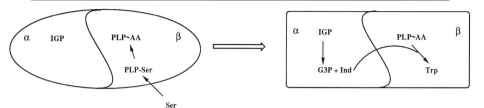

FIG. 5. Diagram illustrating serine triggering of indole channeling. Reprinted with permission from K. S. Anderson *et al., J. Biol. Chem.* **266,** 8020 (1991). Copyright (1991) The American Society for Biochemistry and Molecular Biology.

is channeled efficiently: (1) the reaction of serine at the β site modulates the formation of indole at the α site such that indole is not produced until serine has reacted to form E* ~ AA; (2) the rate of reaction of indole and E* ~ AA is fast and largely irreversible; and (3) the rate of indole diffusion from the α site to the β site is very fast (>1000 sec^{-1}).[41] This mechanism accounts for the fact that indole does not accumulate during a single turnover of conversion of IGP to tryptophan (the $\alpha\beta$ reaction).

This model makes several predictions, which have been tested by kinetic and structural analysis of mutants and alternate substrates.[41,46]

Tests of the Model for α-β Intersubunit Communication in Tryptophan Synthase. The model for α-β intersubunit communication indicates that it is the formation of the aminoacrylate species that leads to activation of the α reaction. When both serine and IGP are added simultaneously to an enzyme in a single enzyme turnover experiment, there is a lag in the cleavage of IGP, which is a function of the reaction of serine to form the aminoacrylate species.[41] Accordingly, amino acids other than serine that can undergo dehydration to form the aminoacrylate such as cysteine should serve as alternate substrates but should lead to a longer lag for the α subunit activation as determined by transient kinetic analysis. Cysteine does indeed serve as an alternate substrate, but a substantially slower activation process was observed.[41,47] Nonetheless, the α and β reactions remain coupled, with no observed accumulation of indole, indicating that efficient channeling is still occurring. Amino acids such as glycine and homoserine that cannot undergo dehydration to form an aminoacrylate do not promote activation.[41]

Additional evidence for the importance of E* ~ AA is provided by the preparation and analysis of a mutant form of tryptophan synthase (β Lys-87 Thr) in which the lysine involved in the covalent Schiff base formation

[46] K. S. Anderson, A. Y. Kim, J. M. Quillen, E. Sayers, X. J. Yang, and E. W. Miles, *J. Biol. Chem.* **270,** 29936 (1995).
[47] I. P. Crawford and J. Ito, *Proc. Natl. Acad. Sci. U.S.A.* **51,** 980 (1964).

with PLP has been replaced by a threonine residue. The β Lys-87 Thr mutant $\alpha_2\beta_2$ complex contains active α subunits and β subunits that retain the ability to bind PLP but are completely inactive enzymatically. This mutant can form stable Schiff base intermediates with L-serine and other amino acids that can be converted to the corresponding aminoacrylate derivative by the addition of ammonia, which partially replaces the deleted ε amino group of β Lys-87. Rapid kinetic analysis shows that the "chemically rescued" mutant exhibits a sixfold increase in the rate of cleavage of IGP to indole relative to the L-serine external aldimine derivative of the mutant, thus establishing the importance of the aminoacrylate in activation of the α reaction.[48]

Based on our working model and transient kinetic analysis, we might predict that if we slow the rate of reaction of indole with the E* ~ AA species in the β reaction, the efficiency will be lost and indole may build up in a single turnover of the $\alpha\beta$ reaction. A residue in the β subunit, Glu-109, has been suggested to play a role in activating indole toward nucleophilic attack on the aminoacrylate intermediate and is supported by X-ray crystallographic studies.[41,49] By mutating β Glu-109 residue to Asp, the rate of the β reaction is decreased by a factor of 300.[41] The rapid chemical quench analysis of the βE109D mutant confirmed that the buildup of indole is observable if the rate of catalysis at the β site is slowed. This observation is also supported by rapid-scanning stopped-flow spectroscopic analysis of this mutant.[50] The rate of catalysis at the βE109D subunit is 3 $\sec^{-1}$ compared with >1000 $\sec^{-1}$ in wild-type tryptophan synthase. The slower reaction at the β site allows a substantial amount of indole to be detected in a single turnover reaction as illustrated in Fig. 6. This is the first time indole has been observed as an actual reaction intermediate in the physiological $\alpha\beta$ reaction of tryptophan synthase.[41]

The importance of rapid diffusion of indole from the α active site through the channel to the β active site was tested by a mutation that blocks the tunnel. According to our model, if the hydrophobic tunnel is blocked or impeded to interfere with the passage of indole from the α site to the β site, indole might be observed in a single enzyme turnover experiment. This hypothesis was tested by mutating one of the hydrophobic residues, β Cys-170 (see Fig. 2), that line the tunnel to a bulkier residue (Phe or Trp). The kinetic analysis reveals that with both the βC170F and the βC170W mutant enzymes, we are able to observe indole in a single

[48] U. Banik, D. M. Zhu, P. B. Chock, and E. W. Miles, *Biochemistry* **34,** 12704 (1995).
[49] S. Rhee, K. D. Parris, C. C. Hyde, S. A. Ahmed, E. W. Miles, and D. R. Davies, *Biochemistry* **36,** 7664 (1997).
[50] P. Brzovic, A. Kayastha, E. Miles, and M. F. Dunn, *Biochemistry* **31,** 1180 (1992).

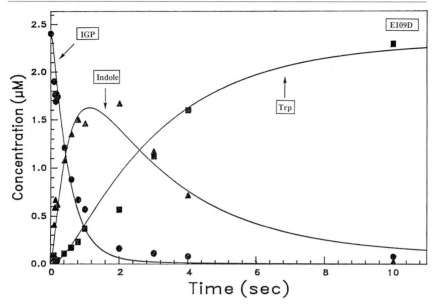

FIG. 6. Single turnover in the $\alpha\beta$ reaction of the E109D mutant. A solution of serine and [^{14}C]IGP was mixed with enzyme to initiate the reaction. The final concentrations were 10 mM serine, 20 μM enzyme, and 2.4 μM [^{14}C]IGP. The disappearance and formation of IGP (●), indole (▲), and tryptophan (■) were monitored. The curves were calculated by the numerical integration using the Kinsim program. Reprinted with permission from K. S. Anderson *et al.*, *J. Biol. Chem.* **266,** 8020 (1991). Copyright (1991) The American Society for Biochemistry and Molecular Biology.

enzyme turnover experiment. These studies indicate that the passage of indole from the α to the β subunit is either blocked or impeded.[46] The X-ray structure of the βC170W mutant has been solved and, as predicted, the indole side chain of the Trp mutant projects into the tunnel, providing evidence of obstruction.[51] Further kinetic analysis by rapid chemical quench and stopped-flow experiments reveals an interesting difference in βC170F and βC170W mutants. Although the Phe mutant indicates a slower rate of formation of E $\sim$ AA (12 sec^{-1}) compared with wild type (45 sec^{-1}), the α reaction is still activated.[46] In the case of βC170W, however, the intersubunit communication is significantly impaired because no activation of the α reaction is noted.[46] We might speculate from these results that the presence of the indole ring of the Trp-170 moiety in the tunnel could mimic the indole intermediate and impede the conformational change necessary

[51] I. Schlichting, X.-J. Yang, E. W. Miles, A. Y. Kim, and K. S. Anderson, *J. Biol. Chem.* **269,** 26591 (1994).

for activating the α reaction. The communication may be mediated by a change in the position of β Phe-280, a residue that may serve as an allosteric gate.[52] The X-ray structures of the wild type and βC170W with bound α ligand, 5-fluoroindole propanol phosphate (5-F-IPP), are in the process of refinement and may provide further insight.

Structural Basis for Efficient Channeling

The kinetic analysis of tryptophan has revealed that the formation of the aminoacrylate species at the β subunit modulates catalysis at the α subunit more than 25 Å away. The most interesting question from a structural perspective is how does this communication work? The key structural components toward understanding this process are the crystal structures of the following tryptophan synthase enzyme complexes: enzyme with no ligand, enzyme with α ligand, enzyme with PLP-activated serine as an aminoacrylate (E* ~ AA), and enzyme with α ligand and PLP-activated serine (E* ~ AA). At present, the unliganded wild-type enzyme at 2.5 Å is available as well as in the presence of monovalent cations (K$^+$ at 2.0 Å and Cs$^+$ at 2.3 Å).[37,53] Use of a substrate analog of IGP, indole-3-propanol phosphate (IPP), revealed the active site of the α subunit.[37] The crystal structures of several liganded forms of the βK87T mutant enzyme have been published that provide clues into the intersubunit communication[49] and two additional wild-type enzyme–ligand complexes with α ligand and PLP-activated serine (E* ~ AA) have become available.[53a] A second approach for deriving structural information using solid-state nuclear magnetic resource (NMR) techniques is introduced.

X-Ray Crystallographic Studies. The three-dimensional structures for four enzyme–ligand complexes of the βK87T mutant enzyme have been determined.[49] These are βK87T-Ser (the external aldimine formed between PLP and serine), βK87T-Ser-IPP, βK87T-Ser-GP (glycerol 3-phosphate, an analog of glyceraldehyde-3-phosphate), and βK87T-Trp (the external aldimine formed between PLP and tryptophan).

The three-dimensional structures for the wild-type enzyme–ligand complexes with α ligand alone and an α ligand with PLP-activated serine (E* ~ AA) have also been solved.[53a] One enzyme–ligand complex (TRPS^{F-IPP}) contains the α site inhibitor, 5-fluoroindole propanol phosphate (F-IPP). The second wild-type enzyme–ligand complex (TRPS$^{F-IPP;A-A}$) con-

[52] R. M. Stroud, *Struct. Biol.* **1**, 131 (1994).

[53] S. Rhee, K. D. Parris, S. A. Ahmed, E. W. Miles, and D. R. Davies, *Biochemistry* **35**, 4211 (1996).

[53a] T. Schneider, E. Gerhardt, M. Lee, P. H. Liang, K. S. Anderson, and I. Schlichting, *Biochemistry* **37**, 5394 (1998).

tains both F-IPP and serine complexed with PLP to form the aminoacrylate (A-A). Some of the structural effects of ligand-induced conformations are highlighted later; however, the reader is referred to the original articles for detailed accounts.[49,53a] Because the structures for the βK87T mutant enzyme–ligand complexes were solved before the wild-type enzyme–ligand complexes, we will describe them in chronological order.

THE FLEXIBLE LOOP REGION OF THE α SUBUNIT. The structure of the α subunit contains an eightfold α/β barrel similar to that observed in several other α/β barrel enzymes, such as triose phosphate isomerase (TIM).[54-56] It has been suggested that the α177–192 flexible region, termed loop 6, closes down over the substrate in a manner similar to that observed for TIM. This region is rather disordered in the unliganded wild-type tryptophan synthase structure. The βK87T mutant structure reveals that this loop 6 region is also disordered in the presence of the α ligand, IPP. However, in the presence of both serine and an α ligand, such as IPP or glycerol-3-phosphate (GP) (an analog of glyceraldehyde-3-phosphate), the flexible loop (residues α179–187) becomes ordered in the βK87T mutant. The conformational changes between the wild-type enzyme and the βK87T-Ser-GP mutant complex are shown in Fig. 7A, see color insert.

It has been suggested previously from mutagenesis studies[57,58] that α Asp-60 and α Glu-49 play catalytic roles in the aldolytic cleavage of IGP.[58,59] Closure of loop 6 isolates the active site of the α subunit from solvent and results in an interaction between the hydroxyl of α Thr-183 and the catalytic residue α Asp-60 that is located in loop 2, which explains the importance of α Thr-183 in catalysis.[60,61] The structure also supports a role for α Asp-60 in catalysis, as the carboxylate of this residue is located 3.0 Å from the nitrogen of the indole ring of IPP. A structural role for α Glu-49 is unclear. Also located in loop 6 is α Arg179, which interacts with neighboring residues, α Ser-180, α Val-182, and α Gly-184, to stabilize the ordered loop structure and may explain why a α Arg-179T mutant has reduced substrate affinity and affects α-β communication.[62,63]

[54] E. Lolis and G. A. Petsko, *Biochemistry* **29**, 6619 (1990).
[55] E. Lolis, T. Alber, R. C. Davenport, D. Rose, F. C. Hartman, and G. A. Petsko, *Biochemistry* **29**, 6609 (1990).
[56] G. Farber and G. A. Petsko, *Trends Biochem. Sci.* **15**, 228 (1990).
[57] L. Shirvanee, V. Horn, and C. Yanofsky, *J. Biol. Chem.* **265**, 6624 (1990).
[58] S. Nagata, C. C. Hyde, and E. W. Miles, *J. Biol. Chem.* **264**, 6288 (1989).
[59] K. Kirschner, A. Lane, and A. Strasser, *Biochemistry* **30**, 472 (1991).
[60] X. J. Yang and E. W. Miles, *J. Biol. Chem.* **267**, 7520 (1992).
[61] C. Yanofsky and I. Crawford, *in* "The Enzymes," p. 1. Academic Press, New York, 1972.
[62] H. Kawasaki, R. Bauerle, G. Zon, S. A. Ahmed, and E. W. Miles, *J. Biol. Chem.* **262**, 10678 (1987).
[63] P. Brzovic, C. C. Hyde, E. W. Miles, and M. F. Dunn, *Biochemistry* **32**, 10404 (1993).

In contrast to observations with the βK87T mutant structure, the presence of an α-site ligand alone (F-IPP) in the wild-type enzyme–ligand complex (TRPS^{F-IPP}) induced loop αL6 closure and was sufficient to trigger simultaneous ordering of loops αL2 and αL6.[53a] Moreover, for the TRPS^{F-IPP} wild-type structure, a region of the N-terminal domain of the β subunit (introduced previously as mobile region βGly93-βGly189[49]) has been defined as the COMM domain, as this domain appears to play an important role in allosteric communication between α and β sites (see Fig. 8A, color insert).

CONFORMATIONAL CHANGES IN THE HYDROPHOBIC TUNNEL. The hydrophobic tunnel, which connects the α and β subunits (Fig. 2), is implicated in the transfer of the intermediate indole from the α site to the β site. One of the residues suggested to act as a gate, and hence may influence the diffusion of indole, is β Phe-280. It has been shown that this residue appears to be directly in the tunnel in refined wild-type structures but moves out of the tunnel in the presence of the monovalent ions, K^+ and Cs^+.[53] The βK87T mutant structure shows density for β Phe-280 in the tunnel as well as α Leu-58, as shown in Fig. 7B, see color insert.

THE α-β SUBUNIT INTERFACE. A comparison of α-β interface residues for wild-type and mutant structures indicates two groups of residues: α53–61 (loop 2) and β161–181, which exhibit substantial changes in position. While the overall calculated surface area between wild type and mutant in the α-β interface is not altered significantly, two residues that are notably reordered are β Lys-167 and β Arg-175.[49] In the IPP-Ser complex, the side chain of β Lys-167 is observed to interact with α Asp-56, whereas β Arg-175 interacts with the backbone carbonyl of α Pro-57. This may be triggered initially by the closure of loop 6 (α179–187) over the IPP-binding site, which allows for the interaction of α Thr-183 and α Asp-60 and subsequent interactions between loop 2 and the β subunit. These large structural changes suggest that these residues may be involved in allosteric linkage between α and β subunits.[64]

The availability of liganded complexes for mutant structures has provided insights into the structural basis of intersubunit communication; however, there may be differences in the α-β intersubunit communication between wild type and the βK87T mutant. Transient kinetic analysis using rapid chemical quench methods has shown that there is only a sixfold activation of the α reaction by the E* $\sim$ AA with mutant[48] whereas there is a 150-fold activation of wild-type tryptophan synthase.[41] These results suggest that the pathways for intersubunit communication may differ in wild type and mutant enzymes. One of the key structural complexes for

[64] P. Pan, E. Woehl, and M. F. Dunn, *Trends Biochem. Sci.* **22,** 22 (1997).

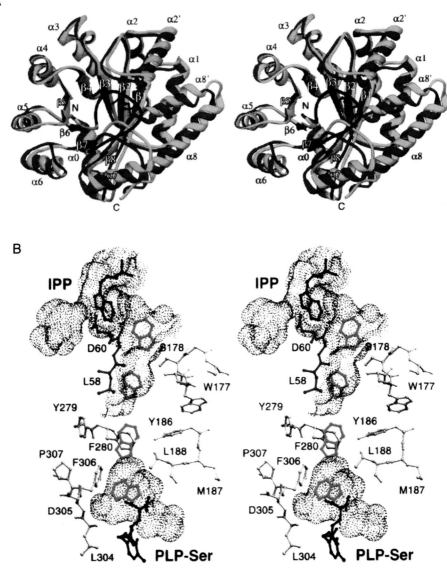

A

B

FIG. 7. (A) Conformational changes between the wild-type (red) and βK87T-Ser-GP mutant complex (yellow) in the α subunit. Loop 6 (residues 179–187 in dark blue) as localized in the βK87T-Ser-GP structure covers the GP (black) bound to the active site. (B) View of the indole tunnel between α and β active sites of the βK87T-Ser-IPP $\alpha_2\beta_2$ complex. Bound IPP and PLP-Ser are indicated in black, selected α subunit residues in red, and β subunit residues in lavender. Four indole molecules can be modeled inside of the tunnel[37] and are represented in green, although there has been no direct observation of indole within the tunnel. Note that the side chain of β Phe-280 is superimposed on the modeled indole at this location and the tunnel in this structure is blocked. Reprinted with permission from S. Rhee *et al., Biochemistry* **36,** 7664 (1997). Copyright (1997) American Chemical Society.

A

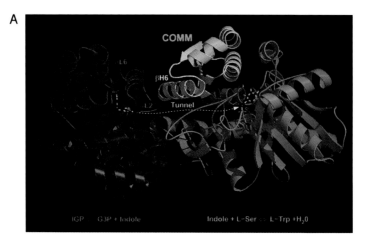

B

C

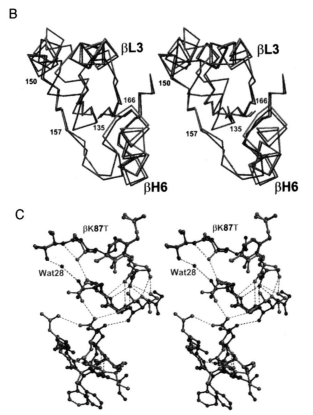

FIG. 8. (A) Schematic representation of the overall fold of the $\alpha\beta$-dimer TRPS^{F-IPP}. F-IPP and PLP bound to the α and β subunits are represented by balls and sticks. The tunnel connecting α and β sites is indicated by a dashed line. The α subunit is shown in red, the β subunit in blue, and the COMM domain in yellow. (B) Cα traces TRPS^{F-IPP} (red) and βK87T-IPP-L-Ser (green) complexes showing regions that exhibit major conformational differences. (C) Ball-and-stick representations of the β site from TRPS$^{F-IPP:A-A}$ (red) and βK87T-IPP-L-Ser (green) complexes.

understanding this process is the wild-type enzyme complexed with α ligand and PLP-activated serine ($E^* \sim AA$). Using a flow cell, the aminoacrylate (A-A) intermediate of the β reaction was generated in the crystal under steady-state conditions in the presence of serine and F-IPP to provide the three-dimensional structure for the wild-type TRPS$^{F-IPP;A-A}$ complex.[53a]

As shown in Fig. 8A, the three-dimensional structure for the wild-type TRPS^{F-IPP} revealed a region of the N-terminal domain of the β subunit, defined as the COMM domain, that seems to play an important role in allosteric communication between α and β sites. Difference distance matrices for atoms indicate that the only significant difference between the polypeptide backbone of wild-type TRPS^{F-IPP} and TRPS$^{F-IPP;A-A}$ complexes is a rigid body displacement of the COMM domain. This rigid body displacement of the COMM domain, on forming the aminoacrylate, is most likely caused by the interaction of the aminoacrylate carboxylate group with loop βL3 that leads to a slight stretching of most of the hydrogen bonds between helix βL6 and loop αL2. The resulting strain could be propaged to the scissile bond of the natural substrate, IGP, at the α site via the hydrogen bond between α Asp-60 oxygen and the indole nitrogen, thereby activating the α reaction.[53a]

Superposition of the structures of the β subunits for the mutant βK87T complexed with IPP and the TRPS$^{F-IPP;A-A}$ complex reveals that although most of the β subunit retains an identical structure in the mutated protein, there are substantial differences in the conformation of the COMM domain in the β135–β150 and β157–β166 regions as illustrated in Fig. 8B, see color insert. However, the differences may also be related differential contacts between the external aldimine and the aminoacrylate, as the mutant does not form the aminoacrylate, as illustrated in Fig. 8C, see color insert.

In summary, X-ray crystallographic studies allow the description of several important features of catalysis for tryptophan synthase in structural terms. Binding of an α site substrate triggers the ordering of two previously disordered loop regions, thereby creating a reaction volume for the α reaction that is secluded from both solvent and the tunnel leading to the β site. Simultaneously, a newly defined domain, the COMM domain, becomes ordered and establishes a pathway by which information can be communicated allosterically from the α site to the β site and vice versa. The actual communication between the two sites involves only a small number of key interactions bridging the intersubunit interface and connecting the COMM domain to the β site, the transmitter between these being rigid body displacements of the entire COMM domain.[53a]

Solid-State NMR Studies. A key step in intersubunit communication is the formation of the aminoacrylate intermediate at the β site, which in turn activates the α subunit. Although one of the key enzyme–ligand complexes to provide information concerning α activation via the β subunit is the E*-AA, until recently it had been difficult to obtain this information by conventional X-ray crystallographic approaches.[53a] This may be due in part to the fact that the aminoacrylate slowly hydrolyzes to pyruvate, resulting in species heterogeneity (see Fig. 3). An alternative approach complementary to X-ray crystallography is the use of solid-state NMR to examine specific ligand–protein interactions.

Solution-state NMR studies are not possible due to the size (143 kDa) and complexity of tryptophan synthase. Solid-state NMR offers an additional advantage over solution-state studies in that it allows selective detection of ligands bound to the enzyme and elimination of signals from high concentrations of substrate in solution necessary to saturate the active site. Also the use of a newly emerging solid-state NMR technique, REDOR, to detect changes in position of selected isotopic nuclei[65,66] may be valuable. This technique was developed to measure weak dipolar coupling constants between isolated spin-1/2 pairs. The dipolar coupling constants are translated directly into internuclear distances. These distances can be determined with accuracy and without interference from surrounding protons. Therefore, select residues in the protein or ligand can be used as sensitive monitors of conformational and dynamic movement.[67,68] The amounts of protein required are comparable to that needed for X-ray crystallographic studies. The preparation of 50–80 mg of pure protein in the form of microcrystals was sufficient for the solid-state NMR studies on tryptophan synthase described later. Two experiments were designed to address issues related to the conformational state of tryptophan synthase in the E* ~ AA state. The first issue is whether the E* ~ AA can be observed directly by solid-state NMR using isotopically labeled serine and the effect of α ligands such as IPP. The second issue is whether there may be residues in the hydrophobic tunnel, such as β Tyr-279 and β Phe-280, that undergo changes in conformation when the E* ~ AA is formed. To address these issues by solid-state NMR, a catalytically active, microcrystalline sample of tryptophan synthase $\alpha_2\beta_1$ containing isotopic labels in the β subunit from [^{19}F]phenyl-

[65] T. Gullion and J. Schaefer, *J. Magn. Res.* **81,** 196 (1989).

[66] T. Gullion and J. Schaefer, *Adv. Magn. Res.* **13,** 57 (1989).

[67] A. W. Hing, N. Tjandra, P. F. Cottam, J. Schaefer, and C. Ho, *Biochemistry* **33,** 8651 (1994).

[68] G. R. Marshall, D. Beusen, K. Kociolek, A. Redlinski, M. Leplawy, S. Holl, R. McKay, and J. Schaefer, *in* "Peptides: Chemistry and Biology, the Proceedings of the 12th American Peptide Symposium, June 16–21, 1991, Cambridge, MA" (J. A. Smith and J. E. Rivier, eds.). Leiden, 1992.

alanine and [^{13}C]tyrosine was prepared.[69,70] A sample of L-[3-^{13}C]serine was used to monitor the formation of aminoacrylate.

In many PLP utilizing enzymes, an aminoacrylate enzyme intermediate between PLP and serine has been postulated.[71,72] Ultraviolet spectroscopic studies have suggested that the aminoacrylate enzyme intermediate is the activated form of serine that condenses with indole to form tryptophan in the β reaction, although direct structural proof for this species is lacking.[44,73] Studies have shown that the most prominent band has a maximum absorbance at 350 nm rather than the anticipated 450–480 nm.[32] A modified version of Fig. 3 illustrating the chemical species that may be involved in this activation process is summarized in Fig. 9.

When serine is not present at the β site, the PLP is bound covalently to β Lys-87 (**I**). On binding, the amino group of serine replaces the ε amino group of the lysine to form an external aldimine species (Ex-Ald, **II**). Subsequent deprotonation forms a quininoid species (**III**) that is, in turn, dehydrated to produce the aminoacrylate species (E-AA, **IV**). This type of species has been observed in other PLP utilizing enzymes by a variety of spectral techniques.[72] For tryptophan synthase, the rate of formation of **IV** is fast and its hydrolytic breakdown, in the absence of indole, to regenerate **I** and pyruvate is relatively slow.[41]

The solid-state NMR experiment to monitor formation of the aminoacrylate in tryptophan synthase (Trp-S) using a doubly labeled microcrystalline sample of enzyme and L-[3-^{13}C]serine is shown in Fig. 10. Figure 10 shows the two regions in the ^{13}C spectra where resonances for aminoacrylates might be expected. Signals from the natural abundance ^{13}C spectra of the protein and from labeled tyrosine residues (156 ppm) are visible in the spectra.

When double-labeled Trp-S is mixed with a solution of L-[3-^{13}C]serine (label chemical shift of 63 ppm), a new ^{13}C resonance due to the serine labeled appears at 28 ppm having an intensity corresponding to a single carbon site (Fig. 10, bottom right). This species is assigned to the bound intermediate methyl tautomer (**V** or **VI**). No resonances due to the label appear in the vinyl region (Fig. 10, bottom left). If both [^{13}C]serine and

[69] L. M. McDowell, M. Lee, R. McKay, K. S. Anderson, and J. Schaefer, *Biochemistry* **35**, 3328 (1996).
[70] L. M. McDowell, M. Lee, J. Schaefer, and K. S. Anderson, *J. Am. Chem. Soc.* **117**, 12352 (1995).
[71] K. Schnackerz, "Progress in Clinical and Biological Research" (A. E. Evangelopoulos, ed.), p. 195. A. R. Liss, New York, 1984.
[72] K. Schnackerz, J. Ehrlich, W. Giesemann, and T. Reed, *Biochemistry* **18**, 3557 (1979).
[73] W. F. Drewe and M. F. Dunn, *Biochemistry* **24**, 3977 (1985).

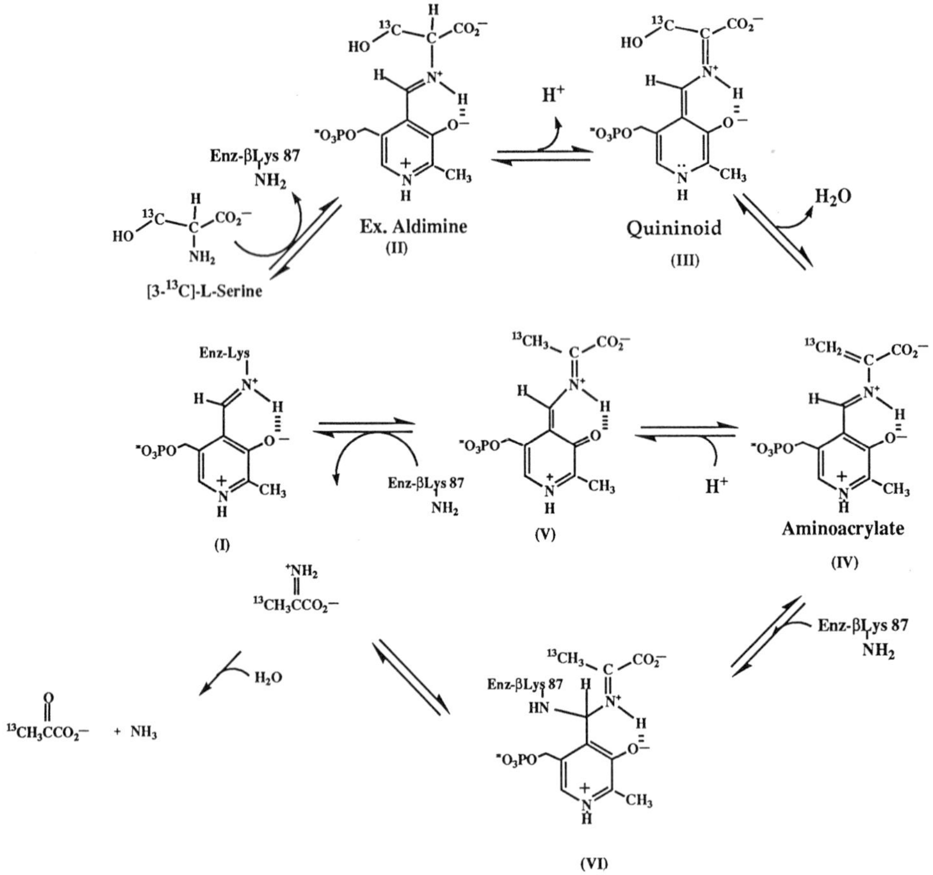

FIG. 9. PLP activation of [^{13}C]serine by tryptophan synthase. Reprinted with permission from L. M. McDowell *et al., J. Am. Chem. Soc.* **117,** 12352 (1995). Copyright (1996) American Chemical Society.

IPP are added to Trp-S, an enhanced resonance appears in the vinyl region at 123 ppm (Fig. 10, top left), and the intensity of the peak at 28 ppm is reduced in intensity (Fig. 10, top right). Both peaks are missing in spectra obtained with unlabeled serine plus IPP (Fig. 10, middle) or when neither serine nor IPP is present (data not shown). Because the methyl region peak is decreased in intensity in the presence of IPP, this indicates that a redistribution of population among aminoacrylate species **IV** and **V** or **VI** has apparently occurred and is consistent with earlier spectroscopic studies.[44,73] The vinylogous species has a ^{13}C chemical shift of 123 ppm and the methyl species has a ^{13}C chemical shift of 28 ppm. These results explain

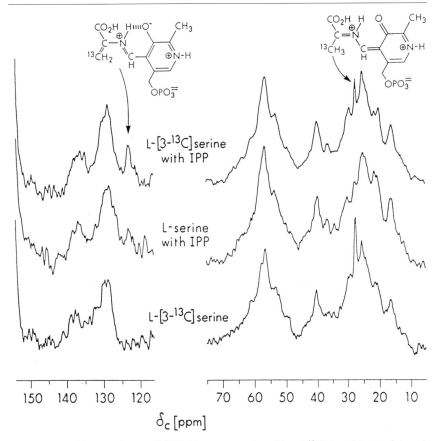

FIG. 10. Fifty-megahertz [^{13}C]NMR spectra of α_2-[ring-4-^{19}F]phenylalanine-[phenol-4-^{13}C]tyrosine-β_2-tryptophan synthase with 5-kHz magic-angle spinning and 80-kHz proton dipolar decoupling. The mother liquor bathing the microcrystals contained 100 mM L-[3-^{13}C]serine and 3 mM indole propanol phosphate (IPP) (top), 100 M L-serine (unlabeled) and 6 mM IPP (middle), or 100 mM L-[3-^{13}C] serine (bottom). The 28-ppm peak in top and bottom spectra is assigned to the methyl ketoamine tautomer (**V** or **VI**) from metabolism of L-[3-^{13}C]serine (Fig. 9). The peak at 123 ppm in the top spectrum may indicate a bound [^{13}C]aminoacrylate and is assigned to **IV** (Fig. 9). Reprinted with permission from L. M. McDowell *et al.*, *J. Am. Chem. Soc.* **117**, 12352 (1995). Copyright (1996) American Chemical Society.

why spectroscopic studies failed to detect an aminoacrylate intermediate at the expected wavelength. Because we now know that there are two aminoacrylate tautomeric forms, X-ray studies may not provide information on the structural species at the β active site region of the protein, which is well resolved. The solid-state NMR resolves these two species (vinylogous aminoacrylate and the methyl tautomer). The equilibrium between the vinylogous species and the methyl tautomer is rationalized on the basis of the highly reactive nature of the aminoacrylate. The equilibrium is only shifted toward the more reactive species when the α site is occupied by ligand and is in a state poised for aldolytic cleavage of IGP to form indole.

A double-labeled sample in which the β subunit of Trp-S contained [^{19}F]Phe and [^{13}C]Tyr was prepared to monitor changes in the hydrophobic region, especially residues β Tyr-279 and β Phe-280, which may occur when serine binds.[69,70] One might predict that a conformational change would result in a change in distance between these two residues, in the absence and presence of serine. Changes in the [^{19}F]F pattern of isotropic shifts in the phenylalanines on the addition of serine were indeed observed, indicating a change in conformation. However, REDOR experiments that determined the change in distance between tyrosine and phenylalanine residues showed that there was less than 1 Å movement. This indicates that the phenylalanine movements were not associated with changes relative to the tyrosine residues in the β subunit.

While the X-ray crystallographic studies provide information on the global conformational changes observed in the protein, the NMR addresses more specific questions, such as residue movements and enzyme intermediates, particularly in cases where disorder in the crystals precludes the interpretation of electron density. The studies described here on tryptophan synthase illustrate the use of solid-state NMR as a complementary technique to X-ray crystallography, which may provide valuable insight in the study of ligand–protein interactions.

Other Enzyme Systems That May Exhibit Substrate Channeling

Structural information has become available for several other enzyme systems that may exhibit channeling behavior. Some of the functional and structural aspects of these systems related to substrate channeling are described.

Bifunctional Thymidylate Synthase–Dihydrofolate Reductase

One of the examples of an enzyme complex suggested to display channeling behavior is the bifunctional enzyme thymidylate synthase–dihydrofolate reductase (TS-DHFR) isolated from the protozoan species *Leishmania major*.[74–76] In several species of parasitic protozoa, the catalytic

[74] R. Grumont, W. Sirawaraporn, and D. V. Santi, Biochemistry **27**, 3776 (1988).
[75] R. Grumont, W. L. Washtien, D. Caput, and D. V. Santi, *Proc. Natl. Acad. Sci. U.S.A.* **83**, 5387 (1986).
[76] T. D. Meek, E. P. Garvey, and D. V. Santi, *Biochemistry* **24**, 678 (1985).

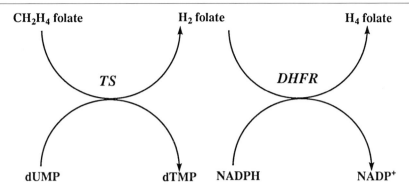

CH$_2$H$_4$ folate   H$_2$ folate   H$_4$ folate

TS   *DHFR*

dUMP   dTMP   NADPH   NADP$^+$

FIG. 11. Catalytic activities of the TS-DHFR bifunctional enzyme.

activities are located on a single polypeptide chain. In the bifunctional enzyme, thymidylate synthase (TS) catalyzes the conversion of dUMP and 5,10-methylene tetrahydrofolate (CH$_2$H$_4$folate) to dTMP and dihydrofolate (H$_2$folate). The dihydrofolate reductase (DHFR) then catalyzes the subsequent reduction of H$_2$folate by NADPH to regenerate tetrahydrofolate (H$_4$folate), which is required for transfer of one carbon unit. The catalytic activities are illustrated in Fig. 11.

This bifunctional enzyme is of considerable clinical significance as the inhibition of either enzyme activity will shut down DNA synthesis. Thus TS-DHFR is a target for antiproliferative and antimicrobial therapeutics. An interesting report on the X-ray structure of the TS-DHFR from *L. major* also revealed the presence of a very different type of channel from that observed with tryptophan synthase.[52,77] The enzyme exists as a symmetrical homodimer with the TS and DHFR sites of a monomeric unit located 40 Å apart, as illustrated in Fig. 12.

In contrast to tryptophan synthase, no obvious "tunnel" is observed, even though the TS- and DHFR-binding sites are too far apart (40 Å) for the direct transfer of H$_2$folate without a major protein conformational rearrangement.[52,77] A novel mechanism has been proposed by Knighton *et al.*[77] Because dihydrofolate has a net charge of -2 and can become more negatively charged by the addition of multiple glutamate moieties, the negatively charged folates might be sensitive to an electrostatic field around the enzyme. The distribution of positively charged residues (Arg, Lys) on the bifunctional enzyme clearly indicates a strongly positive electrostatic "highway" directly across the surface of the protein that links the TS site

[77] D. R. Knighton, C.-C. Kan, E. Howland, C. A. Janson, Z. Hostomska, K. M. Welsh, and D. A. Matthews, *Nature Struct. Biol.* **1,** 186 (1994).

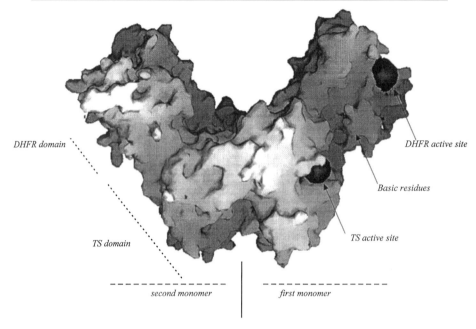

DHFR domain

DHFR active site

Basic residues

TS domain

TS active site

second monomer   *first monomer*

FIG. 12. Structure of the TS-DHFR dimer showing locations of the TS and DHFR domains and active sites.

to the DHFR site, 40 Å away. This highway could shuttle intermediates across the surface of the protein between the two active sites. It is well known that electrostatic funneling occurs in directing substrates toward the active sites of enzymes,[52,78,79] and a detailed modeling study of the TS-DHFR protein has supported the concept of electrostatic channeling.[80]

This proposed mechanism of channeling is very different from the hydrophobic tunnel observed in tryptophan synthase, but the fundamental principles governing channeling may be similar. Steady-state kinetic studies comparing the transient time for monofunctional and bifunctional forms of TS and DHFR provide evidence that substrate channeling of the intermediate, dihydrofolate, is occurring.[76,81] More recent studies indicate that the bifunctional TS-DHFR enzyme from *Toxoplasma gondii* also exhibits chan-

[78] D. R. Ripoll, C. H. Faerman, P. H. Axelsen, I. Silman, and J. L. Sussman, *Proc. Natl. Acad. Sci. U.S.A.* **90,** 5128 (1993).

[79] R. C. Tan, T. N. Truong, J. A. McCammon, and J. L. Sussman, *Biochemistry* **32,** 401 (1993).

[80] A. H. Elcock, M. J. Potter, D. A. Matthews, D. R. Knighton, and J. A. McCammon, *J. Mol. Biol.* **262,** 370 (1996).

[81] K. M. Ivanetich and D. V. Santi, *FASEB J.* **4,** 1594 (1990).

neling behavior.[82] It has also been suggested that the TS and DHFR catalytic domains might communicate through protein conformational changes.[81] Although these steady-state studies indicate that channeling is occurring in the bifunctional TS-DHFR enzyme, the mechanistic details that govern the reaction kinetics for the efficient channeling of dihydrofolate and TS-DHFR domain–domain communication are not understood. An in-depth kinetic analysis of the bifunctional TS-DHFR enzyme using rapid transient methods has been used to provide a quantitative description of the reaction pathway and to identify essential features of reaction kinetics that govern efficient channeling and TS-DHFR active site communication. A detailed kinetic and structural analysis of the TS-DHFR system allows a unique opportunity to compare important aspects of the kinetics of channeling and active site communication with those determined for tryptophan synthase.

Transient Kinetic Analysis to Study Substrate Channeling in the Bifunctional TS-DHFR Enzyme. The bifunctional TS-DHFR enzyme from *L. major* has been suggested to exhibit "electrostatic channeling" between the two active sites.[77,80] A combination of transient kinetic experiments using rapid chemical quench and stopped-flow methods has been used to provide definitive evidence for substrate channeling as well as a more quantitative and detailed mechanistic picture of TS-DHFR active site communication. A brief description of these experiments will be provided. The strategy for investigating substrate channeling was similar to that employed for the study of tryptophan synthase. The first kinetic test of substrate channeling in TS-DHFR was the use of single turnover kinetics[28] to demonstrate that the product from the TS reaction, dihydrofolate, is indeed channeled from the TS active site to the DHFR active site and is not allowed to dissociate from the enzyme. This was done by a series of single enzyme turnover rapid chemical quench experiments as illustrated in Scheme I.

The first experiment established the rate of catalysis of the TS reaction under single turnover conditions designated by the reaction 1 in Scheme I. Previous steady-state kinetic studies suggest that this is an ordered reaction with dUMP binding first, followed by CH_2H_4folate.[83] The bifunctional TS-DHFR enzyme was preincubated with a saturating concentration of unlabeled dUMP and then mixed with a limiting amount of radiolabeled CH_2H_4folate. Under these conditions the rate of catalysis was 2.3 sec^{-1}, similar to the steady-state turnover rate, indicating that either chemical catalysis or substrate binding is rate limiting and not product release. There was no change in the rate of a single turnover performed with enzyme in excess when the enzyme concentration was doubled. This experiment rules

[82] M. Trujillo, R. Donald, D. Roos, D. Greene, and D. Santi, *Biochemistry* **35**, 6366 (1996).
[83] P. Danenberg and K. Danenberg, *Biochemistry* **17**, 4018 (1987).

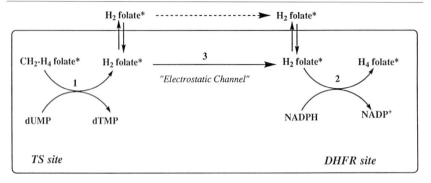

SCHEME I. Rapid chemical quench studies to illustrate channeling in TS-DHFR.

out rate-limiting substrate binding and establishes that chemical catalysis is rate limiting. A similar set of single turnover experiments was carried for examining the DHFR activity of the bifunctional enzyme as designated in reaction 2 in Scheme I. Under conditions in which the enzyme was preincubated with excess unlabeled NADPH and then mixed with a limiting amount of radiolabeled H_2folate, the observed rate of H_4folate formation was 14 sec^{-1}. The same rate was observed if the enzyme was preincubated with radiolabeled H_2folate and then mixed with an excess of unlabeled NADPH. Experiments at a higher enzyme concentration did not result in an increase in the observed rate, indicating that either chemical catalysis or product release was the rate-limiting factor. A pre-steady-state burst experiment showed *no* burst of product, verifying that, under these conditions, chemical catalysis was rate limiting. This result is consistent with studies on the monofunctional DHFR, which have shown that at higher pH the hydride transfer reaction limits the reaction.[84] These two experiments provide two important pieces of kinetic information: (1) the rate of H_2folate formation at the TS site and (2) the rate of reaction of H_2folate from solution at the DHFR site. The next experiment was to measure the rate of formation of H_4folate in the bifunctional TS-DHFR reaction using the radiolabeled CH_2H_4folate substrate as noted by pathway 3 in Scheme I. Under these conditions, *no* H_2folate was observed as an intermediate, even though it was established previously that the rate of reaction of H_2folate from solution is rather slow. The rate of H_4folate formation from CH_2H_4folate (the TS-DHFR reaction) is similar to that measured for the TS reaction. Because the TS and TS-DHFR reaction rates are comparable, this cannot be accounted for if the H_2folate had dissociated from the TS site and rebound to the DHFR site because H_2folate reacts from solution

[84] C. Fierke, K. Johnson, and S. Benkovic, *Biochemistry* **26**, 4085 (1987).

at an observed rate of 14 sec^{-1}. These reaction rates were verified using stopped-flow fluorescence methods. Coenzyme fluorescence energy transfer experiments were used to measure the rate of H_4folate formation and results were consistent with those determined by rapid chemical quench methods.[85,86] Modeling of the rates of each of these reactions using the KINSIM enzyme kinetics simulation program[28,41,87] provided some interesting suggestions regarding the efficient channeling of dihydrofolate. In a single turnover experiment examining the TS-DHFR reaction, no accumulation of H_2folate was observed. The modeling of data indicates that we cannot account for the rate for H_4folate formation and the lack of H_2folate accumulation by simply including a fast rate of transfer (>1000 sec^{-1}) of H_2folate from the TS site to the DHFR site. The reaction kinetics are consistent with a model in which the channeling of H_2folate occurs rapidly and the conversion of H_2folate to H_4folate at the DHFR site is activated by at least an order of magnitude.[85,86] This is best illustrated by Fig. 13. Figure 13 shows the time course for the bifunctional TS-DHFR reaction in which the conversion of radiolabeled CH_2H_4folate to H_4folate was monitored. No H_2folate was detected and the curves shown are predicted according to rate constants in the kinetic model described in Trujillo et al.[82] and Liang and Anderson,[85] which includes a 10-fold activation of DHFR catalysis. The inset in Fig. 13 shows the buildup of H_2folate (dashed line), which is predicted to build up if the H_2folate dissociated from the TS site and rebound at the DHFR site, i.e., two independent catalytic reactions. Accordingly, a very large amount of H_2folate (indicated by the arrow) should be observed along with a lag in the production of H_4folate (dashed line indicated by arrow). These results indicate that TS-DHFR domain–domain interactions are important for the efficient channeling of H_2folate.

The single turnover experiments just described suggest that the rate of DHFR catalysis is enhanced by the TS site, indicating that one of the species involved in TS catalysis may be the chemical switch that triggers this activation. The chemical catalysis of TS involves the formation of a covalent enzyme intermediate that can be trapped using a dead-end inhibitor, 5-fluoro-2'-deoxyuridine monophosphate (FdUMP).[88] Fluorescence studies have indicated that a protein conformational change is associated with the formation of this covalent intermediate.[89] One might hypothesize that the formation of a covalent intermediate at the TS site might be the

[85] P. H. Liang and K. S. Anderson, *Biochemistry* **37,** 12195 (1998).
[86] P. H. Liang and K. S. Anderson, *Biochemistry* **37,** 12206 (1998).
[87] K. S. Anderson, J. A. Sikorski, and K. A. Johnson, *Biochemistry* **27,** 7395 (1988).
[88] D. V. Santi, C. S. McHenry, R. T. Raines, and K. M. Ivanetich, *Biochemistry* **26,** 8606 (1987).
[89] J. Donato, J. Aull, J. Lyon, J. Reinsch, and R. Dunlap, *J. Biol. Chem.* **251,** 1303 (1976).

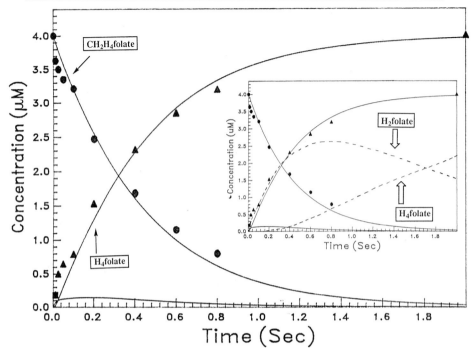

FIG. 13. Single enzyme time course for the TS-DHFR reaction for the bifunctional TS-DHFR. The bifunctional TS-DHFR enzyme (7.5 μM) preincubated with dUMP (500 μM) and NADPH (500 μM) was mixed with [^{3}H]CH$_2$H$_4$folate (4 μM) in the presence of MgCl$_2$ (25 mM) and EDTA (1 mM) at pH 7.8 and 25°. The disappearance of [^{3}H]CH$_2$H$_4$folate (●) and the formation of [^{3}H]H$_4$folate (▲) were monitored. Curves were calculated by the numerical integration using the Kinsim program. (Inset) The amount of H$_2$folate (dashed line, designated by arrow) that would be predicted to build up if the channeling step is removed and the dihydrofolate intermediate dissociates from the TS site and rebinds to the DHFR site. A lag in the formation of H$_4$folate (dashed line, designated by arrow) is also predicted.

species responsible for activating the DHFR reaction. This was tested by measuring the rate of the DHFR reaction in the presence of CH$_2$H$_4$folate and FdUMP using stopped-flow fluorescence methods. Under these conditions the rate of the DHFR is enhanced to 120 sec^{-1} (data not shown). The change in protein conformation that accompanies the covalent intermediate formation during TS chemical catalysis enables DHFR to accept the channeled H$_2$folate and to convert it to H$_4$folate at a much faster rate. The converse experiment was examined in which the rate of formation of covalent intermediate from FdUMP and CH$_2$H$_4$folate was determined by stopped-flow fluorescence. The rate in the absence and presence of NADPH indicates a substantial rate enhancement when the DHFR site is occupied by NADPH and implies reciprocal communication. These results provide direct evidence for the activation of DHFR chemical catalysis by the TS active site as well as domain–domain interaction.

An additional experiment provides corroborative evidence for substrate channeling as illustrated in Scheme II. The underlying principle is similar to an isotope dilution experiment except that measurements are made under single turnover conditions to directly monitor catalysis and partitioning of the intermediate, H₂folate, between the active sites.

In this straightforward pulse-chase type experiment, the enzyme is preincubated with radiolabeled CH₂H₄folate* and NADPH and is then mixed with an excess of cold dUMP and H₂folate. The rationale is as follows: if radiolabeled H₄folate* is formed, then the intermediate H₂folate must have been channeled because if it dissociated from the enzyme it would have been diluted out into the pool of cold excess H₂folate, preventing the observation of radiolabeled H₄folate*. A single turnover experiment was performed in which the bifunctional TS-DHFR enzyme was preincubated with a limiting amount of radiolabeled CH₂H₄folate* and then mixed with an excess of dUMP and unlabeled H₂folate. Indeed, a substantial amount of radiolabeled H₄folate* was formed and no H₂folate accumulated over a period from 10 msec to 8 sec, during which time the CH₂H₄folate* was completely consumed (data not shown),[85,86] thus implying a direct transfer of H₂folate between TS and DHFR active sites. Care should be taken, however, in relying solely on this type of pulse chase experiment as definitive evidence for substrate channeling as the results were more complicated when a similar experiment was conducted in earlier studies with tryptophan synthase.[41] However, this pulse-chase experiment, in combination with the rapid chemical quench and stopped-flow experiments described earlier, provides direct evidence for both substrate channeling in the TS-DHFR enzyme and domain–domain communication.

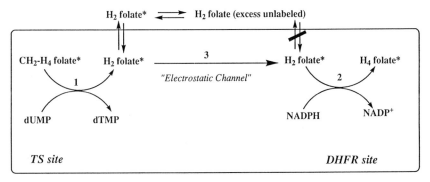

SCHEME II. A pulse-chase experiment to test the channeling behavior of TS-DHFR.

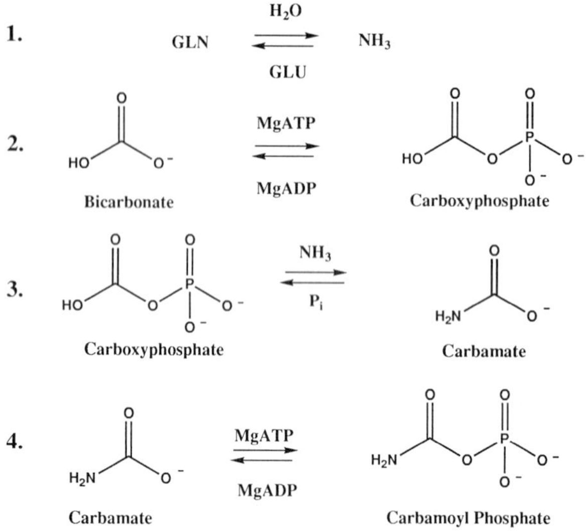

1. $\quad$ GLN $\underset{\text{GLU}}{\overset{\text{H}_2\text{O}}{\rightleftharpoons}}$ NH$_3$

2. Bicarbonate $\underset{\text{MgADP}}{\overset{\text{MgATP}}{\rightleftharpoons}}$ Carboxyphosphate

3. Carboxyphosphate $\underset{\text{P}_i}{\overset{\text{NH}_3}{\rightleftharpoons}}$ Carbamate

4. Carbamate $\underset{\text{MgADP}}{\overset{\text{MgATP}}{\rightleftharpoons}}$ Carbamoyl Phosphate

SCHEME III. Catalytic activities of carbamoyl phosphate synthase.

Carbamoyl Phosphate Synthase

A third enzyme, carbamoyl phosphate synthetase (CPS), has also been suggested to employ substrate channeling. CPS catalyzes the production of carbamoyl phosphate from bicarbonate, glutamine, and two molecules of MgATP. The carbamoyl phosphate is a key biosynthetic intermediate for the formation of pyrimidine nucleotides as well as the urea cycle. Mechanistic studies indicate that four individual catalytic activities are required for the ultimate formation of the carbamoyl phosphate product as illustrated in Scheme III.[90]

In these cases there are obvious advantages for efficient channeling as three discrete, highly reactive, and unstable intermediates—carboxylphosphate, carbamate, and ammonia—may be produced at various stages in the overall catalysis. Several lines of evidence support the existence of these intermediates.[91,92] The CPS crystal structure in the presence of ligands has been solved, which provides a structural basis for substrate channeling.[93] The enzyme contains a small subunit and a large subunit.

Glutamine amidotransferase catalytic activity is associated with the small subunit whereas ATP-dependent phosphorylations of bicarbonate

[90] P. Anderson and A. Meister, *Biochemistry* **35**, 3164 (1996).
[91] F. Raushel, P. Anderson, and J. Villafranca, *Biochemistry* **17**, 5587 (1979).
[92] F. Raushel and J. Villafranca, *Biochemistry* **18**, 3434 (1978).
[93] J. Thoden, H. Holden, G. Wesenberg, F. Raushel, and I. Rayment, *Biochemistry* **36**, 6305 (1997).

and carbamate occur on the large subunit. The large subunit also contains a binding site for ornithine that is an allosteric effector of the catalytic activity. A linear distance of nearly 100 Å separates the glutamine-binding site in the small subunit and the site for carbamoyl phosphate synthesis in the large subunit. Three active sites for glutamine amidotransferase, carboxyphosphate synthetase, and carbamoyl phosphate synthetase activities were identified. Examination of the CPS structure suggests a pathway involving a contour length of >96 Å by which enzymatic intermediates may pass from the small subunit to the ultimate carbamoyl phosphate synthetase active site as illustrated in Fig. 14.[93] Although there is no direct evidence for substrate channeling, the availability of the three-dimensional structure and biochemical studies comparing wild-type and mutant enzymes indicate a mechanism in which two unstable intermediates may be channeled. Accordingly, the ammonia formed from glutamine hydrolysis at the

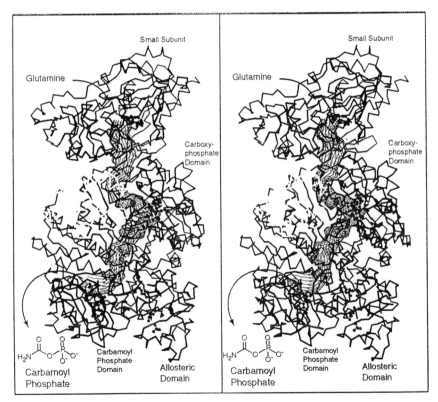

FIG. 14. Putative channel connecting the three active sites in CPS. Reprinted with permission from J. Thoden *et al., Biochemistry* **36,** 6305 (1997). Copyright (1997) American Chemical Society.

small subunit (reaction 1, Scheme III) would be channeled directly to the carboxyphosphate domain of the large subunit. The phosphorylation of bicarbonate at this domain produces carboxyphosphate (reaction 2, Scheme III), which then reacts with the ammonia to produce carbamate (reaction 3, Scheme III). The carbamate could then be channeled to the carbamoyl phosphate domain, also located on the large subunit, for the final phosphorylation by ATP to produce the ultimate product, carbamoyl phosphate (reaction 4, Scheme III).

A detailed kinetic analysis of the reaction pathway and substrate channeling in CPS is currently not available. However, kinetic studies indicate rate enhancements for the complete reaction as compared with the half-reactions, thereby suggesting that there is physical coupling of the chemical reactions at each active site that may be mediated by domain–domain interaction.[94] The modulation of catalytic activity by remote active site occupation is also observed with tryptophan synthase and TS-DHFR and appears to be a more general feature of enzymes involved in substrate channeling.

Malate Dehydrogenase–Citrate Synthase

An example of a sequential pair of enzymes in the Krebs TCA cycle suggested to exhibit substrate channeling are malate dehydrogenase (MDH) and citrate synthase (CS). The sequential enzyme reactions involve the conversion of malate to oxaloacetate (OAA) by malate dehydrogenase and the subsequent conversion of the intermediate OAA to citrate by citrate synthase. A bifunctional form of the enzyme (MDH1/CS1) was produced by preparing a fusion protein from the mitochondrial yeast enzyme in which the CS protein was attached by a short linker to the MDH protein.[95] The catalytic activities of the free enzymes were comparable to that of the fusion protein from steady-state experiments. The lag (transient) time, τ, for the formation of the final products, MDH and NADH, in the fusion protein was shorter than the free enzyme or the theoretical estimate based on the ratio of V_{max} and K_m for the CS part of the coupled reaction.[24] It was also observed that when aspartate aminotransferase was added to serve as a competitor for the OAA intermediate in the coupled MDH/CS reaction that the fusion protein was more efficient at sequestering OAA as compared with the free forms of the enzyme.[95] The crystal structures of the yeast enzymes are not available; however, the pig heart MDH and CS

---

[94] S. Miran, S. Chang, and F. Raushel, *Biochemistry* **30,** 7901 (1991).
[95] C. Lindbladh, M. Rault, C. Hagglund, W. C. Small, K. Mosbach, L. Bulow, C. Evans, and P. A. Srere, *Biochemistry* **33,** 11692 (1994).

enzymes that are homologous are available. Based these structures, a model of the MDH1/CS fusion protein has been built in which the CS and MDH structures are docked such that the carboxy termini of the CS subunit are adjacent to the N termini of the CS.[96] In this model of the fusion protein the CS and MDH active sites are almost 60 Å apart. Brownian dynamics simulations using this model of the fusion protein indicate that an electrostatically based substrate channeling mechanism may be rather efficient.[96] This mechanism may be similar to the electrostatic channeling of dihydrofolate proposed for the bifunctional TS-DHFR enzyme.

Cyclosporin Synthetase and FK506 Synthetase

One of the ultimate challenges for understanding substrate channeling in multifunctional enzyme complexes is provided by the enzymes in some fungi responsible for the biosynthesis of the secondary macrocyclic metabolites such as cyclosporin A (CsA) and FK506. The discovery of the immunosuppressive activities of these two compounds has revolutionized the field of human transplantation surgery.[97]

Cyclosporin A is a cyclic undecapeptide produced by the enzyme cyclosporin synthetase.[98,99] CsA contains unusual amino acids such as D-alanine, $(4R)$-4-[(E)-2-butenyl]-4-methyl-L-threonine (Bmt), and L-2-aminobutryic acid. In order to complete the biosynthesis, the enzyme catalyzes at least 40 reaction steps in an "assembly belt" type mechanism. It activates all constituent amino acids of CsA to thioesters via aminoacyladenylates and carries out specific N-methylation reactions.[100,101] Amazingly, these catalytic activities reside on a single multifunctional polypeptide having a molecular mass of 1400 kDa.

FK506 is a 23 membered macrocyclic polyketide produced by several species of the fungus *Streptomyces*. A multifunctional polypeptide synthase related to fatty acid synthases has been indicated in the biosynthesis of FK506.[97,102] Although a detailed structural and kinetic understanding of these two enzymes is not available, one might speculate that the channeling of intermediates from one reactive center to another with the multifunc-

[96] A. H. Elcock and J. A. McCammon, *Biochemistry* **35,** 12652 (1996).

[97] M. A. Navia, *Curr. Opin. Struct. Biol.* **6,** 838 (1996).

[98] A. Lawen and R. Zocher, *J. Biol. Chem.* **265,** 11355 (1990).

[99] J. Dittmann, R. M. Wenger, H. Kleinkauf, and A. Lawen, *J. Biol. Chem.* **269,** 2841 (1994).

[100] A. Lawen, R. Traber, R. Reuille, and M. Ponelle, *Biochem. J.* **300,** 395 (1994).

[101] G. Weber, K. Schorgendorfer, E. Schneider-Scherzer, and E. Leitner, *Curr. Genet.* **26,** 120 (1994).

[102] H. Motamedi, S. J. Cai, A. Shafiee, and K. O. Elliston, *Eur. J. Biochem.* **244,** 74 (1997).

tional polypeptide could be crucial for the efficient biosynthesis of these compounds.

General Features of Enzyme Complexes That Channel Intermediates

Compelling structural and kinetic evidence exists for the controlled channeling of intermediates between two or more active sites of an enzyme. The structures of TS, TS-DHFR, and CPS reveal that the "channels" are well defined and can either be tunnels through the protein interior or pathways along the surface of the enzyme. Protein residues that line these pathways are representative of the three fundamental interactions known to stablize proteins and protein–ligand interactions: hydrophobic (tryptophan synthase), charged (TS-DHFR), or hydrophilic (CPS). The nature of the residues that line the channel appear instead to be related directly to the physiochemical nature of the intermediate that must be transported, implying that the channel indeed has a functional role, i.e., hydophobic intermediates would travel through hydrophobic tunnels. Although the pathways clearly aid in guiding and facilitating the diffusion, they appear to be far more than simple conduits for intermediate transfer. The channels not only act to direct the complex movement of the intermediates, but also aid in the signaling process between the active sites. The "gating" role of the channel helps maintain the precise synchrony of the reactions at multiple active sites in order to minimize the accumulation of the intermediate. It is becoming clear that the binding of substrate at one of the active sites can apparently signal (toggle-on) or activate the second site before progressing with the rapid channeling of the intermediate. The concept of the active site operating as both a catalytic site and an allosteric site regulating a second site appears to be operating in these channeling enzymes. While there is much that is still not understood regarding the nature of active site communication, it appears most likely to be mediated by standard protein conformational changes, including loop movements and reorganizations, subunit rotations, and secondary structure translations as well as the repositioning of critical protein side chains. Why has nature gone to so much trouble to finely tune these reactions and provide for channeling of intermediates? There are certainly many answers, but clearly control is paramount. The precise synchrony between active sites and the ability to channel between active sites allow for the overall reaction to be well regulated and prevent the buildup of potentially reactive, toxic, or promiscuous intermediates to be released into the cytosol. In a simplistic sense, channeling acts as a molecular compartment and provides for enhanced catalytic efficiency by providing precise active-site communication. While more rigorous experiments are required to confirm the existence of channeling in other enzyme systems, circumstantial evidence is growing that channeling is a general feature of biochemistry and is of fundamental importance to the overall control and catalytic efficiency of multifunctional enzymes and enzyme complexes.

Acknowledgments

The author thanks Dr. Ilme Schlichting and Dr. Thomas Schneider for the preparation of Fig. 8, Dr. Adrian Elcock for the preparation of Fig. 12, and NIH for support of GM45343.

Section II

Intermediates and Complexes in Catalysis

[7] Intermediates and Energetics in Pyruvate Phosphate Dikinase

By DEBRA DUNAWAY-MARIANO

Introduction

Steady-state initial velocity kinetic techniques emerged from the 1960s as a powerful tool for the study of enzyme mechanisms, particularly when used in combination with isotope exchange techniques. For multiple substrate enzymes, intersecting double reciprocal plots (1/v vs 1/[S] at changing fixed [cosubstrate]) were found to be diagnostic of the sequential addition of substrates and release of products while the parallel pattern observed for the ping-pong kinetic mechanism signaled an intervening product release step and thus the formation of a kinetically stable reaction intermediate. Dead-end, product, and substrate inhibition studies became useful applications of the steady-state initial velocity method to determine the order of substrate binding and product release and to identify dead-end complexes formed. The FORTRAN computer programs published in 1979 by Cleland[1] provided a convenient method to fit steady-state rate equations to initial velocity kinetic data and hence to determine the steady-state constants k_{cat}, K_m, and K_i for the enzymatic reaction. Systematic study of the variation of the steady-state kinetic constants with heavy isotope labeling of the substrate or solvent provided a method to probe rate-limiting reaction steps and transition states, whereas variation of the kinetic constants with reaction solution pH allowed one to identify ionizing groups of the substrate or enzyme active site important for binding and catalytic turnover. Finally, the 1980s issued in site-directed mutagenesis as a mechanistic tool. Analysis of steady-state kinetic constants for mutated enzymes became the essential diagnostic tool for the execution of enzyme structure–function studies.

As our understanding of enzyme catalysis has dramatically increased over recent years, so has our desire to probe the individual steps of catalytic turnover. Steady-state kinetic techniques alone no longer satisfy the needs of the modern mechanistic enzymologist. Fast kinetic techniques that allow the experimentalist to measure the rates of binding steps and individual chemical steps have become the focus of the 1990s. Because of the skills and foresight of several enzyme kineticists, the necessary equipment and

---

[1] W. W. Cleland, *Methods Enzymol.* **63,** 84 (1979).

Copyright © 1999 by Academic Press
All rights of reproduction in any form reserved.
0076-6879/99 $30.00

methodology for the measurement of individual steps in enzyme catalysis are widely available. For details, the reader is directed to articles published earlier in this series.[2,3] In general, rapid quench and stopped-flow absorption/fluorescence techniques are used to monitor single turnover reactions on the enzyme. The dead time of the instruments in current use is minimally ca. 2 msec and thus, in practice, rate measurements must conform to a 1000-sec^{-1} limit in order to catch the enzyme before it has completed the act of a single turnover (however, as the minimal sample size needed is decreased with the advent of more sensitive detection methods, the dead time of the instrument will be shortened). Fortunately, for many enzymes the rate constants for substrate and product binding and release and/or for one or more of the chemical steps are within the range that can measured using current technology. Moreover, because enzymes have evolved to bind and stabilize the reaction intermediates tightly, these intermediates can often be detected as they accumulate in the active site of the enzyme during the course of a single turnover. Simple adjustments made in the reaction solution pH or temperature or alterations made in substrate, cofactor, or enzyme structure can, in many cases, facilitate the detection of a reaction intermediate. Single turnover reactions require large quantities of enzyme. If, however, the encoding gene is cloned and overexpressed, the amount of enzyme needed for these experiments is easily obtained.

Thus, in principle it is now possible to identify the chemical pathway and to measure the full set of kinetic constants for catalysis by many enzymes of mechanistic or biological interest. The constants permit construction of reaction coordinate diagrams, the blueprints to enzymatic reaction energetics. The purpose of this article is to show, using pyruvate phosphate dikinase (PPDK) as the example, how transient kinetic techniques (in this case, rapid quench techniques) can be used in conjunction with classical steady-state initial velocity kinetic techniques and isotope exchange and pulse-chase techniques to characterize these features. It is organized into three sections. The first section describes how the chemical pathway of PPDK catalysis was determined, the second section describes how the kinetic mechanism and rate constants were defined, and the third section examines how the Gibbs free energy profile for the reaction was constructed.

[2] C. A. Fierke and G. G. Hames, *Methods Enzymol.* **249**, 3 (1995).
[3] K. A. Johnson, *Methods Enzymol.* **249**, 38 (1995).

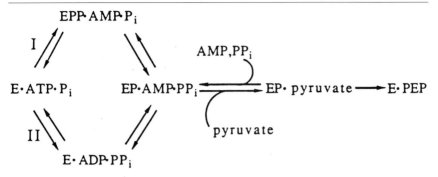

Elucidation of the Chemical Pathway of Pyruvate Phosphate Dikinase
  Catalysis Using Rapid Quench Techniques

The Test Case

In the presence of divalent and monovalent cations (e.g., Mg^{2+} and
NH_4^+), PPDK catalyzes the interconversion of adenosine 5'-triphosphate
(ATP), orthophosphate (P_i), and pyruvate with adenosine 5'-monophos-
phate (AMP), pyrophosphate (PP_i), and phosphoenol pyruvate (PEP).[4]
During catalytic turnover, the β-P of ATP is transferred to pyruvate and
the γ-P of ATP is transferred to P_i. A priori, two reasonable chemical
pathways can be proposed to account for the regiochemistry of the reaction.
Pathway I of Scheme 1 involves the transfer of the β-P of ATP to an
enzyme group to form a pyrophosphorylated enzyme intermediate (E-PP)
and AMP, followed by attack of the P_i at the γ-P of the E-PP to form a
phosphorylated enzyme intermediate (E-P) and PP_i, and, last, attack of
pyruvate at the β-P of the E-P intermediate to form free enzyme (E) and
PEP. Pathway 2, however, involves the initial attack of P_i at the γ-P of
ATP to form adenosine 5'-diphosphate (ADP) and PP_i, followed by transfer
of the β-P of the ADP to an enzyme group to form a phosphorylated
enzyme intermediate (E-P) and AMP, and, last, transfer of the phosphoryl
group from the E-P to pyruvate. Both pathways are consistent with the
observed stereochemistries of the phosphoryl transfers (viz. retention at β-P
and inversion at γ-P).[5] Steady-state kinetic analysis and isotope exchange
studies indicated a Bi(ATP, P_i) Bi(AMP, PP_i) Uni(pyruvate) Uni(PEP)

---

[4] H. G. Wood, W. E. O'Brien, and G. Michaels, *Adv. Enzymol. Relat. Areas Mol. Biol.* **45**,
   85 (1977).
[5] A. G. Cook and J. R. Knowles, *Biochemistry* **24**, 51 (1985).

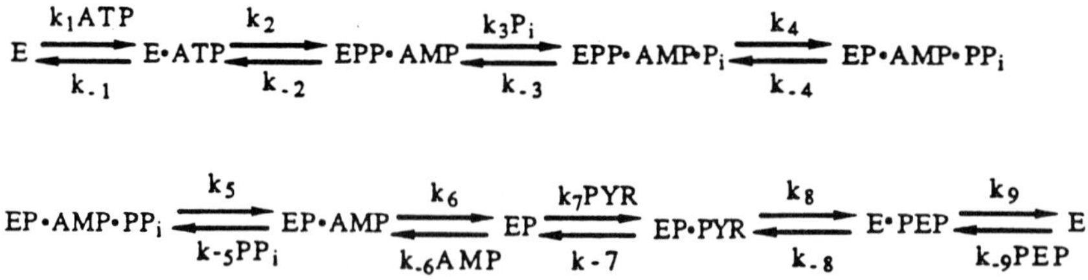

$$E \underset{k_{-1}}{\overset{k_1 ATP}{\rightleftharpoons}} E \cdot ATP \underset{k_{-2}}{\overset{k_2}{\rightleftharpoons}} EPP \cdot AMP \underset{k_{-3}}{\overset{k_3 P_i}{\rightleftharpoons}} EPP \cdot AMP \cdot P_i \underset{k_{-4}}{\overset{k_4}{\rightleftharpoons}} EP \cdot AMP \cdot PP_i$$

$$EP \cdot AMP \cdot PP_i \underset{k_{-5}PP_i}{\overset{k_5}{\rightleftharpoons}} EP \cdot AMP \underset{k_{-6}AMP}{\overset{k_6}{\rightleftharpoons}} EP \underset{k_{-7}}{\overset{k_7 PYR}{\rightleftharpoons}} EP \cdot PYR \underset{k_{-8}}{\overset{k_8}{\rightleftharpoons}} E \cdot PEP \underset{k_{-9}PEP}{\overset{k_9}{\rightleftharpoons}} E$$

SCHEME 2

kinetic mechanism (see Scheme 2), which, in conjunction with the demonstration of enzyme phosphorylation with ATP/P_i or with PEP alone, proved the intermediacy of E-P.[6] The question of how E-P is formed, via pathway I or pathway II remained, however. To distinguish between the two pathways, ^{32}P substrate and rapid quench techniques were used to detect the reaction intermediate (ADP vs E-PP) formed during a single turnover on the enzyme.[7] According to the known regiochemistry of the reaction, the ^{32}P label in [β-^{32}P]ATP should appear in the PP_i product and in the E-PP intermediate, whereas the ^{32}P label of [γ-^{32}P]ATP should appear in the product PEP, in the E-P intermediate, and in either the E-PP or the ADP intermediate. Operating in the reverse direction, PPDK should transfer the ^{32}P label from [^{32}P]PEP to ATP via either the [β-^{32}P]ADP intermediate or the ^{32}P-labeled E-PP intermediate.

Design of the Single Turnover Reaction Experiment

The key to finding a reaction intermediate formed in a single turnover resides in defining the right conditions for its formation and detection. The considerations that were made in selecting the reaction conditions used to detect the intermediates of the PPDK-catalyzed reaction are described. The aim is to follow the disappearance of the radiolabeled substrate, the appearance and disappearance of the intermediate, and the appearance of product in the active site of the enzyme as it undergoes a single turnover. The first decision to be made is the selection of the substrate and substituent to radiolabel. Ideally, the radiolabel will carry over from the reactant to the intermediate and product so that all three species can be monitored. In addition, the key to success of the experiment lies in the sensitivity of detection of the radiolabeled species. Thus, the higher the specific activity of the labeled reactant the better.

[6] H.-C. Wang, L. Ciscanik, W. vonder Saal, J. J. Villafranca, and D. Dunaway-Mariano, *Biochemistry* **27**, 625 (1988).

[7] L. J. Carroll, A. F. Mehl, and D. Dunaway-Mariano, *J. Am. Chem. Soc.* **111**, 5965 (1989).

The time course for the single turnover reaction is measured using a rapid quench instrument to mix enzyme/cofactors with the radiolabeled substrate/unlabeled cosubstrate and then to terminate the reaction after a specified time by mixing with the quench solution. The reader may consult an earlier article for discussion of the selection of instrumentation and quench solutions.[3] In general, the quench solution is designed to act instantaneously to stop catalytic turnover. In practice, we have found that acid quenching works well to stop the reaction and to release the ligands from the enzyme active site. Vortexing the acid-quenched solution with carbon tetrachloride efficiently precipitates the enzyme for separation by filtration or centrifugation. For every time point measured in the construction of the time course, a reaction must be initiated, quenched, and analyzed. Because the amount of intermediate that will be formed in the reaction will be significantly less than the amount of enzyme present, there is a requirement for a considerable quantity of enzyme to execute the single turnover experiment. We have found that rapid quench experiments work best with small volumes (e.g., 40 μl of enzyme solution mixed with 40 μl of substrate solution) and thus it is advantageous to work with as high an enzyme concentration as possible to maximize the amount of intermediate to be detected in the quenched reaction. Ideally, the enzyme concentration used is between 10 and 100 μM. The limit encountered at the high end of this range derives from activity loss caused by protein aggregation and/or mixing problems resulting from the high viscosity of the concentrated enzyme solution.

The amount of radiolabeled substrate used in the experiment is selected by the concentration needed to fill a sufficient number of the enzyme active sites with the level needed to produce an acceptable background of radioactivity in the target fractions of the control reaction (see later discussion). The occupancy of the enzyme active sites should be maximized by using a saturating level of radiolabeled substrate. If the K_d is known, a simple calculation will indicate the ligand concentration needed to saturate the enzyme. If the K_d is small, 1 μM or less, a stoichiometric amount of substrate will do. More typically, however, substrate K_d values are in the range of 10 and 200 μM. Selection of the substrate with the lowest K_d usually provides the most suitable radiolabeled probe.

After the reaction is quenched, the radiolabeled components are separated and analyzed by liquid scintillation counting. The radioactivities present in the substrate, product, and intermediate fractions are summed (the sum should nearly equal the total radioactivity present in the reaction sample subjected to the chromatographic separation), and the sum is then used to calculate the mole fraction of the respective species present in the quenched reaction solution. The mole fractions obtained are multiplied by the concentration of the radiolabeled substrate used in the reaction solution to arrive at the concentrations of the unconsumed substrate, the intermediate, and the product present at the time of the quench. Thus, it is critical that extraneous radioactivity in the product and, particularly, in the intermediate fractions be kept to a minimum. Efficient separations of substrate, intermediate, and product will ensure this. Sometimes, however, the radiolabeled substrate contains small quantities of persistent radiolabeled contaminants (formed during the synthesis of the substrate or resulting from its breakdown during storage) that coincidentally cochromatograph with the target compound (viz. the reaction intermediate). Typically, at the outset, the radiolabeled substrate is subjected to the same separation procedure that will be used for separation of the enzymatic reaction to check for purity. Next, the contents of the "control reaction" are analyzed. The control reactions are prepared by manually mixing the enzyme solution with the quench solution and then adding the substrate solution. The control reaction solution is subjected to the same manipulations used in the separation and analysis of the quenched test reaction, and the radioactivity content of the substrate, intermediate, and product fractions is determined. If a significant level of radioactivity is detected in the intermediate/product fraction obtained from the control reaction, then adjustments must be made in the experimental design. Often the problem can be corrected by employing a concentration of radiolabeled substrate that maximizes the fraction of it that will be enzyme bound and at the same time provides enough enzyme–substrate complex to generate a detectable level of intermediate [with PPDK, a high concentration of enzyme (40 μM) was used with comparatively low concentrations of substrate (5–10 μM) in single turnover experiments]. If, however, this adjustment does not effectively deal with the problem, it may be neccessary to alter the choice of radiolabeled substrate to be used in the experiment.

Execution of the Single Turnover Reaction Experiment

The [γ-^{32}P]ATP used to measure the time course for a single turnover of PPDK is available commercially. However, prior to use in the rapid quench experiments it was advantageous to purify the nucleotide on a 1-ml DEAE Sephadex column using 100 ml of a linear gradient of NH_4HCO_3 (0.1 to 0.6 M) as eluant. The [γ-^{32}P]ATP elutes at ca. 0.3 M NH_4HCO_3. The [β-^{32}P]ATP used in the experiments was prepared from the [γ-^{32}P]ATP by first converting it to [β-^{32}P]ADP with adenylate kinase and AMP/Mg^{2+} (purification on a DEAE-cellulose column) and then phosphorylating the [β-^{32}P]ADP with acetylphosphate in the presence of acetate kinase/Mg^{2+} (purification on a DEAE-cellulose column). The [^{32}P]PEP was synthesized using PPDK/Mg^{2+}, NH$_4^+$ to catalyze the reaction of [β-^{32}P]ATP, P$_i$, and pyruvate (inorganic pyrophosphatase was added to the reaction to drive it to completion via catalysis of PP$_i$ hydrolysis) and by using ion-exchange chromatographic techniques for product purification.[7]

Single turnover reactions were carried out in a three-syringe rapid quench instrument equipped with a thermostatically controlled circulator to maintain a reaction temperature of 25°. The first syringe was loaded with enzyme, buffer, and metal cofactors, the second syringe with [γ-^{32}P]ATP, [β-^{32}P]ATP, or [^{32}P]PEP and cosubstrates in buffer, and the third with an acid (0.6 N HCl) quench solution. Typically 40–45 μl of 80 μM PPDK in 50 mM K$^+$ HEPES (pH 7.0)/5 mM MgCl$_2$/20 mM NH$_4$Cl solution was mixed with an equal volume of substrate in 50 mM K$^+$ HEPES (pH 7.0) and then quenched with 164 μl of the acid solution. Next, 100 μl of CCl$_4$ was added to the quenched solution, followed by vigorous mixing with a vortex to induce precipitation of the enzyme. The precipitated enzyme was pelleted by centrifugation. The supernatent (adjusted to pH 7 with 6 M NaOH and frozen) was saved for chromatographic separation while the enzyme pellet was dried on a piece of tissue paper and then dissolved in 0.4 ml of 10 N H$_2$SO$_4$ at 100° prior to its addition to the 1 ml of liquid scintillation fluid used for counting.

The radiolabeled components contained in an aliquot of the supernatant fraction (a 20-μl aliquot containing ca. 200,000 cpm was first combined with 3 μl of a solution containing 100 mM PEP, 1 mM ADP, and 1 mM ATP before injecting) were separated by chromatography on an ODS C-18 analytical HPLC column using a solution of 2.5% triethylamine, 25 mM KH$_2$PO$_4$, and 5% methanol (pH 6.5) as eluant. The column was eluted at a flow rate of 0.5 ml/min for 9 min and then at a flow rate of 1.5 ml/min for the remainder of the chromatography. Under these conditions the P$_i$/PP$_i$ had a retention time of 5.5 min, PEP 7.5 min, ADP 13.9 min, and ATP 17.3 min. The chromatography was monitored using a UV detector. The 0.25-ml fractions were collected using a microfraction collector and then assayed for radioactivity by liquid scintillation counting.

Results

Time courses for the single turnover reactions catalyzed by PPDK activated with Mg^{2+}/NH$_4^+$ at pH 7.0 are shown in Figs. 1–4.[7,8] Figure 1 illustrates the time course for the formation of the [^{32}P]E-P intermediate from the reaction of 1 μM [^{32}P]PEP with 20 μM PPDK in the absence of cosubstrates. The radioactivity associated with the enzyme fraction minus the background

[8] A. F. Mehl, Y. Xu, and D. Dunaway-Mariano, *Biochemistry* **33**, 1093 (1994).

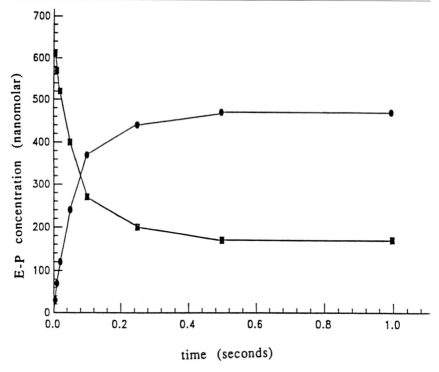

FIG. 1. Time course for the single turnover reaction of PEP in the active site of $Mg^{2+}/$ NH_4^+-activated PPDK at 25°. The reaction contained 1 mM $[^{32}P]$PEP, 20 μM PPDK active sites, 5 mM $MgCl_2$, 20 mM NH_4Cl, and 50 mM K^+ HEPES (pH 7.0): (■) PEP and (●) E-PP. Reprinted with permission from L. J. Carroll *et al.*, *J. Am. Chem. Soc.* **111**, 5965 (1989).

of the control reached a maximum of 40% of the total. Figure 2 shows the time course for the reaction of 4.6 μM $[^{32}P]$PEP, 32 μM PPDK, 20 μM AMP, and 100 μM PP_i. The radioactivity associated with the enzyme fraction (representing E-P and possibly E-PP) reached a maximum level of ca. 40% at 100 msec before declining in concert with the appearance of the $[\beta\text{-}^{32}P]$ATP. $[\beta\text{-}^{32}P]$ADP was not observed in this reaction. The time courses of Figs. 1 and 2 demonstrate clearly the intermediacy of E-P in the overall reaction but do not by themselves distinguish between pathway I and pathway II of Scheme 1.

Figure 3 illustrates the time course for a single turnover of 3.6 μM $[\beta\text{-}^{32}P]$ATP by 32 μM PPDK in the presence of 8.5 mM P_i and 1 mM pyruvate. The level of radiolabeled enzyme minus the background reached a maximum of ca. 15% of the radioactivity originating in the $[\beta\text{-}^{32}P]$ATP. The appearance of the radiolabel in the enzyme fraction precedes the

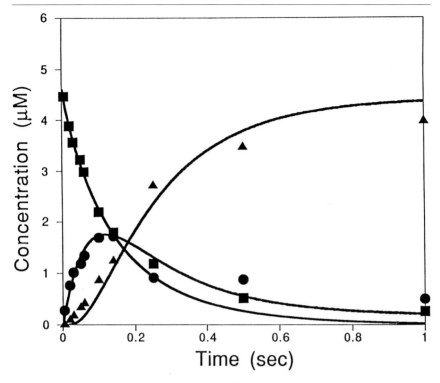

FIG. 2. Time course for a single turnover of [^{32}P]PEP, AMP, and PP$_i$ in the active site of Mg^{2+}/NH$_4^+$-activated PPDK at 25°. The initial reaction mixture contained 4.6 μM [^{32}P]PEP, 32 μM PPDK active sites, 5 mM MgCl$_2$, 10 mM NH$_4$Cl, 20 μM AMP, 100 μM PP$_i$, and 50 mM K$^+$ HEPES (pH 7.0): (▲) ATP, and (■) PEP, and (●) E-P + E-PP. Reprinted with permission from A. F. Mehl *et al.*, *Biochemistry* **33**, 1093 (1994).

appearance of radiolabel in the PEP fraction, thus confirming the intermediacy of E-P. The level of radioactivity detected in the ADP fraction, ranging from 1.5 to 2.2% throughout the reaction, was not in excess of the 2% background determined from the control reactions. Thus, we concluded that ADP is not accumulated during PPDK turnover.

Figure 4 shows the key diagnostic time course for the single turnover of 5.7 μM [γ-^{32}P]ATP with 34 μM PPDK, 4 mM P$_i$, and 1 mM pyruvate. An accumulation of ca. 2% of the radioactivity (minus background) originating in [γ-^{32}P]ATP is observed in the enzyme fraction (in this case only the E-PP species will carry the radiolabel). The observation that the amount of radiolabeled enzyme increased with time (from 0.007 μM at 2.5 msec to 0.07 μM at 15 msec), coincident with the consumption of the [γ-^{32}P]ATP, and then leveled off suggested that we may indeed have found evidence

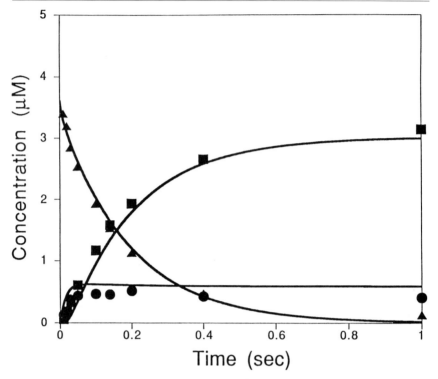

FIG. 3. Time course for a single turnover of $[\beta\text{-}^{32}P]ATP$, P_i, and pyruvate in the active site of Mg^{2+}/NH_4^+-activated PPDK at 25°. The reaction contained 3.6 μM $[\beta\text{-}^{32}P]ATP$, 32 μM PPDK active sites, 3 mM $MgCl_2$, 10 mM NH_4Cl, 8.5 mM P_i, 1 mM pyruvate, and 50 mM K^+ HEPES (pH 7.0): (▲) ATP, (■) PEP, and (●) E-PP + E-P. Reprinted with permission from A. F. Mehl *et al., Biochemistry* **33,** 1093 (1994).

for the E-PP intermediate. As a precaution, a second control study was carried out in which $[U\text{-}^{14}C]ATP$ was reacted with the enzyme to determine how much radiolabel would adhere noncovalently to the enzyme when quenched with acid and precipitated. The amount of nucleotide that remained associated with the enzyme fraction was not dependent on incubation time nor did it exceed the range of 0.008–0.02 μM (0.1–0.5%). Thus, the E-PP species was detected in a small, but seemingly significant amount during catalytic turnover. Moreover, the rate of E-PP appearance was observed to coincide with the loss of $[\gamma\text{-}^{32}P]ATP$, signifying a kinetically competent intermediate.

Because the level of the E-PP intermediate observed during a single turnover of $[\gamma\text{-}^{32}P]ATP$, P_i, and pyruvate was so small, additional experiments were carried out to verify this intermediate. First, the level of

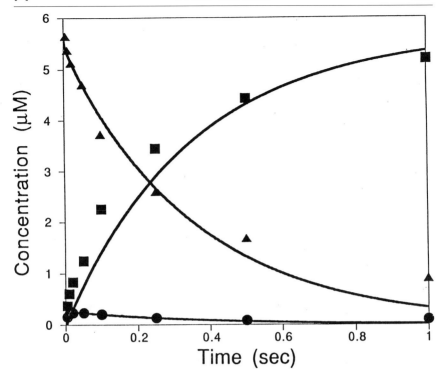

FIG. 4. Time course for a single turnover of $[\gamma\text{-}^{32}P]ATP$, P_i, and pyruvate in the active site of Mg^{2+}/NH_4^+-activated PPDK at 25°. The initial reaction mixture contained 5.7 μM $[\gamma\text{-}^{32}P]ATP$, 34 μM PPDK active sites, 1 mM pyruvate, 4 mM P_i, 2.5 mM $MgCl_2$, 10 mM NH_4Cl, 50 mM K^+ HEPES (pH 7.0): (▲) ATP, (■) PP_i, and (●) E-PP. Reprinted with permission from A. F. Mehl *et al.*, *Biochemistry* **33,** 1093 (1994).

E-PP · AMP formed during the course of a single turnover of $[\gamma\text{-}^{32}P]ATP$ was measured in the presence or absence of the cosubstrate P_i.[9] The results, shown in Fig. 5, demonstrate that the E-PP species is formed in both the presence and the absence of P_i. However, in the absence of P_i, the E-PP persists, whereas in the presence of P_i it reaches a maximum level and then declines rapidly, presumably via reaction to form E-P and $[^{32}P]PP_i$. This observation provides further evidence of the intermediacy of E-PP in the overall reaction.

As mentioned in the Introduction, one can often adjust experimental conditions in such a way to maximize the accumulation of the reaction

[9] S. H. Thrall, A. F. Mehl, L. J. Carroll, and D. Dunaway-Mariano, *Biochemistry* **32,** 1803 (1993).

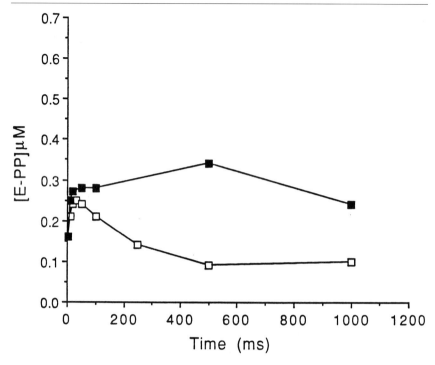

FIG. 5. Time course for a single turnover of $[\gamma\text{-}^{32}P]ATP$ in the active site of Mg^{2+}/NH_4^+-activated PPDK at 25°. The initial reaction mixture contained 6.4 μM $[\gamma\text{-}^{32}P]ATP$, 34 μM PPDK active sites, 1 mM pyruvate, 2.5 mM MgCl$_2$, 10 mM NH$_4$Cl, 50 mM K$^+$ HEPES (pH 7.0), and (■) mM P$_i$ or (□) 3 mM P$_i$. Reprinted with permission from S. H. Thrall *et al.*, *Biochemistry* **32**, 1803 (1993).

intermediate in the enzyme active site during turnover. With PPDK we found[9] that the accumulation of E-PP could be enhanced by switching the divalent cation activator from Mg^{2+} to Co^{2+} or to Mn^{2+}. Figure 6 shows the time course for a single turnover of $[\gamma\text{-}^{32}P]ATP$ measured in the absence and presence of P$_i$ using Mg^{2+}, Co^{2+}, and Mn^{2+} as activators. The high levels of E-PP accumulation seen with these alternate metal ion activators leaves no doubt about the intermediacy of E-PP and hence the chemical pathway of the PPDK reaction.

Elucidation of Gibbs Free Energy Profile of Pyruvate Phosphate Catalysis Using Kinetic and Thermodynamic Techniques

Determination of the free energy profile for PPDK catalysis required the determination of the rate constants for the forward and reverse directions of

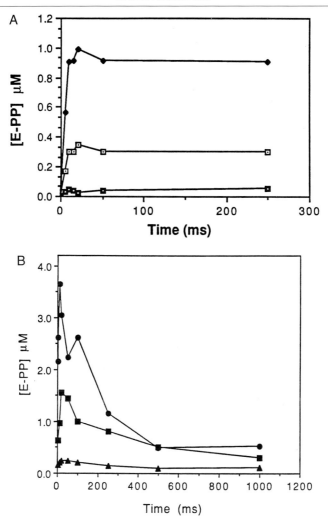

FIG. 6. (A) Time course for a single turnover of $[\gamma\text{-}^{32}P]ATP$ in the active site NH_4^+-activated PPDK as a function of divalent cation activator at 25°. The initial reaction mixture contained 1.5 μM $[\gamma\text{-}^{32}P]ATP$, 40 μM PPDK, 10 mM NH_4Cl, 50 mM K^+ HEPES (pH 7.0), and 2.5 mM (□) $MgCl_2$, (▣) $CoCl_2$, or (◆) $MnCl_2$. (B) Time course for a single turnover of $[\gamma\text{-}^{32}P]ATP$, P_i, and pyruvate in the active site NH_4^+-activated PPDK as a function of divalent cation activator at 25°. The initial reaction mixture contained 6.0 μM $[\gamma\text{-}^{32}P]ATP$, 34 μM PPDK, 2 mM P_i, 1 mM pyruvate, 10 mM NH_4Cl, 50 mM K^+ HEPES (pH 7.0), and 2.5 mM (▲) $MgCl_2$, (■) $CoCl_2$, or (●) $MnCl_2$. Reprinted with permission from S. H. Thrall et al., *Biochemistry* **32**, 1803 (1993).

the three chemical steps and of the six binding steps of the Bi Bi Uni Uni reaction sequence represented in Scheme 2 (18 rate constants total). Our general strategy was to first establish the kinetic mechanism of the reaction. We next measured the equilibrium constants for the substrate/product binding steps and for the interconversion of central complexes and then determined estimates of the rate constants for the binding steps. Single turnover time courses for both directions of the full reaction and partial reactions were then measured to provide estimates of the rate constants governing the chemical steps and to generate reaction profiles for fitting by curve simulation. By using the kinetics simulation program KINSIM, the kinetic model of Scheme 2, the experimentally determined thermodynamic and kinetic constants, and estimates of the rate constants for the chemical steps, curves were simulated to single turnover time course data. The fits of the curves to time course data were optimized by iterative adjustments made in the estimated values of the rate constants until a set of rate constants that satisfied all sets of time course data was obtained. From this final set of rate and thermodynamic constants, and the concentrations of substrates and products present, the Gibbs free energy changes associated with each step of catalysis were calculated. From these values the free energy profile for PPDK catalysis was constructed.

Gibbs free energy profiles are useful in illustrating the energetics of enzyme catalysis under both laboratory and physiological conditions. As will be illustrated later, it is easy to see why the PPDK intermediates accumulate more under one set of experimental conditions than another simply by comparing the two respective energy profiles. In addition, the free energy profile of the physiological enzyme reaction can provide insight into the evolution of catalytic function. A perfectly evolved catalyst will operate at the diffusion limit, the consequence of optimally balanced transition-state and ground-state energies.[10] This property is observed with triose phosphate isomerase of the glycolytic pathway, whose goal is one of maximum catalytic efficiency, developed with the benefit of a long time period for evolution. However, what type of energetics do we expect to see for recently evolved enzymes (the products of directed evolution) or enzymes for which incredible speed is not the primary goal of evolution? What might we expect to see for PPDK, which in some organisms must function to produce ATP and in others operates in the opposite direction of catalysis to produce PEP? As we will see later, the key to success for PPDK appears to be in the balance of energy states that provide for efficient flux in either direction.

[10] W. J. Albery and J. R. Knowles, *Biochemistry* **15**, 5631 (1976).

Finally, we note (but do not describe further in this article) that the free energy profile of an enzyme reaction provides a framework for the dissection of the catalytic mechanism via structure–function studies. The results that strategic changes made in substrate or enzyme structure have on catalytic efficiency can frequently be resolved to the level of the individual rate constants. To visualize how a particular structural change has impacted the energetics of catalysis, the free energy profile for the mutant is constructed and overlaid with that of the natural enzyme.

Kinetic Model

Ideally, a single pathway (or at least a highly preferred pathway) for substrate binding and product release exists for the enzyme under investigation. This simplifies the kinetic analysis to be made. PPDK has three substrates, three products, and three stable enzyme forms. A fully random kinetic mechanism would generate too many enzyme forms and associated steps to deal with. Fortunately, the overall reaction can be broken down into two kinetically independent partial reactions: the E + ATP + P_i ↔ E-P + PP_i + AMP reaction and the E-P + pyruvate ↔ E + PEP reaction.[6] The release of PP_i from E-P · AMP · PP_i followed by AMP is strictly ordered, thus limiting the possible enzyme forms that must be considered. However, steady-state kinetic analysis indicated that ATP followed by P_i binding is the preferred but not the required pathway. Nevertheless, we assumed that ATP binding precedes P_i binding as the major pathway, keeping in mind the relative binding rates and ligand concentrations used in the single turnover experiments. A second simplification made was that turnover of the E · ATP complex to E-PP · AMP precedes P_i binding. Although single turnover experiments had shown that E-PP formation from reaction of the enzyme with ATP occurred at the same rate in the presence and absence of P_i,[9] it was not until we compared fits of data using the two separate kinetic models were we convinced that ATP turnover precedes P_i binding. In the final data analysis, the PPDK kinetic model (Scheme 2) that was used to simulate the curves to the single turnover reaction profiles consisted of a single pathway for substrate binding, catalysis, and product release.

Ligand-Binding Constants

The object is to determine both the rate constant for ligand binding (k_{on}) to the principal enzyme form and the rate constant for ligand dissociation (k_{off}) from the principal enzyme form. If the K_d is known, only one of these values must be determined experimentally, otherwise the K_d ($= k_{off}/k_{on}$) can serve as a check of the accuracy of the k_{on} and k_{off} determinations. A particularly convenient method commonly used to monitor enzyme–ligand binding is to measure changes that occur in the intrinsic fluorescence of the enzyme resulting from complexation. Unfortunately, with PPDK no such changes take place and we thus had to turn to a variety of alternate kinetic and thermodynamic methods. The strategy was to define a range for the kinetic or thermodynamic constants by making independent measurements using different experimental techniques. This range would then be used to provide an estimate of the constant for the simulation of progress curves to fit the single turnover time courses.

The Uni(PEP) Uni(pyruvate) partial reaction gave us the opportunity to determine the K_d values of the $E \cdot PEP$ and E-P$\cdot$pyruvate complexes by using equilibrium exchange techniques, where the rate of $[^{14}C]$pyruvate to $[^{14}C]$PEP exchange was measured as a function of pyruvate concentration at different fixed concentrations of PEP.[8] To obtain estimates of binding constants of the remaining ligands, product inhibition constants, competitive inhibition constants, and equilibrium binding constants of inert structural analogs were measured.[8] For instance, equilibrium dialysis techniques were used in conjunction with Mg$[\alpha$-^{32}P]AMPPNP (in the presence and absence of saturating P$_i$) to obtain a K_d by Scatchard analysis. Three independent measurements gave a range of 50–80 μM. The K_i of MgAMPPNP vs MgATP, however, was determined to be 200 μM. These values compare with the K_i of 200 μM for MgATP product inhibition vs AMP. For the simulations we stayed in the range of 50–200 μM for the K_d of the $E \cdot MgATP$ complex.

The k_{on} values for the substrates were determined by measuring the rate of the single turnover reaction of a given substrate under conditions where the binding rate is slower than the rate of catalytic turnover of the enzyme–substrate complex.[8] It was possible to measure the k_{on} of each of the six substrate ligands using this approach because of the relatively low k_{on} and k_{off} values and high k_{cat} values encountered. In general, to set up an experiment to measure k_{on}, the limiting substrate is used at the lowest concentration possible that will still produce enough product in a single turnover to measure at varying conversion. The enzyme, however, is used in at least a 10-fold excess to the limiting substrate such that the rate of substrate binding is governed by the pseudo first-order rate constant defined by the product $k_{binding}$ {E}, where the $k_{binding} = k_{on}k_{cat}/(k_{off} + k_{cat})$ is less than or equal to the k_{on}, depending on the relative values of k_{off} (for substrate release) and k_{cat} (for substrate turnover). To test whether the binding step is truly rate limiting under the reaction conditions set, the reaction rate is determined at double the enzyme concentration, keeping the limiting substrate concentration the same. The observed rate of product formation will double if the substrate-binding step is indeed rate limiting, otherwise the rate will remain the same. The rate constant for PEP binding

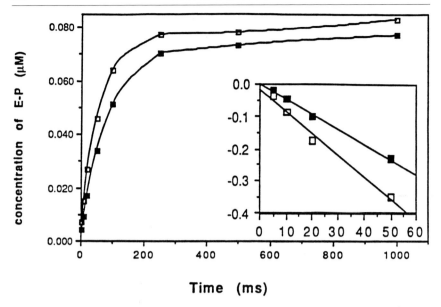

FIG. 7. Time course profile for the formation of [^{32}P]E-P from the reaction of [^{32}P]PEP and Mg^{2+}/NH$_4^+$-activated PPDK during a single turnover (25°). The initial reaction solutions contained 0.11 μM [^{32}P]PEP, 2.5 mM MgCl$_2$, 10 mM NH$_4$Cl, 100 mM, K$^+$ HEPES (pH 7.0), and (■) 18 μM or (□) 36 μM PPDK active sites. Reprinted with permission from A. F. Mehl et al., Biochemistry 33, 1093 (1994).

to Mg^{2+}-activated PPDK was, for example, determined by measuring the time course for the reaction of 0.11 μM [^{32}P]PEP with 18 and 36 μM enzyme (Fig. 7). The observed rates, determined from first-order fits to data, are 12 and 20 sec^{-1}, respectively. The average $k_{binding} = 0.6$ μM^{-1} sec^{-1} was calculated and used as a lower limit for the rate of PEP binding in fitting the reaction profiles with KINSIM. The final k_{on} for PEP that best fit data was 0.75 μM^{-1} sec^{-1} ($k_{off} = 125$ sec^{-1} and $k_{cat} = 300$ sec^{-1}) (Table I).

The k_{off} values for the six PPDK ligands were calculated from the k_{on} and K_d values. A single turnover version of the classic pulse-chase kinetic experiment developed by Rose[11] allowed us to determine experimentally a range for the k_{off} for AMP, an estimate of the k_{on} for AMP, and the K_d for E-P · AMP.[8] If it were technically possible, the same method would have been used to define the range of k_{off} for each of the PPDK substrates. Because a saturating cosubstrate concentration is used to trap limiting radiolabeled substrate on the enzyme, the method was not applicable to the substrates of the Uni(PEP) Uni(pyruvate) partial reaction.

[11] I. A. Rose, Methods Enzymol. 64, 47 (1980).

TABLE I

ESTIMATED AND KINSIM (KINETIC SIMULATIONS)-GENERATED RATE CONSTANTS FOR
Mg^{2+}/NH_4^+-ACTIVATED PPDK[a]

Step	Reaction	k_+ Estimated	k_+ KINSIM	k_- Estimated	k_- KINSIM	K_{eq} Estimated	K_{eq} KINSIM
1	$E + ATP \leftrightarrow E \cdot ATP$	5	4	300	525	50–200	131
2	$E \cdot ATP \leftrightarrow EPP \cdot AMP$		220		1100	0.08	0.2
3	$EPP \cdot AMP + P_i \leftrightarrow EPP \cdot AMP \cdot P_i$	0.03	0.04	100	230	1000–3000	575
4	$EPP \cdot AMP \cdot P_i \leftrightarrow EP \cdot AMP \cdot PP_i$		850		250		3.4
5	$EP \cdot AMP \cdot PP_i \leftrightarrow EP \cdot AMP + PP_i$	8	100	0.3	2.1	30	48
6	$EP \cdot AMP \leftrightarrow EP + AMP$	200	180	5	2.5	40	72
7	$EP + PYR \leftrightarrow EP \cdot PYR$	0.6	1	80	75	14	75
8	$EP \cdot PYR \leftrightarrow E \cdot PEP$		260		300	1	0.87
9	$E \cdot PEP \leftrightarrow E + PEP$	30	125	0.6	0.75	50	170
Full	reaction: $ATP + P_i + PYR \leftrightarrow AMP + PP_i + PEP$ ($k_+ = 5$ sec^{-1}; $k_- = 16$ sec^{-1}; $K_{eq} = 0.003$)						0.006

[a] The reaction sequence refers to the kinetic mechanism shown in Scheme 1. Values are given in μM^{-1} sec^{-1} for substrate on rates and sec^{-1} for substrate off rates. All catalytic steps are in units of sec^{-1}. Equilibrium constants for substrate-binding steps and product release steps are given in units of μM.

Also, the method does not work with a limiting radiolabeled substrate that has a large K_d value, as is the case with P_i ($K_d > 1$ mM), nor does it work for the second substrate (i.e., [^{32}P]PP$_i$) of the ordered Bi(AMP, PP$_i$) Bi(P$_i$, ATP) segment.

For the successful pulse-chase experiment, excess E-P (40 μM) was mixed with [^{14}C]AMP (6 μM) in the rapid quench apparatus and, after a specified number of milliseconds, mixed with a solution containing excess unlabeled AMP (1 mM) and a saturating level of PP$_i$ (5 mM). The solution was quenched within a 3-sec period with acid. The amount of [^{14}C]ATP formed, representing the amount of [^{14}C]AMP trapped on E-P by the PP$_i$, was plotted as a function of the amount of time that the [^{14}C]AMP was allowed to equilibrate with the E-P prior to the addition of the unlabeled AMP and PP$_i$. A first-order fit to data yielded an estimate of $k_{binding} = 8$ μM^{-1} sec^{-1}, which was used to set a minimum for the value of the AMP k_{on}. According to the partition analysis described by Rose,[11] the k_{off} lies between the limits:

$$(k_{cat})K_t/K_m < k_{off} > (k_{cat})K_t/K_m \cdot [\text{E-P} \cdot \text{AMP}]/P\alpha$$

where K_m is the Michaelis constant for PP$_i$ (80 μM), k_{cat} is the steady-state turnover rate of PPDK in the ATP forming direction (16 sec^{-1}), $P\alpha/[\text{E-P} \cdot \text{AMP}]$ is the fraction of AMP trapped at saturating PP$_i$ concentration (0.67), and K_t is the concentration of PP$_i$ required for half-maximal

AMP trapped (450 μM). The latter two values were determined by mixing the PPDK with the [^{14}C]AMP, incubating for 500 msec, and then mixing with the chase containing 1 mM AMP and variable concentrations of PP_i, followed immediately by the acid quench (see the plot of data in Fig. 8). These data place the estimated k_{off} within the range of 90–130 sec^{-1}. The value for k_{off} determined as part of the final simulation is 180 sec^{-1} (Table I).

External and Internal Equilibrium Constants

The equilibrium constant for the PPDK-catalyzed conversion of ATP, P_i, pyruvate with AMP, PP_i, and PEP was determined by using [^{14}C]pyruvate in the equilibration reaction and measuring the ratio of [^{14}C]pyruvate/[^{14}C]PEP present at equilibrium.[8] Highly purified and carefully calibrated samples of substrates were used. The initial reaction contained a catalytic amount of PPDK, 5 mM MgCl$_2$, 20 mM NH$_4$Cl, 0.405 mM [^{14}C]pyruvate, 0.935 mM ATP, and 1.02 mM P$_i$ in 100 mM K$^+$HEPES (pH 7.0, 25°). At equilibrium, 0.093 mM [^{14}C]PEP was present. The concentration of AMP and PP_i was thus 0.093 mM and the concentrations of the ATP, P_i, and

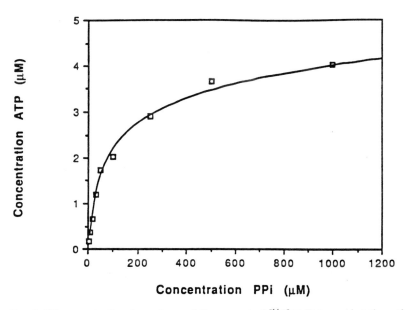

FIG. 8. PP_i concentration dependence of the amount of [^{14}C]AMP trapped at the active site of E-P. The initial solution (80 μl and mixed for 500 msec) contained 6 μM [^{14}C]AMP, 40 μM E-P, 6.3 mM MgCl$_2$, 25 mM NH$_4$Cl, and 50 mM K$^+$ HEPES (pH 7.0, 25°). Reprinted with permission from A. F. Mehl et al., Biochemistry 33, 1093 (1994).

[^{14}C]pyruvate were calculated by subtracting 0.093 from the initial concentrations. The $K_{eq} = 0.003$ was calculated from

$$K_{eq}(obs) = [AMP][PP_i][PEP]/[ATP][P_i][pyruvate]$$

and used to constrain the product of the equilibrium constants in the KINSIM simulations.

An internal equilibrium constant defines the ratio of the rate constants for the forward and reverse directions of a chemical step. Experimentally, it is determined by measuring the ratio of enzyme-bound reactant to enzyme-bound product at equilibrium. For the PPDK partial reaction, the internal equilibrium constant = $[\text{E-PP} \cdot \text{AMP}]/[\text{E} \cdot \text{ATP}]$ was determined from a single turnover time course profile of the reaction of 1 mM [γ-^{32}P]ATP with 40 μM PPDK.[8] The concentration of E-PP present at equilibrium = 3 μM was determined by measuring the ratio of radioactivity present in the precipitated enzyme fraction vs that remaining in the supernatant fraction and multiplying this ratio by the initial [γ-^{32}P]ATP concentration (1 mM). The ratio of $[\text{E-PP} \cdot \text{AMP}]/[\text{E} \cdot \text{ATP}]$ is thus $3/(40 - 3) = 0.08$. We note that the K_d for E $\cdot$ ATP is estimated at 50 μM and, thus, in the presence of 1 mM ATP the enzyme is saturated. In addition, because AMP is not released from the E-PP $\cdot$ AMP complex, we can assume that all of the AMP formed is bound to the enzyme. The simulated internal equilibrium constant determined from the analysis was equal to 0.2.

The internal equilibrium constant = $[\text{E} \cdot \text{PEP}]/[\text{E-P} \cdot \text{pyruvate}]$ was determined in an analogous manner where 40 μM PPDK was equilibrated with an equal, saturating level (1 mM each) of [^{32}P]PEP and pyruvate.[8] The concentration of radiolabeled enzyme formed at equilibrium was found to be 20 μM and hence the internal equilibrium constant is 1.0. The simulated value determined from the analysis is 0.87.

Time Courses for Single Turnover Reactions/Simulations

The time courses used in the simulation of the rate constants for the Mg^{2+}/NH$_4{}^+$-activated, PPDK-catalyzed reaction are shown in Figs. 1–4, presented earlier in the text. Figures 9–11 show analogous single turnover time courses measured for the Co^{2+}/NH$_4{}^+$-activated enzyme to illustrate how the differences in rate/thermodynamic constants resulting from the substitution of the divalent cation activator are reflected in the overall profile. The rate/thermodynamic constants determined from the experimental measurements and the time course simulations are shown in Table I in the case of the Mg^{2+}/NH$_4{}^+$-activated system and in Table II for the Co^{2+}/NH$_4{}^+$-activated system.

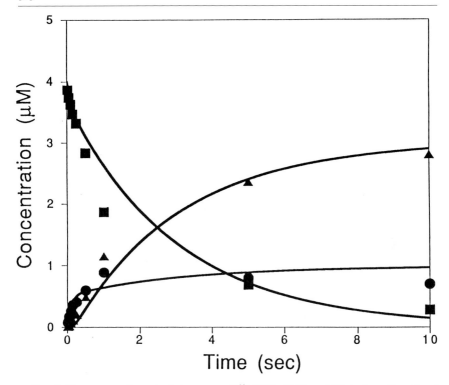

FIG. 9. Time course for a single turnover of [^{32}P]PEP, AMP, and PP$_i$ in the active site of Co^{2+}/NH$_4^+$-activated PPDK at 25°. The initial reaction mixture contained 4.0 μM [^{32}P]PEP, 34 μM PPDK active sites, 2.5 mM CoCl$_2$, 10 mM NH$_4$Cl, 100 μM PP$_i$, 20 μM AMP, and 50 mM K$^+$ HEPES (pH 7.0): (▲) ATP, (■) PEP, and (●) E-P + E-PP. Reprinted with permission from A. F. Mehl et al., Biochemistry **33**, 1093 (1994).

The progress curves drawn to time course data in Figs. 2–4 and 9–11 were simulated using the computer program KINSIM.[12] The general strategy used was to fit the reaction profiles of the individual partial reactions using the estimated values of the rate and thermodynamic constants and to use the values of the rate constants arising from the simulations as starting estimates in the fit to the full reaction.[8] At the outset, it is important to know that the curve fitting is a time-consuming process based on trial and error. The goal is to define the rate constants for the forward and reverse direction of each chemical step. To start, one makes educated guesses at these values (constrained in some cases by the value of the internal equilibrium constant) and uses these along with the other rate

[12] B. A. Barshop, R. F. Wrenn, and C. Frieden, Anal. Biochem. **130**, 134 (1983).

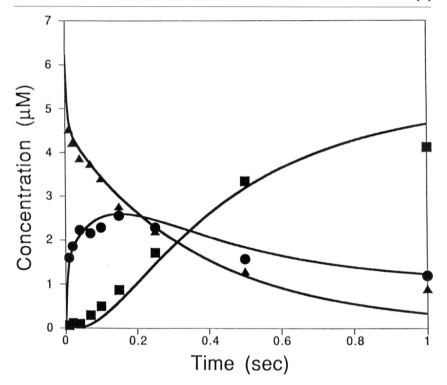

FIG. 10. Time course for a single turnover of $[\beta\text{-}^{32}P]ATP$, P_i, and pyruvate in the active site of Co^{2+}/NH_4^+-activated PPDK at 25°. The reaction mixture contained 6.2 μM $[\beta\text{-}^{32}P]ATP$, 40 μM PPDK active sites, 2.5 mM $CoCl_2$, 10 mM NH_4Cl, 2 mM P_i, 2 mM pyruvate, and 50 mM K^+ HEPES (pH 7.0): (▲) ATP, (■) PEP, and (●) E-PP + E-P. Reprinted with permission from A. F. Mehl *et al., Biochemistry* **33,** 1093 (1994).

and thermodynamic constants, the kinetic scheme, the initial enzyme, and substrate concentrations as inputs to the computer. A progress curve is generated and compared to the experimental data. A decision is made to increase or decrease the value of the rate constant(s), holding all other values constant, and a second generation curve is simulated and overlaid with the time course data. Iterations in the value of one or more rate constant at a time are made until the fit is optimized. The final set of rate constants is evaluated by attempts to simulate progress curves to other single turnover profiles. A good fit is evidence that the rate constants have been well defined. A bad fit means more accurate rate constants are needed.

How accurate are the rate constants derived from trial and error KINSIM fits? Our experience has been that they are as accurate as the experimental data used to define them. The extent to which individual

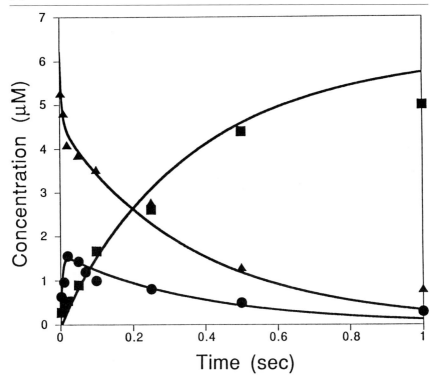

Fig. 11. Time course for a single turnover of [γ-^{32}P]ATP, P$_i$, and pyruvate in the active site of Co^{2+}/NH$_4$$^+$-activated PPDK at 25°. The initial reaction mixture contained 40 μM PPDK active sites, 50 mM K$^+$ HEPES (pH 7.0), 2.5 mM CoCl$_2$, 10 mM, NH$_4$Cl, 6.2 μM [γ-^{32}P]ATP, 2 mM P$_i$, and 2 mM pyruvate: (▲) ATP, (■) PP$_i$, and (●) E-PP. Reprinted with permission from A. F. Mehl *et al., Biochemistry* **33,** 1093 (1994).

rate constants can be varied during the course of the simulation process is relatively small. To illustrate this point, Fig. 12[13] shows the time course for E-P + E-PP formation and consumption in the reaction of E + [^{32}P]PEP + PP$_i$ + AMP ↔ [β-^{32}P]ATP + P$_i$ + pyruvate is overlaid with curves generated using the rate constants listed in Table I. The different curves are based on the use of different values for the rate constants of the E · PEP ↔ E-P · pyruvate step, varied in constant ratio. Note that a threefold change in the values of the two rate constants leads to a substantial deviation in the fit of the simulated curve.

A comparison of the set of single turnover time courses obtained with the Mg^{2+}-activated enzyme with the set obtained with the Co^{2+}-activated

[13] A. F. Mehl, Ph.D. dissertaton, University of Maryland (1991).

TABLE II

ESTIMATED AND KINSIM (KINETIC SIMULATIONS)-GENERATED RATE CONSTANTS FOR
Co^{2+}/NH_4^+-ACTIVATED PPDK[a]

Step	Reaction	k_+ Estimated	k_+ KINSIM	k_- Estimated	k_- KINSIM	K_{eq} Estimated	K_{eq} KINSIM
1	E + ATP ↔ E·ATP	5	5	50	35	10	7
2	E·ATP ↔ EPP·AMP		600		1500	0.5	0.40
3	EPP·AMP + P_i ↔ EPP·AMP·P_i	0.006	0.012	10	150	1000–3000	12,500
4	EPP·AMP·P_i ↔ EP·AMP·PP_i		1000		300		4
5	EP·AMP·PP_i ↔ EP·AMP + PP_i	2	50	0.08	2	30	25
6	EP·AMP ↔ EP + AMP	30	60	1	3	30	20
7	EP + PYR ↔ EP·PYR	0.03	0.07	1.5	3	50	43
8	EP·PYR ↔ E·PEP		12		10	1	1.2
9	E·PEP ↔ E + PEP	0.3	20	0.05	0.15	5	130
Full	reaction: ATP + P_i + PYR ↔ AMP + PP_i + PEP (k_+ = 2 sec^{-1}; k_- = 3 sec^{-1}; K_{eq} = 0.01)						0.033

[a] The reaction sequence refers to the kinetic mechanism shown in Scheme 1. Values are given in μM^{-1} sec^{-1} for substrate on rates and sec^{-1} for substrate off rates. All catalytic steps are in units of sec^{-1}. Equilibrium constants for substrate-binding steps and product release steps are given in units of μM.

enzyme illustrates how changes made in the individual rate constants impact the accumulation of the reaction intermediates. For instance, the time course for a single turnover of [^{32}P]PEP, AMP, and PP_i was determined for each activator under essentially identical conditions. The quantity of E-PP + E-P that accumulates with the Mg^{2+}-activated enzyme far exceeds that which accumulates with the Co^{2+} enzyme principally because the rate of E-P formation from PEP is ca 20-fold faster in the Mg^{2+}-activated enzyme. However, the level of E-PP accumulated on the Co^{2+}-activated enzyme in reaction with [γ-^{32}P]ATP, P_i, and pyruvate is significantly greater than that observed with the Mg^{2+}-activated enzyme. An important factor here is the >10-fold decrease observed in the off rate of ATP from the Co^{2+}-activated enzyme, which allows more of the E·ATP complex to partition forward, thus forming the E-PP·AMP complex rather than returning to E.

From Rate Constants to Energy Profile

The conversion of the set of rate/thermodynamic constants into a Gibbs free energy profile for catalysis is a comparatively simple process. For illustration of the procedure we will first look at the construction of the free energy profile of the E + ATP + P_i ↔ E-P·AMP·PP_i partial reaction shown in Fig. 13. The differences in ground state energies and the difference in transition state energies were accentuated by subtracting 10 kcal/mol from all transition-state energies before entering these values on the profile.

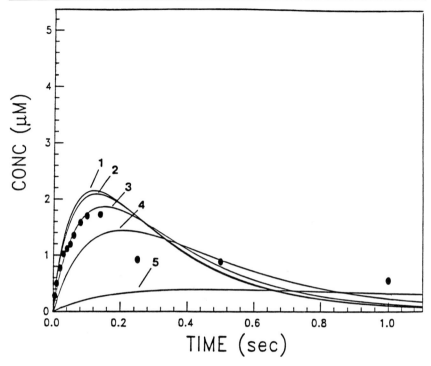

FIG. 12. Time course for the formation of E-P + E-PP (●) from a single turnover of 4.8 μM [^{32}P]PEP, 32 μM PPDK active sites, 20 μM AMP, 100 μM PP$_i$, 5 mM MgCl$_2$, 20 mM NH$_4$Cl, and 50 mM K$^+$ HEPES (pH 7.0, 25°). The curves were generated by simulation (KINSIM) using the kinetic mechanism of Scheme 2 and the rate constants of Table I. The family of curves was generated for a series of forward and reverse rate constants (varied in constant ratio) for the chemical step. E · PEP → E-P · pyruvate: curve 1 (1000 sec^{-1}, 900 sec^{-1}), curve 2 (500 sec^{-1}, 450 sec^{-1}), curve 3 (150 sec^{-1}, 135 sec^{-1}), curve 4 (50 sec^{-1}, 45 sec^{-1}), and curve 5 (5.0 sec^{-1}, 4.5 sec^{-1}). Reprinted with permission from A. F. Mehl, Ph.D. dissertation, University of Maryland (1991).

This profile was constructed for the purpose of illustrating why E-PP accumulates to a much greater extent during turnover of the Co^{2+}-activated enzyme than it does with the Mg^{2+} enzyme under the experimental conditions with 5.9 μM ATP and 2 mM P$_i$.[8] The first step of the process was to adjust the substrate-binding constants (to k_{obs}, K_{obs}) to reflect the concentration of substrate undergoing the binding step. The respective ATP k_{on} (in units of μM^{-1} sec^{-1}) values were multiplied by 5.9 (the dissociation constants are first converted to association constants having units of μM^{-1} and then divided by 5.9). Similarly, values of the k_{on} and the association constant for P$_i$ were multiplied by 2000 μM. The next step was to calculate

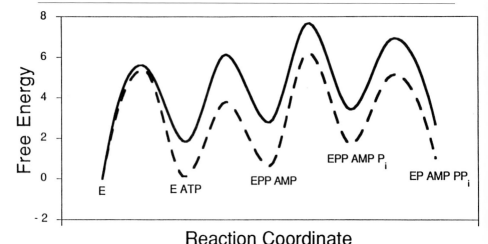

Reaction Coordinate

FIG. 13. Reaction: $P_i \rightarrow$ E-P $\cdot$ AMP $\cdot$ PP$_i$. The free energy change and the apparent free energies of activation for each step were calculated from the constants from Table I (Mg^{2+}) and Table II (Co^{2+}). The concentrations of ATP and PP$_i$ used in the calculations were 5.9 μM and 2 mM, respectively. Ten kcal/mol was subtracted from the energy of activation so that the relative heights of the transition states could be better depicted in the graph. Reprinted with permission from A. F. Mehl *et al., Biochemistry* **33**, 1093 (1994).

the $\Delta G^{\neq} = RT[\ln(kB/Th) - \ln(k)]$ and the $\Delta G = -RT \ln(K_{eq})$ for this step. The calculations were extended to the ensuing steps. By assigning E a free energy value of zero, the $G^{\neq}$ and G for transition-state and ground-states could be plotted. The process of calculation and plotting continued until the end of the reaction coordinate was reached with the E-P $\cdot$ AMP $\cdot$ PP$_i$ complex. The overlay of the free energy profiles obtained with the Mg^{2+}- and Co^{2+}-activated enzymes allows one to visualize the differences in energetics of the two systems that results in a greater accumulation of the E-PP $\cdot$ AMP complex with the Co^{2+}-activated enzyme.

To construct an energy profile that reflects catalysis under physiological conditions, one needs to know the concentrations of the substrate and product ligands under physiological conditions. In the case of PPDK we used the concentrations that had been measured in *Escherichia coli* as an estimate of the values for the host organism, *Clostridium symbiosum*. The Gibbs free energy profiles obtained for the Mg^{2+}- and Co^{2+}-activated PPDK under "physiological conditions" are represented in Fig. 14.[8] The free energy of the E at the end is not zero, which may be disconcerting to some. The reason the energies do not balance at the end is because the contributions from the ligands, ATP in the first instance and PEP in the

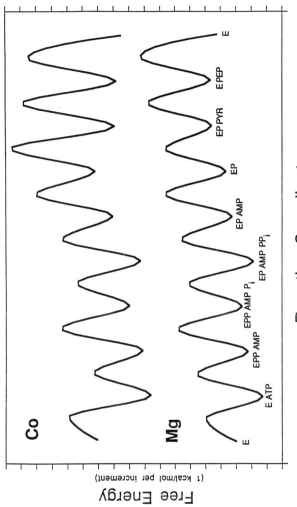

FIG. 14. Reaction-free energy profile for Mg^{2+}/NH_4^+-activated PPDK and Co^{2+}/NH_4^+-activated PPDK. The free energy change and the apparent free energies of activation for each step were calculated from the constants given in Tables I and II. The concentrations of the substrates and products were set equal to the *in vivo* values determined for the bacterium *E. coli* as described in the text. Ten kcal/mol was subtracted from the energy of activation so that the relative heights of the transition states could be better depicted in the graph. Reprinted with permission from A. F. Mehl *et al.*, *Biochemistry* **33**, 1093 (1994).

second, are factored in even though their presence is not implicitly represented in the diagram.

In *C. symbiosum,* PPDK operates in the direction of ATP synthesis, replacing the function of pyruvate kinase. Energetically, the reaction is a gradual downhill slide through balanced transition states and ground states. The ride is substantially "smoother" with the PPDK activated by its physiological divalent cation Mg^{2+} as opposed to our laboratory substitute, Co^{2+}.

By the judicious use of steady-state, pre-steady-state, and equilibrium measurements of substrate and product-binding steps and chemical steps it is possible to resolve all steps in most enzymatic reactions. The detailed reaction energetics forms the basis for understanding the evolution of enzyme catalysis as well as structure–function relationships in enzyme catalysis.

[8] Raman Spectroscopic Studies of the Structures, Energetics, and Bond Distortions of Substrates Bound to Enzymes

By HUA DENG and ROBERT CALLENDER

Vibrational spectroscopy is a powerful tool to study the electronic structure of molecules and the bonds within these molecules. The typical energy range of this spectroscopy is 100–4000 cm^{-1} when expressed in frequency. For comparison, thermal motion, i.e., $kT = 0.58$ kcal/mol at room temperature, is about 204 cm^{-1} in frequency units. Vibrational spectroscopy probes vibrational motions of molecules or molecular bonds ranging from stretch motions of hydrogen bonds in the low end of the spectrum to the stretch motions of various covalent bonds in the medium to high end of the spectrum. Because the resolution of vibrational spectroscopy is on the order of 5 cm^{-1}, band shifts on this order, which correspond to a 0.02-kcal/mol bond energy change or about 0.02% of a typical chemical bond energy, can be detected. For a given chemical bond, the bond energy is directly related with its bond length, which in turn can be correlated empirically with its stretch frequency (cf. references 1 and 2). Therefore, vibrational spectroscopy is well suited for the studies of the chemical bond distortions found in enzymatic catalysis. Although an oversimplification, it can be said that

[1] F. D. Hardcastle and I. E. Wachs, *J. Phys. Chem.* **95,** 5031 (1991).
[2] G. Horvath, J. Illenyi, L. Pusztay, and K. Simon, *Acta Chim. Hung.* **124,** 819 (1987).

Copyright © 1999 by Academic Press
All rights of reproduction in any form reserved.
0076-6879/99 $30.00

the resolution of vibrational spectroscopy picks up where diffraction and multidimensional nuclear magnetic resonance (NMR) techniques leave off, at ca. 0.2 Å, and extends down to much lower lengths.

Although clearly useful as a structural tool, vibrational spectroscopy has been little used for studies of enzymic structure and mechanism until recently because of the problem of spectral crowding. Many vibrational modes contribute to the spectrum of a protein at any given frequency, which results in a spectrum that is very difficult to interpret. Some method must be employed to assign the spectra. For example, it has become standard to perform resonance Raman spectroscopy of chromophores inside proteins where, because of resonance enhancement, the spectrum of the chromophore dominates the observed spectrum (cf. references 3 and 4). Also, there have been many studies involving Fourier transform infrared (FTIR) difference spectroscopy on systems that contain a photolabile chromophore, such as bacteriorhodopsin or the photosynthetic reaction center and some others. Here, protein molecules in films are "switched" from one state to another by light, a feature needed to measure accurately the small differences in the IR spectra (cf. references 5–9). Because both approaches require a "friendly" system, one containing a chromophore, their application is limited to a small set of problems.

It has become apparent that very accurate nonresonance Raman difference spectroscopy has become technically feasible, and it is a general method of probing protein structure.[10–29] In such an experiment, a protein

[3] T. G. Spiro, ed., "Biological Applications of Raman Spectroscopy." Wiley, New York, 1988.

[4] T. G. Spiro, ed., "Biological Applications of Raman Spectroscopy." Wiley, New York, 1987.

[5] J. Breton, E. Nabedryk, and W. W. Parson, *Biochemistry* **31**, 7503 (1992).

[6] W. Mäntele, *Trends Biochem. Sci.* **18**, 197 (1993).

[7] J. Breton and A. Vermeglio, eds., "The Photosynthetic Bacterial Reaction Center II Structure, Spectroscopy, and Dynamics." Plenum Press, New York, 1992.

[8] K. J. Rothschild, *J. Bioenerget. Biomembr.* **24**, 147 (1992).

[9] M. S. Braiman and K. J. Rothschild, *Annu. Rev. Biophys. Biophys. Chem.* **17**, 541 (1988).

[10] R. Callender and H. Deng, *Annu. Rev. Biophys. Biomol. Struct.* **23**, 215 (1994).

[11] R. Callender, H. Deng, D. Sloan, J. Burgner, and T. K. Yue, *Proc. Int. Soc. Opt. Engin.* **1057**, 154 (1989).

[12] D. Chen, K. T. Yue, C. Martin, K. W. Rhee, D. Sloan, and R. Callender, *Biochemistry* **26**, 4776 (1987).

[13] W. L. Peticolas, K. Bajdor, T. W. Patapoff, and K. J. Wilson, *in* "Studies in Physical and Theoretical Chemistry" (J. Stepanek, P. Anzenbacher, and B. Sedlacek, eds.), p. 249. Elsevier, Amsterdam, 1987.

[14] K. T. Yue, H. Deng, and R. Callender, *J. Raman Spec.* **20**, 541 (1989).

[15] M. Kim, H. Owen, and P. R. Carey, *J. Chem. Soc. Faraday Trans.* **93**, 3619 (1997).

[16] D. Manor, G. Weng, H. Deng, S. Cosloy, C. X. Chen, V. Balogh-Nair, K. Delaria, F. Jurnak, and R. H. Callender, *Biochemistry* **30**, 10914 (1991).

[17] H. Deng, J. M. Goldberg, J. F. Kirsch, and R. Callender, *J. Am. Chem. Soc.* **115**, 8869 (1993).

is "tagged" in some way and the difference spectrum between the protein and its modified version measured. For example, a protein spectrum is measured as is that of a protein complexed with ligand. Subtraction of the two yields the spectrum of the bound ligand. Alternatively, isotope editing procedures are very powerful and general. Here, an atom within a bond of interest is labeled with a stable isotope (^{2}H, ^{13}C, ^{15}N, ^{18}O, etc.) and this shifts the frequency of the modes that involve the motion of the labeled atom. Subtraction of labeled and unlabeled protein spectra yields an "isotopically edited" difference spectrum, whereby only those modes involving the labeled atom show up in the spectrum. Water (H_2O or D_2O) generally has a very small Raman cross section so that samples in their biological environment can be studied, and virtually the entire spectrum of a protein or protein/ligand complex can be measured. It is also possible to obtain the spectrum of membrane-bound proteins.[30] Thus, Raman difference spectroscopy can be used on most protein systems, and results are available in a relatively short period of time. Following similar protocols, there has been some success using FTIR difference techniques on small proteins. This work was performed on samples in D_2O in order to move the intense and otherwise masking water absorption band from about 1640 cm^{-1}, the spectral region of interest for these experiments, down to 1200 cm^{-1} in order to study certain protein–ligand interactions.[31-36] This may also develop into

[18] J. M. Goldberg, J. Zheng, H. Deng, Y. Q. Chen, R. Callender, and J. F. Kirsch, *Biochemistry* **32,** 8092 (1993).

[19] H. Deng, J. Burgner, and R. Callender, *J. Am. Chem. Soc.* **114,** 7997 (1992).

[20] H. Deng, J. Zheng, D. Sloan, J. Burgner, and R. Callender, *Biochemistry* **31,** 5085 (1992).

[21] H. Deng, J. Burgner, and R. Callender, *Biochemistry* **30,** 8804 (1991).

[22] H. Deng, J. Zheng, J. Burgner, D. Sloan, and R. Callender, *Biochemistry* **28,** 1525 (1989).

[23] H. Deng, J. Zheng, J. Burgner, and R. Callender, *J. Phys. Chem.* **93,** 4710 (1989).

[24] H. Deng, J. Zheng, J. Burgner, and R. Callender, *Proc. Natl. Acad. Sci. U.S.A.* **86,** 4484 (1989).

[25] H. Deng, D. Manor, G. Weng, C.-X. Chen, V. Balogh-Nair, and R. Callender, *Proc. Int. Soc. Opt. Engin.* **1890,** 114 (1993).

[26] H. Deng, W. J. Ray, J. W. Burgner, and R. Callender, *Biochemistry* **32,** 12984 (1993).

[27] H. Deng and R. Callender, *Comm. Mol. Cell. Biophys.* **8,** 137 (1993).

[28] H. Deng, J. Zheng, A. Clarke, J. J. Holbrook, R. Callender, and J. W. Burgner, *Biochemistry* **33,** 2297 (1994).

[29] H. Deng, A. Y. Chan, C. K. Bagdassarian, B. Estupinan, B. Ganem, R. H. Callender, and V. L. Schramm, *Biochemistry* **35,** 6037 (1996).

[30] T. Spiro, ed., "Biological Applications of Raman Spectroscopy: Raman Spectra and the Conformations of Biological Molecules." Wiley, New York, 1987.

[31] P. J. Tonge, M. Pusztai, A. J. White, C. W. Wharton, and P. R. Carey, *Biochemistry* **30,** 4790 (1991).

[32] P. J. Tonge and P. R. Carey, *in* "Biomolecular Spectroscopy" (R. J. H. Clarke and R. E. Hester, eds.), p. 129. Wiley, New York, 1993.

a general tool, particularly if some technique can be employed to overcome the problem of the strong water (H_2O or D_2O) absorption.

It is clear that the techniques of vibrational spectroscopy can now yield interpretable spectra of specific molecular moieties and particular bonds within protein molecules. Reviews cited earlier discuss resonance Raman and FTIR difference techniques. This article describes briefly Raman difference spectroscopy, this being the newest and most general technique. It then summarizes how the results from a few selected enzymatic groups and systems have been interpreted in order to illustrate the types of questions that may be answered with this structural probe.

Raman Difference Spectroscopy

In Raman spectroscopy, when light of a certain frequency (say ν_L because the light is usually produced by a laser) irradiates a sample, a (small) portion is scattered from the sample, frequency shifted by an amount that corresponds to the frequency of a particular vibrational mode, ν_O. Thus, photons are emitted from the sample with a frequency of $\nu_L-\nu_O$. This is measured by a spectrometer, and because the frequency of the laser is known, the frequency of the vibrational mode may be determined. The universally used unit of the frequency is wave numbers (cm^{-1}), which is the reciprocal of the wavelength of ν_O expressed in centimeters. The spectrum that is plotted is intensity on the y axis (usually in arbitrary units but proportional to the number of detected photons) versus frequency. Each peak in this plot corresponds to the frequency of a vibrational mode. The characteristic time scale associated with vibrational spectroscopy is on the order of the vibrational motion (ca. 10^{-13} sec) or faster so that signal size and its characteristics are not affected by slow motions of the protein, such as the tumbling or rotational motions, that affect NMR spectra. Thus, large proteins can be studied quite easily.

In our work, we adapted a conventional Raman spectrometer to perform difference spectroscopy. The essential feature of this is that the two parent spectra must be very accurate so that small differences between them may be seen without error in the difference spectrum (see references 10, 37,

[33] P. K. Slaich, W. U. Primrose, D. Robinson, K. Drabble, C. W. Wharton, A. J. White, and G. C. K. Robert, *Biochem. J.* **288,** 167 (1992).

[34] A. J. White and C. W. Wharton, *Biochem. J.* **270,** 627 (1990).

[35] A. J. White, K. Drabble, S. Ward, and C. W. Wharton, *Biochem. J.* **287,** 317 (1992).

[36] J. E. Baenziger, K. W. Miller, and K. J. Rothschild, *Biophys. J.* **61,** 983 (1992).

[37] W. Kiefer, *Appl. Spect.* **27,** 253 (1973).

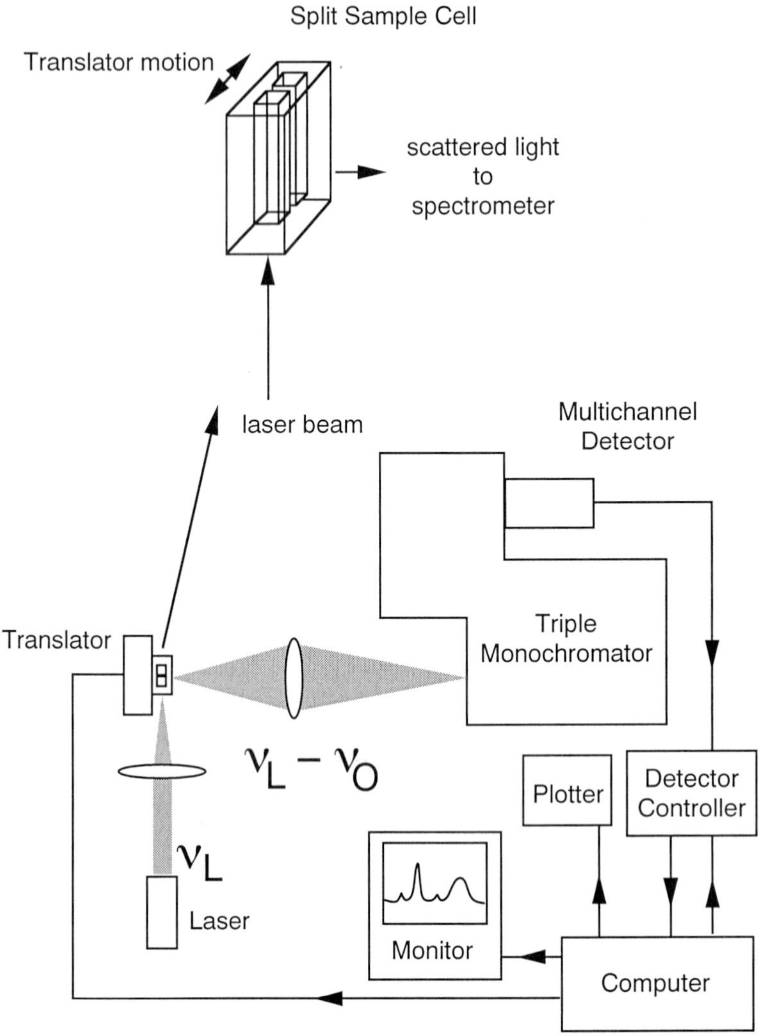

FIG. 1. A schematic of the Raman difference instrument. Laser light is incident on a specially fabricated split cell from the bottom. The Raman-scattered light at ν_L is frequency shifted from the incoming laser light by ν_O, the frequency of a vibrational mode. Raman spectroscopy is an especially useful form of vibrational spectroscopy for biomacromolecules because the water in protein solutions yields a negligible Raman background. The split cell has been designed so as to make the light paths in both halves as equal as possible in order to reduce subtraction artifacts. The Raman-scattered light is collected by a monochromator and multichannel detector combination. Scattered light is collected from one side of the split cell for about 30 min, and this spectrum is stored in the computer. The cell is translated and scattered light is collected from the second side. The two spectra are then subtracted in the computer to form the difference spectrum. Sample concentrations are about 1 mM. This

38). Figure 1 shows our spectrometer system. Defining subtraction fidelity as the ratio of the maximum false signal in the difference spectrum divided by the maximum signal in the parent spectra [$\Delta I_{max}/I(0)$], the instrument depicted in Fig. 1 has a fidelity of about 0.1%.

This method is illustrated by a study[26] of the protein phosphoglucomutase (PGM), which catalyzes the interconversion of glucose 1-phosphate and glucose 6-phosphate. PGM is a ca. 60-kDa protein, and one of our goals was to measure the symmetric stretch frequency of the phosphate group of glucose 1-phosphate when bound to the protein. Figure 2a shows the Raman spectrum for PGM complexed with [^{16}O]glucose 1-phosphate, whereas Fig. 2b shows the spectrum for PGM complexed with [^{18}O]glucose 1-phosphate. Differences between these two spectra involve only the phosphate group of bound glucose 1-phosphate, and it is nearly impossible to observe by eye any changes due to the labeling of the phosphate group. The difference spectrum is shown in Fig. 2c, where the intensity scale is about 100 times that of Fig. 2a and 2b, and the effect of the label is now evident. The positive peak at 977 cm^{-1} in the difference spectrum is the P••^{16}O symmetric stretch of bound glucose 1-phosphate, and the negative peak, shifted down by 42 cm^{-1}, is the P••^{18}O stretch. This experiment demonstrates the feasibility of measuring small molecules or small molecular fragments bound to a ca. 60-kDa protein with a signal to noise of about 10/1.

General Concepts

It is clear from this that it is possible to obtain the ground-state vibrational spectrum of substrates bound to their enzymes. What about transition states, which are so short lived as to make a direct measurement very difficult. Many times a remnant of the bond distortions that take place in the transition state shows up in the ground-state electronic structure. Although such remnants may be small, vibrational spectroscopy is a high-resolution technique so that even such small effects can be monitored. In addition, following the paradigm that the binding sites of enzymes are made to bind tightly to the transitions state(s) of substrates, it is sometimes

[38] D. L. Rousseau, *J. Raman Spect.* **10,** 94 (1981).

means that about 1.2 mg of a 40-kDa protein is presently required. Typically, 100 mW of visible light from either an argon ion or a krypton ion laser is used to excite Raman scattering, and a typical experiment takes 1 to 2 hr. About 1200 cm^{-1} of the Raman spectrum can be detected simultaneously.

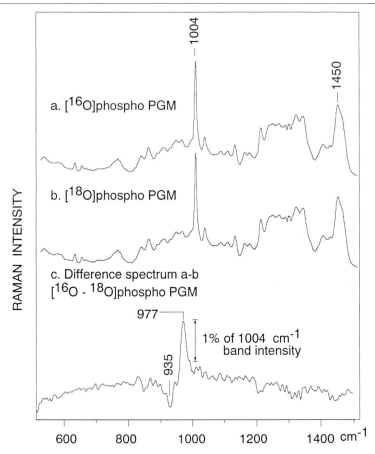

FIG. 2. Raman spectra of phosphoglucomutase (PGM) at pH 7.4 at a temperature of 4°. The approximate time for data accumulation for protein spectra was 2 hr; a 514.5-nm laser was used. (a) The spectrum of PGM complexed with [^{16}O]glucose 1-phosphate; (b) the spectrum of PGM complexed with [^{18}O]glucose 1-phosphate; and (c) the difference spectrum between (a) and (b) where the intensity scale has been expanded by a factor of 100.

possible to find a "transition-state analog." To the extent that the molecule is really a transition state-analog, it is then possible to map out the transition state from measurements of this bound ligand. Last, as discussed in this volume in Chapter 13, kinetic isotope analysis can yield a pretty detailed picture of the vibrational structure at the transition state. Taking all this together, it is now feasible to determine the vibrational structure of the substrate in solution, when bound to the enzyme, and at the transition state, and all this yields a very detailed map of the binding patterns between the protein and its substrate along the reaction pathway.

A very useful concept in interpreting vibrational spectroscopy has been the concept of the local oscillator or local group mode. It is often found that the molecular vibration from a molecule is localized substantially to a characteristic moiety, such as the carboxyl group, the carbonyl group, the phosphate group, or the amino group. This is an approximation in that coupling always exists between the moiety and nearby coordinates. Nevertheless, there has been a long history of employing this approximation along with empirically derived correlations that relate changes in vibrational group frequencies to physically useful parameters such as bond order, atomic charges, bond geometry, ionization state, and the like. This approach can be very useful in specific circumstances.

It has been shown that reasonably accurate force fields for fairly large organic molecules can be calculated by quantum mechanical *ab initio* methods (see, e.g., reference 39). Particularly, the calculated changes of the vibrational modes of a molecule, induced either by conformational change or by interaction with other molecules/ions, are quite accurate.[29,40] This allows specific modeling of enzyme–substrate/protein–ligand interactions at the atomic level. For example, various conformations of a substrate that interacts with other molecules/ions can be modeled in these calculations. The agreement between the calculated vibrational frequencies and the observed ones is the criterion in the model-building process and is an explicit test of the putative interactions. In this way, the vibrational spectroscopic results serve as a very sensitive probe of the substrate–enzyme interactions and the geometry changes of the substrate on its binding to the enzyme. In addition, once a correct model of the active site is found, on the basis of the agreement between calculated and observed vibrational frequencies of the bound substrate, the electronic picture of the substrate can be visualized by its electrostatic potential surfaces calculated with *ab initio* methods.[29,40]

Interpretation Tools and Spectroscopy of Specific Groups

Badger–Bauer Relationships

It is intuitive that force constants, and hence vibrational frequencies, of localized modes will be affected by hydrogen bonding. Such thermodynamic

[39] P. Pulay, G. Fogarasi, G. Pongor, J. E. Boggs, and A. Vargha, *J. Am. Chem. Soc.* **105**, 7037 (1983).

[40] H. Deng, L. Huang, M. Groesbeek, J. Lugtenburg, and R. H. Callender, *J. Am. Chem. Soc.* **98**, 4776 (1994).

correlations have been investigated, and sometimes the relationship is very direct and simple. For example, Badger and Bauer suggested quite some time ago that the enthalpy of formation of a hydrogen bond, ΔH is related linearly to the vibrational frequency shift, $\Delta \nu$, of the O–H stretch frequency of an alcohol. A strong hydrogen bond weakens the O–H bond and the stretch force constant more than a weak hydrogen bond does; this gives rise to a significant decrease in the observed stretching frequency. A series of papers have explored this relationship in a number of ways, both experimentally and theoretically.[41–43] The general methodology of this body of work is to determine the association constants, enthalpies, and binding entropies of donor–acceptor pairs in organic solvents. It has often been found that a linear relationship between binding enthalpy and change in stretch frequency is reasonably well followed over certain energy ranges for particular modes of a hydrogen-bonding X—H$\cdots$Y group. The relationship depends, however, on the nature of the group and how its chemistry may be modified by bonded atoms. Because we are interested in the change in enthalpy, ΔH, and the change in entropy, ΔS, when a substrate binds to a enzymic-binding site, the quantitative exploration and use of such relationships can be very important.

Table I lists the Badger–Bauer (linear) relationships found in various compounds for specific group modes. It should be pointed out that studies to date have suggested that the slope as well as the intercept of the ΔH versus $\Delta \nu$ relationship may vary with the strength of the hydrogen bond. For example, the relatively large intercept value for phenol indicates that the validity of the linear relationship in the low energy region (<4 kcal/mol) may not be justified for this molecule so that a different linear relationship or even a nonlinear relationship should be used to cover a wider region.[44] In general, the empirically measured relationship is over a fairly small range in ΔH (<10 kcal/mol). However, there are exceptions. The relationship between carbonyl stretch frequency and hydrogen bond energy of H_2CO interacting with various cations and water has been investigated by *ab initio* calculations.[43] It was found that the shift in C$=$O stretching frequency in H_2CO fits a linear correlation with the computed interaction energy, ΔE, having a slope of 0.5 kcal/mol for every 1 cm^{-1} of frequency shift, agreeing remarkably well with the experiment (see Table I for acetone). In addition, the linear relationship was found to hold up to an interaction energy of 27

[41] M. Joesten and L. J. Schaad, "Hydrogen Bonding." Dekker, New York, 1974.
[42] R. Thijs and T. Zeegers-Huyskens, *Spectrochem. Acta* **40A**, 307 (1984).
[43] Z. Latajka and S. Scheiner, *Chem. Phys. Letts.* **174**, 179 (1990).
[44] C. N. R. Rao, P. C. Dwivedi, H. Ratajczak, and W. J. Orville-Thomas, *J. Chem. Soc. Faraday Trans.* **71**, 955 (1975).

TABLE I

BADGER–BAUER LINEAR RELATIONSHIPS BETWEEN THE CHANGE IN NORMAL MODE
FREQUENCY, $\Delta\nu$, AND THE HYDROGEN BOND ENTHALPY, ΔH, AND BETWEEN THE BOND OF
THE COMPOUND LISTED AND VARIOUS HYDROGEN BOND DONORS OR ACCEPTORS[a]

Mode	Compound	Slope (a)	Intercept (b)
C=O stretch	Acetone[b]	0.47	−0.11
	Acetophenone[b]	0.31	−0.506
	Benzophenone[b]	0.36	−1.05
	Methylacetate[c]	0.40	−5.488
	Retinal	0.27	
O–H stretch[d]	Phenol	0.0105	3.0
	p-Fluorophenol	0.0103	3.1
	t-Butanol	0.0106	1.7
	Perfluoro-t-butanol	0.0106	3.9
	2,2,2-Trifluro-ethanol	0.0121	2.7
	1,1,1,3,3,3-Hexafluoro-isopropanol	0.0115	3.6
N–H stretch	Pyrrole[e]	0.0123	1.8
	Indole[f]	0.0116	0.8
	N-Methylaniline[g]	0.0394	2.1
Tyrosine ring mode (ν_8)	Tyrosine[h]	0.8	−0.64

[a] The linear relationship is defined by $-\Delta H = a\,\Delta\nu + b$, where H is in kcal/mol and ν in cm^{-1}.
[b] Ref. 42.
[c] Ref. 73.
[d] Reviewed in ref. 44.
[e] Ref. 74.
[f] Ref. 41.
[g] Ref. 75.
[h] Ref. 76.

kcal/mol, at which point the character of the donor/acceptor pair becomes clearly covalent in nature.

Ionization States

Protonation/deprotonation of titratable groups has a profound effect on the distribution of electrons in bonds; hence, the vibrational spectra of molecular moieties in different ionization states are very different. This affects the imidazole group of histidines, the tautomeric state of nitrogens, carboxyl groups, and so forth.[45,46] We give one example that is used in

[45] P. R. Carey, "Biochemical Applications of Raman and Resonance Raman Spectroscopy." Academic Press, New York, 1982.
[46] L.-V. Daimay, N. B. Colthup, W. G. Fateley, and J. G. Grasselli, "The Handbook of Infrared and Raman Characteristic Frequencies of Organic Molecules." Academic Press, San Diego, 1991.

discussion later. The side chain carboxylic acid group of aspartate or gluta-mate, when protonated, has a characteristic $C=O$ stretch band at 1720–1770 cm^{-1} in the Raman spectrum depending on the degree of hydrogen bonding (toward the lower end for very strong H bonding) and effective dielectric constant and a symmetric CO_2^- stretch band at 1360–1450 cm^{-1} when ionized.[46] These are well defined and very characteristic marker bands for the two ionization states of carboxyl.

Phosphate–Vanadate Spectroscopy

A very useful approach to understanding the vibrational spectroscopy of phosphates and vanadates, molecular groups important in the numerous phosphoryl transfer enzymes and transition state analogs thereof, involves empirical ideas of Gordy and Badger that a characteristic stretch frequency is related to the bond strength. This and the concept of network theory, as developed by Brown and Wu,[47] leads to empirical relationships between frequencies and bond order that appear to work particularly well for phos-phates and vanadates. Network theory holds that the sum of bond strengths for each atom in a crystal[47] or in the solution phase[48,49] is equal to its formal oxidation strength. For phosphate and vanadate, this is equal to five. It has been found then that bond order (or perhaps more properly, bond valence), s, bond frequency, ω, and bond length, r, are related by[1,47–49]

$$s = [a \cdot \ln(b/\nu)]^p$$

and

$$s = (r/r_o)^p$$

where a, b, r_o, and p are empirical parameters, r is the length of the PO or VO bond, and ν is the fundamental frequency of the specific phosphate or vanadate group, $n\nu^2 = \nu_s^2 + (n + 1)\nu_a^2$ with ν_s the symmetric, ν_a the asymmetric, and n the degeneracy of the mode. From these relationships, bond lengths and bond orders can be determined from the vibrational spectroscopy of phosphates and vanadates once the empirical constants have been determined. The great beauty of these relationships is that they are relatively independent of the exact geometry or coordination of the central metal ion.

[47] I. D. Brown and K. K. Wu, *Acta Crystallogr.* **B32**, 1957 (1976).
[48] J. W. J. Ray, I. J. W. Burgner, H. Deng, and R. Callender, *Biochemistry* **32**, 12977 (1993).
[49] W. J. Ray, D. C. Crans, J. Zheng, J. W. Burgner, H. Deng, and M. Mahroof-Tahir, *J. Am. Chem. Soc.* **117**, 6015 (1995).

Entropy Determination

The time scale of vibrational spectroscopy is very fast, as mentioned earlier, faster than any averaging time scale of interest here. Thus, the vibrational spectrum of a molecule contains the spectrum of each conformer of a molecule in proportion to its concentration. To the extent that a particular band position depends on conformation, the widths of these vibrational bands are broadened (heterogeneous broadening) or, in extreme cases, separate resolved bands are observed. Heterogeneous band broadening is then a measure of conformer population. The Boltzman definition of entropy is $S = R \cdot \ln \Omega$, where R is the gas constant and Ω is the number of accessible states. Changes in entropy become a counting game, and it is possible to count the number of available states from band widths. This concept has been used in an analysis of the vibrational spectroscopy of lactate dehydrogenase (LDH).

Applications of these interpretation tools in Raman studies of several enzyme systems are described in the following section.

Lactate Dehydrogenase

Lactate dehydrogenase accelerates the oxidation of lactate by NAD^+ to pyruvate and NADH by hydride transfer from the C-4 carbon of NAD to and from the cofactor. Figure 3 shows the active site of LDH with the

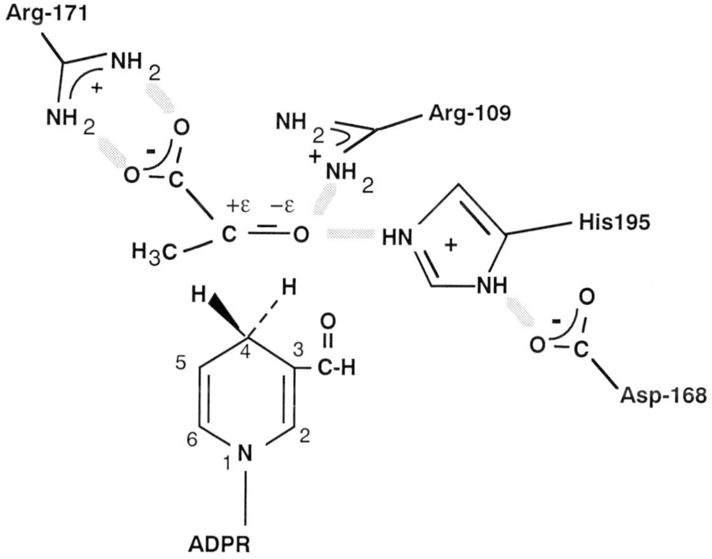

Fig. 3. The active site of LDH.

bound substrate pyruvate and the NAD cofactor as derived from X-ray crystallographic data. A number of important coordinates of NAD^+, NADH, and pyruvate bound to LDHs of different types and also to mutants have been probed by Raman difference spectroscopy and isotope editing techniques, including pyruvate's $C{=}O$ and COO^- group and the carboxamide group and the C-4–H stretch of the NAD cofactor. This is one of the best studied enzymes from the point of view of vibrational spectroscopy.

Energetics of Substrate Binding

The spectroscopy of bound pyruvate is virtually the same in LDHs from pig heart, dog fish, and thermophilic bacteria. This means that the binding patterns of these enzymes are virtually the same given the high degree of resolution inherent in vibrational spectroscopy. The frequency of the carbonyl group of pyruvate shifts downward 35 cm^{-1} from its solution value at ca. 1710 cm^{-1}.[24] This is a large shift and corresponds to a 14- to 17-kcal/ mol favorable interaction enthalpy between the protein and this group from data in Table I (the number is somewhat imprecise because a precise Badger–Bauer relationship has not been performed on pyruvate). This interaction may be understood easily by electrostatic (hydrogen bonding) interactions between these groups and specific charged protein groups, i.e., Arg-109 and His-195 with pyruvate's carbonyl. The estimated binding enthalpy of pyruvate to lactate dehydrogenase is 18.1 kcal/mol,[50,51] which is the sum of any number of interactions between pyruvate and protein, pyruvate with water, and so forth. This value is quite close to the single contribution of pyruvate carbonyl as determined from Raman study.[24] The entropy loss, given by $-T\,\Delta S$, involved in binding oxamate, an isoelectronic substrate analog for pyruvate, is 8.4 kcal/mol at 300 K.[51] Clearly, there is more than sufficient binding enthalpy to overcome the loss of entropy. Raman results pinpoint a main, essentially electrostatic, interaction responsible for the binding.

Stereospecific Hydride Transfer

In LDH, the error in the fidelity of hydride transfer from the pro-R side of the NAD cofactor compared to the pro-S side is smaller than 1 part in 10^8.[52] From a structural point of view, it is known that pro-R enzymes bind the nicotinamide ring in the *anti* conformation about the glycosidic bond (the bond between the N-1 nitrogen of the nicotinamide and the C-1

[50] W. B. Novoa, A. Winer, A. J. Glaid, and G. W. Schwert, *J. Biol. Chem.* **234,** 1143 (1959).
[51] F. Schmid, H.-J. Hinz, and R. Jaenicke, *Biochemistry* **15,** 3052 (1976).
[52] R. D. LaReau, W. Wan, and V. E. Anderson, *Biochemistry* **28,** 3619 (1989).

carbon of the ribose) with the pro-R proton pointing toward the substrate-binding pocket whereas pro-S enzymes bind the ring in the *syn* conformation with, now, the pro-S proton pointing toward the substrate. This prompted the early suggestion that a simple geometrical consideration, one proton is closer to the substrate than the other, is responsible for the stereospecific nature of these enzymes.[53] This concept suggests a stringent requirement. The nicotinamide ring must bind in the correct geometry and remain in the correct geometry once bound, within the error rate of hydride transfer. This means a difference in the equilibrium constant between *syn* and *anti* of 1 part in 10^8, or 10.4 kcal/mol, for LDH. From measurements of the C=O and $-NH_2$ groups of NAD's carboxamide moiety and the use of Badger–Bauer relationships,[21] it has been estimated that the energy difference of the hydrogen-bonding interactions between NADH's amide arm and protein is 9.3–11.3 kcal/mol lower in the *anti* conformation compared to the *syn* conformation, in very good agreement with that estimated by the error rate of the hydride transfer.

Transition-State Stabilization

Lactate dehydrogenase accelerates the hydride transfer rate between NADH and pyruvate by a factor of 10^{14} over the rate found in solution.[54–56] How does this come about? One effect is likely entropic. A decrease in the number of accessible conformational states available to NADH was observed when the substrate analog oxamate binds to the binary complex as judged by the narrowing of certain Raman bands, particularly the C-4–H stretch mode of NADH.[19] From an analysis relating loss of entropy to the widths of vibrational bands as sketched earlier, it has been calculated that an entropy loss corresponding to ~0.7 kcal/mol of free energy is associated with this constraint of the NADH nicotinamide ring, and another ~0.7 kcal/mole is associated with the constraint of the carbonyl of pyruvate.[19] Therefore, at least 1.4 kcal/mol, or about a factor of 10 in rate enhancement, arises from elimination of various nonproductive conformations of nicotinamide ring and pyruvate on binding of the substrate. (It has been shown that entropic effects, related mainly to the loss of translational and rotational degrees of freedom of the substrate at the protein active site, can

[53] J. J. Holbrook, A. Liljas, S. J. Steindel, and M. G. Rossmann, *in* "The Enzymes" (P. D. Boyer, ed.), p. 191. Academic Press, New York, 1975.
[54] J. W. Burgner, II and W. J. Ray, *Biochemistry* **23,** 3636 (1984).
[55] J. W. Burgner, II and W. J. Ray, *Biochemistry* **23,** 3626 (1984).
[56] J. W. Burgner, II and W. J. Ray, *Biochemistry* **23,** 3620 (1984).

contribute to a rate enhancement of up to $10^{6.5}$; only some of it shows up in the vibrational spectroscopy.[57])

A most significant interaction between pyruvate and LDH is the exceptionally strong and favorable electrostatic interaction between pyruvate's carbonyl oxygen and His-195, as noted earlier. This same interaction is likely even stronger in the transition state because the transition state contains substantially more $^+C\text{--}O^-$ character than the ground state. Use of the reaction rate theory then suggests that an Arrhenius plot of k_{cat} versus the shift in carbonyl frequency should yield a straight line, assuming a linear relationship between the energy of the transition state and the frequency of $C{=}O$ stretch, such as the Badger–Bauer rule found for the ground state.[28] We then could write

$$k_{cat} \sim \exp(-\Delta E^{\dagger}/RT) \sim \exp(x \cdot \Delta \nu_{C=O}/RT)$$

where x is an unknown numerical factor that depends on the relationship between the transition-state $C^+ {\bullet\bullet} O^-$ stretch and its enthalpic interaction with the imidazole proton.

Figure 4 shows a plot of the hydride transfer step rate as a function of shift in carbonyl frequency for the wild-type protein, various mutants, and the estimated solution rate.[54] It can be seen that reasonable agreement is obtained except for the R109Q mutant. In the R109Q mutant, the hydride transfer rate is substantially below the value that would be predicted from the observed polarization of the $C{=}O$ bond. This is not unreasonable, as in LDH, a "loop" of the protein involving residues 98–110 closes over the active site, which moves Arg-109 0.8 nm from a position in the solvent to one in the active site in the enzyme/cofactor binary complex. The replacement of arginine by glutamine almost certainly affects loop closure, resulting in an active site structure that is altered substantially geometrically.[19,58] As a consequence, the $C{=}O$ stretch band in the R109Q mutant is very broad, which suggests nonproductive binding modes. It is possible that the coenzyme–substrate distance is increased in some of these conformations, substantially decreasing k_{cat}. The reasonable agreement observed in Fig. 4 can be rationalized by supposing that the enthalpy of the transition state is stabilized to an even greater extent than the ground state so that the net reaction barrier is lowered by the electrostatic interaction. This is true because the carbonyl moiety is substantially more polarized in the transition state, with more negative character on the oxygen. This analysis[28] suggests that about 6 orders of magnitude out of the ca. 14 orders-of-magnitude

[57] M. I. Page and W. P. Jencks, *Proc. Natl Acad. Sci. U.S.A.* **68,** 1678 (1971).
[58] A. R. Clarke, D. B. Wigley, W. N. Chia, and J. J. Holbrook, *Nature* **324,** 699 (1986).

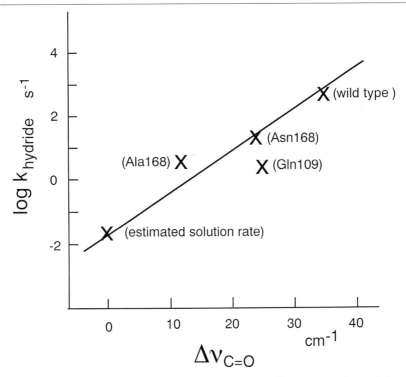

FIG. 4. The log of the rate of hydride transfer step (in sec^{-1}) versus the observed change (solution value minus protein value; see Deng et al.[24] for wild-type determination) in C=O stretching frequency of pyruvate (as NAD-pyruvate adduct). The C=O stretch in solution lies at 1710 cm^{-1}.[24] The solution hydride transfer rate, which has not been measured, is estimated from the ratio of enolization rate of pyruvate in solution and in LDH[55] and taking the transfer rate to be 1000 sec^{-1}. The hydride transfer of the Asp-168 mutants is taken from Clarke et al.[77]

rate enhancement come from stabilization of the polar $^+$C–O$^-$ group in the transition state, in agreement with previous chemical studies.[54-56]

Serine Proteases

Serine proteases catalyze the polypeptide bond cleavage via a two-step, acyl enzyme mechanism. In the first step, concerted formation of a covalent ester linkage between the serine oxygen and the acyl carbon atom of the substrate, with the cleavage of the peptide bond, takes place through a transient, tetrahedral intermediate, resulting in the release of the amide portion of the peptide and the formation of the acylenzyme. In the second

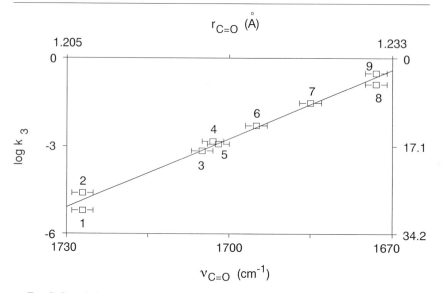

FIG. 5. Correlation between carbonyl stretching frequency, $\nu_{C=O}$, and log k_3, where k_3 is the maximal deacylation rate at pH 10 for the series of acyl-serine protease. Each point represents an acyl enzyme. $\nu_{C=O}$ and log k_3 are also recast in the forms of $r_{C=O}$ (carbonyl bond length) and $\Delta\Delta G$ (free energy of activation for deacylation), respectively.

step, the ester linkage of the acyl enzyme, which is very unstable compared to the original peptide bond of the substrate, undergoes hydrolysis to release the carboxylic portion of the peptide. In extensive work on acyl intermediates of a series of serine proteases and functional mutants, Carey and Tonge[59] (and references therein) have shown that quantitative information about the hydrogen-bonding enthalpy, the deacylation rate, and the C=O bond length can be obtained through empirical relationships based on the C=O stretch frequency shifts observed in acyl enzymes. For the series of acyl enzymes, the 54-cm^{-1} shift of the C=O stretch mode correlates a change of the hydrogen-bonding strength of -27 kJ/mol, which is very close to the change in the measured activation energy of -24 kJ/mol (Fig. 5). The deacylation rate is related directly to the C=O stretch frequency of the acyl enzyme in the ground state. A shift of the C=O stretch mode of the acyl enzyme to higher frequencies is correlated linearly with the logarithm of the decrease of the deacylation rate, log(k_3). As the C=O frequency shifts upward by 54 cm^{-1}, the deacylation rate decreases by 16,300-fold (Fig. 5). The empirical relationship between C=O stretch fre-

[59] P. R. Carey and P. J. Tonge, *Acc. Chem. Res.* **28**, 8 (1995).

quency and $C=O$ bond length based on X-ray crystallographic and vibrational spectroscopic studies on the same crystals of a large series of compounds has been determined.[2] According to this relationship, the $C=O$ stretch frequency difference of 54 cm^{-1} corresponds to a difference of the $C=O$ bond length of 0.025 Å (Fig. 5), about 10% of the change from a $C=O$ double bond to a $C–O$ single bond. Such relationships established that a tetrahedral intermediate is formed at the transition state, as a polarization of the $C=O$ bond, which lowers its stretch frequency, in the enzymic acyl intermediates reflects a ground-state distortion toward the transition state, where the $C=O$ double bond character is substantially reduced.[59]

Dihydrofolate Reductase

Dihydrofolate reductase (DHFR) catalyzes the reduction of 7,8-dihydrofolate (DHF) by NADPH to form 5,6,7,8-tetrahydrofolate and NADP$^+$. Figure 6 depicts the active site showing the overall reaction. The active site of all DHFRs contains an invariant carboxylic acid residue, aspartic acid in bacteria (Asp-27 in the *E. coli* enzyme) and glutamic acid in vertebrates, which is the only ionizable residue in the active site cavity. It has been proposed that DHFR promotes the hydride transfer from NADPH to substrates by facilitating the protonation of N5 of DHF or N8 of folate (see, e.g., Kraut and Matthews[60]) as the transition-state energy of protonated DHF is presumably substantially smaller than that for the neutral form. For the *E. coli* enzyme, the binding of certain inhibitors and the rate of hydride transfer have characteristic titration curves with a pK_a of 6.5, which also suggests that the protonation of some specific groups and concomitant structural changes are responsible for much of the catalytic behavior of DHFR. This conjecture that the protonation of N-5 is important requires that the pK_a of bound dihydrofolate be raised some four units, as the solution pK_a of N-5 is around 2.6.

Vibrational spectroscopy is an ideal tool to determine the pK_a of molecular groups because of its sensitivity to the ionization state. Figure 7 shows the Raman difference experiments of dihydrofolate bound to DHF/NADP$^+$, which is likely a good model for an active Michaelis complex. The 1650- and the 1675-cm^{-1} bands can be assigned to the N-5$=$C-6 stretch of the pteridine ring of DHF for N-5 unprotonated and protonated, respectively, so that these bands serve as characteristic marker bands.[61] As can be seen in Fig. 7, the intensities of these bands are quite pH dependent.

[60] J. Kraut and D. A. Matthews, *in* "Biological Macromolecules and Assemblies: Active Sites of Enzymes" (F. A. Jurnak and A. McPherson, eds.), p. 1. Wiley, New York, 1987.
[61] Y.-Q. Chen, J. Kraut, R. L. Blakley, and R. Callender, *Biochemistry* **33,** 7021 (1994).

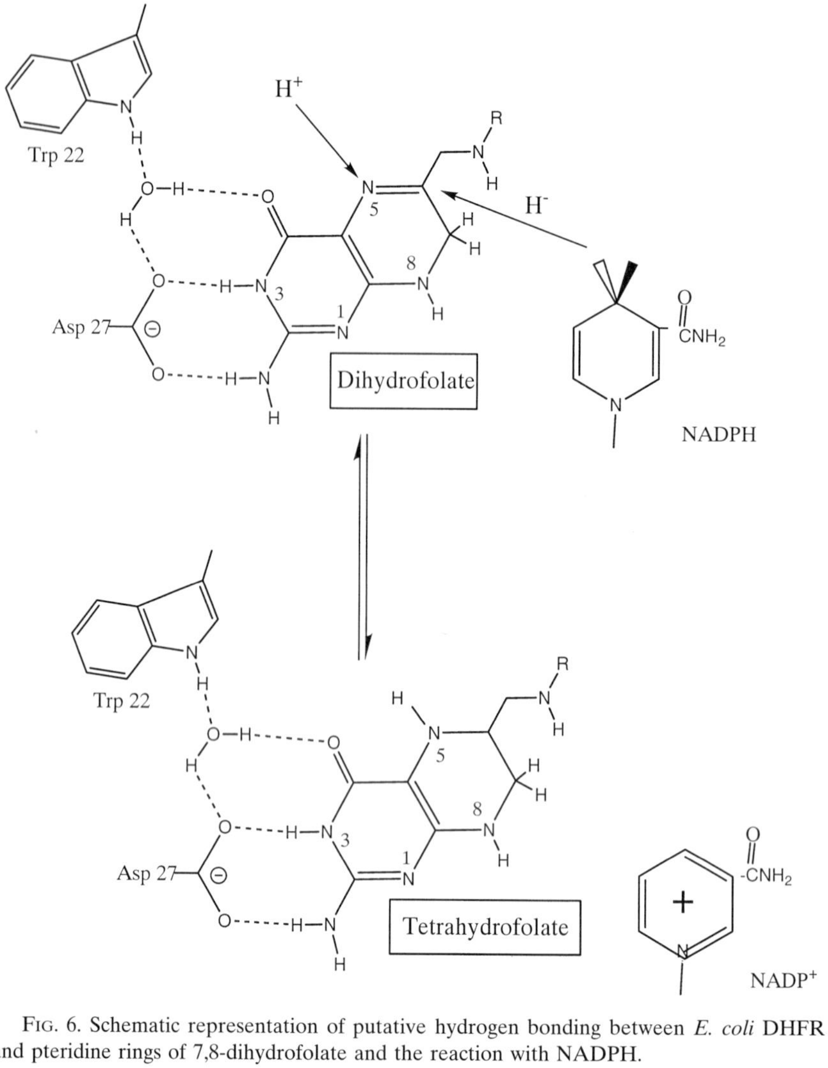

Fig. 6. Schematic representation of putative hydrogen bonding between *E. coli* DHFR and pteridine rings of 7,8-dihydrofolate and the reaction with NADPH.

Using the fact that the intensity of a band is proportional to the concentration of a species, a titration curve was constructed from these data (and data taken at other pH values), and it was found that the pK_a of N-5 is 6.5 for bound DHF. Thus, at physiological pH values, the concentration of protonated DHF bound to dihydrofolate reductase is some four orders of magnitude higher than for the substrate in solution, and this alone can account for a four order of magnitude increase in the reaction rate (adding

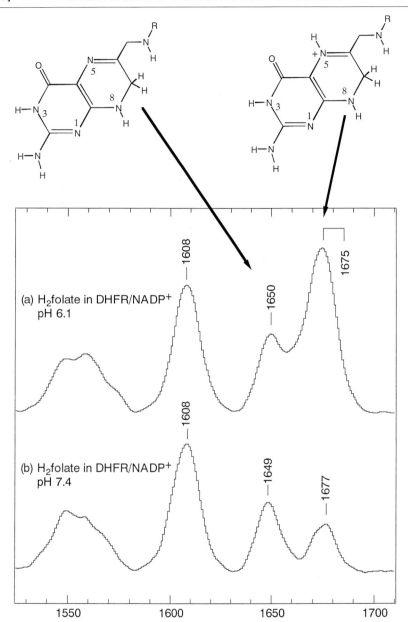

FIG. 7. Difference Raman spectra of dihydrofolate bound in the ternary complex of DHFR with NADP$^+$ (DHFR/NADP$^+$/dihydrofolate = 3.4 : 2.5) at 4° in (a) 25 mM Bis–Tris buffer containing 0.5 M KCl, pH 6.1, and (b) 20 mM Tris buffer containing 0.5 M KCl, pH 7.4. The marker bands, identified by isotope editing, for protonated and neutral dihydrofolate are marked in the diagram.

the reasonable possibility of another four orders of magnitude increase by bringing the reactants together, essentially an entropic factor, would yield a very respectable enzyme).

What structural features of the enzyme are responsible for raising the pK_a of bound DHF by four units? X-ray crystallographic studies[62,63] show that the Asp-27 side chain of *E. coli* DHFR forms hydrogen bonds to the 2-amino group and to N-3 of DHF's pteridine ring rather than N-5 or N-8. Thus Asp-27 cannot be the ultimate proton donor to the key nitrogen of substrate. Instead, an indirect role for the carboxyl group has been proposed.[63–65] An essential step in most versions of the protonation mechanism is that bound pterin substrates undergo enolization from the 4-keto to the 4-hydroxyl tautomer and that the 4-hydroxyl group then becomes part of a proton relay system, possibly involving bound waters, for proton donation to the substrate. The ionization state of this carboxyl and its pK_a were determined by Raman difference spectroscopy by a novel approach.[66] The Raman difference spectrum between wild-type ecDHFR and the Asp-27 to serine mutant (D27S) in the pH range of 5.6–9.0 was taken, and this difference spectrum contained the spectrum of Asp-27. No protonation of the carboxyl group was detected, implying that its pK_a is less than 5.0. However, a pH dependence in the intensity of Raman bands, including bands assigned to Trp-22, in the protein difference spectrum with a pK_a of 6.3 was detected, indicating that the apo enzyme undergoes pH-dependent conformational changes. Because the carboxyl group of Asp-27 at the active site is the only ionizable group in the binding site, other groups, away from the catalytic site, must be responsible for this pH behavior of ecDHFR.

Phosphoryl Transfer Proteins

Phosphoglucomutase catalyzes the reversible transfer of the PO_3^- fragment of a dianionic phosphate ester between the 1- and the 6-oxygens of a glucose molecule in a process that involves two binding and two transfer steps plus a fifth step where the initially formed intermediate rearranges (cf. Ray *et al.*[67] and Ray and Post[68]). In order to assess how PGM promotes

[62] J. T. Bolin, D. J. Filman, D. A. Matthews, R. C. Hamlin, and J. Kraut, *J. Biol. Chem.* **257,** 13650 (1982).
[63] C. Bystroff, S. J. Oatley, and J. Kraut, *Biochemistry* **29,** 3263 (1990).
[64] J. F. Morrison and S. R. Stone, *Biochemistry* **27,** 5499 (1988).
[65] T. Uchimaru, S. Tsuzuki, K. Tanabe, S. J. Benkovic, K. Furukawa, and K. Taira, *Biochem. Biophys. Res. Commun.* **161,** 64 (1989).
[66] Y.-Q. Chen, J. Kraut, and R. Callender, *Biophys. J.* **72,** 936 (1997).
[67] W. J. J. Ray, J. W. Burgner, and C. B. Post, *Biochemistry* **29,** 2770 (1990).
[68] W. J. J. Ray and C. B. Post, *Biochemistry* **29,** 2779 (1990).

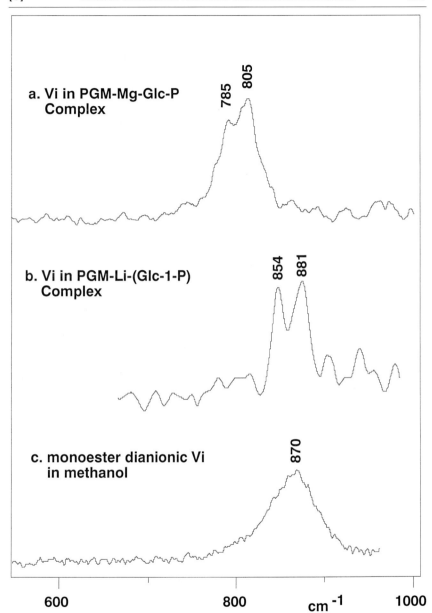

FIG. 8. Raman spectra of (a) vanadate bound in the PGM·Mg·Glc-Vi complex at 4°, (b) vanadate bound in the PGM·Li·Glc-Vi complex at 4°, and (c) monoester dianionic methyl vanadate in methanol.

phosphate transfer, we studied phosphate bound to the protein as well as vanadate because nature has endowed vanadate with sufficient "plastic" properties that make it a good transition-state analog for phosphate bound to PGM.[26,67,68] Classical Raman spectroscopy was chosen to probe the internal bonding of the bound phosphate and a vanadate group in an analog complex closely similar to the normal glucose bisphosphate complex—one where the remainder of the glucose phosphate moiety is bound in the same way as the –Glc-1-P portion of Glc-P_2. In this way, the electronic properties of bonds in the ground state and the transition state were probed.

The symmetric stretching frequency of the P$\cdot\cdot$O bonds within the enzymic phosphate group of muscle phosphoglucomutase was measured by [^{16}O]/[^{18}O] Raman difference spectroscopy.[26] The observed frequency and its shift on isotopic substitution of the phosphate band is characteristic of dianionic phosphate. The P$\cdot\cdot$O stretching frequency is not altered detectably by the binding of the metal ion activators, Mg^{2+}, Zn^{2+}, or Cd^{2+}, or by the subsequent binding of glucose phosphate (see Fig. 2). Hence, a binding-induced distortion of the bonding within the enzymic phosphate group in the ground enzyme–substrate complex cannot serve as a rationale for the large value of k_{cat} observed for the phosphoglucomutase reaction. In contrast, the frequency of the V$\cdot\cdot$O bonds within a vanadate group bound at the same site in the transition-state analog complex involving glucose 1-phosphate 6-vanadate is unusually low for a tetrahedral vanadate (Figs. 8a and 8c). This unusually low V$\cdot\cdot$O stretching frequency is rationalized in terms of the partial bonding of a fifth ligand, HOR′, which decreases the bond order of all three of the nonbridging V$\cdot\cdot$O bonds of the vanadate group, as in RO$-$[VO_3^{2-}]$\cdots_{OR'}^{H}$.[26] This polarization of the vanadate group depends on the interaction of the metal ion activator with the central $-$[VO_3^{2-}]$\cdots$ group, as the replacement of Mg^{2+} by Li^+ largely abolishes the polarization of this group and thus the involvement of the fifth ligand (Fig. 8b). The average bond order of the three nonbridging V$\cdot\cdot$O bonds is changed from 1.37 vu in solution to about 1.20 in the Mg^{2+} complex, according to the bond-order frequency relationship.[1,48,49] According to Brown and Wu's paradigm,[47] there should be a bond-order increase of about 0.5 vu in the bridging and newly formed incoming V$\cdot\cdot$O bond. Thus, to the extent that the vanadate–inhibitor complex mimics the transition state for the normal phosphoglucomutase reaction, the enzymic reaction involves an associative process (Fig. 9). It thus is quite different from the process by which model phosphate ester dianions normally react, as in aqueous solution a dissociative process is almost universal.[69]

---

[69] D. Herschlag and W. P. Jencks, *Biochemistry* **29**, 5172 (1990).

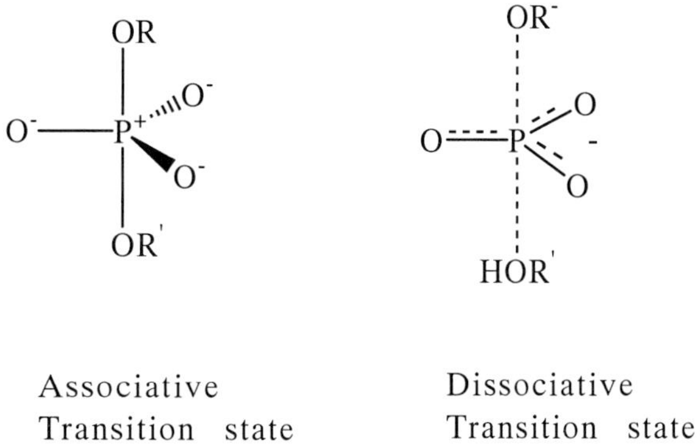

Associative
Transition state

Dissociative
Transition state

FIG. 9. Dissociative and associative transition states. In the dissociative process, the breaking of the P–O bond of the leaving group is essentially complete before the formation of the P–O bond with the incoming group. The sum of the bond orders of the two P–O bonds involved with the incoming and leaving groups decreases substantially in the transition state, whereas the bond orders of the nonbridging P–O bonds increase significantly. Prevailing evidence suggests that this is the reaction mechanism for the hydrolysis of monoester dianionic phosphate model compounds in aqueous solution.[69] In the associative process, the breaking of the P–O bond of the leaving group starts after the formation of the P–O bond with the incoming group. The sum of the bond orders of the two P–O bonds involved with the incoming and leaving groups increases substantially in the transition state, whereas the bond orders of the nonbridging P–O bonds decrease significantly.

Nucleoside Hydrolase

The inosine–uridine nucleoside hydrolase (IU-necleoside hydrolase) catalyzes the hydrolysis of N-glycoside linkage of the commonly occurring purine and pyridine nucleosides. Kinetic isotope studies of this enzyme (see chapter 13 in this volume) have established that the transition state is characterized by an elongated C–N glycosidic bond (see Fig. 10).[70] Studies on this enzyme have brought together several quantitative methods to describe the transition state and to build transition-state analogs *de novo*. The approach combines kinetic isotope studies with molecular modeling and *ab initio* quantum mechanical calculations to describe the electronic structure of the transition state. Inhibitors having features of the transition-state structure are then constructed. From the paradigm that binding sites of enzymes are made to bind the transition state of their substrates tightly, a systematic analysis of the binding constants of these inhibitors yields measurement of how specific features of the electronic structure contribute

[70] B. A. Horenstein, D. W. Parkin, B. Estupinan, and V. Schramm, *Biochemistry* **30,** 10788 (1991).

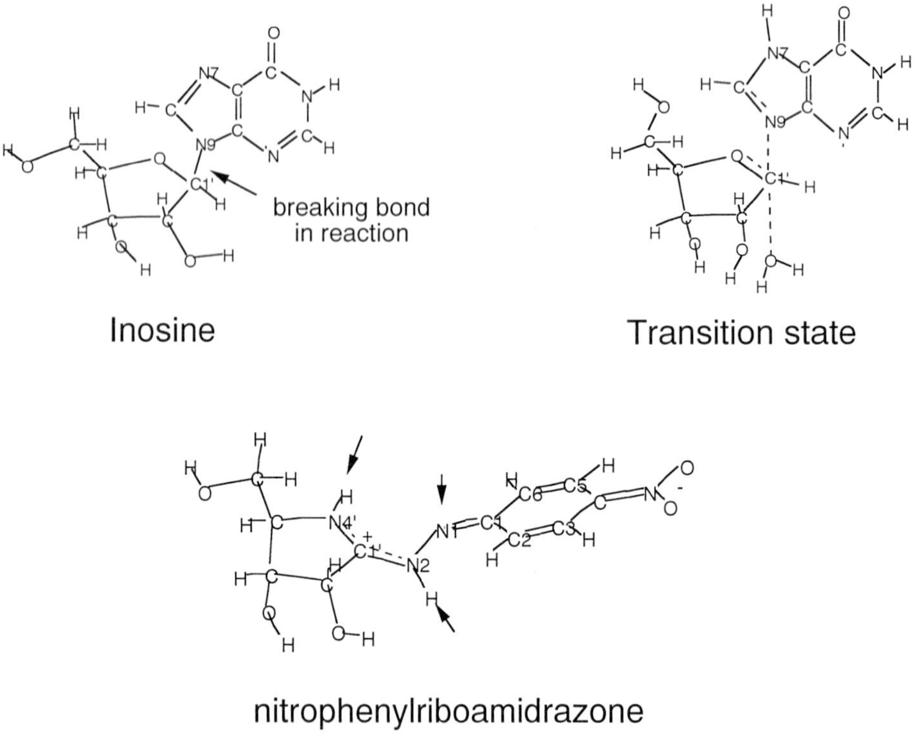

Inosine   Transition state

nitrophenylriboamidrazone

FIG. 10. Molecular structure of the inosine substrate, transition state, and the nitrophenyl-riboamidrazone inhibitor of IU-nucleoside hydrolase.

to the binding and, hence, the transition-state structure.[71,72] Within this context, it is thus clear that the electronic structure of the bound inhibitors needs quantitative characterization, which is feasible if the vibrational spectrum of the bound inhibitor can be measured.

It has been shown, or rather pointed to, how all this can fit together in a quantitative way yielding very detailed insights into the transition state,

[71] B. A. Horenstein and V. L. Schramm, *Biochemistry* **32,** 9917 (1993).

[72] B. A. Horenstein and V. L. Schramm, *Biochemistry* **32,** 7089 (1993).

[73] L. Vanderheyden and J. Th. Zeegers-Huyskens, *J. Mol. Liquids* **25,** 1 (1983).

[74] M. S. Nozari and R. S. Drago, *J. Am. Chem. Soc.* **92,** 7086 (1970).

[75] J. H. Lady and K. B. Whetsel, *J. Phys. Chem.* **71,** 1421 (1967).

[76] P. G. Hildebrandt, R. A. Copeland, T. Spiro, J. Otlewski, M. J. Laskowski, and F. G. Prendergast, *Biochemistry* **27,** 5426 (1988).

[77] A. R. Clarke, H. M. Wilks, D. A. Barstow, T. Atkinson, W. N. Chia, and J. J. Holbrook, *Biochemistry* **27,** 1617 (1988).

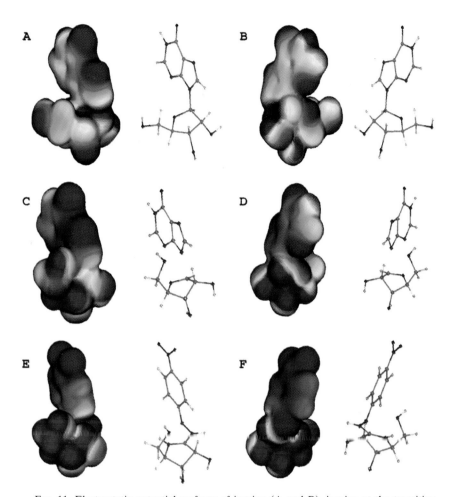

Fig. 11. Electrostatic potential surfaces of inosine (A and B), inosine at the transition state of nucleoside hydrolase (C and D), and the bound nitrophenylriboamidrazone transition-state analog (E and F). The electrostatic potential surfaces of inosine and the N-7 protonated transition state for inosine were determined from a kinetic isotope analysis,[71,72] while the surface for the bound transition-state analog from Raman difference spectroscopy and other spectral measurements.[29] In the surfaces shown here, the geometry of inosine and the transition state were aligned to permit the best match among substrate, transition state, and inhibitor. Each left–right view shows 180° vertical rotational views of the same atomic structure. The color code for the stick figures is oxygen, red; nitrogen, blue; carbon, green; and hydrogen, white. The color codes for the electrostatic potential surfaces are partial negative charge (electron excess) as blue, partial positive charge (electron deficiency) as red, and yellow as neutral.

the design of transition-state analogs, and the general enzymology of a given enzyme.[29] Nitrophenylriboamidrazone binds 2×10^5 more tightly to IU-nucleoside hydrolase than inosine to yield a dissociation constant of 2 nM. The vibrational spectrum of the bound inhibitor was measured by resonance Raman experiments. From this data set and from *ab initio* calculations, the electrostatic interaction between the NO_2 group of the inhibitor and a positively charged residue was characterized. The electrostatic potential surfaces of the bound inhibitor were also derived from the calculations. These surfaces, along with the potential surface of the transition state and that of the ground state of the substrate, are given in Fig. 11, see color insert. It is clear that the surface of the inhibitor resembles that of the transition state and is markedly different from that of the ground-state surface. Although the comparison is not perfect between the surfaces of the inhibitor and that of the transition state, it certainly seems feasible from this study that it is possible to either (1) derive the electronic structure of the transition in detail from studies of transition-state analogs if good ones can be found or (2) design very good inhibitors for a specific enzyme in a systematic way (nice from the point of view of pharmaceuticals) if the transition state is otherwise known.

Acknowledgment

This work is supported by a grant from the National Institutes of Health (GM35183).

[9] Crystallographic Analysis of Solvent-Trapped Intermediates of Chymotrypsin

By GREGORY K. FARBER

Introduction

There are now a number of different crystallographic techniques to trap enzyme-bound intermediates in a crystal lattice. These techniques can be broken down into two very different approaches. The most straightforward technique attempts to use ultrafast data collection methods to obtain complete data sets on the enzymatic time scale.[1] The major problem with this technique is trying to ensure that all molecules in the crystal are synchronized so that they are at the same step of the reaction at the same time.[2]

[1] B. L. Stoddard, *Pharm. Therap.* **70**, 215 (1996).
[2] G. K. Farber, *Curr. Biol.* **5**, 1088 (1995).

Copyright © 1999 by Academic Press
All rights of reproduction in any form reserved.
0076-6879/99 $30.00

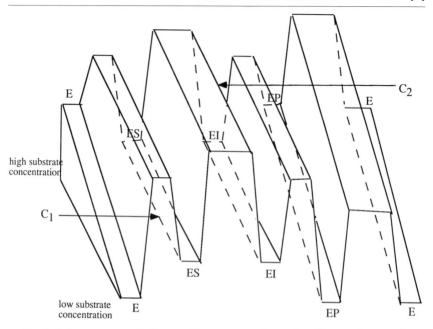

FIG. 1. A free energy profile for an arbitrary enzymatic reaction. The reaction coordinate goes from left to right, the free energy goes up and down, and the effect of changing the substrate concentration is shown in the third dimension. This free energy profile was adapted from the free energy profile of proline racemase.[3]

The second approach to solving this problem involves turning a kinetic problem into a thermodynamic problem. The free energy profile shown in Fig. 1[3] illustrates how this approach works. At low substrate concentrations, it is clear that the apo enzyme has lower free energy than any of the enzyme-bound intermediates. However, as the substrate concentration increases, the free energy of the intermediates becomes lower than the free enzyme. In other words, if the substrate concentration is high enough, all of the active sites in the crystal lattice will be filled with something. Such an experiment is performed by placing an enzyme crystal in a small volume containing substrate and buffer at the appropriate concentrations. The system is allowed to come to equilibrium, and then the crystal is mounted in the capillary tube for X-ray data collection. Throughout data collection, catalysis will continue because the product of the reaction is simply the substrate for the back reaction. Enzymes with unstable or gaseous products would present difficult challenges for this approach.

[3] W. J. Albery and J. R. Knowles, *Biochemistry* **25**, 2572 (1986).

In addition to potential problems with unstable products, there is another limitation on this approach. This thermodynamic solution to trapping intermediates works in protein crystals because the solvent channels in protein crystals are large enough that substrate and product can diffuse between active sites.[4] If the active site of the enzyme is not accessible to the solvent channel or if a large conformational change occurs during the catalytic cycle that disrupts the crystal lattice, this approach will fail. Finally, for enzymes that have substrates that are very large it may not be possible for the substrate to diffuse into the active site.

In the free energy profile shown in Fig. 1, an equilibrium structure obtained at high substrate concentrations would be a mixture of ES and EI. Such a superposition might be very difficult to interpret because S and I are presumably reasonably similar in structure and occupy roughly the same location in the active site. Fortunately, a flow cell experiment often allows these intermediates to be decoupled.[5] The flow cell experiment is very similar to a steady-state kinetic experiment. The enzyme in the crystal is not allowed to settle to equilibrium. Rather, substrate is flowed through the crystal during the experiment. In this experiment, the intermediate that precedes the rate-determining step of the reaction will dominate the electron density map. In the free energy profile shown in Fig. 1, the rate-determining step actually changes as the substrate concentration increases. To trap the ES complex, S would be flowed through the crystal at concentrations lower than C_2. To trap the EI complex, P would be flowed through the crystal at concentrations lower than C_2. Finally, the E $\cdot$ P complex could be trapped by flowing S through the crystal at concentrations higher than C_2. Using a combination of flow cell experiments and equilibrium experiments, it is often possible to obtain structures of all of the important intermediates in an enzymatic reaction.[2,6–8]

However, this protocol outlined for trapping intermediates will not work if water is a substrate in the reaction. It is simply not possible to alter the concentration of water significantly in an aqueous crystallization solution. Fortunately, we have discovered that for many enzyme crystals it is possible to transfer the crystals from their aqueous crystallization buffer to nonaque-

[4] M. W. Makinen and A. L. Fink, *Annu. Rev. Biophys. Bioeng.* **6,** 301 (1977).
[5] B. L. Stoddard and G. K. Farber, *Structure* **3,** 991 (1995).
[6] G. K. Farber, A. Glasfeld, G. Tiraby, D. Ringe, and G. A. Petsko, *Biochemistry* **28,** 7289 (1989).
[7] J. M. Bolduc, D. H. Dyer, W. G. Scott, P. Singer, R. M. Sweet, D. E. Koshland, and B. L. Stoddard, *Science* **268,** 1312 (1995)
[8] K. H. Verschueren, F. Seljee, H. J. Rozeboom, K. H. Kalk, and B. W. Dijkstra, *Nature* **363,** 693 (1993).

ous solvents without cross-linking the crystals before the transfer.[9-11] If these nonaqueous solvents are chosen carefully, it is then possible to either increase or decrease the concentration of water in the active site. The Ringe group at Brandeis has done similar experiments, but has cross-linked the crystals prior to transfer.[12,13] Cross-linking seems to preserve the aqueous conformation of the protein in the organic solvent. This protocol allows the possibility of determining binding sites for particular molecular shapes and could lead to a new method of rational drug design.

In contrast, our technique allows changes to occur in the protein structure in the organic solvent when compared to the aqueous structure. Generally, the changes that we observe are what would be expected. Charged and polar groups that were on the surface of the protein in the aqueous structure move these groups away from the nonpolar environment. We have observed new salt bridges and hydrogen bonds in crystals transferred to the nonpolar solvents.[9,10] These side chain rotations are permitted because the crystals are not cross-linked.

In addition to side chain rearrangements, we have observed significant changes in the water structure on the surface of the protein. Both the location of water molecules and the number of observed water molecules change in organic solvents. In chymotrypsin, it is possible to lower the number of water molecules in the active site by transferring the crystals into hexane. In order to raise the number of waters on the surface of the protein (including the active site), we used a weak-binding ligand that was expected to cause such preferential hydration.[14] For the chymotrypsin system, the ligand was isopropanol.[10] Low concentrations (4%, v/v) of isopropanol caused the number of observable water molecules to double in either an aqueous or a nonaqueous environment.[9,10]

Nonaqueous solvents make it possible to extend the idea of trapping enzyme-bound intermediates using the thermodynamic approach to enzymes where one of the substrates is water. However, it is not easy to measure free energy profiles in such solvents. Almost all enzymes form heterogeneous mixtures when placed in organic solvents, and transport of substrate to the active site can become the rate-determining step in such mixtures, which makes trapping intermediates in organic solvents a more

[9] N. H. Yennawar, H. P. Yennawar, and G. K. Farber, *Biochemistry* **33,** 7326 (1994).
[10] H. P. Yennawar, N. H. Yennawar, and G. K. Farber, *J. Am. Chem. Soc.* **117,** 577 (1995).
[11] Farber, G. K., *Ann. N.Y. Acad. Sci.* **799,** 85 (1996).
[12] P. A. Fitzpatrick, A. C. Steinmetz, D. Ringe, and A. M. Klibanov, *Proc. Natl. Acad. Sci. U.S.A.* **90,** 8653 (1993).
[13] P. A. Fitzpatrick, D. Ringe, and A. M. Klibanov, *Biochem. Biophys. Res. Commun.* **198,** 675 (1994).
[14] S. N. Timasheff, *Biochemistry* **31,** 9857 (1992).

difficult challenge than trapping them by simply adjusting substrate concentrations.

Experimental Procedures

We have tried to transfer four different protein crystals (chymotrypsin, subtilisin, turkey egg white lysozyme, and ribonuclease) into organic solvents. With the exception of ribonuclease, these crystals all behave similarly in their ability to withstand transfer into organic solvents. The crystals are grown in the normal way from aqueous solutions. Once they have reached final size, the crystals are transferred to a quartz capillary, which is chosen so that the crystal will wedge partway down the capillary tube. It is important to choose the correct size of capillary as free flow of solution through the crystal is often critical to success in these experiments. If the capillary is blocked by fibers to keep the crystal in the tube,[15] the flow rate is reduced greatly.

Once the crystal is wedged inside the capillary tube, all of the visible water in the tube is removed. This is done using a combination of filter paper and thin drawn-out glass capillary tubes attached to a mouth pipette. This drying procedure should be completed as quickly as possible to prevent damage to the crystal. Once the crystal has been dried, the organic solvent is added to the top of the capillary tube and fills the whole capillary. Air bubbles that develop near the crystal as the organic solvent is added to the top of the capillary tube can also be removed using the thin drawn-out glass capillaries. The organic solvent should be stored over activated molecular sieves (4 Å, 1/8-in, 4–8 mesh), which will remove all of the water in the solvent. The capillary tube is then washed extensively by pipetting the organic solvent into the top of the tube and allowing it to drain out the bottom. After 10–20 column volumes have washed through the capillary, the tube is placed into a large reservoir of the organic solvent and allowed to equilibrate for 24 hr or longer before data collection. We typically use the glass tubes in which the quartz capillaries are delivered. The amount of water remaining in the lattice can be lowered by refreshing the solvent, as water is slightly soluble in these very nonpolar solvents. If water is a substrate in the reaction, loosely bound water can be also be eliminated by adding the other substrates and allowing the enzyme to catalyze the reaction and eliminate water during each turnover.

Generally, protein crystals are stable in very nonpolar solvents such as hexane or benzene. The crystal lifetime in these nonpolar solvents can be prolonged by including a very small molecule such as carbon disulfide,

[15] G. A. Petsko, *Methods Enzymol.* **114**, 141 (1985).

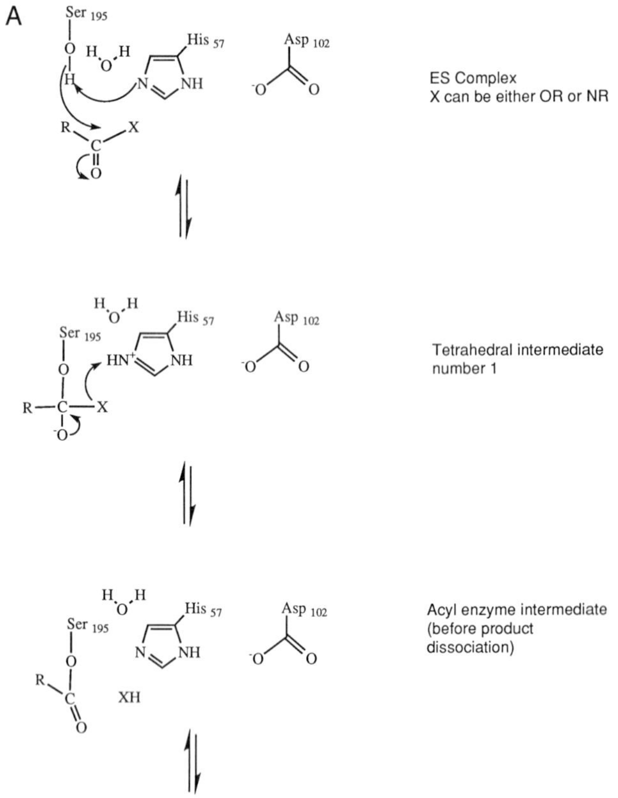

FIG. 2. Structures of the intermediates that occur during the serine protease reaction. Substrates with very good leaving groups may go through a tetrahedral transition state rather than a tetrahedral intermediate.

methanol, or acetonitrile at low concentration in the bulk solvent. After roughly 10 days in large nonpolar solvents, many of the crystals fracture and turn into piles of small crystals. Presumably, this occurs because as water leaves the crystal some of the spaces that were occupied by water are too small for a large molecule such as hexane. This problem is solved to some extent by the small organic molecules. However, crystals soaked in a solution containing a small organic molecule still fracture after roughly 15 days.

Once an appropriate solvent condition has been found, substrates can be added to the crystal in order to trap an enzyme-bound intermediate. As mentioned earlier, due to the problems in measuring accurate free energy profiles in organic solvents, this process has an element of trial and

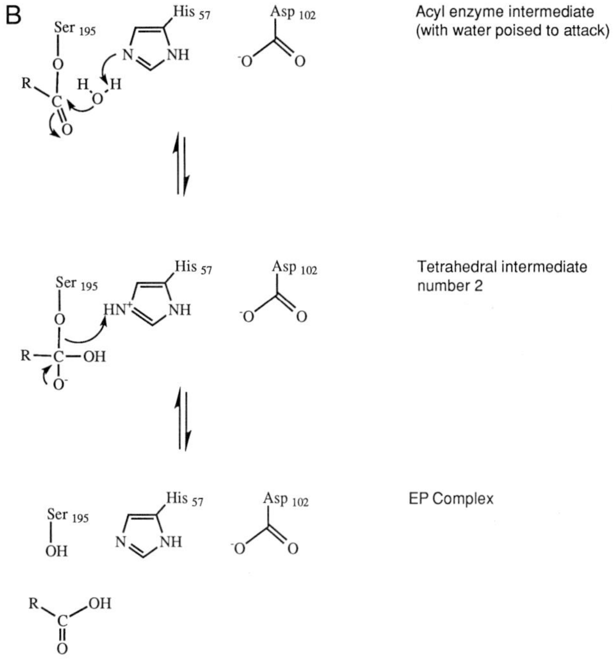

B

Ser 195   His 57   Asp 102   Acyl enzyme intermediate
(with water poised to attack)

Tetrahedral intermediate
number 2

EP Complex

FIG. 2. (*continued*)

error. Most of the intermediates in the reaction catalyzed by a serine protease have now been trapped. This movie of enzymatic catalysis is discussed in detail in the following section.

Chymotrypsin: A Case Study

Although there has been much debate about the chemical mechanism of the reaction catalyzed by serine proteases, there is now general agreement about the number of intermediates and the general structure of these intermediates.[16] An outline of the mechanism is shown in Fig. 2. There are six true intermediates that one could imagine trapping: the native enzyme, the enzyme · substrate complex, the first tetrahedral intermediate, the acyl enzyme intermediate, the second tetrahedral intermediate, and the enzyme · product complex. In aqueous solutions, the tetrahedral intermediates are much less stable than any of the other intermediates. At low pH, it is

[16] L. Polgar, in "Hydrolytic Enzymes" (A. Neuberger and K. Brocklehurst, eds.), p. 159. Elsevier Science, New York, 1987.

possible to trap the acyl enzyme.[17,18] Because this intermediate is in the middle of the reaction, it serves as a good starting point to trap all of the others. The structures of five of these intermediates and the methods used to trap these intermediates are discussed. Following that brief description, the motions that occur during catalysis are discussed.

Structure of the Native Enzyme

The structure used as the apo enzyme is the high-resolution structure of α-chymotrypsin by Tskuada and Blow.[19] The active site can be broken down into several different regions: the specificity pocket, the oxyanion hole, and the catalytic triad. The specificity pocket binds the side chain in the P1 subsite (adjacent to the scissile bond in the substrate). The pocket is defined by residues 189–195, 214–220, and 225–228. We observed that the side chain of Met-192, which forms part of the binding pocket, shows an interesting conformational change during the reaction. The oxyanion hole, which stabilizes the tetrahedral intermediates, is formed by the backbone nitrogens of residues 193 and 195. The infamous catalytic triad is formed by Ser-195, His-57, and Asp-102 (Fig. 3)

Structure of the First Tetrahedral Intermediate

In order to trap the tetrahedral intermediate, we began with the acyl enzyme structure and transferred the crystals into hexane.[9] During the crystallization process, the mother liquor that produced these crystals contained a variety of peptides and amino acids. These small molecules were still present in the solvent channels of the crystal when it was transferred into hexane. Under these conditions, it has been demonstrated that chymotrypsin catalyzes peptide bond synthesis.[9,20,21] Presumably, hexane alters the equilibrium to favor peptide bond synthesis by altering the protonation state of the attacking amino acid and by simply forcing the rather polar amino acids into the active site and away from the very nonpolar solvent. It was not clear to us whether the reaction would stop at the tetrahedral intermediate or continue on to the $E \cdot S$ intermediate. It seemed unlikely that the reaction would proceed completely to the apo enzyme as this would require that the peptide product would preferentially partition into hexane rather than stay in the polar active site. The tetrahedral intermediate

[17] M. M. Dixon and B. W. Matthews, *Biochemistry* **28,** 7033 (1989).
[18] M. Harel, C.-T. Su, F. Frolow, I. Silman, and J. L. Sussman, *Biochemistry* **30,** 5217 (1991).
[19] H. Tskuada and D. M. Blow, *J. Mol. Biol.* **262,** 703 (1985).
[20] G. A. Homandberg, J. A. Mattis, and M. Laskowski, Jr., *Biochemistry* **17,** 5220 (1978).
[21] V. Kasche, G. Michaelis, and B. Galunsky, *Biotechnol. Lett.* **13,** 75 (1991).

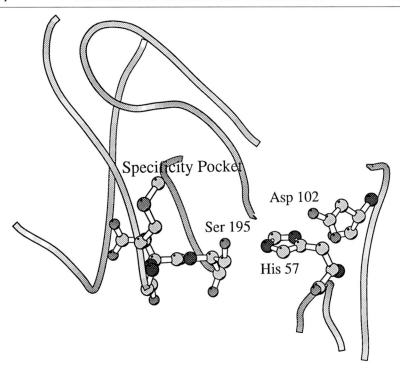

FIG. 3. A view of some of the residues in the active site region of chymotrypsin. Most backbone atoms are shown as a coil. When individual atoms are shown, carbons have the lightest gray shading, oxygen and sulfur atoms are darker, and nitrogen atoms are the darkest. The residues in the catalytic triad are shown as a ball and stick and are labeled. Most of the residues that form the borders of the specificity pocket are shown. The side chain of Met-192 is just below the word "specificity." The oxyanion hole is created by the two nitrogen atoms of residues 193 and 195. These are below the Met-192 side chain. This figure was made using Molscript.[24]

presumably was stabilized by the favorable hydrogen-bonding pattern in the active site for the attacking aspartic acid (Fig. 4).

Structure of the Acyl Enzyme Intermediate

For many years, the acyl enzyme intermediate was known as γ-chymotrypsin.[22] The difference between α- and γ-chymotrypsin was the space group and unit cell dimensions of the resulting crystals. γ-Chymotrypsin is created by taking an enzyme that would otherwise form α-chymotrypsin crystals and either placing it in a Waring blender for a few minutes or

[22] G. H. Cohen, E. W. Silverton, and D. R. Davies, *J. Mol. Biol.* **148,** 449 (1981).

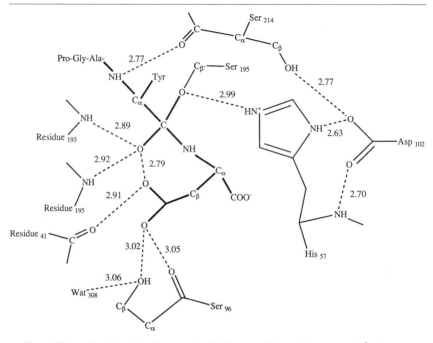

FIG. 4. The active site of the first tetrahedral intermediate of chymotrypsin.[9] The proton-ation state of the attacking side chain and the oxyanion in this nonaqueous solvent are not shown. There is at least one and perhaps as many as three protons distributed among these oxygen atoms.

incubating it at a pH above 7.0 and at raised temperatures for a few hours.[9,10,17,18,22] Either of these treatments seems to cause a fraction of the chymotrypsin molecules in solution to unfold. Because the incubations occur at pH values where chymotrypsin is very active and because the active chymotrypsin concentration is very high, the unfolded proteins are hydrolyzed into a variety of peptides and amino acids by the intact molecules of chymotrypsin.[18] Prior to crystallization, the pH of this mixture of intact protein and peptides is lowered to 5.6 to protonate His-57 and prevents any further reaction. Crystals that grow from this solution have a tetrapeptide attached covalently to the catalytic serine.[17,18] In our hands, the electron density for this tetrapeptide has been modeled as Pro–Gly–Ala–Tyr.[9,10] This tetrapeptide is presumably derived from a -Pro–Gly–Val–Tyr–Ala–Arg- sequence in chymotrypsin with the Val side chain disordered in the crystal and the Tyr–Ala bond cleaved by chymotrypsin. The structure determined in our laboratory[9] is used in the following comparisons. It is very similar to other reported structures of the acyl enzyme.[17,18] The electron density for the carbon bound to the γ oxygen of Ser-195 is unambiguously trigonal planar.

Structure on the Way to the Second Tetrahedral Intermediate

Singer *et al.*[23] have trapped a structure between the acyl enzyme intermediate and the second tetrahedral intermediate using the Laue method of data collection. These experiments were done using trypsin that had been treated with *p*-nitrophenyl guanidinobenzoate to form a covalent bond to the active site serine. This derivatized form of trypsin is analogous to the acyl enzyme in chymotrypsin. It is stable for days at low pH values and hydrolyzes in hours at high pH values. All data for this experiment were collected using a single crystal mounted in a flow cell. The native structure of the *p*-guanidinobenzoyl trypsin was taken with the crystal surrounded by a buffer at pH 5.5. Then the pH of the elution buffer was raised to 8.5, and two data sets were collected both 3 min after the pH jump and 90 min after the jump. In both of the high pH structures, a new water molecule is observed in the active site, which is located in a position that is well suited for attack on the ester formed at the catalytic serine. This water molecule is located in the same position as the attacking Asp group found in the first tetrahedral intermediate.[9] The fact that both of these structures find the attacking group in the same location is strong evidence that both have trapped enzymatically relevant intermediates.

It should be emphasized that because this was a Laue experiment, the system had not settled to equilibrium. The Laue method shows a time dependent snapshot of the structure(s) in the crystal as the reaction proceeds from the acyl enzyme toward the second tetrahedral intermediate. In this experiment, the reaction was sufficiently slow so that the Laue data sets showed a single clear picture of a structure between these two intermediates.

Structure of the Enzyme Product Complex

The final intermediate in the serine protease pathway that has been trapped successfully is the E · P complex prior to dissociation.[10] Once again, the acyl enzyme of chymotrypsin was the starting point for this structure. In order to get from the acyl enzyme to the E · P complex, it is necessary to increase the concentration of water molecules in the active site while still making the solvent sufficiently nonpolar so that the product does not dissociate away from the active site. We have observed that adding a low

[23] P. T. Singer, A. Smalås, R. P. Carty, W. F. Mangel, and R. M. Sweet, *Science* **259,** 669 (1993).
[24] P. J. Kraulis, *J. Appl. Cryst.* **24,** 946 (1991).

percentage of a weak binding ligand to the solution around a crystal causes the number of observable water molecules to increase dramatically.[10] This increased hydration occurs in either aqueous or nonaqueous solutions and has a good theoretical basis.[10,14]

To form the E · P complex, we took crystals of γ-chymotrypsin (the acyl enzyme) and changed the surrounding solvent to a solution of 96% hexane, 4% isopropanol. The number of water molecules observed went from 130 to 307 between the 100% hexane structure and the 96% hexane, 4% isopropanol structure. A similar increase was observed in control experiments done in aqueous solutions.[10] In the active site, the density between Ser-195 and the product that was clear and continuous in both the acyl enzyme and the tetrahedral intermediate structures was broken in the E · P complex. The water molecules present in the acyl enzyme structure are missing in the E · P complex. Presumably, this occurs due to one water being incorporated into the product and the resulting movement of the product toward the solvent. The product does not dissociate completely as the bulk solvent is still rather nonpolar.

Motions that Occur between Intermediates

Once all of these structures were determined, it was straightforward to overlap these structures to observe the motions that occur during catalysis. Superpositions were done using the α carbons of the whole protein. The acyl enzyme of chymotrypsin was chosen as the base model for these superpositions because it is in the middle of the reaction. In the case of trypsin, a few α carbons that clearly did not have any equivalent residues in the chymotrypsin molecule were omitted. In general, when the α carbons of all of these structures are superimposed, the differences between the structures are rather small. Encouragingly, the largest deviations occur in the active site region.

The largest change that occurs between the native enzyme and the first tetrahedral intermediate is in the region of the oxyanion hole. The backbone atoms near nitrogen-193 rearrange to move this nitrogen into a much better position to interact with the substrate oxygen. In the apo enzyme, N-193 is 3.39 Å away from the location of the substrate oxygen in the oxyanion hole. In the tetrahedral intermediate, this distance is 2.89 Å. Interestingly, the distance between the oxygen and the nitrogen that makes up the other side of the oxyanion hole also changes. The N–O distance in the native enzyme would be 2.77 Å. In the tetrahedral intermediate the distance is 2.92 Å.

Changes in the catalytic triad are much smaller than the change in the oxyanion hole. There is almost no change in the position of Ser-195 between these two structures. The imidazole ring of His-57 does rotate a little bit. The largest changes involve the ε carbon and nitrogen. Smaller changes are observed between the δ carbon and nitrogen. The oxygen of Asp-102, which makes a short hydrogen bond to the δ nitrogen of His-57, does move a little bit to track the motions of the imidazole ring. This tracking motion is observed during other steps in the reaction. The other oxygen of the Asp side chain is hydrogen bound to the backbone atoms of residues 57 and 56. These atoms do not seem to move at all during the reaction and serve as an anchor for the triad.

In summary, the oxyanion hole rearranges between the native enzyme and the tetrahedral intermediate by motions involving the backbone near N-193. In the tetrahedral intermediate, both of the hydrogen bonds between the substrate oxygen and the backbone nitrogens are almost identical in length. Because there is no E·S structure, it is unclear whether substrate binding is enough to initiate the changes near N-193 or whether these occur during the transition from the E·S to the first tetrahedral intermediate.

The transition between the first tetrahedral intermediate and the acyl enzyme intermediate involves almost no change in the structure of the protein. Atoms making up the oxyanion hole do not move between these two intermediates. The hydrogen bond distances between the oxygen and the backbone nitrogens do lengthen as the substrate goes from the tetrahedral structure to the trigonal planar intermediate. The distance between oxygen and N-195 is 3.05 in the acyl enzyme, and the bond between substrate oxygen and N-193 has increased to 2.95.

The motions in the catalytic triad are similarly small. The Ser and His side chains do not appear to move at all. The oxygen of the Asp that hydrogen bonds to the His is very slightly different, but this change is so small that it is unlikely to be meaningful. The bond length between the Asp and the His is still very short: 2.64 Å.

The transition from the acyl enzyme toward the second tetrahedral intermediate is where most of the large changes occur. Potentially, there are three sources for these changes: the molecule is now trypsin, not chymotrypsin; the structure is not a true low energy intermediate, rather it is a time point during the chemical reaction; and, most interestingly, this is the location in the reaction that should require the largest conformational changes. This is because the water molecule that is going to be attacked by His-57 cannot occupy exactly the same space as the Ser that was deprotonated in the first half of the reaction. It is therefore necessary that either the Ser move out of the way or that the His ring rotates to attack a water molecule. The Laue structure of trypsin suggests that the later explanation is true.

The largest changes occur in the catalytic triad. As suggested earlier, the motions from the acyl enzyme toward the second tetrahedral intermediate are smallest at Asp-102 and get much larger at His-57 and Ser-195. The imidazole ring of His-57 both translates away from Asp-102 and rotates about the bond between β and γ carbon atoms. The surprising result of this motion is that the distance between the δ nitrogen of the histidine and the oxygen of the Asp side chain has lengthened to 2.86 Å—well outside the range of short-strong hydrogen bonds. The bond between the ε nitrogen and the γ oxygen of Ser-195 has also lengthened significantly to 3.84 Å. This motion makes sense because the His ring is rotating to attack the water molecule in the second half of the reaction.

The hydrolytic water molecule that is observed in both high pH Laue structures is located in the same area as the nitrogen of the attacking aspartic acid in the first tetrahedral intermediate. The fact that these two

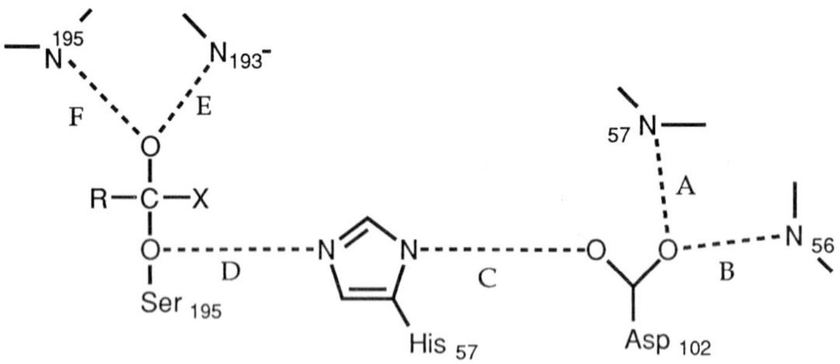

Structure	PDB	A	B	C	D	E	F
apo enzyme	4CHA	2.76	2.78	2.61	2.82		
first tetrahedral	1GMD	2.70	2.82	2.63	2.99	2.89	2.92
acyl enzyme	1GMC	2.87	2.84	2.64	3.14	2.95	3.05
Laue structure		2.94	2.94	2.86	3.84	3.34	3.19
E·P complex	1GHB	2.83	2.89	2.60	3.02		

FIG. 5. Key hydrogen bond distances in the active site of chymotrypsin during a catalytic cycle. PDB lists the four-letter code for the coordinate file used in these comparisons. The Laue structure was generously provided by Dr. Robert Sweet. All distances are given in angstroms.

structures arrive at the same general location for either the attacking water or the attacking amino acid suggests that both structures are relevant to normal catalysis in an aqueous solution.

In the oxyanion hole, the oxygen of the substrate is much closer to the position of the oxygen of the first tetrahedral intermediate than of the oxygen of the acyl enzyme. However, the rearrangement of the backbone atoms that will bring N-195 and N-193 closer to the oxygen has not yet occurred in the Laue structure. Both of these distances are rather long (3.19 and 3.34 Å) in the Laue structure.

The final set of motions to consider involves the formation of the E · P complex. When compared to the preceding structures there are two notable changes. The first involves the γ oxygen of Ser-195. When all five structures (native, tetrahedral intermediate, acyl enzyme, Laue structure, and E · P complex) are superimposed, the largest motion among the active site residues involves this oxygen. In the E · P structure it is 0.95 Å away from the position in the native structure. The positions of the other three structures are intermediate between these two extremes. In addition to this motion, the side chain of Met-192 shows an interesting motion in the E · P complex. In all other structures the terminal methyl is oriented away from the specificity pocket (in trypsin this residue is a Gln but the terminal side chain atoms of the Gln are in a similar location to the methyl group of Met-192). In the E · P complex, this methyl group has rotated into the specificity pocket. Presumably, this torsion helps push the product out of the active site and toward the solvent. All of these motions are summarized in Fig. 5.

Conclusions

The chymotrypsin experiments described in this article demonstrate that the approach of trying to thermodynamically trap the intermediates during an enzymatic reaction can be successful even in a case where one of the substrates is water. The motions that were observed between these intermediates are largely what were expected. The only truly surprising result is the observation of a normal rather than a short hydrogen bond between the His and the Asp in the Laue structure. All of the rest of the observed motions can be explained easily.

These experiments suggest that introducing nonaqueous solvents into a protein crystal causes only minor perturbations and that the resulting structures do have relevance to the reaction that occurs in aqueous solvents. Of course, this will have to be tested with other enzymes, but the approach has worked well for the serine protease reaction.

Section III

Detection and Properties of Low-Barrier Hydrogen Bonds

[10] Nuclear Magnetic Resonance Methods for the Detection and Study of Low-Barrier Hydrogen Bonds on Enzymes

By ALBERT S. MILDVAN, THOMAS K. HARRIS, and
CHITRANANDA ABEYGUNAWARDANA

Introduction

Hydrogen bonds in nature range in strength from very weak ($\Delta G_{\text{formation}} \sim -2$ kcal/mol) to very strong ($\Delta G_{\text{formation}} < -25$ kcal/mol) approaching, in some cases, the strengths of covalent bonds.[1] In weak hydrogen bonds, the proton is bonded covalently to one electronegative atom at a distance of ~1.0 Å, and electrostatically attracted to another electronegative atom at approximately twice the distance (1.7 to 2.0 Å) (Fig. 1). The overall hydrogen bond length, given by the distance between the two electronegative or heavy atoms, is 2.7 to 3.0 Å. The proton has two alternative locations or free energy wells as it can be bonded covalently to either A or B [Eq. (1)]:

$$-\text{A}-\text{H}\cdots\cdots:\text{B}-$$
$$\updownarrow \tag{1}$$
$$-\text{A}:\cdots\cdots\text{H}-\text{B}-$$

As the electronegative atoms A and B approach each other, shortening the hydrogen bond to values between 2.60 and 2.50 Å, the covalent A–H moiety lengthens, the noncovalent moiety H$\cdots$B shortens, the hydrogen bond becomes stronger (-7 kcal/mol $\geq \Delta G_{\text{formation}} \geq -25$ kcal/mol), and the barrier between the two proton wells becomes lower[1-3] (Fig. 1). When the barrier height approaches the zero-point vibrational energy of hydrogen, the proton becomes delocalized between the two wells, while a deuteron would remain localized. This state is referred to as a low-barrier hydrogen bond (LBHB). Specifically, as O–H$\cdots$O hydrogen bond lengths decrease from 2.54 to 2.45 Å, their strengths ($\Delta E_{\text{formation}}$) increase rapidly from -7.8 to -32 kcal/mol,[1] presumably due to exponentially increasing overlap of the proton and oxygen wave functions. As the heavy atoms continue to approach each other, further shortening the hydrogen bond lengths to values between 2.45 and 2.30 Å, the barrier between the wells disappears

---

[1] F. Hibbert and J. Emsley, *Adv Phys. Organ. Chem.* **26**, 255 (1990).
[2] M. M. Kreevoy and T. M. Liang, *J. Am. Chem. Soc.* **102**, 3315 (1980).
[3] Th. Steiner and W. Saenger, *Acta Cryst.* **B50**, 348 (1994).

Copyright © 1999 by Academic Press
All rights of reproduction in any form reserved.
0076-6879/99 $30.00

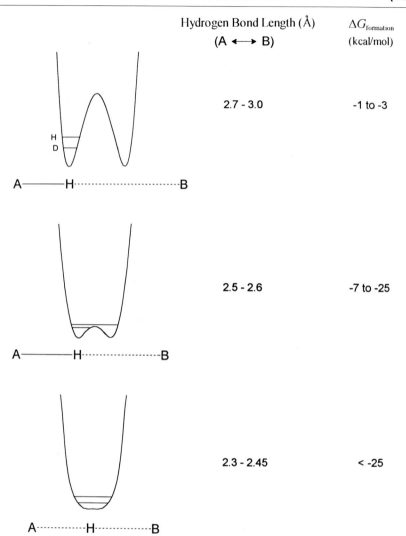

Hydrogen Bond Length (Å) $\Delta G_{\text{formation}}$
(A ⟷ B)   (kcal/mol)

2.7 - 3.0   -1 to -3

2.5 - 2.6   -7 to -25

2.3 - 2.45   < -25

FIG. 1. Potential functions for weak, double-well (upper), strong, low-barrier (middle), and very strong, single-well hydrogen bonds (lower) in which the pK_a values of A–H and B–H are equal. Horizontal lines give the zero-point vibrational energy levels of protium and deuterium. Also shown are the lengths and strengths of such hydrogen bonds.

and the proton (or deuteron) is in a single well (Fig. 1). In such single-well hydrogen bonds, which are very strong ($\Delta G_{formation} < -25$ kcal/mol), the average position of the proton is now equidistant from the two electronegative atoms. Further decreases in hydrogen bond length now decrease the distances between the proton and both heavy atoms. Single-well hydrogen bonds have not been reported in proteins.

While the strength of hydrogen bonds correlates strongly with their lengths (Fig. 1), little or no correlation of strength with hydrogen bond angle, A–H····B, over a range 180 ± 30°, is found experimentally[4] and theoretically.[5] Angle bending beyond ±30° can lead to weakening of strong hydrogen bonds.[6] Classic examples of very strong, single-well hydrogen bonds are $F^-\cdots H^+\cdots F^-$ ($r = 2.27$ Å, $\Delta E_{formation} = -39$ kcal/mol)[7] and $HO^-\cdots H^+\cdots {}^-OH$ ($r = 2.29$ Å, $\Delta E_{formation} = -32$ kcal/mol)[8] in the gas phase. In aqueous environments, due to competition with the solvent, weak hydrogen bonding is expected and is generally found.[9] Thus the average strength of hydrogen bonds contributing to protein structure has been estimated calorimetrically as $\Delta G_{formation} \sim -2 \pm 1$ kcal/mol.[10] Similarly, in nucleic acids the hydrogen bonds responsible for base pairing are comparably weak.[11] Hence strong or low-barrier hydrogen bonds in biological systems with $\Delta G_{formation} \leq -7$ kcal/mol and with lengths ≤2.6 Å are unusual and noteworthy.

Although LBHBs can be detected in small molecules by vibrational, ultraviolet, and nuclear magnetic resonance (NMR) spectroscopies,[1,12,13] they are studied most readily in biological macromolecules by high-resolution proton NMR spectroscopy as this method permits their structural assignments. To directly relate structure to function, LBHBs are best studied in enzymes. LBHBs have been proposed as participants in the mechanisms of action of hydrolase enzymes,[14–19] lyases,[15] a ligase,[15] and several

[4] J. Donohue, in "Structural Chemistry and Molecular Biology" (A. Rich and N. Davidson, eds.), p. 443. Freeman, San Francisco, 1968.

[5] H. Adalsteinsson, A. H. Maulitz, and T. C. Bruice, J. Am. Chem. Soc. 119, 7689 (1996).

[6] C. J. Smallwood and M. A. McAllister, J. Am. Chem. Soc. 119, 11277 (1997).

[7] T. B. McMahon and J. W. Larson, J. Am. Chem. Soc. 104, 5848 (1982).

[8] K. Abu-Dari, K. N. Raymond, and D. P. Freyberg, J. Am. Chem. Soc 101, 3639 (1979).

[9] S. Shan and D. Herschlag, Proc. Natl. Acad. Sci. U.S.A. 93, 14474 (1996).

[10] G. I. Makhatadze and P. Privalov, Adv. Prot. Chem. 47, 307 (1995).

[11] J. D. Watson, N. H. Hopkins, J. W. Roberts, J. A. Steitz, and A. M. Weiner, "Molecular Biology of the Gene," p. 132. Benjamin/Cummings, Menlo Park, CA, 1987.

[12] S. N. Smirnov, N. S. Golubev, G. S. Denisov, H. Benedict, P. Schah-Mohammedi, and H. H. Limbach, J. Am. Chem. Soc. 118, 4094 (1996).

[13] N. S. Golubev, G. S. Denisov, S. N. Smirnov, D. N. Shchepkin, and H. H. Limbach, Z. Physikal. Chem. 196, 73 (1996).

[14] W. W. Cleland, Biochemistry 31, 317 (1992).

isomerases.[15,20–22] NMR spectroscopic evidence for strong hydrogen bonding has been obtained in serine proteases,[16–19,23–25] aspartate aminotransferase,[26,27] ketosteroid isomerase,[28,29] and triosephosphate isomerase[30] (Fig. 2). It is noteworthy that all LBHBs detected thus far on enzymes involve carboxyl groups.[19,27,29,30]

NMR Criteria for Low-Barrier Hydrogen Bonds

Four useful NMR criteria for detecting strong hydrogen bonds in proteins are the proton chemical shift, the fractionation factor, the exchange rate protection factor, and the free energy of formation of the hydrogen bond (Table I). While a strong hydrogen bond is established most directly by measurement of its $\Delta G_{formation}$, such measurements in macromolecules are of necessity indirect because of the multiplicity of interactions in such complex systems. The theoretical basis and methods of measurement of each of these four parameters will be considered.

Proton Chemical Shift and Assignment of Deshielded Proton Resonances

Theory

Correlations of hydrogen bond lengths measured by X-ray diffraction with the chemical shift of the proton in the hydrogen bond have established

[15] W. W. Cleland and M. M. Kreevoy, *Science* **264**, 1887 (1994).

[16] G. Robillard and R. G. Shulman, *J. Mol. Biol.* **71**, 507 (1972).

[17] N. S. Golubev, V. A. Gindin, S. S. Ligai, and S. N. Smirnov, *Biochemistry* (*Moscow*) **59**, 613 (1994).

[18] P. A. Frey, S. A. Whitt, and J. B. Tobin, *Science* **264**, 1927 (1994).

[19] C. S. Cassidy, J. Lin, and P. A. Frey, *Biochemistry* **36**, 4576 (1997).

[20] Y.-K. Li, A. Kuliopulos, A. S. Mildvan, and P. Talalay, *Biochemistry* **32**, 1816 (1993).

[21] Q. Zhao, A. S. Mildvan, and P. Talalay, *Biochemistry* **34**, 426 (1995).

[22] J. A. Gerlt and P. G. Gassman, *J. Am. Chem. Soc.* **115**, 11552 (1993).

[23] J. Markley and I. B. Ibanez, *Biochemistry* **17**, 4637 (1978).

[24] T.-C. Liang and R. H. Abeles, *Biochemistry* **26**, 7603 (1987).

[25] J. L. Markley and W. M. Westler, *Biochemistry* **35**, 11092 (1996).

[26] A. Kintanar, C. M. Metzler, D. E. Metzler, and, R. D. Scott, *J. Biol. Chem.* **266**, 17222 (1991).

[27] E. Mollova, D. E. Metzler, A. Kintanar, H. Kagamiyama, H. Hayashi, K. Hirotsu, and I. Miyahara, *Biochemistry* **36**, 615 (1997).

[28] Q. Zhao, C. Abeygunawardana, P. Talalay, and A. S. Mildvan, *Proc. Natl. Acad. Sci. U.S.A.* **93**, 8220 (1996).

[29] Q. Zhao, C. Abeygunawardana, A. G. Gittis, and A. S. Mildvan, *Biochemistry* **36**, 14616 (1997).

[30] T. K. Harris, C. Abeygunawardana, and A. S. Mildvan, *Biochemistry* **36**, 14661 (1997).

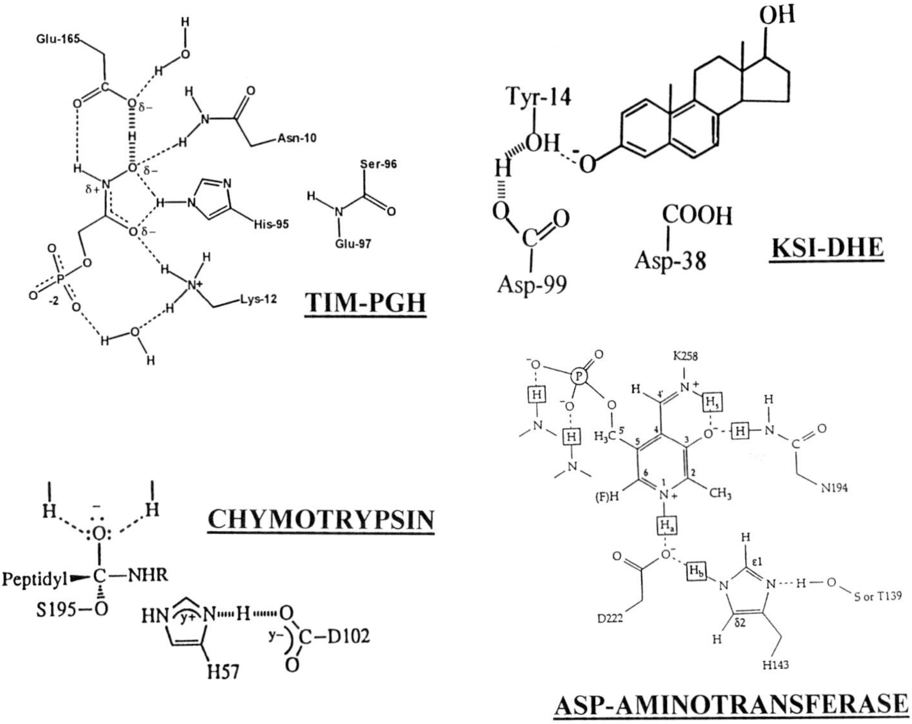

FIG. 2. Examples of low-barrier hydrogen bonds on enzymes based on NMR data. For details on the triosephosphate isomerase complex with phosphoglycolohydroxamic acid (TIM-PGH), see T. K. Harris, C. Abeygunawardana, and A. S. Mildvan, *Biochemistry* **36,** 14661 (1997); for the ketosteroid isomerase complex of dihydroequilenin (KSI-DHE), see Q. Zhao, C. Abeygunawardana, A. G. Gittis, and A. S. Mildvan, *Biochemistry* **36,** 14616 (1997); for chymotrypsin, see C. S. Cassidy, J. Lin, and P. A. Frey, *Biochemistry* **36,** 4576 (1997); and for aspartate aminotransferase, see D. E. Metzler, *Methods Enzymol.* **280,** article [4].

TABLE I
NMR CRITERIA FOR LOW-BARRIER HYDROGEN BONDS

1. Chemical shift: 15 to 20 ppm. $\Delta\delta = (\delta^{obs} - \delta^{intrinsic}) \geq 5$ ppm

2. Fractionation factor: $\phi = \dfrac{[\text{Enz-D}][\text{H}_{solvent}]}{[\text{Enz-H}][\text{D}_{solvent}]} \leq 0.4$

3. Exchange rate protection factor: $\dfrac{k_{intrinsic}}{k_{ex}} \geq 10$

4. Hydrogen bond strength ≤ -7 kcal/mol

that as hydrogen bonds decrease in length from 3.0 to 2.45 Å, i.e., proceed from weak to strong (Fig. 1), the proton becomes more deshielded, i.e., its resonance shifts downfield, approaching a maximum value of ~21 ppm.[31] A proton involved in a strong hydrogen bond should be deshielded by at least 5 ppm from its chemical shift in the absence of hydrogen bonding, as in an organic solvent (Table I). The reason for deshielding is the previously mentioned lengthening of the covalent A–H moiety of the A–H$\cdots$B hydrogen bond. This lengthening attenuates the major shielding provided by the σ bonding electron pair.

At hydrogen bond lengths below 2.45 Å, which have not been detected in proteins, the hydrogen bond becomes very strong and single well, i.e., the proton becomes equidistant from both A and B (Fig. 1). Shortening of the hydrogen bond below 2.45 Å now increases the shielding of the proton by electron pairs from both A and B, resulting in an upfield chemical shift.[1] A biochemically useful correlation of proton chemical shifts, corrected for amino acid type and ring current effects ($\Delta\delta$), with noncovalent (H$\cdots$O) lengths (d) for weak peptide NH$\cdots$O hydrogen bonds in proteins[32] has been fit by Eq. (2), permitting the estimation of unknown NH$\cdots$O distances from chemical shifts.[33]

$$\Delta\delta = \frac{19.2}{d^3} - 2.3 \tag{2}$$

Equation (2) is justified theoretically at least for long, weak hydrogen bonds, as electric field effects, local magnetic anisotropies, and polarization of the electron cloud near the hydrogen atom by the proximity of an oxygen all have a d^{-3} dependence.[32]

A rigorous and highly useful correlation of a wide range of full O–H$\cdots$O hydrogen bond lengths (D), measured by small molecule high-resolution X-ray crystallography, with chemical shifts (δ) determined by solid-state NMR in the same crystals[31] is given in Fig. 3. While the data of Fig. 3 can be fit by an inverse cubic relationship of the form of Eq. (2), which has some theoretical justification for long, weak hydrogen bonds,[32] they are better fit empirically by Eq. (3)[29]:

$$D = 5.04 - 1.16 \ln(\delta) + 0.0447(\delta) \tag{3}$$

In Eq. (3), D refers to the distance between the heavy atoms. The improved fit of the exponential function, especially for short, strong O–H$\cdots$O hydro-

[31] A. McDermott and C. F. Ridenour, in "Encyclopedia of NMR," p. 3820. Wiley, Sussex, England, 1996.
[32] G. Wagner, A. Pardi, and K. Wüthrich, J. Am. Chem. Soc. 105, 5948 (1983).
[33] H. Li, H. Yamada, and K. Akasaka, Biochemistry 37, 1167 (1998).

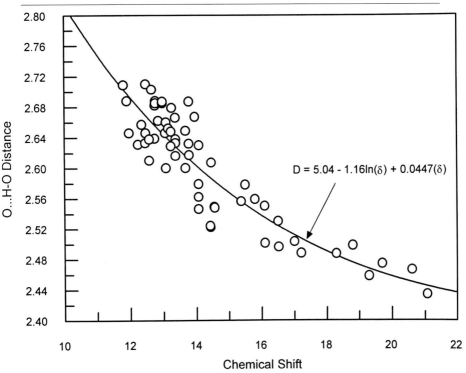

FIG. 3. Correlation of O–H·····O hydrogen bond distances (*D*) from X-ray diffraction with ¹H chemical shifts from solid-state NMR of crystalline amino acids. Data are from A. McDermott and C. F. Ridenour, *in* "Encyclopedia of NMR," p. 3820. Wiley, Sussex, England, 1996.

gen bonds, may reflect incipient orbital overlap with an exponentially decreasing wave function. Figure 3 and Eq. (3) have been very useful in estimating the lengths of strong, low-barrier O–H·····O hydrogen bonds in proteins with errors ≤0.05 Å.[29,30] These distances are summarized in Table II and are compared with distances obtained from fractionation factors and from protein X-ray diffraction. Distances obtained from chemical shifts are more precise and probably more accurate than those determined by protein X-ray diffraction, which, as a rule of thumb, have errors ranging from 0.1 to 0.3 times the resolution, depending on the quality of the data.

Chemical Shift Measurements

The detection and measurement of chemical shifts of resonances of protons that are involved in strong hydrogen bonds generally require high

TABLE II

COMPARISON OF HYDROGEN BOND LENGTHS ON ENZYMES DERIVED FROM PROTON CHEMICAL
SHIFTS (δ) AND FRACTIONATION FACTORS (ϕ) WITH THOSE FROM X-RAY CRYSTALLOGRAPHY

Enzyme complex	Interaction	Hydrogen bond length (Å) from		
		δ	ϕ	X ray
KSI–DHE[a]	D99···Y14	2.49 ± 0.02[b]	2.50 ± 0.01[b]	3.9[c]
	Y14···O–C	2.72 ± 0.02[b]	2.68 ± 0.02[b]	2.61, 2.63[c]
TIM–PGH[d]	E165···HON	2.57 ± 0.05[e]	2.52 ± 0.02[e]	2.68, 3.08[f]
	H95···O=C		2.62 ± 0.01[e]	2.66, 3.12[f]
H[+]–chymotrypsinogen	H57···D102		2.54 ± 0.01[g]	2.62[h]
H[+]–chymotrypsin	H57···D102		2.63 ± 0.01[i]	2.62[h]
Chymotrypsin–BoroPhe[j]	H57···D102		2.63 ± 0.01[i]	2.52–2.59[k]
Subtilisin–BoroPhe[j]	H64···D32		2.60 ± 0.01[i]	2.52–2.59[k]

[a] Ketosteroid isomerase complex of dihydroequilenin.

[b] Q. Zhao, C. Abeygunawardana, A. G. Gittis, and A. S. Mildvan, *Biochemistry* **36,** 14616 (1997).

[c] S. W. Kim, S.-S. Cha, H.-S. Cho, J.-S. Kim, N.-C. Ha, M.-J. Cho, S. Joo, K. K. Kim, K. Y. Choi, and B.-H. Oh, *Biochemistry* **36,** 14030 (1997).

[d] Triosephosphate isomerase complex of phosphoglycolohydroxamic acid.

[e] T. K. Harris, C. Abeygunawardana, and A. S. Mildvan, *Biochemistry* **36,** 14661 (1997).

[f] R. C. Davenport, P. A. Bash, B. A. Seaton, M. Karplus, G. A. Petsko, and D. Ringe, *Biochemistry* **30,** 5821 (1991).

[g] J. L. Markley and W. M. Westler, *Biochemistry* **35,** 11092 (1996).

[h] M. N. G. James, A. R. Sielecki, G. D. Brayer, and L. T. Delbaere, *J. Mol. Biol.* **144,** 43 (1980).

[i] W. P. Huskey, personal communication. For details, see D. Bao, C. A. Kettner, W. P. Huskey, and F. Jordan, submitted for publication (1999).

[j] BoroPhe is methoxysuccinyl-Ala-Ala-Pro-2-amino-3-phenylethylboronic acid.

[k] K. Brady, A. Wei, D. Ringe, and R. H. Abeles, *Biochemistry* **29,** 7600 (1990).

sensitivity and low temperatures to suppress resonance broadening due to proton exchange with solvent, which may cause such signals to disappear at room temperature. Line widths decrease with decreasing temperature, reach a minimum as exchange is slowed, and then start broadening again due to dipolar effects (see later). For example, the proton resonances involved in the LBHBs of ketosteroid isomerase[28,29] and triosephosphate isomerase[30] show narrowest lines at −3.3° and −4.8°, respectively. To avoid sample freezing at subzero temperatures and to permit field/frequency locking in H_2O solutions, miscible deuterated solvents may be added to the sample in amounts too low to affect catalytic activity or protein structure. Typically, 10% (v/v) DMSO-d_6 provides a strong deuterium signal for field frequency locking and permits a temperature of −4.8° without sample freezing.[28–30] High sensitivity is achieved by strong magnetic fields as provided by 400- to 750-MHz spectrometers equipped with proton probes with temperature control to $\pm 0.1°C$. Protein samples should be in the range of 0.1 to 1.0 mM, noting that ≥ 0.3 mM is needed for the accurate determination of fractionation factors (see later).

Temperature Measurement and Calibration

To characterize hydrogen bonds and to obtain accurate kinetic parameters (see later), the sample temperature in the NMR probe needs to be measured accurately. The set temperature of the NMR spectrometer provides only a coarse estimate of the actual sample temperature. To obtain more accurate measurements and to correlate them with the set temperature, a calibration curve is constructed using the standard neat methanol sample.[34] The actual sample temperature $T(°K)$ for the range 250–320°K is given by Eq. (4):

$$T(°K) = 403.0 - 29.48|\Delta\delta| - 23.81(\Delta\delta)^2 \qquad (4)$$

where $\Delta\delta$ is the chemical shift difference between methyl and hydroxyl protons in ppm.

For accurate determination of chemical shifts, a chemical shift external standard is prepared containing 1–2 mM sodium 3-(trimethylsilyl)propionate-2,2,3,3-d_4 (TSP) in D_2O together with DMSO-d_6 in the same amount as in the sample and at the same volume. After the field is locked on the strong DMSO-d_6 signal and the temperature has equilibrated for about 15 min, the frequency of the methyl proton resonance of TSP is obtained and defined as 0.0 ppm. The chemical shift of the HDO resonance in the same sample is recorded. This value is very sensitive to temperature, moving upfield with increasing temperature, and is typically 5.06 ppm downfield from TSP at $-4.8°$, 5.01 ppm at $0°$, and 4.77 ppm at $25°$, providing an internal chemical shift reference for the protein sample. The chemical shift of water with respect to *internal* TSP also provides an independent estimate of the sample temperature, which is closely approximated by Eq. (5):

$$T(°C) = 104.69[5.012 - \delta(H_2O)] \qquad (5)$$

where $\delta(H_2O)$ is the chemical shift of water with respect to internal TSP at pH 5.5.[35,36] Because the chemical shift of TSP itself is sensitive to solvents and to pH changes,[36] Eq. (5) provides a somewhat less accurate measurement of sample temperature than Eq. (4) using neat methanol.

The broad water signal in the protein sample may preclude accurate setting of the carrier frequency, shimming, or chemical shift referencing.

[34] S. S. Raiford, C. L. Fisk, and E. D. Becker, *Anal. Chem.* **51**, 2050 (1979).
[35] L. P. M. Orbons, G. A. van der Marel, and C. Altona, *Eur. J. Biochem.* **170**, 225 (1987).
[36] D. S. Wishart and B. D. Sykes, *Methods Enzymol.* **239**, 363 (1994).

In such cases, the just-mentioned blank sample with identical sample height, buffer, and DMSO-d_6 content made up in D_2O and containing 1–2 mM TSP as a chemical shift standard and can be used for these purposes. Once the shims have been optimized on this blank sample, the carrier frequency can be positioned readily on the residual HDO line, which also provides an accurate chemical shift standard. Use of this HDO frequency both as the carrier and as the chemical shift reference for the sample containing H_2O and protein provides reproduceable chemical shift values for the deshielded protons.

In the protein sample in H_2O, detection of the weak low field signals of the protons involved in strong hydrogen bonds requires suppression of the intense water resonance. This must be accomplished without saturating or otherwise exciting the water resonance in order to avoid the transfer of saturation to the resonances of the hydrogen-bonded protons that could abolish them. Deshielded proton resonances may also show relatively fast transverse relaxation. Therefore the ideal detection pulse is a short, selective excitation pulse that delivers zero net excitation at the water resonance. Among the various pulse schemes designed for this purpose,[37] we have found the binomial 1331 sequence[38] to be satisfactory for one-dimensional (1D) spectra. This sequence provides the best single-scan water suppression (or minimal excitation of water), however, at the expense of a large, frequency-dependent phase correction and severe baseline roll. These problems can be remedied during data processing by linear prediction of the first few data points before Fourier transformation and with baseline flattening corrections. The 90° pulse width is divided into four parts with pulse widths in the ratio $1:3:3:1$. The equal delays (τ_D) between the 1331 pulses are adjusted to provide maximal excitation at the frequency of interest and minimal excitation of the water resonance. This is accomplished by setting $\tau_D = 1/(2\Delta\nu)$, where $\Delta\nu$ is the frequency difference in Hz between the water resonance and the deshielded resonance of interest. The carrier frequency is set at the water resonance. The recycle time, given by the sum of the relaxation delay (~3 sec) and the acquisition time (~0.5 sec), should exceed three times the T_1 of the slowest relaxing proton resonance of interest. Because 1331 pulses produce a frequency-dependent excitation profile across the spectrum, care must be taken when comparing the intensities of various signals. Alternative methods for water suppression that have been used for observing deshielded proton resonances[23,24] are the 1-1 pulse

[37] P. J. Hore, *Methods Enzymol.* **176,** 64 (1989).
[38] D. L. Turner, *J. Magn. Reson.* **54,** 146 (1983).

followed by a 1-1 refocusing pulse[39] (see later) and the jump-return pulse sequence.[40]

Assignment of Deshielded Proton Resonances

The assignment of resonances of protons involved in LBHBs to specific residues and/or ligands is difficult, often requiring multiple approaches. Three approaches that have been found useful will be discussed. Table III summarizes the properties of assigned LBHBs in enzymes.

J Coupling to Assigned Resonances. If the deshielded proton involved in a hydrogen bond is bonded covalently to a nitrogen, selective ^{15}N labeling and the detection of *J* coupling of this proton to ^{15}N would establish the assignment. This approach has been used to assign the deshielded proton resonance in α-lytic protease to that of the NδH of the catalytic base, His-57, which is hydrogen bonded to Asp-102.[41] Other serine proteases (Fig. 2) also show this deshielded resonance. With triosephosphate isomerase, ^{15}N labeling has been used to assign the Nε proton of the active site residue, His-95,[42] and the deshielded NH proton of enzyme-bound phosphoglycolo-hydroxamic acid, an analog of the enediol intermediate[30] (Fig. 2). For these studies, 2D ^{1}H–^{15}N HMQC spectra are most effective because they provide the chemical shifts of both the proton and the attached ^{15}N in the same experiment. In favorable cases, the $^1J_{N–H}$ is also obtained, which gives information on the extent of covalent bonding.[43] In the HMQC experiment (Fig. 4A), suppression of the strong water signal is necessary and is accomplished by the 1-1 pulse sequence together with phase cycling.[39] Pulse-field gradients are included to further remove imperfections in the 180° refocussing pulse.[30] Pulse widths, delays, and gradient strength will vary, depending on the instrumentation and the sample.

Mutagenesis of the Enzyme. When the deshielded proton is not attached to nitrogen, alternative assignment methods must be used. If the hydrogen bond donor can be removed, either by mutagenesis of the enzyme[28,29,44] or by alteration of the ligand of the enzyme,[26,30] the deshielded proton resonance should disappear. This approach has permitted the assignment of

[39] V. Sklenár and A. Bax, *J. Magn. Reson.* **74**, 469 (1987).
[40] P. Plateau and M. Gueron, *J. Am. Chem. Soc.* **104**, 7310 (1982).
[41] W. W. Bachovchin, *Proc. Natl. Acad. Sci. U.S.A.* **82**, 7948 (1985).
[42] P. J. Lodi and J. R. Knowles, *Biochemistry* **30**, 6948 (1991).
[43] E. L. Ash, J. L. Sudmeier, E. C. De Fabo, and W. W. Bachovchin, *Science* **278**, 1128 (1997).
[44] R. W. Wu, S. Ebrahemian, M. E. Zawrotney, L. D. Thornberg, G. C. Perez-Alverado, P. Brothers, R. M. Pollack, and M. F. Summers, *Science* **276**, 415 (1997).

TABLE III

PROPERTIES OF LOW-BARRIER HYDROGEN BONDS ON ENZYMES AS STUDIED BY NMR

Enzyme complex	δ (ppm)	Assignment	Interaction	$\Delta\delta$ (ppm)	ϕ	H-bond strength (kcal/mol)	k_{ex} (sec^{-1})	E_{act} (kcal/mol)	$k_{intrinsic}$ (sec^{-1})	Protection factor
KSI–DHE[a]	18.2	D99-COOH	D99···Y14	6.2	0.34	≥7.1 (diad)	137 (−3.3°)	12.9	≥3500	≥25.5
TIM–PGH[b]	14.9	PGH-NOH	E165···HON	6.2	0.38	—	3900[g] (30°)	8.9	3500	~1[g]
H⁺-chymotrypsinogen[c]	18.1	H57-NδH	H57···D102	4.7	0.4	—	952 (1°)	10.7	8720	9.2
Chymotrypsin–TFK[d]	18.9	H57-NδH	H57···D102	5.5	—	≥7	—	—	—	—
Asp-amino transferase[e]	17.4	Pyr-NH⁺	Pyr-NH⁺···D222	—	—	≥7.6	—	—	—	—
AKB–CoA-ligase[f]	19.1	Pyr-NH⁺	Pyr-NH⁺···COO⁻	—	—	≥3.5	—	—	—	—

[a] Ketosteroid isomerase complex of dihydroequilenin. Q. Zhao, C. Abeygunawardana, P. Talalay, and A. S. Mildvan, *Proc. Natl. Acad. Sci. U.S.A.* **93**, 8220 (1996); Q. Zhao, C. Abeygunawardana, A. G. Gittis, and A. S. Mildvan, *Biochemistry* **36**, 14616 (1997).

[b] Triosephosphate isomerase complex of phosphoglycolohydroxamic acid. T. K. Harris, C. Abeygunawardana, and A. S. Mildvan, *Biochemistry* **36**, 14661 (1997).

[c] J. L. Markley and W. M. Westler, *Biochemistry* **35**, 11092 (1996).

[d] TFK is *N*-acetyl-L-Leu-DL-Phe-CF₃. C. S. Cassidy, J. Lin, and P. A. Frey, *Biochemistry* **36**, 4576 (1997).

[e] E. Mollova, D. E. Metzler, A. Kintanar, H. Kagamiyama, H. Hayashi, K. Hirotsu, and I. Miyahara, *Biochemistry* **36**, 615 (1997).

[f] 2-Amino-3-ketobutyrate–CoA ligase. H. Tong and L. Davis, *Biochemistry* **34**, 3362 (1995).

[g] Rate is that of PGH dissociation.

A

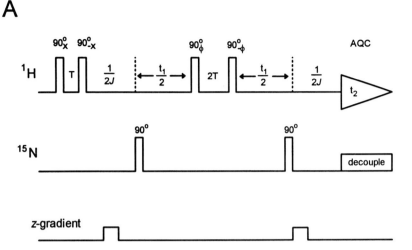

B

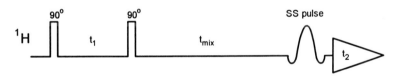

FIG. 4. Pulse sequences for 2D experiments used to assign deshielded proton resonances. (A) ^{1}H–^{15}N HMQC pulse sequence modified from V. Sklenár and A. Bax, *J. Magn. Reson.* **74,** 469 (1987) by the addition of pulse-field gradients. Excitation is provided by the first pair of 90° ^{1}H pulses and refocusing is accomplished by the second pair of 90° ^{1}H pulses at twice the separation. In favorable cases, ^{15}N decoupling during acquisition may be omitted to measure $^1J_{N-H}$. In this scheme the phase is cycled along the x, y, $-x$, and $-y$ axes. (B) SS NOESY pulse sequence [S. H. Smallcomb, *J. Am. Chem. Soc.* **115,** 4776 (1993)]. The duration of the shaped SS pulse was 140 μsec.

deshielded proton resonances of ketosteroid isomerase,[29] triosephosphate isomerase,[30] and aspartate aminotransferase.[26,45,46]

[45] D. E. Metzler, C. M. Metzler, E. T. Mollova, R. D. Scott, S. Tanase, K. Kogo, T. Higacki, and Y. Morino, *J. Biol. Chem.* **269,** 28017 (1994).

[46] D. E. Metzler, C. M. Metzler, R. D. Scott, E. T. Mollova, H. Kagamiyama, T. Yano, S. Kuramitsu, H. Hayashi, K. Hirotsu, and I. Miyahara, *J. Biol. Chem.* **269,** 28027 (1994).

Nuclear Overhauser effects (NOEs) to Assigned Resonances. Assignment of deshielded proton resonances involved in strong hydrogen bonds may be facilitated by the detection of NOEs from these signals to assigned proton resonances of the enzyme.[28–30,45,46] Ambiguities may be resolved by comparison with X-ray structures, if available.[26,30,47] For 2D Nuclear Overhauser effect spectroscopy (NOESY) spectra involving such deshielded proton resonances, we have used the symmetrically shifted or SS-NOESY pulse sequence (Fig. 4B).[48] The SS detection pulse is soft and continuous, which provides an excitation profile analogous to that produced by hard binomial pulses such as the 121 and 1331 pulses. The SS pulse provides much improved phase and baseline properties, which are necessary for 2D experiments.[29,30] Generation of the SS pulse requires pulse-shaping capabilities in the ^{1}H transmitter channel. The exact procedure for generation of the SS pulse varies with the spectrometer, requiring as input parameters the pulse width and frequency separation between the water resonance and those of the deshielded proton resonances of interest. The SS pulse width (pw) (100–160 μsec) is generated to provide maximum excitation at the desired frequency. As with the 1331 pulse, the carrier frequency is set at the water resonance.

Fractionation Factors

Theory

Consider a solvent-exchangeable proton on an enzyme (Eq. 6):

$$\text{Enz-H} + \text{D}_{\text{solvent}} \rightleftharpoons \text{Enz-D} + \text{H}_{\text{solvent}} \qquad (6)$$

The fractionation factor (ϕ) at this position is defined as the equilibrium constant for the exchange of deuterium from the solvent into this position (Eq. 7)[49]:

$$\phi = \frac{[\text{Enz-D}][\text{H}_{\text{solvent}}]}{[\text{Enz-H}][\text{D}_{\text{solvent}}]} \qquad (7)$$

Thus ϕ measures the preference of this site for deuterium over protium. If this proton is hydrogen bonded, the effective force constant of this bond is lower, i.e., it vibrates in a wider, shallower well at a lower zero-point frequency or energy (Fig. 5). The zero-point vibrational frequency or en-

[47] R. C. Davenport, P. A. Bash, B. A. Seaton, M. Karplus, G. A. Petsko, and D. Ringe, *Biochemistry* **30**, 5821 (1991).
[48] S. H. Smallcomb, *J. Am. Chem. Soc.* **115**, 4776 (1993).
[49] K. B. Schowen and R. L. Schowen, *Methods Enzymol.* **87**, 551 (1982).

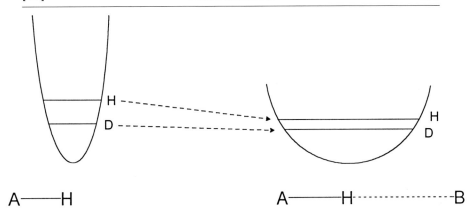

F<small>IG</small>. 5. Effect of hydrogen bonding on the effective force constant and resulting zero-point vibrational energies of a proton and a deuteron.

ergy of a deuteron at this position, because of its 2-fold greater mass, is $\leq\sqrt{2}$-fold less sensitive to this decrease in the effective force constant. Hence the zero-point vibrational energy of a deuteron in a hydrogen bond decreases less than that of a proton. As a result, the presence of a deuteron in the hydrogen bond is relatively disfavored over the presence of a proton, yielding a fractionation factor <1 (Fig. 5). A quantum mechanical explanation for this equilibrium isotope effect has been given by Kreevoy.[50] The Heisenberg uncertainty principle states that the product of the uncertainty in the position and the uncertainty in the momentum of a proton (or deuteron) is constant. When the proton well becomes wider due to hydrogen bonding, the uncertainty in the position of the proton increases. Hence the uncertainty in its momentum, as defined by the height of its zero-point vibrational energy, decreases. A deuteron, with a $\leq\sqrt{2}$-fold lower zero-point energy than a proton, undergoes a corresponding $\leq\sqrt{2}$-fold lower decrease in zero-point energy than a proton does in the wider well of the hydrogen bond. In this wider well, $\leq\sqrt{2}$ has multiplied a smaller number than in the narrow well, causing a numerically smaller lowering of the zero-point energy of the deuteron below that of the proton. The net result is the disfavoring of the deuteron in the wider well, yielding a fractionation factor <1 (Fig. 5). The shorter and stronger the hydrogen bond, the lower the fractionation factor, minimizing at a ϕ value of ~0.16 at an O–H$\cdots$O hydrogen bond distance of 2.39 Å.[2]

At distances shorter than 2.39 Å, where a very strong single-well hydrogen bond exists, the effective force constant now increases with decreasing

[50] M. M. Kreevoy, *J. Chem. Edu.* **41,** 636 (1964).

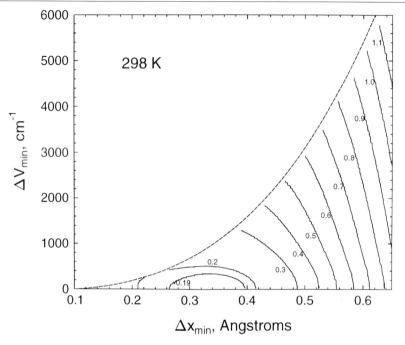

F<small>IG</small>. 6. Estimates of fractionation factors using model one-dimensional, double-minimum potentials of the type used by M. M. Kreevoy and T. M. Liang, *J. Am. Chem. Soc.* **102,** 3315 (1980). Contours show the dependence of H/D fractionation factors at 25° on the energy difference between the wells of the double-minimum potential (ΔV_{min}) and the distance between the wells (Δx_{min}). W. P. Huskey, personal communication. For details, see D. Bao, C. A. Kettner, W. P. Huskey, and F. Jordan, submitted for publication (1999).

distance as the well narrows, increasing the fractionation factor (Fig. 1). Quantum mechanically, the uncertainties in the positions of both the proton and the deuteron now decrease with decreasing hydrogen bond distance as the well narrows. The uncertainties in their momenta will increase, spreading apart their zero-point energies, thereby removing the energetic disfavoring of the deuteron in the hydrogen bond and increasing the fractionation factor, as in the nonhydrogen bonded case. Such very strong, single-well hydrogen bonds have not been reported in proteins.

Using quartic potential functions, Kreevoy and Liang[2] have computed ϕ values of O–H $\cdots$ O hydrogen bonds as a function of distance between the proton wells, and these computations have been expanded and generalized by Bao *et al.*[51] Figure 6 relates the distance between proton wells to

[51] D. Bao, C. Kettner, W. P. Huskey, and F. Jordan, submitted for publication (1999).

the ϕ values and to the energy difference between the two wells. Figure 6 may be used to determine hydrogen bond lengths from ϕ values by adding two covalent bond lengths, typically 2.00 Å, to the distance between the wells obtained graphically. Table II compares hydrogen bond distances determined from ϕ values using Fig. 6 with those obtained from chemical shifts using Fig. 3 and Eq. (3). The agreement is seen to be excellent.

Measurement of Fractionation Factors

As described by Loh and Markley,[52,53] fractionation factors for solvent exchangeable proton resonances are determined by integration of the corresponding ^1H NMR resonances at equilibrium in mixed H_2O–D_2O solutions. Quantitative comparisons between peak intensities in multiple samples are best achieved by normalizing the peak area of the exchangeable proton resonance to the peak area of an isolated nonexchangeable proton signal such as an upfield methyl resonance.[28,29] However, such a reference peak may not be available for high molecular weight proteins because of extensive overlap in the upfield region of the spectrum. In such cases, precise measurements can be achieved by ensuring that the protein concentrations in each of the mixed H_2O–D_2O solutions are constant and that identical NMR tubes are used for each sample. We have used Wilmad Cat. No. 535-PP 5-mm tubes containing 0.6-ml sample volumes. Careful sample preparation also requires that the mole fractions of H_2O and D_2O are precise and that the solvent exchange of the proton resonance of interest has reached equilibrium before the measurements are made.

Initially, a stock solution of the protein at its maximal concentration, e.g., 3.8 mM triosephosphate isomerase sites, in appropriate nonvolatile buffers, ligands, and salts is made in H_2O. Next, two stock solutions of the buffer, ligands, and salts at the same concentrations and pH as in the protein solution are made: (a) one in H_2O and (b) the other in D_2O. Solution (b) is made by first adjusting the pH in H_2O and then lyophilizing and redissolving the solutes at least three times in 99.96% D_2O (Aldrich, Cat. No. 19234-1). A constant volume of the protein solution (typically 72 μl or 12% of the final volume) is added to a weighed Eppendorf tube and weighed. To this solution is added the desired volume (0 to 528 μl) of solution (a) and the weight is recorded. Solution (b) is now added in an amount necessary (528 to 0 μl) to reach a total volume of 600 μl, assuming additivity of volumes, and the weight is again recorded. The additivity of volumes may be checked at the end of the experiment by measurement of the protein

[52] S. N. Loh and J. L. Markley, in "Techniques in Protein Chemistry IV" (R. Angeletti, ed.), p. 517. Academic Press, New York, 1993.
[53] S. N. Loh and J. L. Markley, Biochemistry 33, 1029 (1994).

concentration in each sample. The mole fraction of H_2O in D_2O (X) is calculated from the weights of H_2O and D_2O in each mixture, using Eq. (8):

$$X = \frac{\dfrac{\text{wt } H_2O}{18.015}}{\dfrac{\text{wt } H_2O}{18.015} + \dfrac{\text{wt } D_2O}{20.028}} \tag{8}$$

To each of these solutions, a fixed volume [67 μl or 10% (v/v)] of DMSO-d_6 is added for field frequency locking and to permit NMR measurements at $-4.8°$.

For proteins that survive lyophilization, it may be more convenient to prepare samples by freeze drying and redissolving them into H_2O–D_2O mixtures. Because some of the protein is lost during lyophilization, an internal intensity standard from the protein is necessary. If a high concentration protein sample cannot be obtained from which samples of equal protein concentration can be made, an internal standard[28,29] or concentric intensity standard in a capillary is necessary for each sample.[52] For the internal standard, well-resolved nonexchangeable methyl resonances of the protein may be used. If they are not resolved completely, spectral simulation is necessary.[28,29]

The samples are incubated for a time long enough for complete equilibration of all of the exchangeable proton resonances of interest. Because of their facile exchange with solvent (see later), 1 hr at 4° is frequently sufficient for protons in LBHBs. Complete equilibration is checked by repeated measurements, after 24 hr, and after a week, if necessary. Even this length of time is not sufficient for the complete equilibration of certain slowly exchanging amide protons.

NMR spectra of H_2O–D_2O mixed samples are obtained as described earlier with the 1331 pulse sequence to avoid excitation of the water resonance and the transfer of saturation. In the absence of an internal intensity standard, each sample is acquired and processed with identical parameters such as numbers of transients, receiver gain, and vertical scale, regardless of the mole fraction of H_2O, resulting in a lower signal/noise at lower mole fractions of H_2O. Hence the number of transients should be adjusted to give adequate signal/noise ($>6:1$) of the resonance of interest at the lowest mole fraction of H_2O. Resonance integration further increases the signal/noise as it sums all of the points of the measured resonance. The protein concentration in each tube may be checked and the measured resonance intensities corrected accordingly. In the presence of a nonexchangeable internal intensity standard, the number of transients can be adjusted to equalize the signal/noise for all solutions. The intensities of each resonance may be determined accurately by spectral integration when the baseline is flat. When the baseline is irregular, integration may be accomplished by cutting and weighing of resonances or, most accurately, by spectral simulation of the resonances of interest. Simulation is accomplished with the program FELIX 2.0 (Biosym Technologies, Inc.) by visually adjusting the chemical shift, width, and height of a Lorentzian signal to optimally fit each measured resonance. The integrated area is computed rapidly and displayed. Spectral simulation is essential when either the measured resonances or those of the internal standard are not resolved completely.[28-30] When an internal or concentric capillary[52] intensity standard is available, peak intensities (I) are normalized to this standard. The maximum signal intensity at 100% H_2O (I_{max}) is defined as 1.0.

The fractionation factors are derived directly by fitting NMR data to either of the following rearranged forms of Eq. (7):

$$I = \frac{I_{max}(X)}{\phi(1 - X) + X} \tag{9a}$$

$$\frac{1}{I} = \frac{\phi}{I_{max}} \left(\frac{1 - X}{X} \right) + \frac{1}{I_{max}} \tag{9b}$$

in which I is the observed peak intensity at a given mole fraction (X) of H_2O, I_{max} is the maximal peak intensity when $X = 1.0$, and $1 - X$ is the mole fraction of D_2O. When plotted as I versus X, Eq. 9a yields a straight line when $\phi = 1.0$ and a hyperbola for $\phi \neq 1$, convex up when $\phi < 1$ and convex down when $\phi > 1$.[52] Such curves can be fit by nonlinear least-squares regression to yield ϕ and, if necessary, I_{max} (Fig. 7). Equation (9a) can also be expressed in linear form [Eq. (9b)] and plotted as $1/I$ versus $(1 - X)/X$, yielding a slope of ϕ/I_{max}, a y intercept of $1/I_{max}$, and an x intercept of $-1/\phi$, permitting evaluation of ϕ and I_{max} by linear regression.[53] Because the linear plot exaggerates the errors at low mole fractions of H_2O, the nonlinear analysis using Eq. (9a) is preferred. Table III summarizes the ϕ values of protons in LBHBs on enzymes.

Proton Exchange Rates and Protection Factors

Theory

Strongly hydrogen-bonded protons are expected to be slowed in their exchange rate with solvent protons by at least an order of magnitude. This slowing is expressed quantitatively by the protection factor (P.F.), which is given by Eq. (10),

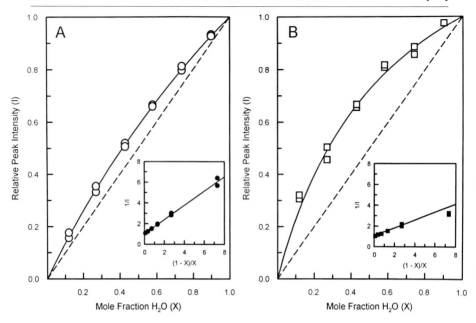

FIG. 7. Determination of the fractionation factors of downfield proton resonances in the complex of triosephosphate isomerase with phosphoglycolohydroxamic acid, an analog of the enediol intermediate. Plot of the relative peak intensity of (A) the His-95 NεH proton signal in the triosephosphate isomerase complexed with phosphoglycolohydroxamic acid and of (B) the NOH proton signal of enzyme-bound phosphoglycolohydroxamic acid as a function of the mole fraction H_2O in the aqueous solution. The curve is described by Eq. (9a) using (A) $\phi = 0.71 \pm 0.02$ and (B) $\phi = 0.38 \pm 0.06$; the dashed line is drawn for $\phi = 1$. The inset shows the same data plotted in linear form and the line is described by Eq. (9b) using (A) $\phi = 0.69 \pm 0.02$ and (B) $\phi = 0.38 \pm 0.06$ [T. K. Harris, C. Abeygunawardana, and A. S. Mildvan, *Biochemistry* **36**, 14661 (1997)].

$$\text{P.F.} = \frac{k_{\text{intrinsic}}}{k_{\text{ex}}} \qquad (10)$$

where k_{ex} is the pseudo-first-order rate constant for exchange with solvent of the hydrogen-bonded proton and $k_{\text{intrinsic}}$ is the pseudo-first-order rate constant for exchange of the same proton in the absence of hydrogen bonding. Within a protein, the exchange mechanisms are described by Eq. (11)[54,55]:

$$\text{A-H} \cdots \text{B} \underset{k_2}{\overset{k_1}{\rightleftharpoons}} [\text{A-H} + \text{B}] \xrightarrow[\text{D}_2\text{O}]{k_3} \text{A-D} + \text{B} \qquad (11)$$

[54] A. Hvidt and S. O. Nielsen, *Adv. Prot. Chem.* **21**, 288 (1966).
[55] W. Qiwen, A. D. Kline, and K. Wüthrich, *Biochemistry* **26**, 6488 (1987).

In Eq. (11), $k_{\text{intrinsic}} = k_3$, and k_1 and k_2 are first-order rate constants for breakage and formation, respectively, of a hydrogen bond within a protein. Assuming the [AH + B] species to be in a steady state, one obtains

$$k_{\text{ex}} = \frac{k_1 k_3}{k_1 + k_2 + k_3} \tag{12}$$

From Eqs. (10) and (12) we can write

$$\text{P.F.} = \frac{k_1 + k_2 + k_3}{k_1} \tag{13}$$

Two limiting proton exchange mechanisms have been described for Eq. (11).[54,55] In the EX1 mechanism $k_3 \gg k_2$, k_1 resulting in rate-limiting hydrogen bond breakage (k_1), i.e.,

$$k_{\text{ex}} = k_1 \tag{12a}$$

$$\text{P.F.} = \frac{k_3}{k_1} \tag{13a}$$

Note that in the EX1 mechanism P.F. is equal to the intrinsic rate constant divided by the rate constant for breakage of the hydrogen bond. In the EX2 mechanism $k_2 \gg k_3$, k_1 resulting in rate-limiting proton exchange gated by preequilibrium hydrogen bond breakage and formation.

$$k_{\text{ex}} = \frac{k_1}{k_2} k_3 \tag{12b}$$

$$\text{P.F.} = \frac{k_2}{k_1} \tag{13b}$$

In the EX2 mechanism, P.F. is equal to the equilibrium constant for formation of the hydrogen bond.

The intrinsic rate constant for exchange of the exposed proton (k_3) is catalyzed by H_3O^+ and OH^-, resulting in a first-order dependence of k_3 on $[H_3O^+]$ and $[OH^-]$. Hence, in an EX1 mechanism, k_{ex} should be independent of pH [Eq. (12a)], whereas in an EX2 mechanism, k_{ex} should be strongly dependent on pH [Eq. (12b)]. While both mechanisms thus permit a protection factor >1, only the EX1 mechanism predicts a pH-independent exchange rate constant. The absence of a pH dependence of the exchange rate of the protons in the LBHBs detected in ketosteroid isomerase[28,29] and triosephosphate isomerase[30] indicates EX1 mechanisms. Table III summarizes the protection factors of protons involved in LBHBs on enzymes.

Determination of Proton Exchange Rates and Protection Factors

The proton exchange rate with solvent is determined from the temperature dependence of the width of its resonance in ^{1}H NMR spectra. Initially, thermal stability studies monitoring catalytic activity should be performed to indicate a safe temperature range for each protein. Next, the NMR sample (600 μl) is prepared in H_2O containing ≥ 0.7 mM protein sites, the appropriate buffer, ligands, and salt and 10% (v/v) DMSO-d_6 (67 μl) for field frequency locking and to permit NMR measurements at $-4.8°$. The probe temperatures are calibrated with the methanol standard using Eq. (4), and NMR spectra at various temperatures over a wide temperature range, e.g., -4.8 to $43°$,[30] are obtained as described earlier using the 1331 pulse sequence to avoid excitation of the water resonance and transfer of saturation. After being placed in the probe, the sample is allowed to equilibrate at the designated temperature for at least 15 min before acquiring spectra. Because incipient thermal denaturation of the protein may occur at high temperatures, it is preferable to begin acquiring spectra at the lowest temperature and then to proceed to higher temperatures.

The line widths of the exchangeable proton resonances of interest are determined at each temperature by taking the width at half-height of each resonance measured in Hz. This is accomplished by resonance simulation with the program FELIX 2.0 (Biosym Technologies, Inc.) as described earlier or simply by hand using a ruler and pencil. The temperature dependences of the line widths ($\Delta \nu_{1/2}$) are then analyzed by Arrhenius plots of $\ln(1/T_{2\text{obs}})$ versus (temperature)$^{-1}$ according to Eqs. (14) to (17) (Fig. 8)[56]:

$$\frac{1}{T_{2\text{obs}}} = \pi\Delta\nu_{1/2} = k_{\text{ex}} + \frac{1}{T_{2\text{d}}} \qquad (14)$$

In Eq. (14), k_{ex} is the exchange rate constant and $1/T_{2\text{d}}$ is the dipolar contribution to the line width. These two terms have unequal and opposite temperature dependences. Thus, k_{ex} is a chemical rate constant that typically increases with temperature with a high activation energy ($E_{\text{ex}} > 5$ kcal/mol) [Eq. (15)]:

$$\ln(k_{\text{ex}}) = -\frac{E_{\text{ex}}}{RT} + C_{\text{ex}} \qquad (15)$$

In contrast, $1/T_{2\text{d}}$ decreases with increasing temperature as it is inversely related to the dipolar correlation time, which has a low activation energy ($E_{\text{d}} \leq 4$ kcal/mol) [Eq. (16)]:

[56] T. J. Swift and R. E. Connick, *J. Chem. Phys.* **37**, 307 (1963).

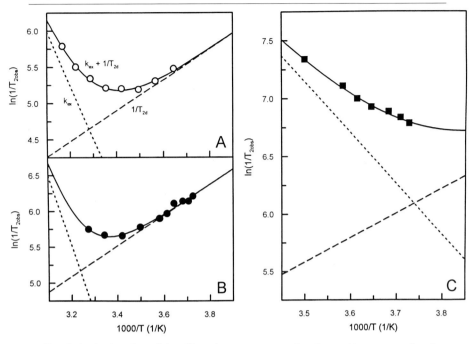

FIG. 8. Arrhenius plots of the effect of temperature on the observed transverse relaxation rate ($1/T_{2\text{obs}}$) of downfield proton resonances in free triosephosphate isomerase and triosephosphate isomerase complexed with phosphoglycolohydroxamic acid (PGH). (A) His-95 NεH proton signal ($\delta = 13.1$ ppm) in free triosephosphate isomerase. (B) His-95 NεH proton signal ($\delta = 13.5$ ppm) in triosephosphate isomerase complexed with PGH. (C) PGH-NOH proton signal ($\delta = 14.9$ ppm) in enzyme-bound PGH. The dotted line represents the exchange contribution (k_{ex}) to $1/T_{2\text{obs}}$, and its slope and y intercept are given by the activation energy (E_{ex}) and Arrhenius coefficient (C_{ex}), respectively [Eq. (15)]. The dashed line represents the dipolar contribution ($1/T_{2\text{d}}$) to $1/T_{2\text{obs}}$, and its slope and y intercept are given by E_{d} and C_{d}, respectively [Eq. (16)]. The curve through data represents the composite results for both k_{ex} and $1/T_{2\text{d}}$ contributions to $1/T_{2\text{obs}}$ and is described by Eq. (17). Values used to fit data were $E_{\text{ex}} = 14.7 \pm 0.9$ kcal/mol, $C_{\text{ex}} = 28.8 \pm 1.5$, $E_{\text{d}} = -4.3 \pm 0.8$ kcal/mol, and $C_{\text{d}} = -2.36 \pm 0.30$ (○); $E_{\text{ex}} = 19.1 \pm 1.9$ kcal/mol, $C_{\text{ex}} = 36.2 \pm 1.5$, $E_{\text{d}} = -4.3 \pm 0.6$ kcal/mol, and $C_{\text{d}} = -1.74 \pm 0.20$ (●); and $E_{\text{ex}} = 8.9 \pm 0.3$ kcal/mol, $C_{\text{ex}} = 22.7 \pm 0.5$, $E_{\text{d}} = -4.3$ kcal/mol, and $C_{\text{d}} = -1.89$ (■) [T. K. Harris, C. Abeygunawardana, and A. S. Mildvan, *Biochemistry* **36**, 14661 (1997)].

$$\ln\left(\frac{1}{T_{2\text{d}}}\right) = -\frac{E_{\text{d}}}{RT} + C_{\text{d}} \tag{16}$$

The terms C_{ex} and C_{d} are the intercepts of the two lines at infinite temperature given by Eqs. (14) and (16). From Eqs. (14) to (16), we may write Eq. (17),

$$\ln\left(\frac{1}{T_{2obs}}\right) = \ln\left[\exp\left(-\frac{E_{ex}}{RT} + C_{ex}\right) + \exp\left(-\frac{E_d}{RT} + C_d\right)\right] \quad (17)$$

which is fitted to an Arrhenius plot of $\ln(1/T_{2obs})$ as a function of (temperature)$^{-1}$ (Fig. 8).

From the values of E_{ex} and C_{ex}, the exchange rate constant k_{ex} can be calculated at any temperature. This value may then be used, together with E_{ex}, to evaluate ΔG^*, ΔH^*, and $-T\Delta S^*$ of the exchange process (in kcal/mol), using Eqs. (18), (19), and (20), which apply in solution at constant pressure:

$$\Delta H^* = E_{ex} - RT \quad (18)$$

$$\Delta G^* = -RT\ln\left(\frac{hk_{ex}}{k_B T}\right) = -(1.98 \times 10^{-3})T\ln\left(\frac{k_{ex}}{T} 4.80 \times 10^{-11}\right) \quad (19)$$

$$-T\Delta S^* = \Delta G^* - \Delta H^* \quad (20)$$

In Eq. (19), R is the gas constant, k_B is the Boltzman constant, and h is Planck's constant.

Once the pseudo-first-order rate constant for exchange with the solvent of the hydrogen-bonded proton, k_{ex}, has been determined, the protection factor is calculated from Eq. (10) using the pseudo-first-order rate constant for exchange of the same proton in the absence of hydrogen bonding, $k_{intrinsic}$. While, in principle, the T_2 method may be used to determine $k_{intrinsic}$,[57] in practice,[58] $k_{intrinsic}$ is often too fast to measure by this method. In these cases, $k_{intrinsic}$ may be calculated using Eq. (21)[58,59]:

$$k_{intrinsic} = \sum_i \frac{k_D[\text{catalyst}]_i \, 10^{\Delta pK_a}}{10^{\Delta pK_a} + 1} \quad (21)$$

In Eq. (21) $k_{intrinsic}$ of an exchangeable site A is equal to the sum of the contributions to the exchange rate at site A given by each potential acid and base catalyst in the solution (i.e., H_3O^+, OH^-, H_2O, and buffers). The concentration of each potential acid and base catalyst is multiplied by the second-order rate constant, k_D ($\sim 10^{10}$ M^{-1} sec^{-1} at 25°), for diffusion-controlled complex formation between the exchangeable site A and the catalyst. The remaining factors in Eq. (21) describe the fraction of successful exchanges that occur on encounter between site A and each catalyst. These

[57] E. Liepinsh, G. Otting, and K. Wüthrich, *J. Biomol. NMR* **2**, 447 (1992).
[58] K. Wüthrich, *in* "NMR of Proteins and Nucleic Acids," p. 23. Wiley, New York, 1986.
[59] M. Eigen, *Agnew. Chem. Intl. Ed. Engl.* **3**, 1 (1964).

factors depend on the differences (ΔpK_a) between the pK_a of the proton acceptor and the pK_a of the proton donor (i.e., $\Delta pK_a = pK_a^{\text{acceptor}} - pK_a^{\text{donor}}$). For example, when calculating the ΔpK_a for OH^--catalyzed exchange, OH^- is the proton acceptor and the exchangeable site A–H is the proton donor, whereas for H_3O^+-catalyzed exchange, the site A is the proton acceptor and H_3O^+ is the proton donor. In this calculation, all of the potential acid and base catalysts must be considered and the concentration and the pK_a of each catalyst and of site A must be known. In aqueous solution under standard conditions, the base catalysts are OH^- ($pK_a = 15.57$), H_2O ($pK_a = -1.7$), and buffer bases. The acid catalysts are H_3O^+ ($pK_a = -1.7$), H_2O ($pK_a = 15.57$), and buffer acids. The concentration of H_2O is 55.5 M and the concentrations of OH^- and H_3O^+ are calculated from the pH at which the rate of exchange is measured. The pK_a values of A–H and of buffers are used to determine the concentrations of their acidic and basic components.

Free Energy of Formation of the Hydrogen Bond

Theory

Measurement of the $\Delta G_{\text{formation}}$ of a hydrogen bond would provide the most direct evidence for its strength. However, on an enzyme where multiple interactions of many types exist, and where unusually strong hydrogen bonds may be part of a complex network, the measurement of $\Delta G_{\text{formation}}$ of a particular hydrogen bond is difficult and usually indirect. The effect of mutation of an amino acid residue, abolishing its ability to donate or to accept a hydrogen bond, on a rate constant or an equilibrium constant in which the hydrogen bond participates would provide indirect information relevant to its strength. Thus in the catalytic diad mechanism of ketosteroid isomerase (Fig. 2) in which Asp-99 donates a strong hydrogen bond to Tyr-14, which, in turn, donates a hydrogen bond to the dienolic intermediate,[29] mutation of Asp-99 to Ala decreases k_{cat} by $10^{3.7}$-fold,[44] indicating a 5.0 kcal/mol weaker binding of the transition state. Mutation of Tyr-14 to Phe, which disrupts both hydrogen bonds of the catalytic diad, decreases k_{cat} by $10^{4.7}$-fold, weakening transition state binding by 6.4 kcal/mol.[60] The Y14F mutation also accelerates greatly the dissociation of the dienolic intermediate from the enzyme by a factor of $\geq 10^{5.6}$-fold, corresponding to ≥ 7.6 kcal/mol weaker binding of the intermediate in the absence of the catalytic

[60] A. Kuliopulos, P. Talalay, and A. S. Mildvan, *Biochemistry* **29**, 10271 (1990).

diad.[61] These indirect kinetic approaches yield estimates of $\Delta G_{\text{formation}}$ of the strong hydrogen bond ranging from -5.0 to ≤ -7.6 kcal/mol.

With the serine protease trypsin, mutation of Asp-102 of the catalytic triad (Fig. 2) to Asn decreases k_{cat} by $10^{4.0}$-fold, corresponding to a 5.4-kcal/mol weakening of the hydrogen bond between this residue and His-57, the catalytic base.[62] Complete loss of the corresponding Asp $\cdots$ His hydrogen bond in the D32A mutant of subtilisin results in a $10^{6.2}$-fold decrease in k_{cat}, corresponding to a $\Delta G_{\text{formation}}$ of -8.4 kcal/mol.[63] This latter value, based on the complete loss of the hydrogen bond, likely provides a better estimate of its strength. Similarly the D222A mutant of aspartate aminotransferase (Fig. 2), which abolishes a strong hydrogen bond between Asp-222 and the pyrimidine nitrogen of pyridoxal phosphate,[26] results in a 10^5-fold decrease in catalysis corresponding to a $\Delta G_{\text{formation}}$ of -6.8 kcal/mol.[64]

In principle, the strength of a hydrogen bond can be estimated by the decrease in pK_a of the acceptor or by the increase in the pK_a of the donor resulting from the existence of the hydrogen bond. More generally, any change in pK_a resulting from the formation of a hydrogen bond should reflect the strength of the hydrogen bond. When the proton involved in the hydrogen bond is detectable by NMR, its pK_a shift (ΔpK_a) or a limiting value for ΔpK_a can often be obtained by pH titration monitoring the chemical shift of the proton resonance. Thus with chymotrypsin, the increase in pK_a of the Nε proton of His-57 from a value of 6.8 to 12.0 as the strong hydrogen bond forms between Nδ of His-57 and the carboxylate oxygen of Asp-102 (Fig. 2) yields a $\Delta G_{\text{formation}}$ of -6.6 kcal/mol.[19] This value is similar to those estimated from the effects of mutation of the catalytic Asp residue. With the ketosteroid isomerase complex of the anion of dihydro-equilenin, an analog of the dienolic intermediate[28,29] that accepts a hydrogen bond from the catalytic diad (Fig. 2), no change in the chemical shift of the hydrogen-bonded deshielded protons of the catalytic diad was detected over the pH range 4.3 to 9.0. Because the pK_a of free dihydroequilenin is 9.0, the pK_a of bound dihydroequilenin is at least one pK unit lower than 4.3. The resulting $\Delta pK_a \geq 5.7$ corresponds to $\Delta G_{\text{formation}} \leq -7.0$ kcal/mol for the hydrogen bonds in the catalytic diad,[28,29] which is comparable to the values obtained by mutagenesis of Tyr-14.

[61] L. Xue, P. Talalay, and A. S. Mildvan, *Biochemistry* **30**, 10858 (1991).
[62] C. S. Craik, S. Roczniak, C. Largman, and W. J. Rutter, *Science* **237**, 909 (1987).
[63] P. Carter and J. Wells, *Nature* **332**, 564 (1988).
[64] J. J. Onuffer and J. F. Kirsch, *Prot. Engin.* **1**, 413 (1994).

pH Titration of Deshielded Exchangeable Proton Resonances

NMR spectra are obtained and analyzed as described previously for measurement of the chemical shift (δ) with protein concentrations ≥ 0.1 mM. Two identical samples are prepared at near neutral pH, and the chemical shifts of the deshielded resonances are recorded. In one sample, the pH is decreased by the addition of 1- to 2-μl volumes of 1 M HCl to 600-μl samples with rapid mixing to minimize protein precipitation. If necessary, the sample is centrifuged to remove turbidity. The pH is measured before and after determination of the chemical shift and the process is repeated. In the other sample the pH is increased by adding 1- to 2-μl volumes of 1 M KOH or NaOH and δ is measured. The reversibility of the effects of pH on δ may be checked with one or both samples. If changes in δ with pH are observed, titration data are fit by nonlinear least-squares methods to Eq. (22):

$$\delta(\text{ppm})^{\text{app}} = \frac{\delta_1 + \delta_2(10^{\text{pH}-\text{p}K_a})^n}{1 + (10^{\text{pH}-\text{p}K_a})^n} \qquad (22)$$

yielding four parameters, the limiting chemical shifts at low (δ_1) and high pH (δ_2), the pK_a, and the Hill coefficient (n).

Low-Barrier Hydrogen Bonds on Enzymes Detected by NMR

LBHBs on enzymes detected by NMR, which meet some or all of the criteria listed in Table I, are summarized in Fig. 2 and their properties are given in Table III. It is noteworthy that all appear to involve carboxyl groups, emphasizing that Asp and Glu residues on enzymes can play a unique role in catalysis by forming unusually strong hydrogen bonds.

Acknowledgments

We are grateful to W. Phillip Huskey for permission to reproduce Fig. 6 and data in Table II prior to publication and for helpful discussions, to Ann E. McDermott for permission to reproduce Fig. 3, and to Qinjian Zhao and Paul Talalay for valuable collaboration.

[11] Hydrogen Bonding in Enzymatic Catalysis: Analysis of Energetic Contributions

By Shu-ou Shan and Daniel Herschlag

Overview

Enzymes typically form hydrogen bonds with substrate moieties that undergo electronic rearrangement in the course of reaction. This article reviews results that suggest that these hydrogen bonds can be strengthened in the transition state to a greater extent within the context of the enzymatic active site than in aqueous solution. Comparisons in model systems demonstrate this differential strengthening effect and suggest that it may provide substantial rate enhancements for enzymatic reactions relative to solution reactions.[1-3] Despite their likely importance, quantitative dissection of the role of these "reaction center" hydrogen bonds in enzymatic catalysis is blurred because the energetic contributions of these hydrogen bonds are intertwined with those of hydrogen bonds and other interactions that fold the enzyme and position the bound substrate. This interconnection between binding interactions and rate enhancement is a fundamental property of enzymatic catalysis.[4]

The Hydrogen Bond: General Energetic Considerations

When a hydrogen attached to a heteroatom is near van der Waals' contact with another heteroatom, a hydrogen bond is typically inferred. Although energy cannot be read from structure, the preponderance of such contacts, with preferred distances and angles, suggests an energetic preference for such interactions.[5-11] From an energetic standpoint, when a

[1] S. Shan and D. Herschlag, *Proc. Natl. Acad. Sci. U.S.A.* **93,** 14474 (1996).
[2] S. Shan and D. Herschlag, *J. Am. Chem. Soc.* **118,** 5515 (1996).
[3] S. Shan, S. Loh, and D. Herschlag, *Science* **272,** 97 (1996).
[4] W. P. Jencks, *Adv. Enzymol.* **43,** 219 (1975).
[5] For a detailed, critical review of hydrogen bonds beyond the scope of the presentation herein, see G. A. Jeffrey and W. Saenger, "Hydrogen Bonding in Biological Structures," Berlin, Springer, 1991.
[6] F. Hibbert and J. Emsley, *Adv. Phys. Org. Chem.* **26,** 255 (1990).
[7] W. P. Jencks, *in* "Catalysis in Chemistry and Enzymology," 2nd ed., p. 323. Dover, New York, 1987.
[8] S. N. Vinogradov, *in* "Molecular Interactions" (H. Ratajczak and W. J. Orville-Thomas, eds.), Vol. 2, p. 179. Wiley, New York, 1980.
[9] E. N. Baker and R. E. Hubbard, *Prog. Biophys. Mol. Biol.* **44,** 97 (1984).

Copyright © 1999 by Academic Press
All rights of reproduction in any form reserved.
0076-6879/99 $30.00

complex is formed in which the predominant identifiable interaction is that between a heteroatom-bonded hydrogen atom of one molecule and a heteroatom of a second species, such as that between ROH and F^- in the gas phase [Eq. (1); $\Delta H° = -32$ kcal/mol[12]], we attribute the driving force for complex formation to the hydrogen bond:

$$\text{EtOH} + F^- \underset{\Delta H° = -32 \text{ kcal/mol}}{\overset{K^{HB}}{\rightleftharpoons}} \text{EtOH} \cdot F^- \qquad (1)$$

However, the above picture of the hydrogen bond is deceptively simple. Whereas the equilibrium formation of the hydrogen-bonded EtOH·F^- complex [Eq. (1)] would appear to represent one of the most straightforward cases in which hydrogen bonding is the clear driving force for complex formation, F^- also has large association energies for the formation of complexes that would not typically be described as hydrogen bonding.[12,13] For example, the substantial enthalpy value of $\Delta H° = -30$ kcal/mol for forming the $CF_3-CHF_2 \cdot F^-$ complex [Eq. (2)][12] suggests that F^- interacts strongly with dipolar species in the gas phase, presumably because of its high charge density and the absence of electrostatic screening from solvent. Such observations caution against the view that all of the enthalpy change from formation of the EtOH·F^- complex [Eq. (1)] should necessarily be considered a simple measure of "hydrogen bond strength."

$$CF_3-CHF_2 + F^- \underset{\Delta H° = -30 \text{ kcal/mol}}{\overset{K'}{\rightleftharpoons}} CF_3-CHF_2 \cdot F^- \qquad (2)$$

To analyze the energetics of molecular processes quantitatively, such as the equilibria of Eqs. (1) and (2), an overall process is often broken down into energetic contributions from components that can be considered as "transferable properties," but this appears to be problematic for hydrogen bonding. A property is "transferable" if the same energy, to a reasonable approximation, is obtained for that component regardless of surroundings. Strictly speaking, a change in free energy (or enthalpy) is always a property of an entire system, not an isolated component of the system. In practice, an overall process can often be dissected using components that are, to a reasonable approximation, transferable; this facilitates understand-

[10] R. Taylor and O. Kennard, *Acc. Chem. Res.* **17**, 320 (1984).
[11] M. Etter, *Acc. Chem. Res.* **23**, 120 (1990).
[12] J. W. Larson and T. B. McMahon, *J. Am. Chem. Soc.* **105**, 2944 (1983).
[13] S. A. Sullivan and J. L. Beauchamp, *J. Am. Chem. Soc.* **96**, 1160 (1976).

ing of the complex system under investigation.[14,15] For example, additivity holds to a first level of approximation for covalent bonds and allows reasonable calculations of enthalpies of formation for many compounds. However, complexation via hydrogen bonding appears to be highly sensitive to molecular context and solvent (see later). It appears that hydrogen bonds cannot be treated as simple transferable properties.

Analyses of hydrogen bonds in this article rely on discrete comparisons. This is both because of the difficulty in assigning "absolute" or transferable energies to hydrogen bonds, as described earlier, and because we are, in fact, interested in the *relative* energetics of hydrogen bonds. We would like to know how the energy of a hydrogen bond changes in a transition state relative to a ground state, and further, we would like to compare this change in energy on an enzyme to the change in solution. We suggest that much of the confusion about hydrogen bond and other energetic features of biological systems has arisen because comparisons are often made implicitly, rather than explicitly. Conversely, explicit comparisons, such as those described later, represent powerful analytical tools to probe aspects of hydrogen bonds that may be important in enzymatic catalysis.

Can Strengthened Transition-State Hydrogen Bonding Contribute to Enzymatic Catalysis?[16]

The Catalytic Problem

Charge rearrangement typically occurs in the course of a reaction. This is shown in Scheme 1 using the triosephosphate isomerase (TIM) reaction as an example. As the reaction proceeds from the ground state to the transition state, negative charge develops on the carbonyl oxygen of the substrate. Hydrogen bonds from His-95 and Lys-12 of TIM to the carbonyl

---

[14] S. W. Benson, *in* "Thermochemical Kinetics," 2nd ed., p. 24. Wiley, New York, 1976.

[15] S. W. Benson, F. R. Cruickshank, D. M. Golden, G. R. Haugen, H. E. O'Neal, A. S. Rogers, R. Shaw, and R. Walsh, *Chem. Rev.* **69,** 279 (1969).

[16] Phrases such as "strengthening of hydrogen bonds" and "increase in hydrogen bond strength" are used in the text to refer to increases in the equilibrium constant for the formation of hydrogen-bonded complexes caused by changing local electrostatics [e.g., the charge buildup on the substrate carbonyl oxygen in the transition state for the TIM reaction (Scheme 1)]. Based on simple electrostatic considerations, an increase in "hydrogen bond strength" is expected when there is an increase in electron density on the hydrogen bond acceptor and/or a decrease in electron density on the hydrogen bond donor, with all other factors being equivalent. It is hoped that the suggestion of comparison in these phrases provides a sufficient reminder that equilibria for hydrogen bond formation need to be compared, as hydrogen bond energies cannot be defined as simple transferable properties.

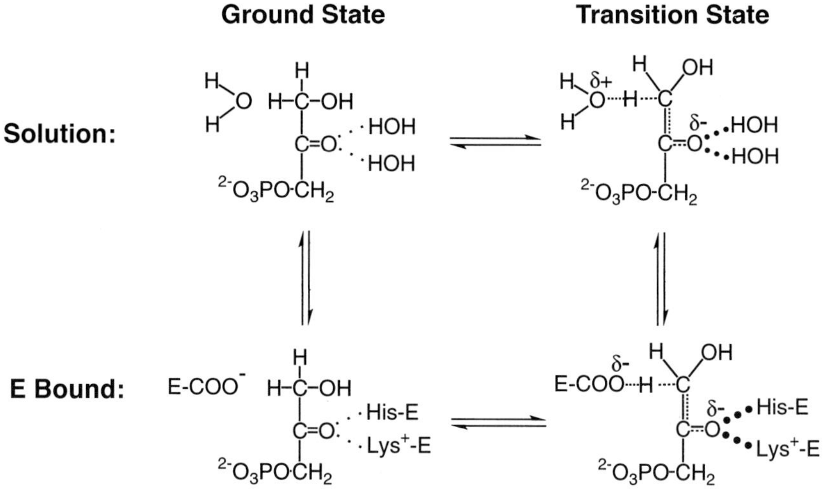

Ground State   **Transition State**

Solution:

E Bound:

SCHEME 1

oxygen[17] (Scheme 1, bottom) become stronger as a result of this charge buildup, as depicted by the darker dots in the transition state. However, in the solution reaction, the hydrogen bonds from water to the substrate also become stronger in the transition state (Scheme 1, top reaction). Enzymes, of course, need to catalyze reactions relative to reactions in aqueous solution. Thus, the strengthening of hydrogen bonds at the active site must be *greater than* the strengthening of the corresponding hydrogen bonds to water.

It is well established that enzymes can achieve some of their catalysis by positioning active site moieties with respect to bound substrate[4,18] (see also later). For example, in the active site of TIM, hydrogen bond donors and acceptors and the general acid and base are positioned in the active site with respect to bound substrate so that the entropic cost of arranging interactions that stabilize the transition state is overcome. Here we address a conceptually distinct question about the potential catalytic role of hydrogen bonds: After accounting for effects from positioning, are there differences between the active site and aqueous solution that allow a larger increase in hydrogen bond strength on the enzyme than in aqueous solution accom-

[17] Z. Zhang, S. Sugio, E. A. Komives, K. D. Liu, J. R. Knowles, G. A. Petsko, and D. Ringe, *Biochemistry* **33,** 2830 (1994).
[18] M. I Page and W. P. Jencks, *Proc. Natl. Acad. Sci. U.S.A.* **68,** 1678 (1971).

panying the electronic rearrangements in going from the ground state to the transition state?

Potential Solutions to the Catalytic Problem

The following sections describe model studies that suggest that this strengthening of hydrogen bonds to enzymatic groups in the course of a reaction can be greater than the corresponding strengthening of hydrogen bonds to water. The general approach is to probe hydrogen-bonding equilibria with homologous series of compounds. The change in pK_a of these compounds as substituents is varied and is used as a mimic of the change in charge distribution of the substrate functional groups in going from the ground state to the transition state.

Stronger Hydrogen Bond Donors and Acceptors at Enzymatic Active Sites Can Provide Greater Strengthening of Hydrogen Bonds Accompanying Charge Rearrangement. Remarkably, simple complexes with only a single hydrogen bond, such as that between a phenolate and a protonated amine [Eq. (3)], can be observed in aqueous solution despite the competition for hydrogen bonding by 55 M water.[19]

$$\text{Ph}{-}O^- \cdot HOH + H_2O \cdot {}^+H{-}NR_3 \rightleftharpoons$$

$$\text{Ph}{-}O^- \cdot {}^+HNR_3 + H_2O \cdot HOH \text{ (aq)} \quad (3)$$

While it is intuitive that the increased negative charge density on the phenolate ion relative to the oxygen of water would increase the strength of the electrostatic interaction with a protonated amine donor,

$$\left[\text{Ph}{-}\underline{O}^- \cdot {}^+HNR_3 \text{ vs } H_2\underline{O} \cdot {}^+HNR_3 \text{ in Eq. (3)} \right],$$

the interaction with a water donor would also be stronger:

$$\text{Ph}{-}\underline{O}^- \cdot HOH \text{ vs } H_2\underline{O} \cdot HOH \text{ in Eq. (3)}.$$

The ability to form the $\text{Ph}{-}\underline{O}^- \cdot {}^+HNR_3$ complex indicates that the increase in hydrogen bond strength is dependent on both the donor and the acceptor, i.e., the increase in hydrogen bond strength upon changing

[19] N. Stahl and W. P. Jencks, *J. Am. Chem. Soc.* **108**, 4196 (1986).

the acceptor from H_2O to RO^- is larger with the stronger hydrogen bond donor, $^+HNR_3$, than it is with HOH.[19,20]

How does this relate to hydrogen bonding in enzymatic catalysis? Consider the reaction catalyzed by TIM (Scheme 1) in which the carbonyl oxygen accumulates negative charge in going from the ground state ($C=O$) to the transition state ($C{\doteq}O^{\delta-}$). For a given hydrogen bond donor, the hydrogen bond strengthens in the transition state relative to the ground state because of this increased charge density on the acceptor. What happens with different hydrogen bond donors, i.e., water for the solution reaction vs Lys-12^+ for the enzymatic reaction (Fig. 1, top and bottom, respectively)? The positively charged Lys-12^+ is a stronger hydrogen bond donor than water. This causes a larger strengthening of the hydrogen bond to the carbonyl oxygen in going from the ground state to the transition state on TIM (Fig. 1, $\Delta\Delta G_E^{HB}$) than in aqueous solution (Fig. 1, $\Delta\Delta G_{soln}^{HB}$), with all other factors being equal.

WHAT IS THE BASIS FOR THIS EFFECT? The larger increase in hydrogen bond strength with a stronger donor is predicted from a simple Coulombic model for electrostatic interactions between hydrogen bond donors and acceptors.[21] Equations (5a) and (5b) represent this Columbic model for

---

[20] The situation is analogous when the hydrogen bond donors and acceptors are switched in the descriptions throughout the text. This can be seen from the symmetry in Eq. (5) and the Hine equation [Eq. (8)] in the Appendix.

[21] It has been suggested that covalent character in hydrogen bonds between donors and acceptors with matched pK_a values could result in large hydrogen bond energies in nonaqueous environments, thereby making a substantial contribution to enzymatic catalysis (e.g., Refs. 22–25). Indeed, it was these hypotheses that stimulated the experiments carried out in our laboratory.[1–3] The results and analyses suggest that hydrogen bonds may indeed contribute more to enzymatic catalysis than previously thought,[1–3] as suggested by these proposals. Nevertheless, data obtained thus far provide no indication of a large special energetic contribution arising from covalent character of hydrogen bonds (see also Refs. 2, 3, 26–31), and the energetic effects observed in model studies can be accounted for by simple electrostatic effects. Thus, a simple electrostatic model of hydrogen bonding is used in the text. In addition, the simplicity of this model facilitates an intuitive understanding of many of the factors that affect the energetics of hydrogen bonding. Nevertheless, structural and spectroscopic data strongly suggest that covalent interactions are also involved in certain hydrogen bonds in model and enzymatic systems, including TIM (Refs. 26–40; see also Refs. 5–7 and references therein). The nature of hydrogen bonds in enzymatic active sites and the physical properties of these hydrogen bonds that are important for their *energetic* behavior remain important areas for future investigation.

[22] W. W. Cleland, *Biochemistry* **31**, 317 (1992).

[23] W. W. Cleland and M. M. Kreevoy, *Science* **264**, 1887 (1994).

[24] J. A. Gerlt and P. G. Gassman, *Biochemistry* **32**, 11943 (1993).

[25] J. A. Gerlt and P. G. Gassman, *J. Am. Chem. Soc.* **115**, 11552 (1993).

[26] B. Schwartz, D. G. Drueckhammer, K. C. Usher, and S. J. Remington, *Biochemistry* **34**, 15459 (1995).

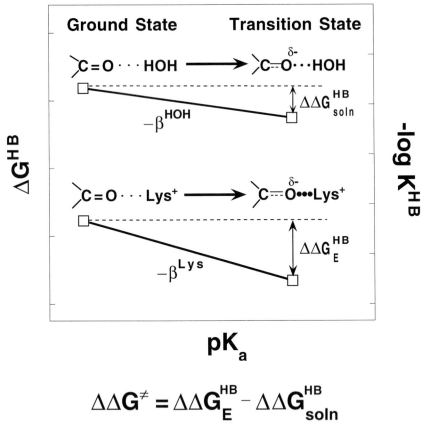

$$\Delta\Delta G^{\neq} = \Delta\Delta G_{E}^{HB} - \Delta\Delta G_{soln}^{HB}$$

FIG. 1. The potential rate enhancement that can be obtained by a stronger hydrogen bond donor at the enzyme active site than water using the TIM reaction as an example.

the equilibria of Eqs. (4a) and (4b), respectively. In Eqs. (5a) and (5b), q^+ is the partial positive charge on the donor proton, q_1^- and q_2^- are the partial negative charges on the acceptor atoms in Eqs. (4a) and (4b), respectively; the constant "C" describes the interaction between the donor and the

[27] K. C. Usher, S. J. Remington, D. P. Martin, and D. G. Drueckhammer, *Biochemistry* **33**, 7753 (1994).
[28] B. Schwartz and D. G. Drueckhammer, *J. Am. Chem. Soc.* **117**, 11902 (1995).
[29] Z. Wang, H. Luecke, N. Yao, and F. A. Quiocho, *Nat. Struct. Biol.* **4**, 519 (1995).
[30] Y. Kato, L. M. Toledo, and J. Rebek, Jr., *J. Am. Chem. Soc.* **118**, 8575 (1996).
[31] G. Zundel and M. Eckert, *J. Mol. Struct.* **200**, 73 (1989).
[32] J. Tobin, S. Whitt, C. Cassidy, and P. Frey, *Biochemistry* **34**, 6919 (1995).

acceptor and depends on the distance between the partial charges (r) and the effective dielectric (ε_{eff}) of the media [Eq. (5c), see next section for a description of ε_{eff}; see Refs. 41–43 for a description of Coulomb's law]. Equation (5) indicates that the electrostatic interaction becomes stronger as the charge density on the donor or acceptor increases, i.e., if q_2^- is more negative than q_1^-, then the value of ΔH_2^{HB} will be more negative than ΔH_1^{HB}; this is represented by a negative $\Delta\Delta H^{HB}$ in Eq. (6).

Now consider the TIM reaction in solution and on the enzyme (Fig. 1). The more negative value of q_2^- relative to q_1^- is analogous to the increased negative charge density for the substrate carbonyl oxygen in the transition state relative to the ground state (Fig. 1; $C\dot{=}O^{\delta-}$ vs $C=O$). The hydrogen bond donor to the carbonyl oxygen on the enzyme (Lys-12$^+$) is stronger than water. This corresponds to an increase in q^+ (i.e., $q_{lys}^+ >$ q_{HOH}^+) and results in a more negative value of $\Delta\Delta H^{HB}$ on the enzyme than in solution [Eqs. (7a) and (7b); $\Delta\Delta H_E^{HB}$ is more negative than $\Delta\Delta H_{soln}^{HB}$]. Thus, the enzymatic hydrogen bond contributes more toward

$$X_1 \text{—} \overset{q_1^-}{\underset{}{O}} + \text{H-NR}_3 \quad \underset{}{\overset{\Delta H_1^{HB}}{\rightleftharpoons}} \quad X_1 \text{—} \overset{q_1^-}{\underset{}{O}} \cdot \text{H-NR}_3 \quad (4a)$$

$$X_2 \text{—} \overset{q_2^-}{\underset{}{O}} + \text{H-NR}_3 \quad \underset{}{\overset{\Delta H_2^{HB}}{\rightleftharpoons}} \quad X_2 \text{—} \overset{q_2^-}{\underset{}{O}} \cdot \text{H-NR}_3 \quad (4b)$$

$$\Delta H_1^{HB} = C \times q^+ q_1^- \quad (5a)$$
$$\Delta H_2^{HB} = C \times q^+ q_2^- \quad (5b)$$

$$C \propto \frac{1}{\varepsilon_{eff}r} \quad (5c)$$

[33] P. Frey, S. Whitt, and J. Tobin, *Science* **264**, 1927 (1994).

[34] C. S. Cassidy, J. Lin, and P. Frey, *Biochemistry* **36**, 4576 (1997).

[35] T. K. Harris, C. Abeygunawardana, and A. S. Mildvan, *Biochemistry* **36**, 14661 (1997).

[36] Q. Zhao, C. Abeygunawardana, P. Talalay, and A. S. Mildvan, *Proc. Natl. Acad. Sci. U.S.A* **93**, 8220 (1996).

[37] Q. Zhao, C. Abeygunawardana, A. G. Gittis, and A. S. Mildvan, *Biochemistry* **36**, 14616 (1997).

[38] J. L. Markley and W. M. Westler, *Biochemistry* **35**, 11092 (1996).

[39] E. T. Mollova, D. E. Metzler, A. Kintanar, H. Kagamiyama, H. Hayashi, K. Hirotsu, and I. Miyahara, *Biochemistry* **36**, 615 (1997).

[40] H. Tong and L. Davis, *Biochemistry* **34**, 3362 (1995).

[41] F. Daniel and R. A. Alberty, *in* "Physical Chemistry," 4th ed., p. 192. Wiley, New York, 1975.

[42] G. M. Barrow, *in* "Physical Chemistry," 4th ed., p 544. McGraw-Hill, New York, 1979.

[43] P. W. Atkins, *in* "Physical Chemistry," 4th ed., p. 649. Freeman, New York, 1990.

preferentially stabilizing the transition state than the hydrogen bond in solution.

$$\Delta\Delta H^{\mathrm{HB}} = \Delta H_2^{\mathrm{HB}} - \Delta H_1^{\mathrm{HB}} = C \times q^+(q_2^- - q_1^-) = C \times q^+ \Delta q^- \qquad (6)$$

$$\Delta\Delta H_{\mathrm{E}}^{\mathrm{HB}} = C \times q_{\mathrm{lys}}^+ \Delta q^- \qquad (7a)$$

$$\Delta\Delta H_{\mathrm{soln}}^{\mathrm{HB}} = C \times q_{\mathrm{HOH}}^+ \Delta q^- \qquad (7b)$$

QUANTITATIVE ESTIMATES FOR THE CATALYTIC CONTRIBUTION FROM THIS EFFECT. What is the magnitude of the possible catalytic effect stemming from the stronger hydrogen bond donor at the enzymatic active site relative to water? This can be estimated using the physical organic approach of linear free energy relationships. A brief introduction to linear free energy relationships is included because this approach is also important in the next section of this article.[44]

To construct a linear free energy relationship, the charge density on a hydrogen bond donor (or acceptor) is varied in a homologous series of compounds via electron-donating and electron-withdrawing substituents, as shown in Scheme 2 for substituted phenols. Changes in the pK_a of the $-\mathrm{OH}$ group provide a measure of the changes in the charge distribution at this position caused by the substituents. The equilibrium for hydrogen bond formation (K^{HB}) for the series of homologous compounds, such as phenolate ions, with a given hydrogen bond donor can be determined experimentally. Plots of the log of the equilibrium constant for hydrogen bond formation (log K^{HB}) vs pK_a are typically linear and the slopes are referred to as Brønsted coefficients, α values for hydrogen bond donors and β values for hydrogen bond acceptors (Figs. 2A and 2B, respectively).

Thus, the Brønsted slopes, α and β, describe the dependence of hydrogen bond formation on the charge distribution of the donor or acceptor (Figs. 2A and 2B, respectively) relative to the dependence of the protonation equilibrium. Small slopes of α or $\beta \approx 0.1$–0.2 have been observed for such correlations in water and other protic solvents (Table I). The small slopes are consistent with the idea that substituents have smaller effects on the largely electrostatic hydrogen bond (log K^{HB}) than on covalent bond

[44] For more complete treatments of linear free energy relationships, see, for example, Refs. 45–48.

[45] J. Hine, in "Physical Organic Chemistry," 2nd ed., p 81. McGraw-Hill, New York, 1972.

[46] F. A. Carey and R. J. Sundberg, in "Advanced Organic Chemistry Part A: Structure and Mechanisms," 3rd ed., p. 196. Plenum, New York, 1990.

[47] J. Shorter, "Correlation Analysis in Chemistry." Plenum, New York, 1978.

[48] O. Exner, "Correlation Analysis of Chemical Data." Plenum, New York, 1988.

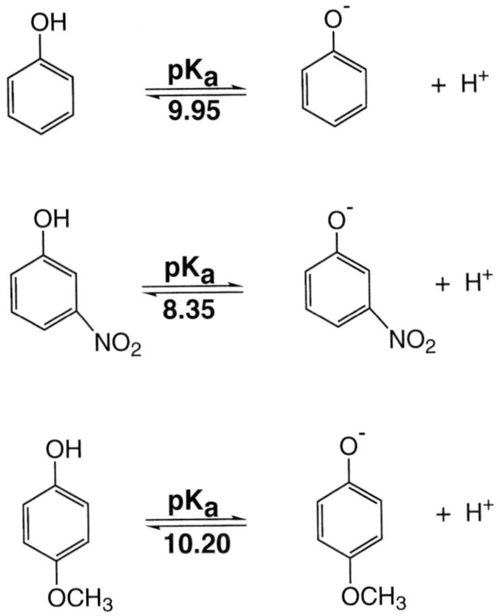

SCHEME 2

formation with a proton (pK_a).[19,49,50] Note that an α or β value of 1 would be obtained if hydrogen bond formation had the same dependence on substituents as protonation to give the phenol.

To quantitatively assess the possible catalytic effect stemming from the stronger hydrogen bond donor at the enzymatic active site relative to water, the Brønsted slope of a series of acceptors with one donor can be compared to the Brønsted slope for the same series of acceptors with a different donor. The different donors are meant to model the different hydrogen bond donors in the enzymatic and solution reactions (e.g., Lys-12$^+$ in TIM and water in solution). It has been observed that the value of β increases when the linear-free energy relationship is obtained with a compound that is a stronger hydrogen bond donor (i.e., lower pK_a) (Fig. 2B).[51] This is precisely what is predicted from Eqs. (7a) and (7b). Equations (7a) and (7b) describe this effect in terms of different changes in the predicted energy for hydrogen bond donors with different partial charges, whereas

[49] L. H. Funderburk and W. P. Jencks, *J. Am. Chem. Soc.* **100**, 6708 (1978).
[50] M. E. Rothemberg, J. P. Richard, and W. P. Jencks, *J. Am. Chem. Soc.* **107**, 1340 (1985).
[51] M. H. Abraham, P. P. Duce, J. J. Morris, and P. J. Taylor, *J. Chem. Soc. Faraday Trans. 1* **83**, 2867 (1987).

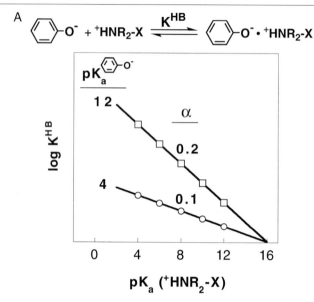

FIG. 2. Theoretical Brønsted plots of the equilibrium for H bonding (K^{HB}) between protonated amines and phenolate anions in water according to the Hine equation. Parameters used in the calculation are $\tau = 0.013$, $pK_a^{HOH} = 16.04$, and $pK_a^{H_2O} = -1.26$.[19,45] (A) Dependences of log K^{HB} on the pK_a values of the hydrogen bond donor ($^+HNR_2$-X) described by the Brønsted slope α. A steeper Brønsted slope is obtained with a stronger hydrogen bond acceptor, with $pK_a = 12$ (□), than with a weaker hydrogen bond of $pK_a = 4$ (○). (B) Dependencies of log K^{HB} on the pK_a values of hydrogen bond acceptor

described by the Brønsted slope β. The β value is larger with a more acidic hydrogen bond donor with $pK_a = 4$ (○) than the β value with a weaker donor with $pK_a = 12$ (□).

the increase in the Brønsted slope describes the same effect in terms of the observed equilibrium constant for hydrogen bond formation with hydrogen bond donors that have different pK_a values.

The change in the value of β with a stronger hydrogen bond donor can be estimated using the Hine equation, which quantitatively describes hydrogen bonding between solutes in water as a competition between solute · solute interactions and solute · solvent interactions, based on an electrostatic model for hydrogen bonding.[19,45,52,53] The reader is referred

[52] J. G. Kirkwood and F. H. Westheimer, *J. Chem. Phys.* **6**, 506 (1938).
[53] J. Hine, *J. Am. Chem. Soc.* **94**, 5766 (1972).

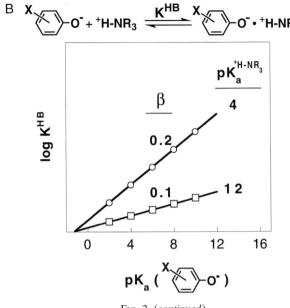

FIG. 2. (*continued*)

to the Appendix for a detailed description of the Hine equation and the calculation of the rate enhancement from the stronger hydrogen bond donor, Lys-12$^+$ on the enzyme, relative to water. This analysis gives an estimated rate enhancement of 10-fold, assuming that the active site environment is the same as aqueous solution. This estimate presumably represents a conservative lower limit because the effect would be accentuated in an environment of low effective dielectric, as the enzymatic active site is thought to provide (see later). The next section describes that a low effective dielectric in an enzymatic active site can cause the increase in hydrogen bond strength in going from the ground state to the transition state to be larger than the corresponding increase in aqueous solution, whether or not the enzyme provides a stronger hydrogen bond donor than water.

   A Low Dielectric Environment Can Increase the Change in Hydrogen Bond Strength Accompanying Charge Rearrangement. The electrostatic model for hydrogen bonding presented in the last section predicts that decreasing the "effective dielectric" of the media can also increase the strengthening of hydrogen bonds that accompanies changes in the charge distribution of donor or acceptor atoms [$\Delta\Delta H \propto 1/\varepsilon_{\text{eff}}$; Eqs. (7) and (5c)]. The term "effective dielectric" is used to describe the effect of the particular

TABLE I

PARTIAL LIST OF OBSERVED BRØNSTED SLOPES FOR HYDROGEN BONDING IN
AQUEOUS AND ORGANIC SOLVENT

Hydrogen bond	Solvent	Brønsted slope $(\alpha$ or $\beta)^a$
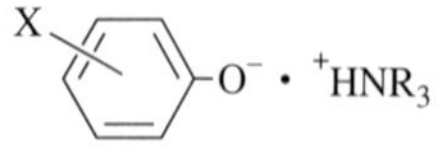 X—C₆H₃—O⁻ · ⁺HNR₃	Water	$\beta = 0.10^b$
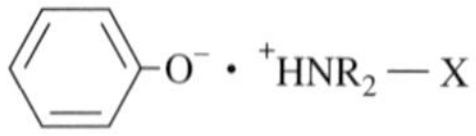 C₆H₅—O⁻ · ⁺HNR₂—X	Water	$\alpha = 0.15^b$
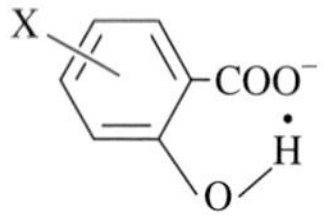 X—C₆H₃—COO⁻ · H—O (ortho)	Water Dimethyl sulfoxide	$\beta = 0.05^c$ $\beta = 0.73^c$
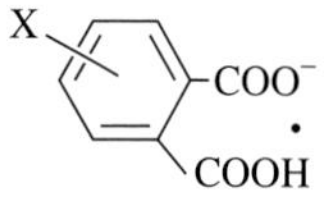 X—C₆H₃—COO⁻ · COOH (ortho)	Dimethyl sulfoxide	$\beta = 0.90^d$
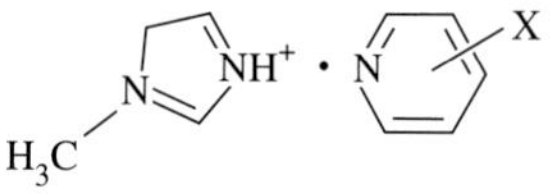 methylimidazole-NH⁺ · N-pyridine—X	Acetonitrile	$\beta = 0.15^{e,f}$
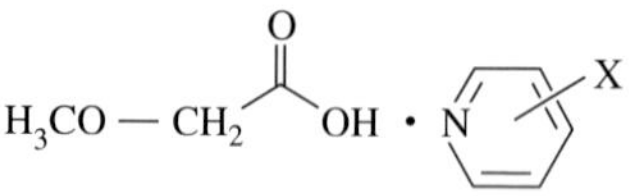 H₃CO—CH₂—C(O)—OH · N-pyridine—X	Acetonitrile	$\beta = 0.28^{g,f}$
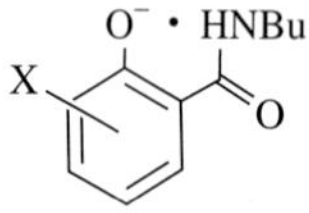 salicyl O⁻ · HNBu, C=O, X	Dimethyl acetamide	$\beta = 0.14^h$
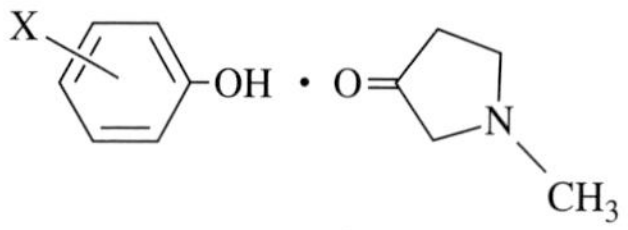 X—C₆H₃—OH · O=pyrrolidinone-N-CH₃	1,1,1-Trichloroethane	$\alpha = 0.41^i$
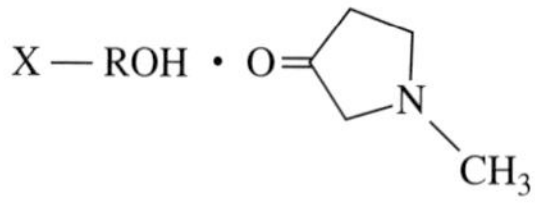 X—ROH · O=pyrrolidinone-N-CH₃	1,1,1-Trichloroethane	$\beta = 0.24^i$
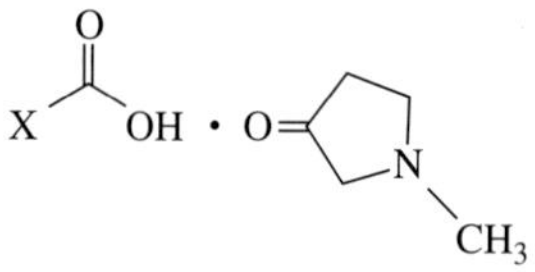 X—C(O)—OH · O=pyrrolidinone-N-CH₃	1,1,1-Trichloroethane	$\beta = 0.40^i$

(*continued*)

TABLE I (*continued*)

Hydrogen bond	Solvent	Brønsted slope $(\alpha \text{ or } \beta)^a$
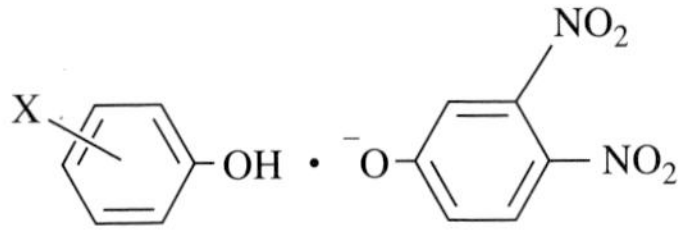 X—⟨benzene⟩—OH · $^-$O—⟨benzene with NO$_2$, NO$_2$⟩—NO$_2$	Tetrahydrofuran	$\alpha = 0.65^d$
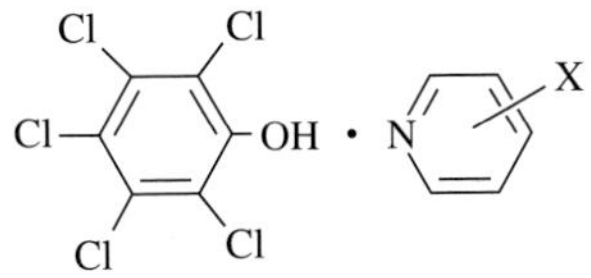 Cl, Cl, Cl—⟨benzene⟩, Cl, Cl—OH · N⟨pyridine⟩—X	Carbon tetrachloride	$\alpha = 1.2^{j,f}$
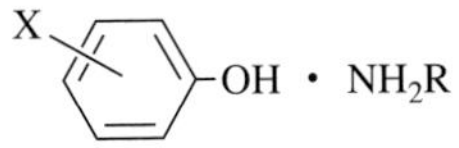 X—⟨benzene⟩—OH · NH$_2$R	Carbon tetrachloride	$\beta = 0.30^{j,f}$
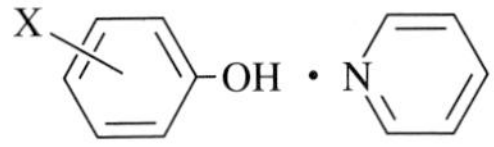 X—⟨benzene⟩—OH · N⟨pyridine⟩	Carbon tetrachloride	$\alpha = 0.53^{k,f}$
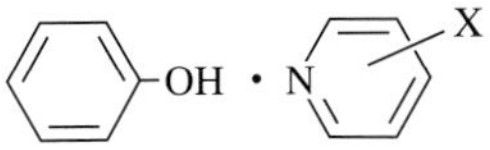 ⟨benzene⟩—OH · N⟨pyridine⟩—X	Carbon tetrachloride	$\beta = 0.13^{l,f}$
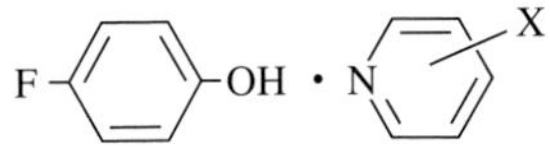 F—⟨benzene⟩—OH · N⟨pyridine⟩—X	Carbon tetrachloride	$\beta = 0.20^{m,f}$
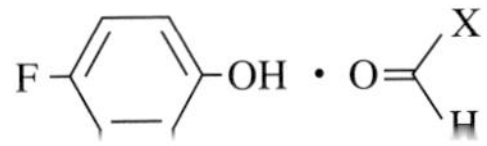 F—⟨benzene⟩—OH · O=C⟨X, H⟩	Carbon tetrachloride	$\beta = 0.23^{m,f}$
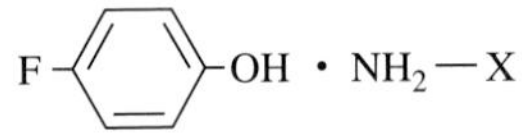 F—⟨benzene⟩—OH · NH$_2$—X	Carbon tetrachloride	$\beta = 0.31^{m,f}$
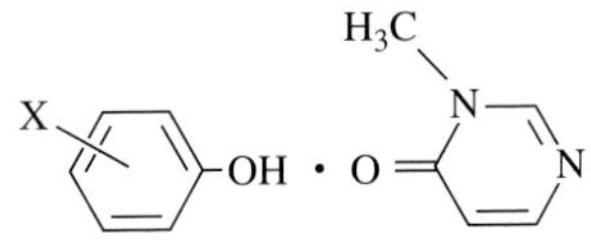 X—⟨benzene⟩—OH · O=⟨pyrimidine with H$_3$C—N⟩	Benzene	$\alpha = 0.33^n$
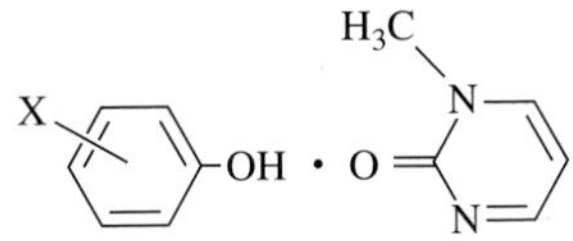 X—⟨benzene⟩—OH · O=⟨pyrimidine with H$_3$C—N⟩	Benzene	$\alpha = 0.44^n$
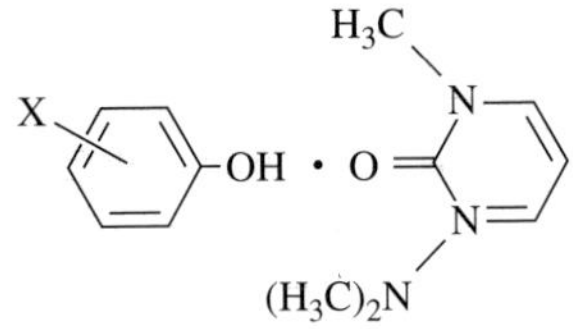 X—⟨benzene⟩—OH · O=⟨pyrimidine with H$_3$C—N, (H$_3$C)$_2$N⟩	Benzene	$\alpha = 0.69^n$

(*continued*)

TABLE I (continued)

Hydrogen bond	Solvent	Brønsted slope (α or β)[a]
	Benzene	$\alpha = 0.30$[n]
	Benzene	$\alpha = 0.28$[n]

[a] α and β values are slopes of plots of log K for hydrogen bond formation against the *aqueous* pK_a values of the hydrogen bond donor or acceptor. In cases with multiple potential hydrogen bond donors and acceptors, complexes are drawn for simplicity of representation, although the functional groups involved in the hydrogen bond have not been determined in general. The number of data points (n) and correlation coefficient (r) for each Brønsted plot are indicated in the parentheses after each reference. The range of β values observed in various model studies is not rationalized easily and suggests that we do not yet have a complete account of the factors that affect the energetics of hydrogen bonding. Tables I and II include all the examples of Brønsted slopes for hydrogen bonding of which we are aware. We would appreciate communication of additional examples for inclusion in future compilations.

[b] Reference 19 ($n = 8$, $r > 0.99$ for the α value; $n = 11$, $r = 0.77$ for the value of β).

[c] Reference 1 ($n = 11$, $r > 0.99$).

[d] Reference 3 ($n = 6$, $r > 0.99$ for the hydrogen bond in phthalate monoanions; $n = 14$, $r > 0.99$ for the hydrogen bond between substituted phenols and 3,4-dinitrophenolate).

[e] Z. Pawlak, G. Zundel, and J. Fritsch, *Electrochim. Acta* **29**, 391 (1984) ($n = 3$, $r = 0.92$).

[f] Individual equilibrium constants for hydrogen bond formation from the cited reference were used to construct Brønsted plots and obtain the α and β values listed.

[g] G. Albrecht and G. Zundel, *Z. Naturforsch.* **39a**, 986 (1984) ($n = 6$, $r = 0.94$).

[h] W. L. Mock and C. Y. Chua, *J. Chem. Soc. Perkin Trans.* **2**, 2069 (1995) ($n = 8$, $r > 0.99$).

[i] Reference 52 ($n = 7$, $r > 0.99$ for the hydrogen bond with substituted alcohols; $n = 13$, $r = 0.90$ for hydrogen bonding with substituted phenols; $n = 8$, $r = 0.97$ for hydrogen bonding with substituted carboxylic acids).

[j] G. Albrecht and G. Zundel, *J. Chem. Soc. Faraday Trans. 1* **79**, 553 (1983) ($n = 3$, $r = 0.98$ for the hydrogen bond with pentachlorophenol; $n = 4$; $r = 0.98$ for the hydrogen bond with substituted phenols).

[k] J. Rubin, B. Z. Senkowski, and G. S. Panson, *J. Phys. Chem.* **68**, 1601 (1964) ($n = 8$; $r = 0.99$).

[l] J. Rubin and G. S. Panson, *J. Phys. Chem.* **69**, 3089 (1965) ($n = 8$, $r = 0.99$).

[m] R. W. Taft, D. Gurka, L. Joris, P. von R. Schleyer, and J. W. Rakshys, *J. Am. Chem. Soc.* **91**, 4801 (1969) ($n = 7$, $r > 0.99$ for the hydrogen bond with substituted pyridines; $n = 4$; $r > 0.99$ for the hydrogen bond with aldehydes; $n = 3$, $r > 0.99$ for the hydrogen bond with amines).

[n] O. Kasende and Th. Zeegers-Huyskens, *J. Phys. Chem.* **88**, 2636 (1984) ($n = 13$, $r = 0.98$–0.99).

local solvent or environment on the electrostatic interaction between nearby charged or dipolar groups. Thus, unlike the bulk dielectric constant, which represents the ability of a solvent to screen electrostatic interactions between groups distant from one another, the effective dielectric depends on the molecular interactions surrounding the hydrogen bond donors and acceptors. As described in this section, the low effective dielectric of enzymatic interiors may allow active site hydrogen bonds with substrate moieties undergoing charge rearrangement to provide a substantial catalytic contribution.[54–60] This effect has been demonstrated in a model study in which the dependence of the equilibrium for hydrogen bond formation (K^{HB}) on the pK_a of the hydrogen bonding groups was compared directly in water and in an organic solvent for the same compounds, a series of substituted salicylate monoanions.[1] The method used in this study to isolate the equilibrium for hydrogen bond formation is first described, as this approach may be applicable for addressing other aspects of hydrogen bonding systems.[61] Results of the model study, the implications for enzymatic catalysis, and relevant enzymatic results are then described.

METHOD FOR LINEAR FREE ENERGY ANALYSIS OF HYDROGEN BOND STRENGTH FOR INTRAMOLECULAR HYDROGEN BONDS. Equilibria for formation of the hydrogen bond between the adjacent carboxylate and hydroxyl groups in salicylate monoanions (Structure I) were determined according to Scheme 3. Formation of this hydrogen bond stabilizes the salicylate monoanion relative to the neutral salicylic acid, resulting in a decrease in the observed pK_a (pK_a^{obsd}) of this compound. Although the value of (pK_a^{obsd}) in water is simply determined using a standard pH electrode, determinations of pK_a values in organic solvents are more involved. Fortu-

[54] A. Warshel and S. T. Russell, *Quart. Rev. Biophys.* **17,** 283 (1984).
[55] G. King, F. S. Lee, and A. Warshel, *J. Chem. Phys.* **95,** 4366 (1991).
[56] B. Honig and A. Nicholls, *Science* **268,** 1144 (1995).
[57] M. K. Gilson and B. Honig, *Nature* **330,** 84 (1987).
[58] K. A. Sharp and B. Honig, *Annu. Rev. Biophys. Biophys. Chem.* **19,** 301 (1990).
[59] The analyses described in this article use model systems to isolate the energetic effects from hydrogen bonds to groups undergoing charge rearrangement. Solvents with low dielectrics and/or low effective dielectrics have been used to model the expected low effective dielectric environment of enzyme active sites, as described in the text. Nevertheless, it should be recognized that the overall energetic interactions with a transition state must be favorable on an enzyme relative to the interactions in aqueous solution in order for the enzyme to provide catalysis. Thus, any potential penalty from imbedding a transition state within a low effective dielectric active site must be overcome by other favorable factors such as additional hydrogen bonds, additional electrostatic interactions, and positioning effects (Refs. 2, 4, 60, and references therein).
[60] A. Warshel, *J. Biol. Chem.* **273,** 27035 (1998).
[61] P. Luo and R. L. Baldwin, *Biochemistry* **36,** 8413 (1997).

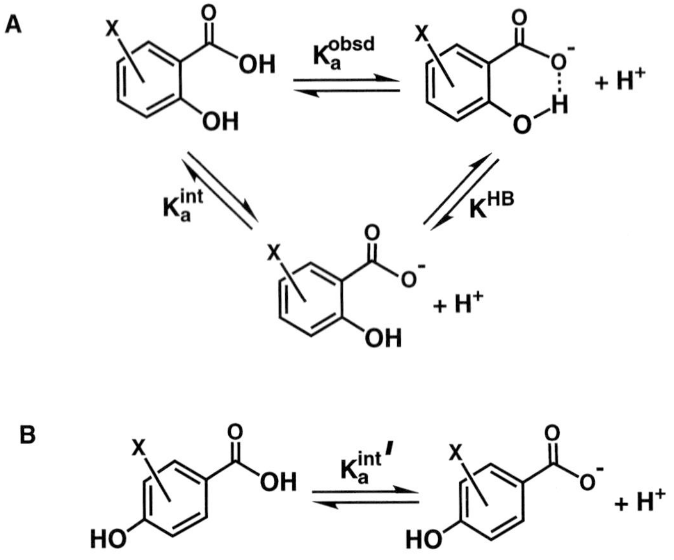

$$K^{HB} = K_a^{obsd} / K_a^{int}$$

$$K_a^{int} \approx K_a^{int\,'}$$

SCHEME 3

nately, indicator scales have been established for pK_a values in dimethyl sulfoxide (DMSO) by Bordwell and co-workers, allowing pK_a values to be obtained spectrophotometrically from the absorbance of proton indicator dyes (Refs. 62–64 and references therein).

Electron-donating and electron-withdrawing substituents on the salicylic acid ($-X$ in Scheme 3) change the charge distribution of the hydrogen bond donor and acceptor. The substituents alter the ability of the hydroxyl and carboxylate groups to donate and accept a hydrogen bond, resulting

[62] W. S. Matthews, J. E. Bares, J. E. Bartness, F. G. Bordwell, F. J. Cornforth, G. E. Drucker, Z. Margolin, R. J. McCallum, G. J. McCollum, and N. R. Vanier, *J. Am. Chem. Soc.* **97**, 7006 (1975).
[63] F. G. Bordwell, R. J. McCallum, and W. N. Olmstead, *J. Org. Chem.* **49**, 1424 (1984).
[64] F. G. Bordwell, *Acc. Chem. Res.* **21**, 456 (1988).

in a change in the equilibrium for hydrogen bonding (K^{HB}); this is the effect that we want to follow. However, the substituents also have a second effect on (pK_a^{obsd}) of the carboxylic acid group, altering its "intrinsic" pK_a (pK_a^{int}). The intrinsic pK_a is a hypothetical equilibrium that represents the pK_a the carboxylic acid group would have in the absence of the hydrogen-bonding interaction. Thus, the effect on pK_a^{int} reflects the "classical" substituent effect and can be approximated by the pK_a of the corresponding compound in which the hydroxyl group is *para* to the carboxylic acid instead of *ortho* (Scheme 3B, $pK_a^{int} \approx pK_a^{int'}$). *Ortho* and *para* substituents have similar inductive and resonance effects, but the *para* positioning prevents intramolecular hydrogen bonding between the $-COO^-$ and $-OH$ groups. As depicted in Scheme 3, the difference between the intrinsic and observed pK_a of the $-COOH$ group provides a measure of the equilibrium for formation of the hydrogen bond[1,65]:

$$K^{HB} = \frac{K_a^{obsd}}{K_a^{int}} \qquad \text{and} \qquad \log K^{HB} = pK_a^{int} - pK_a^{obsd}$$

RESULTS AND IMPLICATIONS. Measuring the equilibrium constant for hydrogen bond formation for salicylate monoanions with a series of substituents in water and in DMSO allowed the determination of linear-free energy relationships for hydrogen bonding in these two different environments (Fig. 3). The dependence of the hydrogen-bonding equilibrium on ΔpK_a, the difference between the pK_a values for the hydrogen bond donor ($-OH$) and acceptor ($-COO^-$), is steeper in DMSO (Fig. 3, circles) than

---

[65] It should be noted that this method does not provide a perfect measure of these hydrogen-bonding equilibria. This is both because of the difficulties in isolating the "absolute" or "intrinsic" energy for hydrogen bonds, as discussed in the text, and because the following simplifying assumptions were made in this analysis: (1) *Ortho*- and *para*-OH groups have similar "intrinsic" effects on deprotonation of the $-COOH$ group; (2) the steric effect on the $-COOH$ pK_a from the *ortho*-OH group is minimal; and (3) the inductive effects from the *ortho*-OH group and added substituents ($-X$ in Scheme 3) are independent and therefore additive. These assumptions could lead to small deviations in the estimated "intrinsic" pK_a values of the hypothetical nonhydrogen-bonded species (Scheme 3A, bottom species) and thus small effects on the K^{HB} values obtained. Nevertheless, control experiments suggest that these assumptions hold to a first approximation (Refs. 1 and 2 and references therein). More importantly, the analysis was performed for a structurally homologous series of compounds so that deviations caused by these simplifications are expected to largely cancel, allowing the *change* in the energetics of these hydrogen bonds to be determined reliably. Thus, these approximations are not expected to affect the conclusions from this study.

in water (Fig. 3, squares), with Brønsted slopes of $\beta = 0.73$ and 0.05, respectively.[66]

The steeper Brønsted slope in DMSO than in water suggests that decreasing the effective dielectric of the media can provide a greater strengthening of hydrogen bonds that accompanies an increase in the charge density on the donor/acceptor groups. Although we know of no other direct comparison between hydrogen bonds in aqueous and nonaqueous environments, steep Brønsted slopes for both intramolecular and intermolecular hydrogen bonds are often observed in nonaqueous solvents and in the gas phase (Tables I and II), suggesting that the observation described earlier is general.

Enzymes may provide an active site environment with a lower effective dielectric than water because protein interiors contain many nonpolar side chains that cannot stabilize isolated charges effectively and because the charged and polar groups that are present are typically involved in networks of interactions that limit their ability to rearrange and stabilize isolated charges (see Refs. 1, 54–58, and references therein for more detailed discussions). Analogous to the enzymatic active site, DMSO molecules are not effective at stabilizing negative charges: their methyl groups prevent close

---

[66] In Fig. 3, a common pK_a scale in water is used to allow direct comparison of the magnitude of changes in the equilibrium for hydrogen bond formation in DMSO and in water for the same series of compounds. Using a common ΔpK_a scale allows simple representation of the transition state and ground state in terms of ΔpK_a values that remain constant, regardless of solvent. This is illustrated in the following example. Consider two compounds in Fig. 3 with ΔpK_a values of 0 and 5 in water, respectively. This five-unit difference in the aqueous ΔpK_a value of the two compounds ($\Delta\Delta pK_a$) is analogous to a change in the pK_a value of a substrate group in the course of a reaction. Using the Brønsted slopes of $\beta_{DMSO} = 0.73$ and $\beta_{water} = 0.05$ in Fig. 3 that are based on the aqueous pK_a scale, the difference in the strength of the two hydrogen bonds in DMSO and in water can be simply compared as $\Delta\Delta\log K^{HB} = \Delta\Delta pK_a \times (\beta_{DMSO} - \beta_{water}) = 5 \times (0.73 - 0.05) = 3.4$. However, because the pK_a scale in DMSO is 2.4-fold larger than that in water, the Brønsted slope for the hydrogen bond based on such a pK_a scale would be smaller, with $\beta'_{DMSO} = 0.30$; for the same reason, the change in the ΔpK_a value would be larger, with $\Delta\Delta pK'_a = 12$ for the same two compounds. The difference in the strength of the two hydrogen bonds in DMSO and water is then compared as $\Delta\Delta\log K^{HB} = \Delta\Delta pK'_a \times \beta'_{DMSO} - \Delta\Delta pK_a \times \beta_{water} = 12 \times 0.30 - 5 \times 0.05 = 3.4$. The same result is obtained using both scales, as it should, but the Brønsted slopes obtained from a common pK_a scale in water allows the change in hydrogen bond strength for the two compounds to be compared directly in different sovents, without the necessity to separately correct for the change in the pK_a scale in going from water to DMSO. It should also be noted that the slope of 0.73 in DMSO does not represent the apparent degree of proton transfer in the hydrogen bond, as described earlier for the β values in water. Such interpretations require the use of the pK_a scale for the solvent in which the hydrogen-bonding equilibrium is being determined and have sometimes been made incorrectly in the literature.

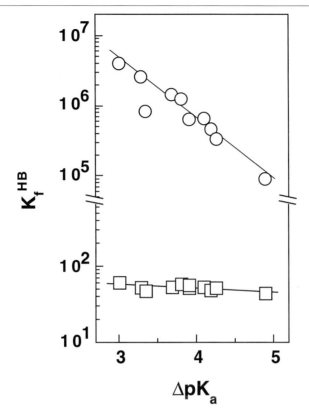

FIG. 3. A greater strengthening of the hydrogen bond accompanying changes in the charge density of donor/acceptor groups in DMSO than water. The equilibrium for formation of the hydrogen bond (K^{HB}) in a series of substituted salicylate monoanions is plotted against ΔpK_a, the difference in the pK_a value of the hydrogen bond donor and acceptor. To allow direct comparison of the magnitude of changes in K^{HB} accompanying changes in the charge distribution of the donor/acceptor groups, a common pK_a scale in water was used for the hydrogen bond in both media. Adapted from S. Shan and D. Herschlag, *Proc. Natl. Acad. Sci. U.S.A* **93,** 14474 (1996), with permission.

approach of the partially positively charged sulfur to anions and their large size limits rearrangement of the molecules in the solvation shell. DMSO may therefore be a reasonable, albeit crude, mimic of the active site environment for interactions of negatively charged species with hydrogen bond donors (see also Ref. 1 and references therein). Data from model studies and the apparent low effective dielectric of enzymatic active sites suggest that hydrogen bonds can be strengthened in the transition state to a greater

TABLE II
PARTIAL LIST OF OBSERVED BRØNSTED SLOPES FOR HYDROGEN
BONDING IN THE GAS PHASE

Hydrogen bond	Brønsted slope (α or β)[a]
$X-OH \cdot F^-$ (X=H, R)	$\alpha = 0.51$[b]
	0.50[c]
$HOH \cdot {}^-O-X$ (X=H, R)	$\beta = 0.28$[d]
$X-OH \cdot {}^-OCC-R$ (X=H, R)	$\alpha = 0.34$[d]
$X-OH \cdot Cl^-$ (X=H, R)	$\alpha = 0.17$[e,f]
$X-R_2NH^+ \cdot OH_2$	$\alpha = 0.26$[g,f]
$X-R_2NH^+ \cdot NH_3$	$\alpha = 0.30$[g,f]
$CH_3NH_3^+ \cdot \underset{\underset{H}{\vert}}{O}-X$	$\beta = 0.17$[g,f]
$CH_3NH_3^+ \cdot O{=}{<}^{R_1}_{R_2}$	$\beta = 0.51$[g,f]
$X-C_6H_4-NH^+ \cdot H_2O$	$\alpha = 0.15$[g,f]
$X-OH_2^+ \cdot OH_2$	$\alpha = 0.34$[g,f]
$R_1R_2OH^+ \cdot OH_2$	$\alpha = 0.31$[g,f]

(continued)

extent within the context of the enzymatic active site than in aqueous solution.[59]

This mechanism may provide substantial rate enhancements relative to solution reactions. The potential catalytic effect from the interaction of His-95 with the substrate carbonyl oxygen in the TIM reaction is outlined in Fig. 4, and a crude estimate for the magnitude of this effect is obtained as follows.[67] As the reaction proceeds from the ground state to the transition state, the pK_a of the carbonyl oxygen increases by $\sim$10 units.[68] If the Brønsted slopes of 0.7 and 0.05 observed for DMSO and water in model

[67] The stronger hydrogen bond donating ability of Lys-12$^+$ and the lower effective dielectric of the enzyme active site could both contribute to a steeper Brønsted slope for the $C{=}O \cdot$ Lys-12$^+$ hydrogen bond. For discussion of the catalytic contribution of hydrogen bonds arising from the lower effective dielectric of the enzyme active site, the $C{=}O \cdot$ His^{95} hydrogen bond is chosen instead because the neutral histidine and water have more similar pK_a values ($\sim$14 and 16, respectively).

[68] A. J. Kresge, *Pure Appl. Chem.* **63,** 213 (1991).

TABLE II (*continued*)

[a] α and β values are slopes of plots of the enthalpy for hydrogen bond formation (ΔH^{HB}) against the enthalpy for gas phase deprotonation (ΔH^{acid}) of the hydrogen bond donor (for α values) or acceptor (for β values). The number of data points (n) and correlation coefficient (r) for each Brønsted plot are indicated in the parentheses after each reference. Significant deviations from a linear relationship are observed in some correlation of ΔH^{HB} vs ΔH^{acid}. A cutoff of $r \geq 0.90$ was used in selecting data sets. In addition, a slope of >1 was also observed for one set of data in reference g (not shown). these deviations suggest that factors in addition to simple electrostatics affect the energetics of complex formation in these data sets.

[b] Reference 12 ($n = 12$, $r > 0.99$).

[c] J. E. Mihalick, G. G. Gatev, and J. Brauman, *J. Am. Chem. Soc.* **118**, 12424 (1996) ($n = 13$, $r > 0.99$).

[d] M. Mautner and L. W. Sieck, *J. Am. Chem. Soc.* **108**, 7525 (1986) ($n = 4$, $r > 0.99$ for the $HOH \cdots O-X$ hydrogen bond; $n = 7$, $r = 0.99$ for the $X-OH \cdots {}^-OOC-R$ hydrogen bond).

[e] R. Yamdagni and P. Kebarle, *J. Am. Chem. Soc.* **93**, 7139 (1971) ($n = 4$, $r = 0.96$).

[f] Individual enthalpies for hydrogen bond formation from the cited reference were used to construct Brønsted plots and to obtain the α and β values listed.

[g] M. Meotner, *J. Am. Chem. Soc.* **106**, 1257 (1984) ($n = 9$, $r = 0.99$ for the $X-R_2NH^+ \cdot OH_2$ hydrogen bond; $n = 5$, $r = 0.98$ for the $X-R_2NH^+ \cdot NH_3$ hydrogen bond; $n = 7$, $r = 0.89$ for the hydrogen bond of methylamine with substituted alcohols; $n = 4$, $r = 0.96$ for the hydrogen bond of methylamine with ketones; $n = 8$, $r = 0.93$ for the hydrogen bond of substituted pyridines with water; $n = 3$, $r = 0.99$ for the hydrogen bond of substituted hydronium ions with water; $n = 5$, $r = 0.96$ for the hydrogen bond of substituted oxonium ion with water).

studies were to hold for the enzymatic and solution hydrogen bonds, respectively (Fig. 4, β_{soln} and β_E in the top and bottom reactions, respectively), this interaction would provide a rate enhancement of $\sim 10^6$-fold [rate enhancement $= 10^{\Delta pK_a \times (\beta_{soln} - \beta_E)} = 10^{10 \times (0.7 - 0.05)} \approx 10^6$; $\Delta\Delta G^{\ddagger} = \Delta\Delta G^{E} - \Delta\Delta G^{soln} \approx 9$ kcal/mol (Fig. 4)]. This estimate certainly oversimplifies features of the active site and ignores the potential change in β value in going from the formally neutral carbonyl oxygen in the ground state to the partially charged enolate-like oxygen in the transition state. Also, DMSO is an imperfect mimic of the active site environment, as noted earlier. Nevertheless, the analysis suggests that the steeper Brønsted slopes in environments of low effective dielectric and the large change in the charge

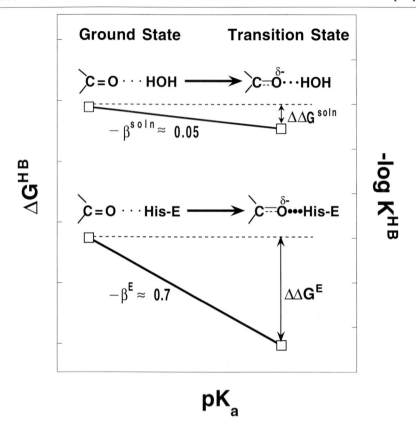

$$\Delta\Delta G^{\neq} = \Delta\Delta G^{E} - \Delta\Delta G^{soln}$$

Fig. 4. Potential catalytic contribution from a greater strengthening of hydrogen bonds accompanying charge redistribution in a low dielectric enzyme active site than in aqueous solutions depicted for the example of the TIM reaction. Reproduced from S. Shan and D. Herschlag, *Proc. Natl. Acad. Sci. U.S.A* **93,** 14474 (1996), with permission.

distribution on substrate groups in the course of a reaction allow a substantial amount of catalysis to be obtained from hydrogen-bonding interactions.[59]

DO ENZYMES EXHIBIT THE LARGE INCREASE IN HYDROGEN BOND STRENGTH UPON CHARGE REARRANGEMENT THAT IS SUGGESTED FROM RESULTS IN MODEL SYSTEMS? An enzymatic study that addresses this question used a series of unnatural amino acids, fluoro-substituted tyrosines, to obtain a Brønsted β value for a Tyr · Glu$^-$ hydrogen bond in staphylococcal nuclease.[69] The effect of the fluoro-substituted tyrosines (F_n-Tyr) on the stability of the folded enzyme was determined. Analysis of the results in a manner analogous to that outlined in Scheme 3, Fig. 4, and the previous discussion suggests that the β value is 0.35–0.75 larger for the F_n-Tyr^{OH} · Glu$^-$ hydrogen bond on the enzyme than the F_n-Tyr^{OH} · OH$_2$ hydrogen bond in solution,[1] consistent with the expectation from the model studies. An analysis of the binding of a series of phenolate ions to the 3-oxo-Δ^5-steroid isomerase active site suggested a smaller increase in the β value, of ~0.1–0.2, for the hydrogen bond(s) at the active site relative to solution.[70] Clearly, more quantitative studies will be required in order to establish the magnitude and range of perturbation of β values on proteins. Advances in the ability to incorporate unnatural amino acids into proteins should allow more studies of this type to be carried out in the near future and to be correlated with structural effects.[71–73]

Rate Enhancements from Multiple Hydrogen Bonds at Sites of Charge Rearrangement. The catalytic contribution of a single hydrogen bond with a substrate group undergoing charge rearrangement in the course of a reaction was discussed in the previous section. However, enzymatic active sites have many groups that interact with substrates, and typically there are multiple interactions with the groups that undergo charge re-arrangement. The two hydrogen bonds from His-95 and Lys-12 to the substrate carbonyl oxygen in the active site of TIM provide one example[17] and the two hydrogen bonds in the oxyanion hole of serine proteases provide another example (Ref. 74 and references therein; for some additional examples, see Refs. 75–80 and references therein). Hence the question arises: What are the energetic consequences of multiple hydrogen bonds? A model study[2] is described that suggests that the energetic effects of multiple hydrogen bonds can be large and additive for donor and acceptor groups that are prepositioned with respect to each other, as occurs in an enzymatic active site.

[69] J. S. Thorson, E. Chapman, and P. G. Shultz, *J. Am. Chem. Soc.* **117**, 9361 (1995).
[70] I. P. Petrounia and R. M. Pollack, *Biochemistry* **37**, 700 (1998).
[71] T. W. Muir and S. B. H. Kent, *Curr. Opin. Biotechnol.* **4**, 420 (1993).
[72] S. B. H. Kent, *Annu. Rev. Biochem.* **57**, 957 (1988).
[73] J. F. Parsons and R. N. Armstrong, *FASEB J.* **10**, A1386 (1996).
[74] W. Bode and R. Huber, *Eur. J. Biochem.* **204**, 433 (1992).
[75] E. A. First and A. R. Fersht, *Biochemistry* **34**, 5030 (1995).
[76] R. Wolfenden, *Pharmacol. Ther.* **60**, 235 (1993).
[77] G. L. Kenyon, J. A. Gerlt, G. A. Petsko, and J. W. Kozarich, *Acc. Chem. Res.* **28**, 178 (1995).
[78] M. Karpusas, D. Holland, and S. J. Remington, *Biochemistry* **30**, 6024 (1991).
[79] Z.-X. Xia and F. S. Mathews, *J. Mol. Biol.* **212**, 837 (1990).
[80] J. E. Wedekind, R. R. Poyner, G. H. Reed, and I. Rayment, *Biochemistry* **33**, 9333 (1994).

The monoanion of 2,6-dihydroxybenzoic acid (**II**), in which the carboxylate group can be hydrogen bonded to both of the *ortho*-hydroxyl groups, was used to mimic multiple hydrogen-bonding interactions at an enzymatic active site.[2] The combined energetic effect of the two hydrogen bonds in DMSO was isolated by the same approach outlined in Scheme 3 for the hydrogen bond in salicylate monoanions. After accounting for inductive effects, a stabilization from the two hydrogen bonds (K^{HB}, Scheme 3) of 10^{10}-fold or $\Delta\Delta G^{HB} = 14.4$ kcal/mol in DMSO was obtained. This is close to the value of 15.8 kcal/mol expected for the additive effect of two $-OH \cdot COO^-$ hydrogen bonds. The value of 15.8 kcal/mol was obtained from the estimated stabilization from a single hydrogen bond in salicylate monoanion (**I**) of 7.9 kcal/mol.

The near additivity of the energetics of hydrogen bonds in **II** suggests that there is no substantial charge redistribution at the donor and acceptor groups upon formation of one hydrogen bond. An additive effect would not be expected for hydrogen bonds that are predominantly covalent in nature, as extensive electronic rearrangement in a group upon forming a covalent bond would be expected to weaken the ability of this group to form a second hydrogen bond. Thus, the nearly additive hydrogen bond energies in the 2,6-dihydroxybenzoate monoanion (**II**), combined with the comparisons presented in the previous section, suggest that large energetic contributions can be obtained from hydrogen bonds that are largely electrostatic in nature in nonaqueous environments and, presumably, in enzyme active sites (see also Refs. 1–3 and references therein for detailed discussions).[21]

I   II   III

Can Geometrical Changes in Going from the Ground State to the Transition State Lead to More Favorable Transition-State Hydrogen Bonding? Geometrical effects on the energetics of hydrogen bonds can be demonstrated by comparing the energetics of hydrogen bonds in salicylate (**I**) vs phthalate (**III**) monoanions. Despite the greater acidity of the carboxylic acid group in **III** relative to the hydroxyl group in **I** and the expectation of a stronger hydrogen bond from the carboxylic acid, the hydrogen bonding in phthalate monoanion (**III**) is 2.7 kcal/mol less favorable than that in salicylate monoanion (**I**). This suggests that the geometry of the carboxylic acid and carboxylate groups in **III** is suboptimal for hydrogen bonding.[1,3]

Analogously, geometrical constraints within an $E \cdot S$ complex might be used to weaken hydrogen-bonding interactions in the ground state; geometrical changes in the course of the reaction might allow the hydrogen bond to be strengthened in the transition state, thereby providing a rate enhancement. For example, it has been suggested that the $C=O$ bond is too short for optimal hydrogen bonding in the oxyanion hole of serine proteases, so that optimal hydrogen bond distances are only achieved when the $C=O$ bond is lengthened in the transition state.[4,81–84]

Catalysis via geometrical destabilization of hydrogen bonds in the ground state is conceptually distinct from catalysis via strengthening of hydrogen bonds accompanying charge rearrangement in the course of a reaction. In practice, however, these two mechanisms are difficult to dissect because geometrical changes typically occur in concert with charge re-arrangement. In addition, even though geometrical effects are observed in model compounds, the geometrical constraints imposed by the noncovalent interactions of an enzyme are presumably less severe than those imposed by covalent interactions in model compounds. The extent to which enzymes can "recognize" the geometrical changes of the substrate and thereby pref-erentially stabilize the transition state remains to be determined.

Interconnections between Binding Interactions and the Differential Strengthening of Hydrogen Bonds

The previous sections focused on hydrogen bonds with substrate moi-eties undergoing charge rearrangement in the course of a reaction and suggested that the increased strengthening of hydrogen bonds in the enzy-matic active site relative to aqueous solution may provide substantial rate enhancements. However, two features of the enzymatic active site are required to achieve catalysis via this mechanism. First, the enzymatic and substrate hydrogen bond donors and acceptors must be aligned with respect to one another. Second, the enzymatic active site must provide a low effec-tive dielectric environment for these hydrogen bonds.

[81] R. J. Henderson, *J. Mol. Biol.* **54,** 341 (1970).
[82] A. Ruhlmann, D. Kukla, P. Schwager, K. Bartels, and R. Huber, *J. Mol. Biol.* **77,** 417 (1973).
[83] D. M. Blow, J. Janin, and R. M. Sweet, *Nature* **249,** 54 (1974).
[84] R. M. Sweet, H. T. Wright, J. Janin, C. H. Chothia, and D. M. Blow, *Biochemistry* **13,** 4212 (1974).

Fixation of active site groups with respect to each other and with respect to bound substrate can also lower the entropic barrier in going from the ground state to the transition state,[4,18] a mechanism that was not discussed earlier. For example, in the active site of TIM, hydrogen bond donors and acceptors and the general acid and base are positioned in the active site with respect to the bound substrate.[17] This minimizes the reorganization of solvent that confers a substantial entropic barrier in nonenzymatic reactions.[85]

Catalysis via lowering the entropic barrier of a reaction is conceptually distinct from catalysis via the greater strengthening of hydrogen bonds accompanying charge rearrangement in the enzymatic active site relative to aqueous solution. However, these mechanisms are inextricably linked in practice because both are effected through the rigidity and precise alignment of the active site groups with respect to each other and with respect to the substrate groups. Positioning of active site residues within the folded enzyme both aligns functional groups for interactions with substrates and lowers the local dielectric.[1,4,18,54–58] Conversely, the "catalytic" hydrogen bonds to groups undergoing charge rearrangement in the course of reaction also contribute to rigidifying the active site and positioning the substrate with respect to other catalytic groups.

This interconnection between binding interactions and interactions at positions involved in chemical transformation represents a fundamental property of enzymatic catalysis (Refs. 4 and 86 and references therein). However, this interconnection also limits our ability to experimentally dissect the catalytic contributions from each mechanism. The following example illustrates these interconnections.

In the active site of TIM, Tyr-208 is hydrogen bonded to a residue in the flexible loop of TIM. Although Tyr-208 is not in direct contact with the substrate, mutation of this residue to phenylalanine results in >2000-fold reduction in k_{cat}/K_m.[87,88] This mutation presumably disrupts the network of interactions within the active site, thereby imparing the interaction of catalytic residues with the substrate (Fig. 5). For example, the active site general base, Glu-65, may be misaligned with respect to the substrate reaction center. This mutation may also impair the interaction of the substrate phosphate group with a number of residues in the flexible loop that are used to position the substrate. It is also possible that the active site becomes more solvated when the closure of the loop is impaired, resulting

[85] E. Grunwald and C. Steel, *J. Am. Chem. Soc.* **117,** 5687 (1995).
[86] G. J. Narlikar and D. Herschlag, *Annu. Rev. Biochem.* **66,** 19 (1997).
[87] N. S. Sampson and J. R. Knowles, *Biochemistry* **31,** 8482 (1992).
[88] N. S. Sampson and J. R. Knowles, *Biochemistry* **31,** 8488 (1992).

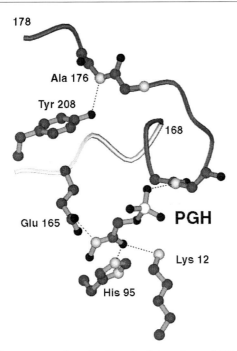

Fig. 5. Schematic representation of active site interactions of TIM complexed with a transition-state analog, phosphoglycolohydroxamate (PGH).[17] The active site base, Glu-165, interacts with N1 of PGH, which represents the site of proton abstraction in the substrate. A loop consisting of residues 168–178 (dark ribbon) folds down over the active site on substrate binding and interacts with the phosphate of PGH.[17,87,88] To show the connection between this loop and Glu-165, the backbone of the loop is extended to residue 165 (lighter ribbon). Tyr-208 is important for closure of this loop via hydrogen bonding with the backbone of Ala-176 (see text). Hydrogen bonds from His-95 and Lys-12+ to the carboxyl oxygen of PGH discussed in the text are also included. Oxygen atoms are shown in black, nitrogen atoms in light gray, carbon atoms in dark gray, and the phosphorus atom of PGH is shown in white.

in increased effective dielectric of the active site. This could lessen the catalytic contribution of hydrogen bonds with substrate moieties undergoing charge relocalization, such as those from His-95 and Lys-12.

The comparative approach described in the previous sections allows probing of the potential energetic contribution of hydrogen bonds that stems from electrostatic properties. These effects are probed more readily in model systems than on enzymes because of the limitations described earlier. Nevertheless, the application of linear free energy relationship to proteins using unnatural amino acids may minimize structural perturbations, thereby providing an improved, although still imperfect, means to probe aspects of the catalytic contribution of hydrogen bonds. The ability to introduce precise and conservative modifications is crucial for preserving the role of these hydrogen bonds in positioning while systematically varying the electrostatic properties of the hydrogen bond. It will also be important to obtain structural and spectroscopic information on these modified enzymes to complement and expand on the results from energetic and mechanistic analyses and to better understand the physical nature of hydrogen bonds.[21]

Summary

Enzymes can provide catalysis by increasing the strengthening of hydrogen bonds to groups undergoing charge rearrangement in the course of reaction *relative* to the strengthening of the hydrogen bonds in the corresponding solution reactions. This can be accomplished by using hydrogen bond donors and acceptors that are stronger than water and by lowering the effective dielectric relative to that in aqueous solution. We suggest that these electrostatic effects are of general significance in enzymatic catalysis.

The effective dielectric is lowered by the overall "rigidity" of the folded enzyme, which facilitates the formation of active site interactions, and by the fixation of active site functional groups within the enzyme · substrate complex. This underscores the fundamental interconnection of catalytic mechanisms in enzymatic catalysis.

Appendix

The Hine equation [Eq. (8)] describes quantitatively hydrogen bonding between solutes in water as a competition between solute · solute interactions and solute · solvent interactions [Eq. (9)], based on an electrostatic model for hydrogen bonding.[19,45,52,53] The Brønsted slopes, α and β, are obtained from the Hine equation by taking the derivative of log K^{HB} with respect to the pK_a of the hydrogen bond donor or acceptor, respectively, as shown in Eqs. (10a) and (10b). As noted in the text, the α and β values are slopes that describe the observed dependence of log K^{HB} on the pK_a of the hydrogen bond donor and acceptor. As shown in Eq. (11), the *change* in these slopes (i.e., the change in the dependence of log K^{HB} on the acceptor with different hydrogen bond donors and the converse) is described in the Hine model by the interaction coefficient, τ. The term "log (2×55)" in the Hine equation [Eq. (8)] is a statistical correction to account for the competition for hydrogen bonds from 55 M water relative to a 1 M standard state for the solute molecules.

This description can be related to the Coulombic model of Eqs. (4) and (5) because the pK_a values in Eq. (8) are analogous to the partial charges on donors and acceptors in Eq. (5). The interaction coefficient τ is analogous to the constant in the Coulombic equation [cf. "C" in Eq. (5) and τ in Eq. (8)] and is dependent on the distance between the groups and the dielectric of the environment surrounding the hydrogen bonding groups:

$$\log K^{HB} = \tau(pK_a^{HOH} - pK_a^{HA})(pK_a^{B:} - pK_a^{H_2O}) - \log(2 \times 55) \qquad (8)$$

$$H_2O:\cdot HA + B:\cdot HOH \overset{K^{HB}}{\rightleftharpoons} B:\cdot HA + H_2O:\cdot HOH \qquad (9)$$

$$\alpha = \frac{\partial \log K^{HB}}{\partial pK_a^{HA}} = \tau(pK_a^{B:} - pK_a^{H_2O}) \qquad (10a)$$

$$\beta = \frac{\partial \log K^{HB}}{\partial pK_a^{B:}} = \tau(pK_a^{HOH} - pK_a^{HA}) \qquad (10b)$$

$$\tau = \frac{\partial \alpha}{\partial pK_a^{B:}} = \frac{\partial \beta}{\partial pK_a^{HA:}} \qquad (11)$$

For the reaction catalyzed by TIM (Fig. 1), the increase in the Brønsted slope for the hydrogen bond with the active site Lys-12$^+$ relative to the hydrogen bond with water ($\Delta\beta$) can be calculated as

$$\Delta\beta = \beta^{Lys} - \beta^{HOH} = \tau \times (pK_a^{HOH} - pK_a^{Lys}) = 0.013 \times (16 - 9) \approx 0.1.$$

This calculation uses the value of $\tau = 0.013$ observed in aqueous solution[19,50,51] and assumes that the relative strength of the hydrogen bonds from Lys$^+$ and water (HOH) can be estimated from the difference in their proton affinities (i.e., pK_a values). This value of $\Delta\beta = \beta^{Lys} - \beta^{HOH} \approx 0.1$ and the change of ~10 units in the pK_a of the carbonyl oxygen in going from the ground state to the transition state[68] allows estimation of the difference in the magnititude of the increase in the hydrogen-bonding equilibrium in the course of the enzymatic reaction relative to that in solution as

$$\begin{aligned}
\Delta\Delta\log K^{HB} &= \Delta\log K_E^{HB} - \Delta\log K_{soln}^{HB} \\
&= \beta^{Lys} \times \Delta pK_a^{GS \to TS} - \beta^{HOH} \times \Delta pK_a^{GS \to TS} \\
&= (\beta^{Lys} - \beta^{HOH}) \times \Delta pK_a^{GS \to TS} \\
&= \Delta\beta \times \Delta pK_a^{GS \to TS} \\
&= 0.1 \times 10 = 1
\end{aligned}$$

This gives an estimated rate enhancement of ~10 fold [rate enhancement = $10^{\Delta\Delta\log K^{HB}} = 10$; $\Delta\Delta G^{\ddagger} = \Delta\Delta G^E - \Delta\Delta G^{soln} = -2.303\, RT \times \Delta\Delta\log K^{HB} = $

-1.4 kcal/mol (Fig. 1)]. As noted in the text, this effect may be accentuated by the low effective dielectric of the enzymatic active site.

Acknowledgments

We thank J. Brauman and G. Narlikar for stimulating discussions, I. Petrounia and R. M. Pollack for communicating unpublished results, and the Herschlag laboratory for comments on the manuscript.

[12] Application of Marcus Rate Theory to Proton Transfer in Enzyme-Catalyzed Reactions

By A. Jerry Kresge and David N. Silverman

Introduction

Although the great variety of chemical substances and the wide diversity of their reactions preclude a generally valid relationship between rate and equilibrium constants, it would seem not unreasonable to expect such a correlation to exist for the same reaction of a sufficiently homogeneous family of substrates. Marcus rate theory fulfills this expectation in a particularily simple and intuitively satisfying way by relating the free energy of activation of a chemical reaction, $\Delta G^{\ddagger}$, to its overall standard free energy change, ΔG°, using only one other parameter, $\Delta G_{o}^{\ddagger}$, commonly called the intrinsic barrier. This basic relationship of Marcus theory is shown in Eq. (1):

$$\Delta G^{\ddagger} = (1 + \Delta G^{\circ}/4\Delta G_{o}^{\ddagger})^2 \Delta G_{o}^{\ddagger} \tag{1}$$

Marcus first developed his theory for electron transfer reactions,[1] but it has been applied widely to proton transfer as well.[2-18] It can be derived

[1] R. A. Marcus, *J. Chem. Phys.* **24**, 966 (1956); *Annu. Rev. Phys. Chem.* **15**, 155 (1964).
[2] R. A. Marcus, *J. Phys. Chem.* **72**, 891 (1968).
[3] A. O. Cohen and R. A. Marcus, *J. Phys. Chem.* **72**, 4249 (1968).
[4] M. M. Kreevoy and D. E. Konasewich, *Adv. Chem. Phys.* **21**, 241 (1971).
[5] M. M. Kreevoy and S.-W. Oh, *J. Am. Chem. Soc.* **95**, 4805 (1973).
[6] A. I. Hassid, M. M. Kreevoy, and T.-M. Liang, *Symp. Faraday Soc.* **10**, 69 (1976).
[7] M. C. Rose and J. Stuehr, *J. Am. Chem. Soc.* **93**, 4350 (1971).
[8] W. J. Albery, A. N. Campbell-Crawford, and J. S. Curran, *J. Chem. Soc. Perkin II*, 2206 (1972).
[9] A. J. Kresge, S. G. Mylonakis, Y. Sato, and V. P. Vitullo, *J. Am. Chem. Soc.* **93**, 6181 (1971).
[10] A. J. Kresge, *Chem. Soc. Rev.* **2**, 475 (1973).

Copyright © 1999 by Academic Press
All rights of reproduction in any form reserved.
0076-6879/99 $30.00

from a number of different models for proton transfer, including hypotheses based on solvent polarization,[19] interesecting parabolas,[20] and the Leffler principle and Hammond postulate.[21] The quadratic form of Eq. (1) is, in fact, the simplest relationship between $\Delta G^{\ddagger}$ and $\Delta G°$ that is consistent with the variable transition state (VTS) principle,[22-26] which expects transition states of exoergic reactions to be reactant-like and those of endoergic reactions to be product-like. This may be seen by examining the first derivative of Eq. (1) with respect to $\Delta G°$, Eq. (2). This derivative, commonly called α, expresses the relative effect that perturbations on the system have on $\Delta G^{\ddagger}$ and $\Delta G°$:

$$d\Delta G^{\ddagger}/d\Delta G° = (1 + \Delta G°/4\Delta G_0^{\ddagger})/2 \equiv \alpha \qquad (2)$$

The VTS principle holds that these perturbations have a smaller effect on $\Delta G^{\ddagger}$ when the transition state is reactant-like than when it is product-like, and it thus requires the derivative α to increase with increasing $\Delta G°$. The simplest relationship consistent with this requirement is a linear dependence of α on $\Delta G°$, which means that the dependence of $\Delta G^{\ddagger}$ on $\Delta G°$ must be at least quadratic. A still simpler, linear relationship between $\Delta G^{\ddagger}$ on $\Delta G°$ will not do, for that would require α to be independent of $\Delta G°$, at variance with the VTS principle.

Intrinsic Barriers

Intrinsic barriers are values of $\Delta G^{\ddagger}$ at $\Delta G° = 0$ and as such they represent the purely kinetic components of reaction barriers, free of any thermody-

[11] A. J. Kresge, *Acc. Chem. Res.* **8,** 354 (1975).

[12] W. K. Chwang, R. Eliason, and A. J. Kresge, *J. Am. Chem. Soc.* **99,** 805 (1977).

[13] A. J. Kresge, D. S. Sagatys, and H. L. Chen, *J. Am. Chem. Soc.* **99,** 7228 (1977).

[14] A. J. Kresge and W. K. Chwang, *J. Am. Chem. Soc.* **100,** 1249 (1978).

[15] P. Pruszynski, Y. Chiang, A. J. Kresge, N. P. Schepp, and J. Wirz, *J. Phys. Chem.* **90,** 3760 (1986).

[16] Y. Chiang, A. J. Kresge, J. A. Santaballa, and J. Wirz, *J. Am. Chem. Soc.* **110,** 5506 (1988).

[17] C. E. Bannister, D. E. Margerum, J. T. Raycheba, and L. F. Wong, *Symp. Faraday Soc.* **10,** 78 (1972).

[18] C. J. Schlesener, C. Amatore, and J. K. Kochi, *J. Phys. Chem.* **90,** 3747 (1986).

[19] E. D. German, R. R. Dogonadze, A. M. Kuznetsov, V. G. Levich, and Y. I. Kharkats, *J. Res. Inst. Catal. Hokkaido Univ.* **19,** 99, 115 (1971).

[20] G. W. Koeppl and A. J. Kresge, *J. Chem. Soc. Chem. Commun.,* 371 (1973).

[21] J. R. Murdoch, *J. Am. Chem. Soc.* **94,** 4410 (1972).

[22] R. P. Bell, *Proc. R Soc. A* **154,** 414 (1936).

[23] M. G. Evans and M. Polanyi, *Trans. Faraday Soc.* **32,** 1340 (1936).

[24] J. E. Leffler, *Science* **117,** 340 (1953).

[25] G. S. Hammond, *J. Am. Chem. Soc.* **77,** 334 (1955).

[26] W. P. Jencks, *Chem. Rev.* **85,** 511 (1985).

namic influence. A trivial example of the influence of thermodynamics on reaction velocity is provided by endoergic or "uphill" processes. As shown in Fig. 1, the barrier to such a reaction, $\Delta G^{\ddagger}$, must be at least as great as the endoergicity, ΔG°, and if the latter is large, the reaction will be slow even when it is an intrinsically fast process.

A less obvious influence of thermodynamics on reaction rate may be illustrated by the family of increasingly exoergic reactions shown in Fig. 2. According to the VTS principle, the transition state for such a series will become more reactant like as the process becomes more exoergic; this causes the reaction barrier to decrease and finally to vanish completely. These reactions therefore become faster as they become more exoergic: they acquire a thermodynamic drive that they do not have at $\Delta G^{\circ} = 0$.

The kinetic component of endoergic reactions is also sped up by their endoergicity. The effect in this case, however, is apparent less readily because the reaction barrier here includes ΔG° in addition to the kinetic component. When ΔG° is subtracted from $\Delta G^{\ddagger}$, the remainder, which is the kinetic part of the barrier, decreases with increasing endoergicity. This may be seen by considering the reactions of Fig. 2 in reverse. When ΔG° is subtracted from $\Delta G^{\ddagger}$ for each of these reverse reactions, the remainder is in fact equal to the barrier for that reaction in the forward direction, which decreases as the reverse reaction becomes more endoergic.

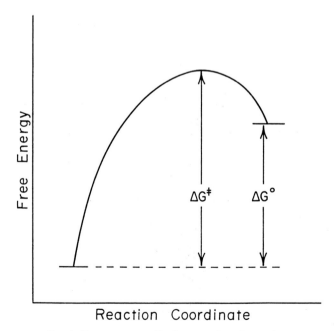

FIG. 1. Free energy profile for an endoergic reaction.

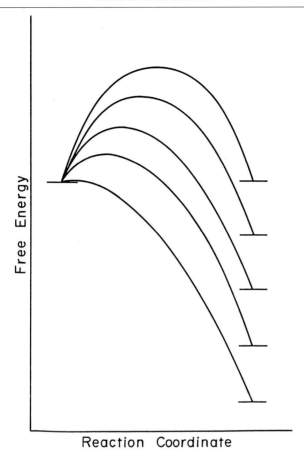

FIG. 2. Free energy profiles for a family of increasingly exoergic reactions.

The kinetic component of a reaction barrier thus decreases with increasing energy imbalance in either the exoergic or endoergic direction. This thermodynamic influence must be eliminated before a reaction can be judged to be intrinsically fast or intrinsically slow. Marcus theory provides a means of doing this by showing how observed barriers may be converted into intrinsic barriers. Particularily illuminating examples of the use of Marcus theory for this purpose have been provided by Bernasconi.[27-30]

[27] C. F. Bernasconi, *Tetrahedron* **41,** 3219 (1985).
[28] C. F. Bernasconi, *Acc. Chem. Res.* **20,** 301 (1987).
[29] C. F. Bernasconi, *Adv. Phys. Org. Chem.* **27,** 119 (1991).
[30] C. F. Bernasconi, *Acc. Chem. Res.* **25,** 9 (1992).

Inverted Region

Although attractive in its simplicity, the quadratic form of Eq. (1) produces what may seem to be unrealistic behavior. This equation gives diminishing values of $\Delta G^{\ddagger}$ with decreasing ΔG° until the exoergic limit $\Delta G^{\ddagger} = 0$ is reached at $\Delta G^{\circ} = -4\,\Delta G_{0}^{\ddagger}$. Beyond this point, $\Delta G^{\ddagger}$ takes on positive values again and grows larger with a continuing decrease in ΔG°. This has the seemingly unrealistic effect of having reaction rates speed up, reach a fast reaction limit, and then slow down again in an "inverted region" as the exoergicity is increased continuously.

This apparent shortcoming is removed in a more elaborate relationship between $\Delta G^{\ddagger}$ and ΔG° provided by Marcus[1] and also by a simple hyperbolic equation relating these two variables deduced by Lewis et al.[31] It is not clear, however, that these alternative formulations are any more realistic than the simple quadratic expression, for inverted region behavior has now been discovered.[32,33] Nevertheless, all of the examples found so far have been for electron transfer reactions, and whether proton transfer is capable of inverted region behavior is still an open question.

Brønsted Relation

The derivative α of Eq. (2) may be identified with the exponent α of the Brønsted relation. This relation[34,35] correlates general acid or general base catalytic coefficients with the acid or base strengths of the catalysts. The relationship for acid catalysis is shown as Eq. (3):

$$k_{HA} = G_{A}(K_{a}^{HA})^{\alpha} \tag{3}$$

in which k_{HA} is the rate constant for catalysis by the acid HA and K_{a}^{HA} is the acidity constant of this acid; G_{A} is a simple proportionality constant. Although this relation ostensibly correlates rate and equilibrium constants of two different reactions, i.e., the catalytic process and the acid ionization reaction, it also compares rate and equilibrium constants of the same process, the catalyzed reaction, and so is equivalent to Eq. (1). This may be seen by considering a particular example to which the Brønsted relation

[31] E. S. Lewis, C. C. Shen, and R. A. More O'Ferrall, *J. Chem. Soc. Perkin II*, 1084 (1981).
[32] G. L. Closs, L. T. Calcaterra, N. J. Green, K. W. Penfield, and J. R. Miller, *J. Phys. Chem.* **90**, 3673 (1986).
[33] Numerous other examples are cited in N. Liang, J. R. Miller, and G. L. Closs, *J. Am. Chem. Soc.* **112**, 5353 (1990).
[34] J. N. Brønsted and K. Pedersen, *Z. Phys. Chem.* **108**, 185 (1924).
[35] R. P. Bell, "The Proton in Chemistry," 2nd ed., Chap. 10. Cornell Univ. Press, Ithaca, NY, 1973.

would apply: the simple proton transfer from catalyst to substrate shown
as Eq. (4):

$$S + HA \underset{K}{\overset{k_{HA}}{\rightleftharpoons}} SH^+ + A^- \tag{4}$$

The equilibrium constant for this reaction, K, is equal to the acidity constant
of the proton donor, K_a^{HA}, divided by the acidity constant of the protonated
substrate, K_a^{SH}: $K = K_a^{SH}/K_a^{HA}$. Substitution of the value for K_a^{HA} provided
by this relationship into Eq. (3) followed by taking logarithms gives Eq. (5):

$$\log k_{HA} = \log G_A + \alpha \log K + \alpha \log K_a^{SH} \tag{5}$$

Brønsted relations are generally constructed by holding the substrate constant and varying the catalyst HA. Under these conditions, K_a^{SH} does not
change along the reaction series. Differentiation of Eq. (5) with respect to
$\log K$ then gives d $\log(k_{HA})$/d $\log(K) = \alpha$, and replacement of the rate and
equilibrium constants by the appropriate free energy quantities gives $\alpha =$
$d\Delta G^{\ddagger}/d\Delta G°$ [Eq. (6)]:

$$\frac{d \log k_{HA}}{d \log k} = \alpha = \frac{d\Delta G^{\ddagger}}{d\Delta G°} \tag{6}$$

   Brønsted exponents are generally regarded as being measures of transition-state structure, with values near zero denoting reactant-like transition
states and values near unity denoting product-like transition states.[36] It
may be seen that this follows from Eq. (2) coupled with the VTS principle.
When the reaction under study is very exoergic and $\Delta G°$ approaches its
downhill limit, α given by Eq. (2) approaches zero, signifying a reactant-
like transition state; and when the reaction is very endoergic and $\Delta G°$
approaches its uphill limit, α given by Eq. (2) approaches unity, signifying
a product-like transition state.
   It follows from these considerations that Brønsted exponents will vary
with $\Delta G°$, and therefore with K_a^{HA}, and Brønsted plots should consequently
be curved. Curvature of Brønsted plots, however, is often difficult to detect,
and its existence has even been questioned.[37,38] This difficulty, however, is
perfectly understandable, for most reactions for which Brønsted plots have
been constructed are intrinsically slow and Eq. (2) requires α for intrinsically
slow reactions to change only gradually. For example, in a system with
$\Delta G_0^{\ddagger} = 10$ kcal mol^{-1}, a change in α of 0.1, which is often the limit of

[36] A. J. Kresge, in "Proton Transfer Reactions" (E. F. Caldin and V. Gold, eds.), Chap. 7.
    Chapman and Hall, London, 1975.
[37] D. S. Kemp and M. L. Casey, *J. Am. Chem. Soc.* **95,** 6670 (1973).
[38] C. D. Johnson, *Tetrahedron* **36,** 3461 (1980).

detectability, requires a change of 8 kcal mol^{-1} in $\Delta G°$ or 6 pK units in K_a^{HA}. It is often difficult to find a set of suitably homogeneous catalysts whose acidity constants span this pK range, and most Brønsted plots are based on pK_a ranges considerably less than this.

Brønsted exponents commonly lie within the range zero to one, but a number of examples outside these limits are now known.[39–41] It is generally recognized that these "anomalous exponents" are the result of the simultaneous operation of multiple, unsynchronized interaction mechanisms, i.e., changes made in the system in order to alter $\Delta G^{\ddagger}$ and $\Delta G°$, and thus create a family of reactions, exert their influence in more than one way, and these multiple interactions develop at different rates as the system moves from the initial to the final state.[27–30,36,42] The simple Marcus theory expressions of Eqs. (1) and (2) assume only a single interaction mechanism, but modifications have been developed that take care of situations with two or more different interactions.[43–45]

Work Terms

It is common practice to regard the proton transfer process as a three-step event consisting of an encounter of reactants [Eq. (7)] and a separation of products [Eq. (9)] in addition to proton transfer [Eq. (8)]:

$$S + HA \overset{w^r}{\rightleftharpoons} S \cdot HA \tag{7}$$

$$S \cdot HA \overset{\Delta G°}{\rightleftharpoons} SH^+ \cdot A^- \tag{8}$$

$$SH^+ \cdot A^- \overset{-w^p}{\rightleftharpoons} SH^+ + A^- \tag{9}$$

The free energy changes involved in forming encounter complexes are called work terms, with w^r being the work required to form a reactant complex from separated reactants and w^p being the work required to form a product complex from separated products. Because w^r and w^p are expected to be more or less constant along a reaction series and to be unaffected by changes in the free energy of reaction, Marcus theory is taken to apply

[39] F. G. Bordwell, W. J. Boyle, Jr., J. A. Hautala, and K. C. Yee, *J. Am. Chem. Soc.* **91**, 4002 (1969).
[40] M. Fukuyama, P. W. K. Flanagan, F. T. Williams, Jr., L. Frainier, S. A. Miller, and H. Schechter, *J. Am. Chem. Soc.* **92**, 4689 (1970).
[41] F. G. Bordwell and W. J. Boyle, Jr., *J. Am. Chem. Soc.* **93**, 511 (1971); **94**, 3907 (1972).
[42] A. J. Kresge, *J. Am. Chem. Soc.* **92**, 3210 (1970); *Can. J. Chem.* **52**, 1897 (1974).
[43] E. Grunwald, *J. Am. Chem. Soc.* **107**, 125, 4710, 4715 (1985).
[44] E. Grunwald, *Prog. Phys. Org. Chem.* **17**, 31 (1990).
[45] J. P. Guthrie, *Can. J. Chem.* **64**, 1283 (1996).

only to the proton transfer step, with $\Delta G^{\ddagger}$ being the barrier to reaction in that step and ΔG° being the free energy change for that step. The observed reaction barrier, $\Delta G^{\ddagger}_{obs}$, is then equal to $\Delta G^{\ddagger}$ calculated by Eq. (1) plus w^r, whereas the observed free energy of reaction, ΔG°_{obs}, is equal to $\Delta G^{\circ} + w^r - w^p$. Recasting Eq. (1) in terms of observed quantities then leads to Eq. (10):

$$\Delta G^{\ddagger}_{obs} = w^r + (1 + [\Delta G^{\circ}_{obs} - w^r + w^p]/4\Delta G^{\ddagger}_{o})^2 \, \Delta G^{\ddagger}_{o} \qquad (10)$$

Equation (10) is usually fitted to experimental data on the assumption that w^r, w^p, and $\Delta G^{\ddagger}_{o}$ are constant along a reaction series. This makes $\Delta G^{\ddagger}_{obs}$ a quadratic function of ΔG°_{obs} whose three coefficients are themselves functions of w^r, w^p, and $\Delta G^{\ddagger}_{o}$. In early applications of Marcus theory to proton transfer reactions, the three quadratic coefficients were first determined by a least-squares fit of $\Delta G^{\ddagger}_{obs}$ to ΔG°_{obs}, or by its equivalent, a fit of $\log(k_{obs})$ to ΔpK_a, where ΔpK_a is the difference between the pK_a of the protonated substrate and that of the catalyst: $\Delta pK_a = pK^{SH}_a - pK^{HA}_a$; w^r, w^p, and $\Delta G^{\ddagger}$ were then calculated from these quadratic coefficients.[8] However, with the current widespread use of desk-top computers and the ready availability of efficient nonlinear least-squares software, it is a simple matter to fit Eq. (10) directly and to obtain w^r, w^p, and $\Delta G^{\ddagger}_{o}$ as parameters determined by this fit. Sometimes values of ΔG°_{obs} are not available because pK^{SH}_a is not known. In that situation, $\Delta G^{\ddagger}_{o}$ and w^r may still be obtained by a fit of $\Delta G^{\ddagger}_{obs}$ to ΔG°_{HA}, the free energy of ionization of the catalyst, or from its equivalent, a fit of $\log(k_{obs})$ to pK^{HA}_a, but w^p cannot be evaluated. The reason for this may be seen by recasting Eq. (10) as Eq. (11):

$$\Delta G^{\ddagger}_{obs} = w^r + (1 + [\Delta G^{\circ}_{HA} - \Delta G^{\circ}_{SH} - w^r + w^p]/4\Delta G^{\ddagger}_{o})^2 \, \Delta G^{\ddagger}_{o} \qquad (11)$$

in which ΔG°_{obs} is replaced by its equivalent, the difference in free energies of acid ionization of the catalyst and the substrate: $\Delta G^{\circ}_{obs} = \Delta G^{\circ}_{HA} - \Delta G^{\circ}_{SH}$. The quantities w^p and ΔG°_{SH} have the same functional form in this expression and therefore cannot be separated; they can only be evaluated as a composite sum and, because ΔG°_{SH} is not known, w^p cannot be obtained.

Many of the analyses of proton transfer reactions by these methods have produced work terms considerably greater than the small values expected for encounter complex formation. This suggests that the first and third steps of the three-step scheme of Eqs. (7–9) involve more than just the simple coming together or separation of reactants and products. This three-step scheme, in fact, might better be regarded as just a formalism that separates the free energy changes in the system into two categories: those whose magnitude depends strongly on the pK_a difference between proton donor and proton acceptor and those whose magnitude depends only weakly, or not at all, on this pK_a difference. Possible phenomena that might fall into the latter category are the reorganization of solvent molecules and orientation of reactants into positions required for proton transfer to occur.

There is evidence also that intrinsic barriers may not be constant along a reaction series, and modifications of the simple Marcus equation have been developed to take this into account.[7,46]

Isotope Effects

Marcus theory has also been used to correlate kinetic isotope effects on proton transfer reactions. According to an idea first put forward by Melander[47] and Westheimer,[48] primary hydrogen isotope effects are expected to vary systematically with transition-state structure and have maximum values for reactions with central transition states midway between reactants and products; these maximum values should then fall off regularly and approach unity as the transition state becomes either reactant-like or product-like. An expression consistent with this hypothesis relating isotope effects to free energy of reaction [Eq. (12)] can be derived from Marcus rate theory[13,49]:

$$\ln(k_H/k_D) = \ln(k_H/k_D)_{max} [1 - (\Delta G^\circ_{obs}/4\Delta G^{\ddagger}_0)^2] \qquad (12)$$

An exactly equivalent relationship can also be derived by using α ($= d\Delta G^{\ddagger}/d\Delta G^\circ$) as a measure of transition-state structure and relating k_H/k_D to α in the simplest way consistent with the Melander–Westheimer principle, i.e., by a quadratic expression.[50] Although it is not clear that the quadratic form of these expressions is the correct relationship between $\ln(k_H/k_D)$ and ΔG°_{obs}, these equations do give results that are consistent with those obtained by the rate-equilibrium correlation of Eq. (10).[13,49]

The Melander–Westheimer hypothesis requires isotope effects to have maximum values when the free energy of reaction for the proton transfer step, ΔG°, is zero, which will occur when observed free energies of reaction, ΔG°_{obs}, are zero only if the work terms are equal: $w^r = w^p$. This will not generally be the case, and the isotope effect maximum will then occur at a value of ΔG°_{obs} offset from zero by the difference between w^p and w^r. Thus, although individual values of w^r and w^p cannot be obtained from a

[46] J. W. Bunting and D. Stefanidis, *J. Am. Chem. Soc.* **110**, 4008 (1988); **111**, 5834 (1989).
[47] L. Melander, "Isotope Effects on Reaction Rates," p. 24. Ronald Press, New York, 1960.
[48] F. H. Westheimer, *Chem. Rev.* **61**, 265 (1961).
[49] A. J. Kresge, *in* "Isotope Effects on Enzyme-Catalyzed Reactions" (W. W. Cleland, M. H. O'Leary, and D. B. Northrop, eds.), p. 37. Univ. Park Press, Baltimore, MD, 1977.
[50] A. J. Kresge, *J. Am. Chem. Soc.* **102**, 7797 (1980).

fit of experimental data using Eq. (12), their difference can be evaluated by determining where the isotope effect maximum lies.

An expression has also been derived linking k_H/k_D to $\Delta G^{\ddagger}$ for use in situations where ΔG^{o}_{obs} is not available, and another expression taking explicit account of the secondary isotope effect accompanying hydrogen transfer from the hydronium ion has been developed as well.[13]

Applications to Enzymatic Reactions

Although considerable literature has accrued on the application of Marcus theory to nonenzymatic, bimolecular proton transfers in solution, there has been rather little application of these ideas to the many examples of proton transfer in enzymatic reactions. Marcus theory has been the context in which Gerlt and Gassman[51,52] have explained their proposal for rapid enzyme-catalyzed proton abstraction from carbons adjacent to carbonyl or carboxylic acid groups and Hawkinson *et al.*[53] for catalysis by 3-oxo-Δ^5-steroid isomerase. In both of these cases, the catalysis was explained in terms of the effect of the enzyme on the intrinsic barrier, $\Delta G^{\ddagger}_{o}$, as well as the thermodynamic component, the free energy of reaction ΔG^{o}_{obs}; the function of the enzyme in significant part is to lower ΔG^{o}_{obs} for the formation of an intermediate. However, the most complete example of the application of Marcus theory to an enzymatic reaction, in a manner that elucidates both the theory and the catalysis, pertains to carbonic anhydrase.

Through the example of carbonic anhydrase, this article discusses how Marcus rate theory can be made applicable to proton transfer in enzyme-catalyzed reactions and the meaning of the resulting parameters of Marcus theory. The great efficiency of hydration of CO_2 catalyzed by carbonic anhydrase, one of the best examples of rapid enzymatic catalysis, is ultimately dependent on rate-limiting proton transfer steps. This article reviews how this catalysis is explained in terms of Marcus parameters, the intrinsic barrier $\Delta G^{\ddagger}_{o}$, and the work functions that contribute to the observed free energy of reaction. Not only does this procedure characterize the catalysis by breaking it into its kinetic and thermodynamic components, but it allows comparison to Marcus parameters for uncatalyzed, bimolecular proton transfer reactions in solution. Moreover, these Marcus parameters describe entire series of proton transfer reactions, e.g., between normal nitrogen and oxygen acids and bases, and are independent of the thermodynamics of a single particular reaction in that series.

[51] J. A. Gerlt and P. G. Gassman, *J. Am. Chem. Soc.* **115,** 11552 (1993).
[52] J. A. Gerlt and P. G. Gassman, *Biochemistry* **32,** 11943 (1993).
[53] D. C. Hawkinson, R. M. Pollack, and N. P. Ambulos, *Biochemistry* **33,** 12172 (1994).

Design of Experiments Utilizing Carbonic Anhydrase

Catalysis by Carbonic Anhydrase

The hydration of CO_2 catalyzed by carbonic anhydrase II is a good starting point for the application of Marcus theory to enzymatic proton transfers because its maximal velocity is dominated by proton transfer processes. For this and other well-studied isozymes of carbonic anhydrase, the catalysis occurs in two stages.[54] In the first stage, the enzyme with zinc-bound hydroxide at the active site reacts with CO_2 to produce bicarbonate; the ligand position of the departed bicarbonate is then replaced by a water molecule [Eq. (13)]. This stage of the catalysis is characterized by a maximal value of k_{cat}/K_m of $1.5 \times 10^8 \ M^{-1} \ sec^{-1}$.[55] The second stage for isozyme II comprises the proton transfer steps that regenerate the zinc-bound hydroxide and complete the catalysis [Eq. (14)]. These include an intramolecular proton transfer from the zinc-bound water molecule to His-64 and subsequently the transfer of a proton to buffer in solution, designated B in Eq. (14). For the case in which the buffer concentration is sufficiently large ($>10 \ mM$), the intramolecular proton transfer is rate limiting and is characterized by a maximal value of k_{cat} at $1.4 \times 10^6 \ sec^{-1}$.[55] It has been estimated that the intramolecular transfer at all pH values is the most sensitive or rate-limiting step in the maximal velocity; the solvent hydrogen isotope effect on k_{cat} is dominated to an extent greater than 95% by the isotope effect on this intramolecular proton transfer[56]:

$$CO_2 + EZnOH^- \rightleftharpoons EZnHCO_3^- \overset{H_2O}{\rightleftharpoons} EZnH_2O + HCO_3^- \quad (13)$$
$$His\text{-}64\text{-}EZnH_2O + B \rightleftharpoons H^+His\text{-}64\text{-}EZnOH^- + B$$
$$\rightleftharpoons His\text{-}64\text{-}EZnOH^- + BH^+ \quad (14)$$

This proton transfer pathway was first suggested by Steiner *et al.*[57] based on solvent hydrogen isotope effects; the role of His-64 in this proton shuttle was subsequently supported by site-specific mutagenesis[58] as well as many less direct methods.[54] Histidine-64 in human carbonic anhydrase II is located in the active-site cavity with its side chain extended into the cavity and its imidazole ring about 8 Å from the zinc.[59,60] The imidazole ring of His-64

[54] D. N. Silverman and S. Lindskog, *Acc. Chem. Res.* **21,** 30 (1988).
[55] R. G. Khalifah, *J. Biol. Chem.* **246,** 2561 (1971).
[56] R. S. Rowlett, *J. Prot. Chem.* **3,** 369 (1984).
[57] H. Steiner, B.-H. Jonsson, and S. Lindskog, *Eur. J. Biochem.* **59,** 253 (1975).
[58] C. K. Tu, D. N. Silverman, C. Forsman, B.-H., Jonsson, and S. Lindskog, *Biochemistry,* **28,** 7913 (1989).
[59] A. E. Eriksson, T. A. Jones, and A. Liljas, *Prot. Struct. Funct. Genet.* **4,** 274 (1988).
[60] K. Håkansson, M. Carlsson, L. A. Svensson, and A. Liljas, *J. Mol. Biol.* **227,** 1192 (1992).

has no apparent interactions with other residues in the cavity, thus it has considerable mobility in its side chain conformation. An array of apparently hydrogen-bonded water molecules extends from the zinc to the imidazole ring of His-64, but there is no hydrogen bond to this side chain itself.[59] It is apparent that the intramolecular proton transfer proceeds through such an array of water bridges, although it is not apparent that the water structure observed in the crystal provides the kinetic pathway. Evidence for the utilization of the water in the pathway was obtained from solvent hydrogen isotope effects,[61] and more recently by a combination of X-ray crystallographic and kinetic comparisons.[62,63]

Design of the Marcus Experiment

Application of Marcus theory to carbonic anhydrase was actually performed on the least efficient of the carbonic anhydrases, human isozyme III. This isozyme has a 56% amino acid identity with the more efficient isozyme II.[64] Moreover, the backbone structures of the human isozyme II and bovine isozyme III (partial structure shown in Fig. 3) are nearly superimposable, especially in the vicinity of the active site.[65] The crystal structure of human CA III has not been reported; however, there is an 86% amino acid identity between human and bovine forms of CA III, and their structures are assumed to be very similar. Human isozyme III has no apparent proton shuttle residue; position 64 is a lysine (Fig. 3) that has no significant participation in the catalysis.[66] The replacement of this residue by histidine activated isozyme III is nearly 10-fold in a manner consistent with intramolecular proton transfer.[66] The advantage of using isozyme III in this manner is that the introduction of His at position 64 by mutagenesis causes activation of catalysis, a result easier to explain than observing a decrease in catalysis for which partial or complete denaturation of the enzyme is a possibility.

The parameters of Marcus theory may not remain constant when proton donors or acceptors of different chemical type are being compared, i.e., interpretation of Marcus theory requires that the intrinsic kinetic barrier and the work functions remain constant in a series of reactions. In establish-

[61] K. S. Venkatasubban and D. N. Silverman, *Biochemistry* **19,** 4984 (1980).
[62] J. E. Jackman, K. M. Merz, and C. A. Fierke, *Biochemistry* **35,** 16421 (1996).
[63] L. R. Sconick and D. W. Christianson, *Biochemistry* **35,** 16429 (1996).
[64] R. E. Tashian, *Bioessays* **10,** 186 (1989).
[65] A. E. Eriksson and A. Liljas, *Prot. Struct. Funct. Genet.* **16,** 29 (1993).
[66] D. A. Jewell, C. K. Tu, S. R. Paranawithana, S. M., Tanhauser, P. V. LoGrasso, P. J. Laipis, and D. N. Silverman, *Biochemistry* **30,** 1484 (1991).

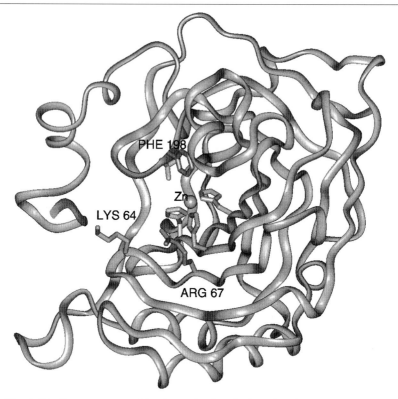

FIG. 3. Backbone structure of bovine carbonic anhydrase III showing residues near the zinc, from the crystal structure of Eriksson and Liljas.[65] The three ligands of the zinc are His-94, His-96, and His-119; a fourth ligand, water, is not shown.

ing free energy relationships with carbonic anhydrase, the pertinent free energies of reaction are obtained from the values of pK_a of the donor and acceptor groups, i.e., His-64 and the zinc-bound water molecule. To obtain a range of values with these pK_a's and yet not alter the chemical nature of the donor and acceptor groups, it was necessary to alter other residues in the active site that influenced these pK_a values. In particular, it was discovered that changing the residue at position 198 in isozyme III (Fig. 3) had a significant effect on the pK_a of the zinc-bound water.[67] The pK_a of this water is near 5 in wild-type carbonic anhydrase III, is 6.9 in the mutant containing the replacement Phe-198 $\rightarrow$ Leu, and is 9.2 in the mutant containing the replacement Phe-198 $\rightarrow$ Asp. The source of this effect is in

[67] P. V. LoGrasso, C. K. Tu, X. Chen, S. Taoka, P. J. Laipis, and D. N. Silverman, *Biochemistry,* **32,** 5786 (1993).

part electrostatic and in part probably due to changes in the interaction of nearby residue Thr-199 in its hydrogen bonding with the zinc-bound water.[68] Importantly, preparation of the appropriate double mutants of isozyme III and measurement of maximal values of k_{cat}/K_m for catalysis of CO_2 hydration showed that the changes at the active site caused by mutations at positions 64 and 198 were generally additive, indicating a noninteracting or indifferent interaction between these sites.[69] Such studies were also extended to interactions at a second proton donation site, residue 67, and residue 198.[70] However, it was difficult to make these comparisons for the pK_a of the zinc-bound water molecule because enzyme denaturation at pH <5.5 precluded measurement of the low value of this pK_a (near 5) for the wild-type enzyme.

Measurement of Rates and Values of pK_a

Accurate measurements of rates of intramolecular proton transfer between His-64 and the zinc-bound water molecule must exclude contributions from intermolecular proton transfer between buffers in solution and the zinc-bound water. Two procedures are applicable for this purpose. One is to use a buffer of large size for pH control in steady-state experiments; the size or possibly the charge on these buffers minimizes their insertion into the active-site cavity for proton transfer. The dianionic sulfonic acid buffers known by their acronyms MOPS, HEPES, and so on have been useful for this purpose.[58] The alternative choice emphasized here is to carry out experiments at chemical equilibrium in which pH control is not a problem and to measure catalysis by the rate of exchange of ^{18}O between CO_2 and solvent water.[71] In this alternative, experiments are carried out in the absence of buffer. Here the measured variable is the catalytic rate at which ^{18}O is depleted from CO_2, measured by mass spectrometry. This rate is used to obtain the rate of release of ^{18}O-labeled water from the enzyme [Eq. (16)], which is dependent on the intramolecular proton transfer.[58,71] It is to be noted that in the ^{18}O exchange method as described in Eq. (16), His-64 is the proton donor and the ^{18}O-labeled zinc-bound hydroxide is the proton acceptor.

$$HOCO^{18}O^- + EZnH_2O \rightleftharpoons EZn^{18}OH^- + CO_2 + H_2O \qquad (15)$$

[68] X. Chen, C. K. Tu, P. V. LoGrasso, P. J. Laipis, and D. N. Silverman, *Biochemistry* **32**, 7861 (1993).

[69] C. K. Tu, X. Chen, P. V. LoGrasso, D. A. Jewell, P. J. Laipis, and D. N. Silverman, *J. Biol. Chem.* **269**, 23002 (1994).

[70] X. Ren, C. K. Tu, P. J. Laipis, and D. N. Silverman, *Biochemistry* **34**, 8492 (1995).

[71] D. N. Silverman, *Method Enzymol.* **87**, 732 (1982).

$$H^+His\text{-}64\text{-}EZn^{18}OH^- \overset{k_B}{\rightleftharpoons} His\text{-}64\text{-}EZn^{18}OH_2$$

$$\overset{H_2O}{\rightleftharpoons} His\text{-}64\text{-}EZnH_2O + H_2^{18}O \qquad (16)$$

In order to establish a free energy relationship, estimates of the free energy of reaction are obtained from values of the pK_a for the proton donor and acceptor. Values of the pK_a of the zinc-bound water in carbonic anhydrase are obtained independently using a number of experiments, all of which give values in good agreement. The pH dependence of k_{cat}/K_m for the hydration of CO_2 and for the hydrolysis of 4-nitrophenyl acetate gives this pK_a directly, as does the measurement of the pH dependence of the visible absorption spectrum of Co(II)-substituted carbonic anhydrase, which is a catalytically active enzyme.[72] In addition, the pH dependence of the exchange of ^{18}O between CO_2 and water is dependent on the pK_a of the zinc-bound water, values that agree with those determined at steady state.[73] Estimates of the pK_a of the proton shuttle residue His-64 are somewhat more difficult to obtain. The pK_a of His-64 can be estimated from the pH dependence of k_{cat} for the hydration of CO_2 in the presence of excess buffer (>10 mM), provided that there is not intermolecular proton transfer between the buffer and the zinc-bound water molecule.[74] An estimate of this pK_a can also be obtained from the pH dependence of the catalyzed rate of exchange of ^{18}O between CO_2 and water; this rate also reflects the pK_a of the proton shuttle.[73] Finally, the most direct manner of determining this pK_a is to measure the pH dependence of the proton nuclear magnetic resonance signal from the imidazole ring of His-64, as was done by Campbell *et al.*[75] for isozyme II. Ren *et al.*[70] made a similar measurement for His-67 in isozyme III.

Marcus Parameters Describing Catalysis by Carbonic Anhydrase

We describe briefly four independent experiments that all support the following conclusion: the intramolecular proton transfer in the hydration of CO_2 catalyzed by carbonic anhydrase proceeds with an intrinsic free energy of activation $\Delta G_o^{\ddagger}$ that is small (≈ 1.5 kcal/mol) and with a work function w^r that is large (≈ 10 kcal/mol). This implies that the value of $\Delta G_o^{\ddagger}$ for this intramolecular proton transfer is comparable to that found for the bimolecular proton transfer between oxygen and nitrogen acids and

[72] I. Simonsson and S. Lindskog, *Eur. J. Biochem.* **123**, 29 (1982).
[73] D. N. Silverman, C. K. Tu, X. Chen, S. M. Tanhauser, A. J. Kresge, and P. J. Laipis, *Biochemistry* **32**, 10757 (1993).
[74] J. E. Coleman and S. Lindskog, *Proc. Natl. Acad. Sci. U.S.A.* **70**, 2505 (1973).
[75] I. D. Campbell, S. Lindskog, and I. A. White, *J. Mol. Biol.* **98**, 597 (1975).

bases in solution[11] and that the rate of the proton transfer in carbonic anhydrase is dominated by the work function, which is discussed further in the next section.

Intramolecular Proton Transfer Involving His-64

Figure 4 shows the Brønsted plot describing k_B of Eq. (16), the rate constants for intramolecular proton transfer in mutants of human carbonic anhydrase III containing His-64; these data were obtained from the catalyzed rates of ^{18}O exchange between CO_2 and water in the absence of buffers.[73] Figure 4 also contains a data point describing the rate constant for proton transfer in catalysis by the wild-type isozyme III; this wild type does not have a His-64, but its relatively small rate constant for proton transfer (2×10^3 sec^{-1}) emphasizes that each mutant containing His-64 is activated relative to the wild type. Figure 4 also contains three points that represent rates of proton transfer for mutants that do not have histidine at residue 64, but are in solutions containing large and saturating concentrations (100–150 mM) of imidazole buffer. This emphasizes the observation

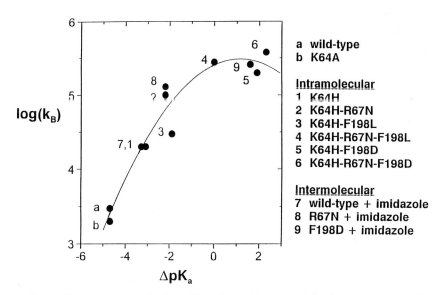

FIG. 4. Brønsted plot of the logarithm of the rate constant for intramolecular proton transfer k_B (sec^{-1}) versus ΔpK_a [pK_a(zinc-bound water) − pK_a(donor group)]. The entries are wild type and mutants of carbonic anhydrase III as indicated. Values of k_B and pK_a were obtained by measurements of catalyzed ^{18}O exchange at 25°. The solid line is a least-squares fit of the Marcus equation [Eq. (10)] to all of the data resulting in $\Delta G_o^{\ddagger} = 1.4 \pm 0.3$ kcal/mol, w^r = 10.0 ± 0.2 kcal/mol, and w^p = 5.9 ± 1.1 kcal/mol. Reprinted with permission from D. N. Silverman *et al.*, *Biochemistry* **32**, 10757 (1993).

that the saturable enhancement of catalysis by proton transfer from imidazole in solution is equivalent to that of His-64. The solid line in Fig. 4 is a least-squares fit of the Marcus equation [Eq. (10)] to all the data, resulting in $\Delta G_0^{\ddagger} = 1.4 \pm 0.3$ kcal/mol, $w^r = 10.0 \pm 0.2$ kcal/mol, and $w^P = 5.9 \pm 1.1$ kcal/mol.

Kinetic Isotope Effects in Proton Transfer Involving His-64

Deuteron transfer is also expected to be described adequately by the Marcus theory. Hence, measurement of the solvent hydrogen isotope effects on the rate constant for intramolecular hydron transfer in mutants of carbonic anhydrase III is an approach for an independent analysis. These isotope effects (Fig. 5) were obtained from observations of the catalyzed exchange of ^{18}O using some of the same mutants as shown in Fig. 4.[73]

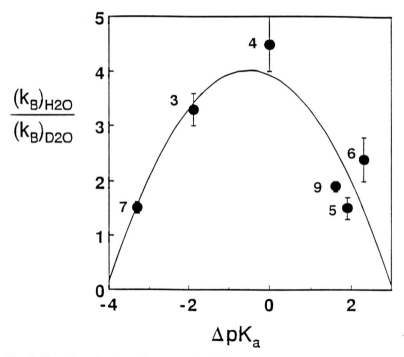

FIG. 5. The solvent hydrogen isotope effect of the rate constant for proton transfer as a function of ΔpK_a [pK_a(zinc-bound water) $-$ pK_a(donor group)]. The entries are wild type and mutants of carbonic anhydrase III numbered as in Fig. 4. Data are means and standard errors of three to five measurements. The solid line is a least-squares fit to Eq. (12) with $\Delta G_0^{\ddagger} = 1.3 \pm 0.3$ kcal/mol, $[(k_B)_{H_2O}/(k_B)_{D_2O}]_{max} = 4.7 \pm 0.7$, and $w^r - w^P = 0.6 \pm 0.5$. Reprinted with permission from D. N. Silverman et al., Biochemistry 32, 10757 (1993).

These data were fit to Eq. (12), which expresses the dependence of the solvent hydrogen isotope effects on ΔG°_{obs}, as presented by Kresge et al.[13] who discuss the assumptions inherent in this derivation. The fit of this expression to the data of Fig. 5 gave $\Delta G^{\ddagger}_{o} = 1.3 \pm 0.3$ kcal/mol with $[(k_B)_{H_2O}/(k_B)_{D_2O}]_{max} = 4.7 \pm 0.7$. From the shift of the maximum of Fig. 5 away from $\Delta G^{\circ} = 0$, the term $w^r - w^p = 0.6 \pm 0.5$ kcal/mol is obtained. This is different from the value of $w^r - w^p$ near 4 kcal/mol obtained from the data of Fig. 4. The source of this difference is not clear but may reflect experimental uncertainties or assumptions used to derive Eq. (12) that are not fully applicable to the enzymic catalysis.

Intramolecular Proton Transfer Involving His-67

The position of Arg-67 in human carbonic anhydrase III has several similarities with Lys-64. Both side chains extend into the active-site cavity with few or no interactions with other residues of the protein (Fig. 3). The α carbons are approximately equidistant from the zinc, with residues 67 and 64 at 9.4 and 9.7 Å, respectively, in the structure of bovine carbonic anhydrase III.[65] Histidine at position 67 in this isozyme is a proton donor in the catalysis with efficiency approximately equivalent to His-64. The Brønsted plot is presented by Ren et al.[70] and is described adequately by Marcus theory with $\Delta G^{\ddagger}_{o} = 1.3 \pm 0.3$ kcal/mol, $w^r = 10.9 \pm 0.1$ kcal/mol, and $w^p = 5.9 \pm 1.1$ kcal/mol. Although values of $\Delta G^{\ddagger}_{o}$ for proton transfer from His-64 and His-67 are indistinguishable, the work function w^r is greater by about 0.9 kcal/mol for proton transfer from His-67. Thus, at its maximum, the rate constant for proton transfer from His-67 to the zinc-bound hydroxide in mutants of isozyme III is smaller by fourfold than the analogous rate constant for proton transfer from His-64.

Intermolecular Proton Transfer Involving Imidazole and Lutidine Buffers

This approach utilizes the capacity of imidazole and other buffers of small size to replace the proton transfer function of His-64 in human carbonic anhydrase II. Here stopped-flow measurements were made by Taoka et al.[76] on rates of proton transfer between buffer in solution and the mutant of human carbonic anhydrase II containing the replacement His-64 → Ala (H64A HCA II). In catalysis of CO_2 hydration by this mutant in the presence of large and saturating concentrations of imidazole and lutidine buffers (see legend to Fig. 6), the maximal values of k_{cat} are comparable with or somewhat smaller than k_{cat} for wild-type HCA II.[76] The solvent hydrogen isotope effect on k_{cat} catalyzed by H64A HCA II is as great as 3.6 (Fig.

[76] S. Taoka, C. K. Tu, K. A. Kistler, and D. N. Silverman, J. Biol. Chem. 269, 17988 (1994).

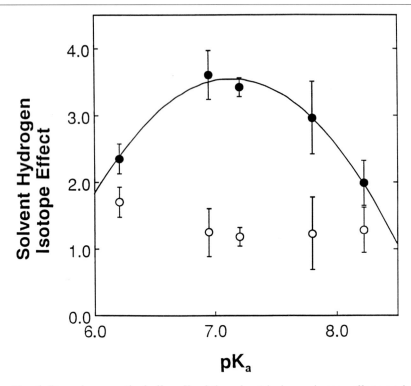

FIG. 6. Dependence on the buffer pK_a of the solvent hydrogen isotope effects on k_{cat} and k_{cat}/K_m for the hydration of CO_2 catalyzed by H64A HCA II. Filled symbols are $(k_{cat})_{H_2O}/(k_{cat})_{D_2O}$ and open symbols are $(k_{cat}/K_m)_{H_2O}/(k_{cat}/K_m)_{D_2O}$. The following buffers were used: 3,5-lutidine (pK_a 6.2), imidazole (pK_a 6.9), 1-methylimidazole (pK_a 7.2), 4-methylimidazole (pK_a 7.8), and 1,2-dimethylimidazole (pK_a 8.2). Temperature was 25°, and the total ionic strength of solution was maintained at 0.1 M by the addition of Na_2SO_4. Buffer concentrations were 100 mM except for 3,5-lutidine, which was 75 mM. In each case the pH of the measurement was in the plateau region of k_{cat}, i.e., at pH > pK_a. Data are given as the mean and standard deviation from three to five experiments. The solid line is a fit of the Marcus rate theory to the data with $\Delta G_0^{\ddagger} = 0.6 \pm 0.3$ kcal/mol and work terms $w^r - w^p = 0.9 \pm 0.5$ kcal/mol. Reprinted with permission from S. Taoka et al., J. Biol. Chem. **269**, 17988 (1994).

6). The fit of Marcus theory as applied to isotope effects using Eq. (12)[13] is shown in Fig. 6 for the intermolecular proton transfer between buffers in solution and H64A HCA II. This fit yields the values $\Delta G_0^{\ddagger} = 0.6 \pm 0.3$ and work terms $w^r - w^p = 0.9 \pm 0.5$ kcal/mol. The observation of a solvent hydrogen isotope effect near unity for k_{cat}/K_m is consistent with a body of evidence supporting a direct nucleophilic attack of zinc-bound hydroxide on CO_2, i.e., the series of steps indicated in Eq. (13) does not involve a rate-contributing proton transfer.[54]

Interpretation of Marcus Parameters for Carbonic Anhydrase

Intrinsic Kinetic Barrier $\Delta G_o^{\ddagger}$

Of the two types of information provided by the Marcus theory, kinetic and thermodynamic, it is the intrinsic kinetic barrier that is the most straightforward to interpret for intramolecular proton transfer in carbonic anhydrase. This is in part because the intrinsic kinetic barrier obtained is comparable in magnitude to $\Delta G_0^{\ddagger} \approx 2$ kcal/mol that characterizes a body of data describing bimolecular proton transfer between normal oxygen and nitrogen acids and bases.[11] Experimental and theoretical studies of hydrogen-bonded reactants confirm that barriers to proton transfer are small if the donor and acceptor are aligned properly.[77] This low intrinsic barrier is consistent with little charge delocalization in these bimolecular proton transfers between nitrogen and oxygen acids and bases, in contrast to aromatic protonation, e.g., for which $\Delta G_0^{\ddagger} \approx 10$ kcal/mol and which involves a high degree of charge delocalization.[11]

The significance of this low intrinsic kinetic barrier for carbonic anhydrase is that the proton shuttle residue His-64 is able to adopt a position in the active-site cavity that optimizes the proton transfer rate with the zinc-bound water or hydroxide and that this optimum mimics in efficiency that for bimolecular proton transfer, even though the donor and acceptor are constrained in their spatial positioning by attachment to the protein backbone. Further support for this conclusion is the observation that free and unrestrained imidazole buffer in solution is a proton donor that, at sufficiently large concentrations, can support catalysis by H64A HCA II that is as large as $k_{cat} = 10^6$ sec^{-1}, the value observed with wild-type HCA II containing the proton shuttle His-64. This is also consistent with the crystallographic picture of the side chain of His-64 extending into the active-site cavity with an apparently large degree of mobility. Thus carbonic anhydrase has evolved a proton shuttle mechanism that provides an intrinsic kinetic barrier that is about as efficient as the very rapid bimolecular proton transfers between comparable donors and acceptors in solution.

Thermodynamic Work Functions w^r and w^p

These thermodynamic work functions are contributions to the free energy of reaction catalyzed by the enzyme. Even in nonenzymatic bimolecular proton transfers, their significance and interpretation are uncertain, and these uncertainties are magnified when Marcus theory is applied to enzymic

---

[77] S. Scheiner, *in* "Proton Transfer in Hydrogen-Bonded Systems" (T. Bountis, ed.), p. 29. Plenum Press, New York, 1992.

catalysis. The classic description of w^r in bimolecular proton transfer is the work that must be done to bring the reactants together and align not only the reactants but also the solvent for facile proton transfer. However, this statement has not been susceptible to experimental support. In a study of model hydride transfers between analogs of NAD^+, the work functions w^r and w^p are not identifiable with any specific feature of the potential energy surface,[78] i.e., w^r and w^p do not appear to represent the free energy of formation of any identifiable structure or local energy minimum that could be obtained from the reactants. As this computation showed, there does not have to be any clear separation between the parts of the reaction that are sensitive to the overall spontaneity and those parts that are not. A significant value of w^r does not imply the presence of a high energy intermediate nor does it exclude such an intermediate.

The most certain statement that can be made about the work function w^r is that it is the free energy that must be subtracted from ΔG° so that the observed free energy of activation can be made to fit to the Marcus equation [Eq. (10)]. In this most severe view of the application of Marcus theory, w^r represents an inadequacy of the theory to explain observations, i.e., w^r represents a function that must alter the free energy of reaction to make Marcus theory fit the experimental results.

In view of these considerations, the application of Marcus theory to carbonic anhydrase indicates that a large part of the observed free energy of activation is unrelated to changes in the free energy of reaction, i.e., the sizable work function $w^r \approx 10$ kcal/mol dominates the expression for $\Delta G^\ddagger_{obs}$. This work function does allow us to be somewhat more specific in answering the ironic question of why carbonic anhydrase is so slow. In principle and observation, an intramolecular proton transfer can occur as fast as a molecular vibration, close to 10^{12} sec^{-1}. The fastest of the isozymes of carbonic anhydrase manages $k_{cat} = 10^6$ sec^{-1}, and the reason it is not faster lies in the work function.

It is perhaps easier to comment on what is not represented in w^r for catalysis by carbonic anhydrase. It probably does not contain an energy to reorient the side chain of His-64, as this side chain appears to be largely unhindered in its mobility about two dihedral angles. Moreover, imidazole in solution can substitute for His-64, giving nearly identical kinetic parameters for catalysis (Fig. 4); hence, it is unlikely that an energy to orient the imidazole or histidine side chain is involved. The work function probably does not contain an energy to desolvate the donor and acceptor groups as these are separated by about 8 Å, with proton transfer occurring through intervening hydrogen-bonded water molecules. The most likely source of

[78] Y. Kim, D. G. Truhlar, and M. M. Kreevoy, *J. Am. Chem. Soc.* **113,** 7837 (1991).

the large value of w^r is the construction of an appropriate, hydrogen-bonded water bridge involving at least two water molecules that spans the gap between the imidazole ring of His-64 and the zinc-bound water and allows facile proton transfer. It is also worth pointing out that this proton transfer alternates with CO_2 binding and HCO_3^- dissociation in the catalytic pathway [Eqs. (13) and (14)], and the water in the active-site cavity is likely to be reoriented thoroughly by such processes in each catalytic cycle. It is interesting that proton transfer from His-67 is slightly less efficient by about 1 kcal/mol than that from His-64 in both isozyme III[70] and isozyme II.[79] Moreover, this decreased efficiency for R67H isozyme III does not appear in the intrinsic kinetic barrier but in w^r. This is interpreted to mean that His-67 has an intrinsic rate constant for proton transfer as efficient as His-64 once it has achieved the appropriate alignment of intervening water molecules, but that achieving such an appropriate alignment requires more energy from position 67. Liang et al.[79] investigated the efficiency of proton transfer by placing a histidine residue at four sites (residues 62, 65, 67, and 200) within the active-site cavity of H64A HCA II. None of these sites could sustain catalysis at the rate of the wild-type enzyme; apparently, isozyme II of carbonic anhydrase has evolved the most efficient proton shuttle site, as efficient as that sustained by free imidazole in solution.

The rate of intramolecular proton transfer from donor groups such as His-64 to the zinc-bound hydroxide in carbonic anhydrase is dominated by the work term w^r, which appears to be dependent on the formation of an appropriate water bridge in the active site. It would be expected perhaps that the rate-limiting proton transfer in carbonic anhydrase would be influenced significantly by an entropic contribution. However, observations of Ghannam et al.[80] indicate that k_{cat} for the hydration of CO_2 catalyzed by human carbonic anhydrase II, which is determined by this intramolecular proton transfer, has a very small entropy of activation and is dominated by an enthalpic term.

[79] Z. Liang, B.-H. Jonsson, and S. Lindskog, *Biochim. Biophys. Acta* **1203,** 142 (1993).
[80] A. F. Ghannam, W. Tsen, and R. S. Rowlett, *J. Biol. Chem.* **261,** 1164 (1986).

Section IV

Transition-State Determination and Inhibitors

[13] Enzymatic Transition-State Analysis and Transition-State Analogs

By VERN L. SCHRAMM

Introduction

Kinetic isotope effects (KIE) permit experimental access to the transition-state (TS) structure of enzymatic reactions and are the only method currently available to obtain detailed TS information. The theory and some methods of this approach have been discussed in two previous volumes of this series, as well as in reviews and monographs.[1–6] The application of isotope effects to establish TS structures for enzymatic reactions has evolved significantly since its last treatment in this series (1995).[2] Some of the practical and theoretical advances during that period will be the focus of this section. In this and the following sections, three major advances since the mid 1990s are treated. This article deals with the methods for application of systematic KIE and methods to measure KIE in several complicated enzymatic systems. These include methods for (1) the enzymatic synthesis of substrates with specific and stereospecific labels in nucleotides, nucleosides, NAD$^+$, RNA, and DNA; (2) KIE analysis for enzymes with suppressed experimental KIE because of commitment factors; (3) the analysis of isotope effects in complex systems of protein covalent modification; (4) measuring KIE in large substrates typified by RNA and DNA; (5) the measurement of binding isotope effects and their influence on observed kinetic isotope effects; and (6) application of TS information for the design of TS inhibitors. In article [14], recently developed methods are summarized for the computational modeling to match intrinsic KIE to bonding models of the TS.[7] Computational approaches are increasingly accessible through the implementation of electron density functional theory, applied in the

[1] W. W. Cleland, *Methods Enzymol.* **87**, 625 (1982).
[2] W. W. Cleland, *Methods Enzymol.* **249**, 341 (1995).
[3] W. W. Cleland, M. H. O'Leary, and D. B. Northrop, "Isotope Effects on Enzyme-Catalyzed Reactions." University Park Press, Baltimore, MD, 1977.
[4] P. F. Cook, ed., "Enzyme Mechanisms from Isotope Effects." CRC Press, Boca Raton, FL, 1991.
[5] R. D. Gandour and R. L. Schowen, eds., "Transition States of Biochemical Processes." Plenum Press, New York, 1978.
[6] L. Melander and W. J. Saunders, Jr., "Reaction Rates of Isotopic Molecules." Wiley New York, 1980.
[7] P. Berti, *Methods Enzymol.*, Chapter [14] (this volume) (1999).

Copyright © 1999 by Academic Press
All rights of reproduction in any form reserved.
0076-6879/99 $30.00

Gaussian computational chemistry programs, to estimate the bonding of nonequilibrium chemical states, which provide the models for the TS.[8,9] These are matched to the KIE using current adaptations of bond-energy bond-order vibrational analysis programs.[10-12] Article [15] demonstrates the relationship of the TS structure to the TS binding energy, and the prediction of inhibitor binding energy by comparing its features to that of experimentally determined TS.[13] These three articles demonstrate that the combined use of KIE, computational chemistry, and TS inhibitor design have become powerful tools in understanding enzymatic TS. Knowledge of the TS is proving helpful in the design of TS inhibitors and has resulted in the most powerful inhibitors known for several different enzymes.[14-18]

Nature of Enzymatic Transition States

Transition-state lifetimes are measured on the time scale of single bond vibrations, as the restoring force for the bond of interest is lost and is replaced by a translational motion, a negative restoring force.[10] Direct observations of the TS are problematic because of the short time scale and have only been accomplished by laser spectroscopy or by supercooling of

[8] R. G. Parr and W. Yang, "Density Functional Theory of Atoms and Molecules." Oxford Univ. Press, New York, 1989.

[9] M. J. Frisch, G. W. Trucks, H. B. Schlegel, P. M. W. Gill, B. G. Johnson, M. A. Robb, J. R. Cheeseman, T. Keith, G. A. Petersson, J. A. Montgomery, K. Raghavachari, M. A. Al-Laham, V. G. Zakrzewski, J. V. Ortiz, J. B. Foresman, J. Cioslowski, B. B. Stefanov, A. Nanayakkara, M. Challacombe, C. Y. Peng, P. Y. Ayala, W. Chen, M. W. Wong, J. L. Andres, E. S. Replogle, R. Gomperts, R. L. Martin, D. J. Fox, J. S. Binkley, D. J. Defrees, J. Baker, J. P. Stewart, M. Hend-Gordon, C. Gonzalez, and J. A. Pople, "Gaussian 94, Revision C.2." Gaussian, Inc., Pittsburgh, 1995.

[10] L. B Sims and D. E. Lewis, in "Isotopes in Organic Chemistry" (E. Buncel and C. C. Lee, eds.), p. 161. Elsevier, New York, 1984.

[11] L. B. Sims, G. W. Burton, and D. E. Lewis, "BEBOVIB-IV, QCPE No. 337." Quantum Chemistry Program Exchange, Department of Chemistry, University of Indiana, Bloomington, 1977.

[12] W. P. Huskey, *J. Am. Chem. Soc.* **118,** 14 (1993).

[13] B. B. Braunheim and S. D. Schwartz, *Methods Enzymol.,* Chapter [15].

[14] V. L. Schramm, *Annu. Rev. Biochem.* **67,** 693 (1998).

[15] R. W. Miles, P. C. Tyler, R. H. Furneaux, C. K. Bagdassarian, and V. L. Schramm, *Biochemistry* **37,** 8615 (1998).

[16] X.-Y Chen, T. M. Link, and V. L. Schramm, *Biochemistry* **37,** 11605 (1998).

[17] C. Li, P. C. Tyler, R. H. Furneaux, G. Kicska, Y. Xu, C. Grubmeyer, M. E. Girvin, and V. L. Schramm, *Nature Struct. Biol.* **6,** in press (1999).

[18] V. L. Schramm B. A. Horenstein, and P. C. Kline, *J. Biol. Chem.* **269,** 18259 (1994).

molecules in the gas phase.[19,20] Solution chemistry is largely intractable to direct TS observation because of the influence of the solvent on the spectroscopic methods to observe transient bonded states and also because of participation of the solvent in the organization or stabilization of the TS. In the usual construct of TS theory, the lifetime is within a single bond vibration or approximately 10^{-13} sec. In TS investigations, we are concerned with the geometric and electronic features of this metastable structure. Many enzymes form unstable intermediates, but the TS is not considered to be an intermediate as intermediates have lifetimes of many bond vibrations. Intermediates in enzymatic reactions are often more closely related to the TS than substrates and therefore bind tightly, as do analogs of the tightly bound intermediate.[21] Examples of this phenomenon are the sp^3-hybridized analogs of serine proteases that mimic the serine adduct at the carbonyl carbon.[22]

Transition-state theory for enzymatic reactions proposes that the rate enhancement imposed by enzymes is due solely to the tight binding or "stabilization" of the activated complex.[23,24] Theories that do not invoke tight binding of the TS complex have also been proposed.[25] Knowledge of the TS for an enzymatic reaction can therefore provide information to design stable analogs as TS inhibitors. It is frequently impossible to estimate the rate of uncatalyzed biological reactions, but in the examples where enzymatic and nonenzymatic rates can be measured (in hydrolytic and decarboxylation reactions, for example), it is well documented that the enzyme enhances the reaction rate by factors of 10^{10} to 10^{18}.[22,26] Transition-state constructs for enzyme-catalyzed reactions propose that the energetics of catalysis arise from binding the TS tighter than the substrate by a factor equivalent to the enzymatic rate enhancement. This means that the TS is bound 10^{10} to 10^{18} times tighter than the substrate in the Michaelis complex. In summaries of TS and multisubstrate analogs, it is apparent that inhibitors with affinities of 10^{-9} to 10^{-11} represent the majority of these inhibitors, with more tightly bound analogs being rare.[21] Because Michaelis complexes have typical dissociation constants of 10^{-3} to 10^{-7} M, enzymatic TS com-

[19] K. Liu, J. C. Polanyi, and S. Yang, *J. Chem. Phys.* **98**, 5431 (1993).

[20] N. F. Scherer, L. R. Khundkar, R. B. Bernstein, and A. H. Zewail, *J. Chem. Phys.* **87**, 1451 (1987).

[21] J. F Morrison and C. T. Walsh, *Adv. Enzymol. Relat Areas Mol. Biol.* **61**, 201 (1988).

[22] A. Radzicka and R. Wolfenden, *Methods Enzymol.* **249**, 284 (1995).

[23] L. Pauling, *Am. Sci.* **36**, 50 (1948).

[24] K. B. J. Schowen, *in* "Transition States of Biochemical Processes" (R. D. Gandour and R. L. Schowen, eds.), p. 225. Plenum Press, New York, 1978.

[25] W. R. Cannon, S. F. Singleton, and S. J. Benkovic, *Nature Struct Biol.* **3**, 821 (1996).

[26] A. Radzicka and R. Wolfenden, *Science* **267**, 90 (1996).

plexes are bound hypothetically with dissociation constants of 10^{-13} to 10^{-25} M that correspond to -18 to -35 kcal/mole binding energy. These dissociation constants are astounding compared to those for known enzymatic inhibitors and indicate that many orders of magnitude of inhibitory potential are available beyond current inhibitor design methods. It may be possible to design improved inhibitors, provided that we have accurate information on the nature of the enzymatic TS and are capable of synthesizing closely related chemical mimics.

Transition-State Binding Energy

Where does TS binding energy come from? The current proposal is that the enzyme conforms to the TS configuration in which amino acid side chains and the peptide backbone form hydrogen bonds to the TS complex, which are energetically more favorable than those in the Michaelis complex.[27–30] Hydrogen bond donor-acceptor pairs that contribute equally to electron sharing of an exchangeable hydrogen atom may form low-barrier hydrogen bonds that facilitate proton transfers and may be more energetic than average hydrogen bonds. The catalytic site becomes more constrained, water is excluded, the site becomes more hydrophobic, and the multiple hydrogen bonds shorten in the hydrophobic environment and lead to the TS.[28,31] Enzymes with substrates of modest size (e.g., the size of glucose) can accommodate 15 or more hydrogen bonds from the enzyme.[32] Methods to detect low-barrier hydrogen bonds in enzymatic complexes are summarized in article [10] of this volume.[30] Some enzymes make use of substrate neighboring group participation by conformational distortion of the bound substrate to provide catalytic groups that participate in TS formation.[33–35] At a relatively modest H-bond energy for optimally aligned bonds, the energy from 15 modest hydrogen bonds of only 1.4–3 kcal/mol can generate -23 to -45 kcal/mol to achieve the TS. A pertinent and recent example of this effect is a TS inhibitor bound to malarial hypoxanthine–guanine

[27] W. W. Cleland, *Biochemistry* **31,** 317 (1992).
[28] S.-O. Shan and D. Herschlag, *Proc. Natl. Acad. Sci. U.S.A* **93,** 14474 (1996).
[29] J. A. Gerlt and P. G. Gassman, *J. Am. Chem. Soc.* **115,** 11552 (1993).
[30] A. S. Mildvan, T. K. Harris, and C. Abeygunawardana, *Methods Enzymol.,* Chapter [10], (this volume) (1999).
[31] S.-O. Shan and D. Herschlag, *Method Enzymol.,* Chapter [11] (this volume) (1999).
[32] A. E. Aleshin, C. Zeng, H. D. Bartunik, H. J. Fromm, and R. B. Honzatko, *J. Mol. Biol.* **282,** 345 (1998).
[33] B. A. Horenstein, D. W. Parkin, B. Estupiñán, and V. L. Schramm, *Biochemistry* **30,** 10788 (1991).
[34] C. R. Muchmore, J. M. Krahn, J. H. Kim, H. Zalkin, and J. L. Smith, *Prot. Sci.* **7,** 39 (1998).
[35] M. Degano, S. C. Almo, J. C. Sacchettini, and V. L. Schramm, *Biochemistry* **37,** 6277 (1998).

phosphoribosyltransferase. The complex demonstrates five hydrogen bonds with proton chemical shifts greater than 12 ppm downfield, none of which appear in a similar complex using substrate analogs.[17] Based on the single criterion of the temperature dependence of the proton nuclear magnetic resonance (NMR) chemical shift for these bonds,[36] four of the five are low-barrier hydrogen bonds. Release of the TS binding energy following the reaction and on the time scale of catalysis is largely unexplored, but requires the loss of the hydrogen bonds that were aligned optimally at the TS. During the several milliseconds for the typical catalytic cycle, the enzyme collides with substrate, binds it into the weakly interacting Michaelis complex, increases the binding/catalytic energy to form the TS, relaxes the tight-binding energy of the TS, forms the weakly bound product complex (Michaelis complex for the reverse reaction), and dissociates the products.

Inhibitor design that takes advantage of the TS structure and binding energy, relies on accurate estimates of the electron distribution at the TS as deduced by the analysis of intrinsic KIE.[37] This article emphasizes the procedure of measuring sufficient intrinsic kinetic and equilibrium isotope effects to construct a complete bonding and electronic structure of the TS for reactions not usually accessible by these methods.

Chemistries Amenable to Transition-State Analysis

Kinetic isotope effects arise from changes in atomic vibrational states between the reactants and the TS.[38] In theory, all reactions can be probed by KIE, but in practice, electron transfers result in intrinsic KIE too small to be quantitated. In reactions where the atom of interest remains in the same bonding environment in the substrate and at the TS, there will be no KIE. Atoms that become vibrationally less constrained in the TS give normal KIE ($k_{normal}/k_{heavy} > 1$), with the heavier isotopic substrate reacting more slowly than that with natural abundance atoms. Conversely, atoms more constrained at the TS cause inverse KIE ($k_{normal}/k_{heavy} < 1$), with the heavy isotopically labeled substrate reacting more rapidly. These generalizations for the interpretation of kinetic isotope effects for some of the common reactions in biochemistry are summarized in Fig. 1. The atomic motions responsible for these effects have been described in the original accounts of Beigeleisen and Wolfsberg[38] and have been reviewed in

[36] M. Garcia-Viloca, R. Gelabert, A. Gonzalez-Lafont, M. Moreno, and J. M. Lluch, *J. Am. Chem. Soc.* **120,** 10203 (1998).

[37] C. K. Bagdassarian, V. L. Schramm, and S. D. Schwartz, *J. Am. Chem. Soc.* **118,** 8825 (1996).

[38] J. Bigeleisen and M. Wolfsberg, *Adv. Chem. Phys.* **1,** 15 (1958).

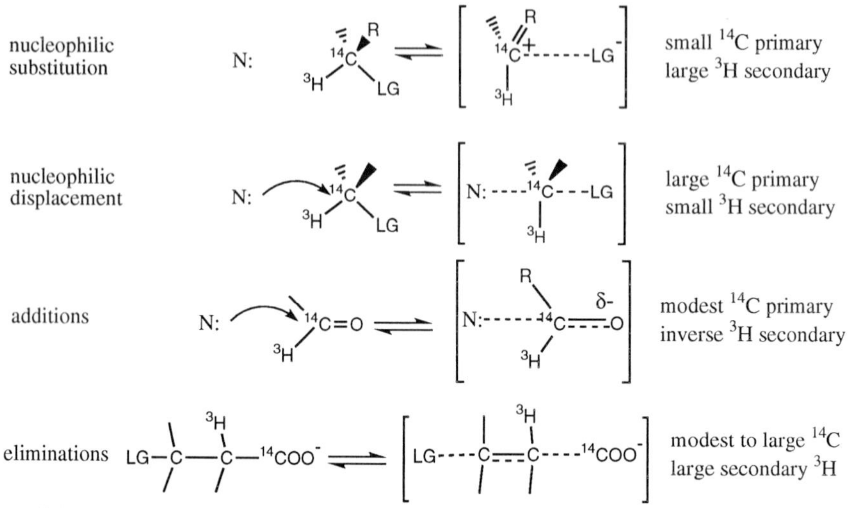

nucleophilic substitution	small ^{14}C primary large ^{3}H secondary
nucleophilic displacement	large ^{14}C primary small ^{3}H secondary
additions	modest ^{14}C primary inverse ^{3}H secondary
eliminations	modest to large ^{14}C large secondary ^{3}H

FIG. 1. Some common enzymatic reactions in biochemical pathways. The KIE expected for the primary ^{14}C and a secondary ^{3}H isotopic label are indicated. The pattern of these two simple KIE are sufficient to distinguish the mechanisms, and additional isotope effects are capable of providing quantitative information of bond orders at the transition state for each type of reaction.

Melander and Saunders,[6] Suhnel and Schowen,[39] and Huskey.[40] The small size of kinetic isotope effects from heavy atom reactants (^{14}C = 1.09, ^{15}N = 1.04, and ^{18}O = 1.07), the necessity for the synthesis of labeled substrates, and the difficulty in establishing intrinsic KIE has prevented the application of KIE methods to most enzymatic reactions. Isotope effects from ^{13}C and ^{18}O in decarboxylases and phosphotransferases are small because the bond order to individual oxygen atoms changes only by a fraction of a bond order at the TS. Despite these problems, remote labeling and isotope ratio mass spectrometry techniques have permitted accurate measurements of these KIE.[41] Finally, the synthesis of isotopically labeled substrates must be accessible.

The most versatile method of measuring kinetic isotope effects is the competitive radiolabeled method. Two substrates are prepared, one with the labeled atom at a site expected to experience bonding changes at the

[39] J. Suhnel and R. L. Schowen, in "Enzyme Mechanism from Isotope Effects" (P. F. Cook, ed.), p. 3. CRC Press, Boca Raton, FL, 1991.
[40] W. P. Huskey, in "Enzyme Mechanism from Isotope Effects" (P. F. Cook., ed.), p. 37. CRC Press, Boca Raton, FL, 1991.
[41] P. M. Weiss, in "Enzyme Mechanism from Isotope Effects" (P. F Cook., ed.), p. 291. CRC Press, Boca Raton, FL, 1991.

TS and a second substrate with a different labeled atom at a site remote from the bond-breaking site, in a position expected to remain vibrationally unchanged at the TS.[42] The double-label method depends on the ability to synthesize labeled substrates.[43] With good techniques in dual-label radioactive counting, it is possible to measure isotope effects with standard errors of ±0.002. Thus isotope effects of 1.01 (1% difference in reaction rates between labeled and unlabeled substrates) can be measured with confidence. Isotope ratio mass spectrometry is more accurate by an order of magnitude, routinely to ±0.0002. However, it is less versatile because the spectrometer measures only volatile samples (N_2, O_2, and CO_2, for example) and requires more material for the analysis. Direct mass spectrometry methods provide accuracy similar to that of the dual-label radioactive counting method.[44,45] Most of the atoms of interest in biochemical reactions are labeled readily with ^3H or ^{14}C and can be incorporated into substrates that also contain ^2H, ^{15}N, and ^{18}O. Double-label techniques provide the opportunity to measure virtually any of these kinetic isotope effects by the analysis of radioactive counting in samples of $^3H/^{14}C$. These isotopes can be counted with accuracies to 0.2% with small amounts of radiolabel, typically one μCi or less per experiment. Counting times decrease and accuracy improves with increased amounts of radioactivity.[46]

Experimental Procedure for Enzymatic Transition-State Analysis and Inhibitor Design

Steps commonly used in this laboratory for the analysis of an enzymatic TS and the design of a TS inhibitor are to[14] (1) select an enzyme expected to give significant experimental KIE; (2) synthesize substrates with isotopic labels at every position where bond changes are expected at the TS and at positions remote from the bonds where chemistry occurs; (3) measure experimental KIE; (4) establish the commitment factors for the enzyme, correct to intrinsic KIE; (5) match intrinsic KIE to the bond orders of a truncated TS state structure consistent with the KIE; (6) establish the molecular electrostatic potential surface for substrate and TS; (7) design inhibitors with molecular electrostatic potential surfaces similar to the TS; and (8) synthesize and test the inhibitors.

[42] F. W. Dahlquist, T. Rand-Meier, and M. A. Raftery, *Biochemistry* **8,** 4214 (1969).
[43] D. W Parkin, H. B. Leung, and V. L. Schramm, *J. Biol. Chem.* **259,** 9411 (1984).
[44] P. J Berti and V. L Schramm, *J. Am. Chem. Soc.* **119,** 12069 (1997).
[45] P. J. Berti, S. R. Blanke, and V. L Schramm, *J. Am. Chem. Soc.* **119,** 12079 (1997).
[46] D. W. Parkin, *in* "Enzyme Mechanism from Isotope Effects" (P. F. Cook, ed.), p. 269. CRC Press, Boca Raton, FL, 1991.

This article considers the practical steps in obtaining intrinsic KIE and interpretation of data to establish a TS structure that provides a practical guide to the design of TS inhibitors. Examples of inhibitor design projects using this experimental approach are described.

Synthesis of Isotopically Labeled Substrates: Nucleoside and Nucleotide Metabolism

Purine Nucleotides and Nucleosides

A technical barrier to systematic KIE analysis is the requirement for the synthesis of as many as 10 different substrates, each labeled specifically with isotopes and usually where none are commercially available. Our approach to this problem is to use enzymatic pathways that synthesize the desired molecules *in vivo*. Enzymatic syntheses are specific and efficient and are increasingly accessible with the easy cloning and expression of most enzymes of bacterial and yeast metabolism. Once the conditions for synthesis are established, the same reaction mixtures, with different labeled precursors, can be used to produce the labeled substrates necessary for TS analysis. An example of the synthesis of a double-labeled $[9\text{-}^{15}N, 5'\text{-}^{14}C]$ATP illustrates this method (Fig. 2). Using the same reaction mixture, or with modifications of the mixture, the labeled ATP molecules summarized in Table I have been synthesized.[43,47] Synthesis from a common precursor (glucose in the case of ATP) provides the benefit of commercially available labeled molecules as well as well-known enzymatic steps for the incorporation of additional isotopic labels. For example, the addition of excess phosphoglucomutase and ^2H_2O results in the incorporation of deuterium at the 2 position of glucose because of the solvent exchange catalyzed by the enzyme. ATP labeled in the indicated positions can be produced in yields of 50–90% from labeled precursors using procedures similar to that of Fig. 2. The synthetic procedures and the purification of the product nucleotides are made more efficient by combining the desired ^3H and ^{14}C precursors in the same reaction mixture to provide the product with the desired ratio of ^3H and ^{14}C product required for KIE measurements. For example, a combination of $[2\text{-}^{14}C]$glucose and $[6\text{-}^3H]$glucose in the same ATP synthesis mixture yields the $[1'\text{-}^{14}C]$ATP and $[5'\text{-}^3H]$ATP pair that is needed to measure the primary $[^{14}C]$KIE.

The labeled ATP molecules of Table I also provide a convenient starting material for the syntheses of other purine nucleotides and nucleosides. The conversions of ATP to dATP, ADP, dADP, AMP, IMP, dAMP, adenosine,

[47] D. W. Parkin and V. L. Schramm, *Biochemistry* **26,** 913 (1987).

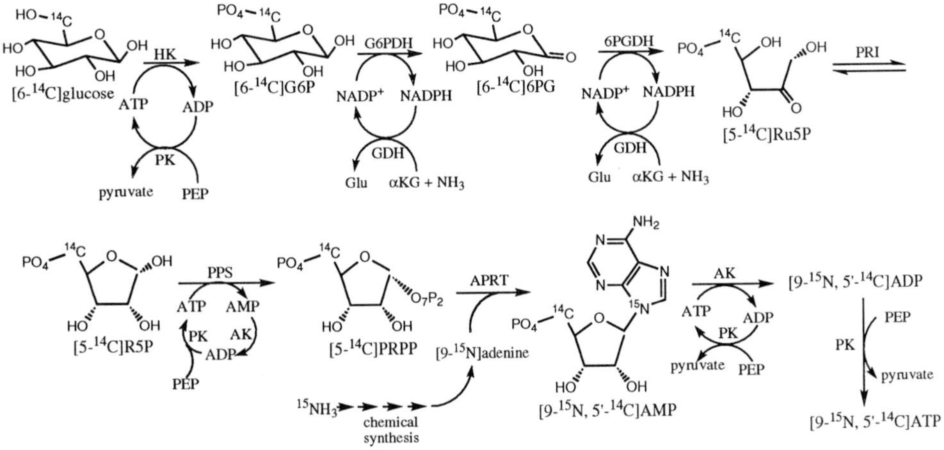

FIG. 2. Synthesis of labeled ATP using coupled enzymatic reactions with cycling of cofactors [D. W. Parkin, H. B. Leung, and V. L. Schramm, *J. Biol. Chem.* **259**, 9411 (1984)]. Reaction mixtures (typically 1 ml) contain 50 mM potassium phosphate, 3.2 mM [9-^{15}N]adenine, 3 mM MgCl$_2$, 50 mM KCl, 5 mM dithiothreitol, 10 mM α-ketoglutarate (α-KG), 1 mM ATP, 0.1 mM NADP$^+$, 10 mM P-enolpyruvate, 1 mM [6-^{14}C]glucose, 0.1–2 U/ml hexokinase (HK), 2 U/ml pyruvate kinase (PK), 0.1–7 U/ml glucose 6-phosphate dehydrogenase (G6PDH), 0.5 U/ml glutamate dehydrogenase (GDH), 0.1–1 U/ml 6-phosphogluconate dehydrogenase (6PGDH), 6 U/ml phosphoriboisomerase (PRI), 0.5–2 U/ml adenylate kinase (myokinase, AK), 0.2–1.3 U/ml PRPP synthetase (PPS), 0.5 U/ml adenine phosphoribosyltransferase (APRT), and ammonium ion added as ammonium sulfate from the commercial preparations of enzymes stored in ammonium sulfate. With the exception of hexokinase, enzymes are mixed in the appropriate ratio and added to the otherwise complete reaction mixture. The reaction is initiated by the addition of hexokinase, incubated at 37° and is monitored for ATP formation by the luciferase reaction or by HPLC. Reactions are complete in 1–8 hr, depending on the quantity of coupling enzymes. Tubes are placed in boiling water for 3 min to stop the reaction. Chromatography on semipreparative reverse-phase HPLC eluting with 50 mM triethylammonium acetate, pH 6.0, containing 5% methanol provides ATP purified sufficiently for RNA synthesis or for conversion to other nucleotides and nucleosides.

d-adenosine, inosine, and d-inosine have all been accomplished in high yields using the steps shown in Fig. 3. Labeled AMP nucleotides used to establish the transition-state structure of AMP nucleosidase were synthesized by the combination of steps in Figs. 2 and 3 (Table II).[47] It is also possible to use the known stereochemistry of these reactions to incorporate additional isotopic labels. For example, ribonucleotide triphosphate reductase (RTR of Fig. 3) is known to catalyze the reduction at C-2' with the retention of configuration.[48] Reactions conducted in the presence of ^{2}H$_2$O

[48] T. J. Batterham, R. K. Ghambeer, R. L. Blakley, and C. Brownson, *Biochemistry* **6**, 1203 (1967).

TABLE I

SYNTHESIS OF ATP FOR KINETIC ISOTOPE EFFECTS

Labeled substrate	Labeled ATP product	Changes from Fig. 1
[1-^{3}H]Ribose	[1'-^{3}H]ATP	Ribokinase replaces first four enzymes
[2-^{3}H]Ribose 5-PO$_4$	[2'-^{3}H]ATP	First four enzymes eliminated
[5-^{3}H]Glucose	[4'-^{3}H]ATP	
[6-^{3}H]Glucose	[5'-^{3}H]ATP	
[6-^{3}H]Glucose	[1'-^{2}H, 5'-^{3}H]ATP	Hexosephosphate isomerase in ^{2}H$_2$O
[6-^{3}H]Glucose	[2'-^{2}H, 5'-^{3}H]ATP	Phosphoribose isomerase in ^{2}H$_2$O
[2-^{14}C]Glucose	[1'-^{14}C]ATP	
[6-^{14}C]Glucose	[5'-^{14}C]ATP	
[6-^{14}C]Glucose	[9-^{15}N, 5'^{14}C]ATP	[9-^{15}N]Adenine replaces adenine
[2-^{14}C]Glucose	[1'-^{2}H, 1'-^{14}C]ATP	Hexosephosphate isomerase in ^{2}H$_2$O
[2-^{14}C]Glucose	[9-^{15}N, 1'-^{14}C]ATP	[9-^{15}N]Adenine replaces adenine

or ^{3}H$_2$O therefore incorporate isotopic hydrogen specifically in the pro-R position from which the hydroxyl is lost. It is possible to take advantage of this stereochemical specificity by producing [2'-^{2}H]ATP or [2'-^{3}H]ATP using the methods of Fig. 2, followed by conversion to dATP in the presence of normal water, to produce isotopic hydrogen specifically in the pro-S position. The availability of pro-R and pro-S isotopes by this synthetic method permits measurement of the stereochemical dependence of isotope effects for the group of enzymes that uses nucleosides, deoxynucleosides, nucleotides, and deoxynucleotides as substrates.

Any of the intermediates from Fig. 2 can be synthesized simply by truncating the reactions at the desired point. A special case is the production of 5-phosphoribosyl-1-pyrophosphate (PRPP), which is used as a substrate for at least a dozen enzymes of nucleotide and amino acid synthesis. The instability of this substrate requires that it be purified rapidly by DEAE ion-exchange chromatography and used promptly to prevent nonenzymatic breakdown. Another approach for production of the unstable PRPP is to include hypoxanthine and hypoxanthine–guanine phosphoribosyltransferase (HGPRT) as a coupling enzyme following the step that produces PRPP. This results in IMP as the end product, which can be purified and stored. When PRPP is required, the reverse reaction—IMP + MgPPi → hypoxanthine + PRPP—can be accomplished, with hypoxanthine removal effected with xanthine oxidase to bring the reaction to completion. Coupling the reaction to remove hypoxanthine is necessary because of the unfavorable K_{eq} for PRPP production.[49]

[49] Y. Xu, J. C. Eads, J. C. Sacchettini, and C. Grubmeyer, *Biochemistry* **36,** 3700 (1997).

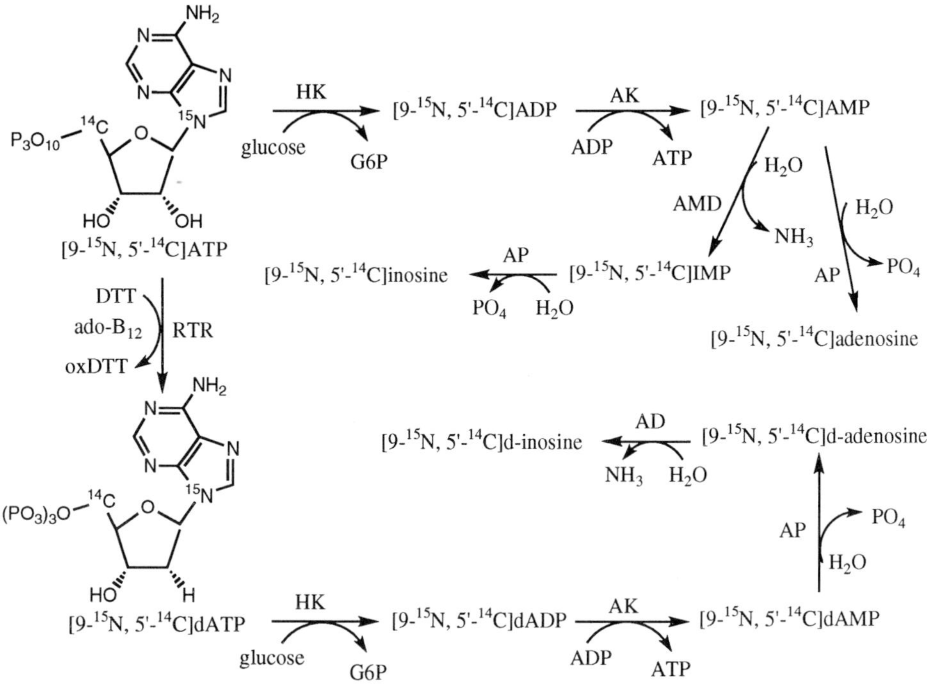

FIG. 3. Synthesis of other nucleotides and nucleosides from ATP [D. W. Parkin, H. B. Leung, and V. L. Schramm, *J. Biol. Chem.* **259,** 9411 (1984); B. A. Horenstein, D. W. Parkin, B. Estupiñán, and V. L. Schramm, *Biochemistry* **30,** 10788 (1991)]. Reaction mixtures for dATP production contained 50 mM potassium phosphate (pH 7.7), 100 mM sodium acetate (pH 7.2), 25 mM dithiothreitol (DTT), 0.1–1 mM labeled ATP, and 2 U/ml ribonucleotide triphosphate reductase (RTR). After incubation under argon, the anaerobic mixture was made to 20 μM in adenosylcobalamin (ado-B$_{12}$) in the dark. The reaction was incubated in the dark, under argon, for 3 hr at 37°, placed at 95° for 2–3 min, cooled, and the labeled dATP purified by semipreparative HPLC as indicated in the legend to Fig. 2. Labeled ATP or dATP can be converted to monophosphates in the same reaction mixtures used to form them. Following denaturation of the ATP-forming enzymes by heating, the cooled reaction mixtures are made to 6 mM glucose, 25 U/ml hexokinase (HK), and 200 U/ml myokinase (AK). After a few minutes at 37°, the reactions are stopped by 3 min in boiling water. Conversions of AMP and dAMP to adenosine or d-adenosine can occur in the same reaction mixtures, after the previous enzymes are denatured. The reactions are made to 2 U/ml alkaline phosphatase (AP), incubated for 30 min at 37°, and boiled for 3 min. In the same reaction mixtures, the addition of 2 U/ml adenosine deaminase (AD) will convert adenosine or d-adenosine to the corresponding inosines in 30 min at 37°. Adenosine monophosphate deaminase (AMD) can be used in the same conditions to convert AMP to IMP, and IMP can be converted to inosine by alkaline phosphatase as indicated. Purification of the monophosphates or the nucleosides and deoxynucleosides can be accomplished by chromatography on Sephadex G-10 in 20 mM acetic acid, followed by HPLC C-18 reverse phase eluted with 0.1 M ammonium acetate, pH 5.0, containing 5% methanol.

TABLE II
Substrates for KIE Measurements: AMP Nucleosidase

Substrate	Kinetic isotope effect ^{3}H
[1'-^{3}H]AMP + [5'-^{14}C]AMP	α secondary ^{3}H
[2'-^{3}H]AMP + [5'-^{14}C]AMP	β secondary ^{3}H
[4'-^{3}H]AMP + [5'-^{14}C]AMP	Remote secondary ^{3}H
[5'-^{3}H]AMP + [5'-^{14}C]AMP	Remote secondary ^{3}H
[1'-^{2}H, 5'-^{3}H]AMP + [5'-^{14}C]AMP	α secondary ^{2}H
[2'-^{2}H, 5'-^{3}H]AMP + [5'-^{14}C]AMP	β secondary ^{2}H
[1'-^{14}C]AMP + [5'-^{14}C]AMP	Primary ^{14}C
[9-^{15}N, 5'-^{14}C]AMP + [5'-^{3}H]AMP	Primary ^{15}N
[1'-^{2}H, 1'-^{14}C]AMP + [5'-^{3}H]AMP	Dual primary ^{14}C and α secondary ^{2}H
[9-^{15}N, 1'-^{14}C]AMP + [5'-^{3}H]AMP	Dual primary ^{14}C and primary ^{15}N

The synthesis of molecules in addition to those just discussed can be inferred directly from the enzymatic reaction schemes. The addition of other isotopes in the starting materials, such as ^{2}H- or ^{13}C-labeled sugars or nitrogenous bases with additional labels, would lead to the expected products. The end product ATP can also be used as a precursor for other nucleosides, nucleotides, and the 2'-deoxynucleosides and -nucleotides. The example of Fig. 4 demonstrates the conversion of the [5'-^{14}C]2'-deoxyribosyl group of [5'-^{14}C]deoxyinosine (Fig. 3) to thymidine, a 2'-deoxypyrimidine nucleoside. A direct extension of this method would provide other pyrimidine nucleosides, nucleotides, and the 2'-deoxy compounds.

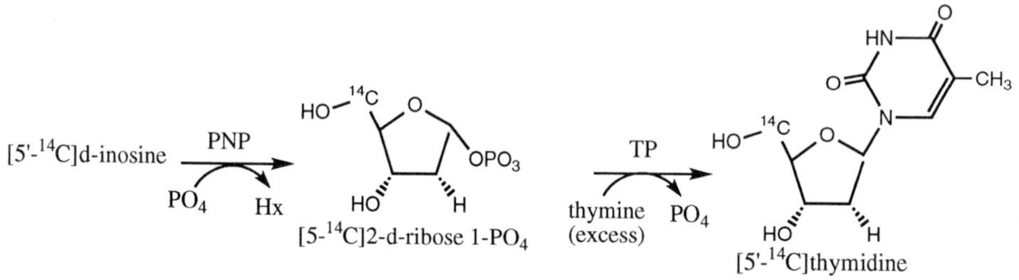

Fig. 4. Synthesis of labeled thymidine from d-inosine (M. Khalil and V. L. Schramm, unpublished experiments, 1998). Reaction mixtures contained 2 mM potassium phosphate, pH 7.7, 30 μM d-inosine, 10 mM thymine, 0.4 U/ml purine nucleoside phosphorylase (PNP), 1 U/ml thymidine phosphorylase (TP), and 0.1 U/ml xanthine oxidase [converts hypoxanthine (Hx) to uric acid]. The product is purified by HPLC C-18 reverse phase, eluted with 50 mM ammonium acetate, pH 5.0.

Nicotinamide Adenine Dinucleotide and Related Nucleotides

NAD$^+$ is a central metabolite used in hydride transfer and was one of the first substrates used to measure enzymatic KIE. The large KIE for deuteride (^{2}H$^-$) transfer reactions is measured easily by direct initial rate techniques, a consequence of the relatively large mass increase from ^{1}H to ^{2}H.[50] NAD$^+$ plays additional roles in metabolism that are of increasing interest. These include the ADP-ribosylation reactions, which are catalyzed by the ADP-ribosylating bacterial toxins; the formation and use of poly-ADPribose in DNA repair reactions and in maintaining chromatin structure; and the formation of cyclic ADPribose as a signal causing calcium release from the endoplasmic reticulum.[51-53] The reaction mechanisms of these enzymes involve scission of the N1-ribosidic bond to nicotinamide followed by transfer of the resulting ADP-ribose to the acceptor molecule. Investigation of the TS for these reactions requires labeling the NMN$^+$-ribose group of NAD$^+$. An example of the synthesis of [1'-^{3}H]NAD$^+$, with the labels being incorporated in the NMN$^+$-ribose, is provided in Fig. 5.[54] Synthesis of NAD$^+$ is accomplished in two steps. In the first, all reagents needed for nicotinic acid adenine dinucleotide (NAAD$^+$) synthesis are combined to convert labeled precursors to this end product. The reaction is stopped by brief heating, the NAAD$^+$ is purified from the reagents of the first step, and the NAD$^+$ synthetase reaction is accomplished in a second incubation (Fig. 5). In this procedure, a component of NAAD$^+$ synthesis inhibits NAD$^+$ synthetase and must be removed prior to NAD$^+$ formation. The same precursors described in Table I can be used to synthesize a similar group of specifically labeled NAD$^+$ molecules (Table III). By combining the synthesis of specifically labeled ATP and that of NAD$^+$, it is possible to synthesize NAD$^+$ with specific atomic labels in any one or more of the nonexchangeable atoms in the entire molecule. The specificity of the synthetic enzymes permits obvious extensions of the methods, e.g., the conversion of labeled NAD$^+$ molecules to nicotinamide guanine dinucleotide (NGD$^+$) and the production of labeled NMN$^+$ molecules by hydrolyzing the phosphodiester of labeled NAD$^+$ (Fig. 6). Both NGD$^+$ and NMN$^+$ are

[50] J. S. Blanchard and K. K. Wong, *in* "Enzyme Mechanism from Isotope Effects" (P. F. Cook, ed.), p. 341. CRC Press, Boca Raton, FL, 1991.

[51] K. Aktories, ed., "Current Topics in Microbiology and Immunology: ADP-Ribosylating Toxins." Springer-Verlag, Berlin, 1992.

[52] M. Miwa., O. Hayaishi, S. Shall, M. Smulson, and T. Sugimura, eds., "ADP-Ribosylation, DNA Repair and Cancer." Japan Scientific Societies Press, Tokyo, 1983.

[53] H. C Lee, T. F Walseth, G. T. Bratt, R. N. Hayes, and D. L. Clapper, *J. Biol. Chem.* **264,** 1608 (1989).

[54] K. A. Rising and V. L. Schramm, *J. Am. Chem. Soc.* **116,** 6531 (1994).

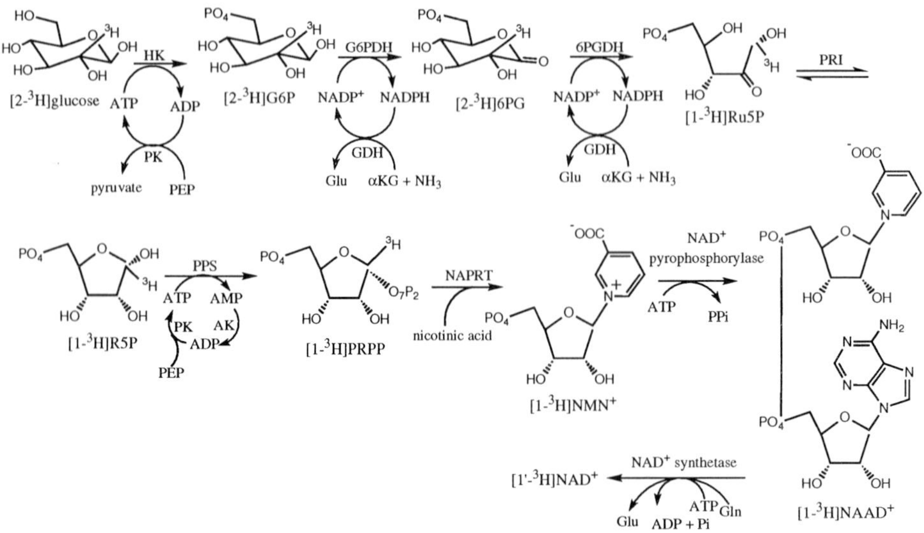

Fig. 5. Synthesis of isotope-labeled NAD^+ [K. A. Rising and V. L. Schramm, *J. Am. Chem. Soc.* **116**, 6531 (1994)]. Reaction mixtures of 1 ml contained 50 mM potassium phosphate, pH 7.5, 3 mM MgCl$_2$, 50 mM KCl, 5 mM dithiothreitol, 2 mM ATP, 10 mM P-enolpyruvate, 10 mM α-ketoglutarate, 0.1 mM NADP$^+$, 2 mM nicotinic acid, pH 6.2, and 1 mM glucose containing 20–100 μCi [^{3}H] or [^{14}C]glucose. Enzymes (as supplied) were diluted into 50 mM potassium phosphate and added to give final concentrations of 0.1 U/ml hexokinase, 2 U/ml pyruvate kinase, 0.1 U/ml glucose 6-phosphate dehydrogenase, 0.5 U/ml glutamate dehydrogenase, 0.1 U/ml 6-phosphogluconate dehydrogenase, 6 U/ml phosphoriboisomerase, 0.5 U/ml myokinase, 0.25 U/ml NAD$^+$ pyrophosphorylase, 0.2 U/ml PRPP synthetase, and 0.02 U/ml nicotinate phosphoribosyltransferase (NAPRT). All enzymes, except hexokinase, were mixed and added together. The other mixture components were added in the order listed. The reaction was initiated by the addition of hexokinase. Synthesis was monitored for 4 hr at 37° for NAAD$^+$ production by periodic HPLC analysis on a C-18 reverse-phase column (7.8 × 300 mm) eluted with 0.1 M ammonium acetate, pH 5.0. The reaction was terminated by placing the sealed reaction mixture (in a plastic vial) in a heating block at 120° for 1.5 min. The sample was centrifuged and the supernate subjected to HPLC using the same method to provide NAAD$^+$. Purified NAAD$^+$ was freeze dried and reconstituted in a reaction mixture of 0.5 ml containing 50 mM potassium phosphate, pH 7.5, 50 mM KCl, 3 mM MgCl$_2$, 5 mM dithiothreitol, 4 mM ATP, 20 mM glutamine, and 0.2 U/ml yeast NAD$^+$ synthetase. Incubation for 2 hr at 37° converted NAAD$^+$ to NAD$^+$. At the end of the incubation, the entire reaction mixture was injected directly onto the HPLC column (7.8 × 300 mm) and eluted as indicated earlier. In this HPLC system, NAD$^+$ is the last component eluted from the column. Fractions containing NAD$^+$ are freeze dried and stored in 50% ethanol at −70°. Under these conditions the labeled NAD$^+$ is stable (>99%) for at least 12 months. Synthesis of [2′-^{3}H]NAD$^+$ was intitiated with [2-^{3}H]R5P. In this synthesis, only the components for PPS and the following steps are used. [2-^{3}H]R5P is prepared by the phosphoribose isomerase (PRI) exchange of ^{3}H$_2$O into an equilibrium mixture of R5P and Ru5P.

TABLE III
Synthesis of NAD$^+$ for Kinetic Isotope Effects

Labeled substrate	Labeled NAD$^+$ product	Changes from Fig. 4
[2-^{3}H]Glucose	[1'-^{3}H]NAD$^+$	
[2-^{3}H]Ribose 5-PO$_4$	[2'-^{3}H]NAD$^+$	First four enzymes eliminated
[5-^{3}H]Glucose	[4'-^{3}H]NAD$^+$	
[6-^{3}H]Glucose	[5'-^{3}H]NAD$^+$	
[2-^{14}C]Glucose	[1'-^{14}C]NAD$^+$	
[6-^{14}C]Glucose	[5'-^{14}C]NAD$^+$	
[5-^{18}O]Glucose	[4'-^{18}O]NAD$^+$	
[6-^{14}C]Glucose	[1-^{15}N, 5'-^{14}C]NAD$^+$[1-^{15}N]nicotinamide	
[2-^{14}C]Glucose	[1-^{15}N, 1'-^{14}C]NAD$^+$[1-^{15}N]nicotinamide	
[8-^{14}C]ATP	[8-^{14}C]NAD$^+$ NMN$^+$ + [8-^{14}C]ATP $\rightarrow$ [8-^{14}C]NAD$^+$	
[5-^{18}O]Glucose	[4'-^{18}O, 8-^{14}C]NAD$^+$[8-^{14}C]ATP	

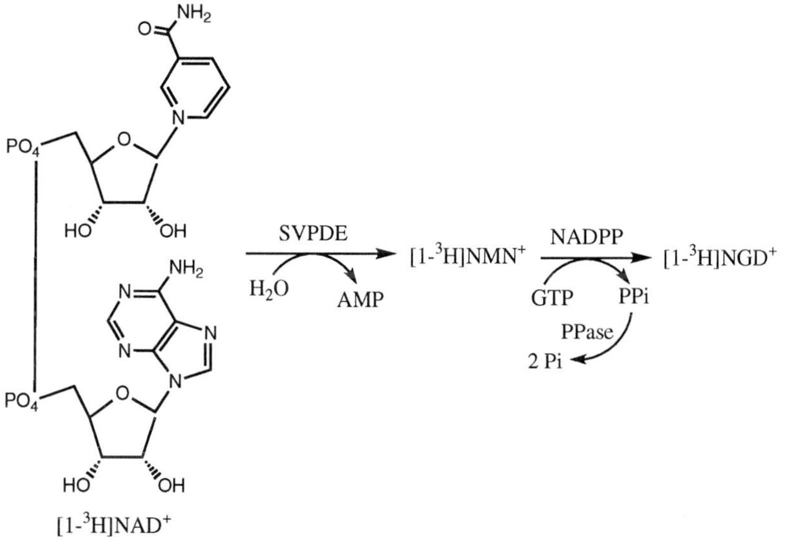

FIG. 6. Synthesis of labeled NGD$^+$ from NAD$^+$ [A. S. Sauve, C. Munshi, H. C. Lee, and V. L. Schramm, *Biochemistry* **37**, 13239 (1998)]. The reaction mixture for the synthesis of NAD$^+$ (legend to Fig. 5) was treated with snake venom phosphodiesterase (SVPDE) (~0.5 U/ml for 30 min 37°) to produce NMN$^+$, heated to 110° in a heating block for 40 sec, centrifuged, and purified on C-18 reverse-phase HPLC with 0.1% trifloroacetic acid containing 1% methanol. After freeze drying, the NMN$^+$ was made to 2 mM with 50 mM potassium phosphate, pH 7.5, containing 5 mM MgCl$_2$, 4 mM GTP, 1 U NAD$^+$ pyrophosphorylase (NADPP), and 0.1 U pyrophosphatase (PPase). After 6 hr at 37°, another addition of enzymes was followed by a further 6-hr incubation at 37°. The reaction was terminated by 20 sec at 110°, centrifuged, and the supernate purified on preparative C-18 HPLC eluted with 50 mM ammonium acetate, pH 5.0.

substrates for CD38, an NAD^+, NMN^+ nicotinamide N-ribohydrolase and ADP-ribosyl cyclase ($NAD^+ \rightarrow$ cyclic ADP-ribose + nicotinamide).[55]

RNA and DNA Oligonucleotides

Isotopically labeled nucleotides and deoxynucleotides (Figs. 2 and 3) serve as substrates for incorporation into oligonucleotide polymers. Figure 7 demonstrates the synthesis of RNA and DNA polymers using [9-^{15}N, 5'-^{14}C]ATP and [9-^{15}N, 5'-^{14}C]dATP for incorporation into oligonucleotide sites coding for adenine nucleotides. RNA synthesis uses T7 RNA polymerase and a double-stranded DNA primer. Under these conditions the product RNA begins with an uncoded GTP, followed by faithful transcription of the remaining DNA primer to the final base. The product RNA terminates with a free 3'-hydroxyl group and contains a triphosphate at the 5' terminus. The primer is not consumed in the reaction, and T7 polymerase is capable of rebinding and synthesis of another RNA strand, displacing the previous RNA strand. Thus, small amounts of the DNA primer can produce large amounts of the labeled RNA product. This process has been used to produce unlabeled stem-loop RNA structures for the adenine depurination reaction catalyzed by the ricin A-chain,[56] as well as labeled RNA for KIE measurements with this reaction (X.-Y. Chen and V. L. Schramm, unpublished results, 1998).

Synthesis of labeled DNA oligonucleotides uses a hairpin primer-template, the desired labeled and unlabeled deoxynucleotide triphosphates, and Klenow fragment DNA polymerase (Fig. 7). The addition of the first nucleotide to the primer results in a covalent phosphodiester bond that must be specifically removed if it is intended to remove the primer-template from the finished product. This is accomplished by the synthesis of a primer-template that incorporates 2'-deoxynucleotides at every position except at the 5'-terminal, where a uridine nucleotide is incorporated. This introduces a site for specific chemical cleavage by base (0.3 M NaOH) without hydrolysis of the DNA primer or the DNA product. Base hydrolysis of the ribonucleotide phosphodiester bond in the product DNA/RNA hybrid yields the replicated DNA strand and regenerates the 3'-phosphorylated primer-template that can be reused following dephosphorylation at the 3' terminus with alkaline phosphatase. The replicative synthesis by this process is stoichiometric, but with the method described here, it can be used in a repetitive fashion following recycling of the primer-template. The methods of Fig. 7 have been used to synthesize stem-loop RNA and DNA containing most of the labeled ATP and dATP described in Table I. For the specific stem-

[55] A. S. Sauve, C. Munshi, H. C. Lee, and V. L. Schramm, *Biochemistry* **37**, 13239 (1998).
[56] X.-Y. Chen, T. M. Link, and V. L. Schramm, *Biochemistry* **37**, 11605 (1998).

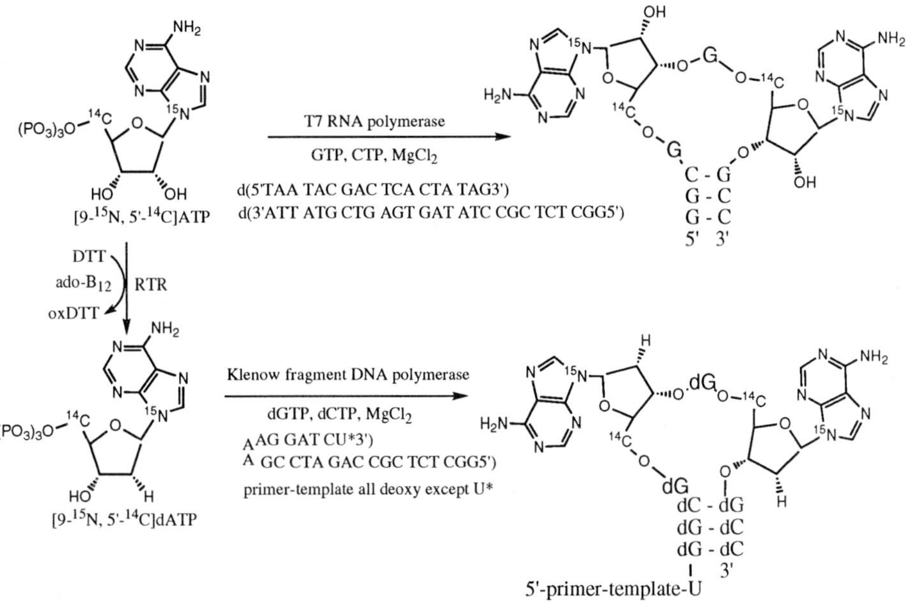

FIG. 7. Synthesis of labeled stem-loop RNA and DNA from labeled nucleotides and deoxynucleotides. Reaction mixtures of 0.4 ml for RNA synthesis contained 0.22 mM ATP, including the desired ^{3}H and ^{14}C, 4.5 mM GTP, 4.5 mM CTP, 22 mM MgCl$_2$, 25 mM NaCl, 2 mM spermidine, 5 mM dithiothreitol, 40 mM Tris–HCl, pH 8.0, 0.2 μM DNA primer-template, 40 U Rnasin (Promega), and 6000 U T7 RNA polymerase. Reactions were incubated at 37° overnight, made to 50 mM EDTA and 0.1 M sodium acetate, and the RNA precipitated with 2.5 volumes of ethanol followed by cooling to −70°. RNA 10-mers were isolated by 24% denaturing polyacrylamide gel electrophoresis, located by UV shadowing, eluted into 1 M ammonium acetate, and desalted on a C-18 Sep-Pac column from Waters. Reaction mixtures of 1.0 ml for DNA synthesis contained 10 mM Tris–HCl, pH 7.5, 5 mM MgCl$_2$, 50 μM hairpin primer-template, 3 mM dGTP, 3 mM dCTP, 50 μM dATP containing the desired ^{3}H and ^{14}C, and 100 U 3'-5' exonuclease-deficient mutant of Klenow fragment DNA polymerase I (U.S. Biochemical). Note that the 3'-terminal U in the primer is the ribosyl nucleotide. After incubation at 37° overnight, the mixture was made to 0.3 N NaOH, followed by incubation at 55° for 2 hr to cleave at the uridine site. The mixture was neutralized with 0.36 M acetic acid, and the DNA was precipitated with ethanol and isolated as described for RNA.

loop RNA and DNA of Fig. 7, the labeled nucleotides are incorporated into both adenylate positions of the oligonucleotide.

Purification of Isotopically Labeled Substrates

Readers will realize that these complex mixtures of enzymes, cofactors, and buffers represent an empirical approach to isotopically labeled synthesis. The complexity of the reaction mixtures and the incomplete conversion of isotopically labeled intermediates present a purification challenge to yield the desired product free of reaction mixture components that would interfere with subsequent KIE determinations. The sequential application of two or three rapid purification steps has provided efficient purifications for all of the small molecule substrates. Following denaturation of the enzymes by brief treatment near 100°, or their removal by ultrafiltration, the first step for nucleosides and nucleotides is molecular exclusion chromatography on Sephadex G-10 or G-25 in 20 mM acetic acid to remove denatured protein and salts. This is followed by anion-exchange chromatography on DEAE Sephadex A-25 developed with gradients of acetic acid, ammonium bicarbonate, or ammonium acetate. In many cases, these two steps are adequate. If not, the product is freeze dried to remove salts followed by HPLC chromatography. NAD^+ is more labile and is purified directly in a single step by chromatography of the reaction mixtures on HPLC using a reverse-phase C-18, 7.8 $\times$ 300-mm column, developed with 0.1 M ammonium acetate, pH 5.0. The stainless-steel components of HPLC systems catalyze the hydrolysis of NAD^+ to ADPribose, resulting in an approximately 5% hydrolysis during the purification.[54] NAD^+ hydrolysis is avoided by the use of coated HPLC systems designed to prevent solvent–metal contact.

Purification of DNA and RNA oligonucleotides is accomplished by precipitation of the product oligonucleotide from reaction mixtures with 2.5 volumes of ethanol, cooling on dry ice, centrifugation, ethanol wash, and chromatography on 24% denaturing polyacrylamide gel electrophoresis. The appropriate product is located with UV shadowing using known oligonucleotides as chromatography standards. Segments containing the desired oligonucleotides are excised, eluted into 1 M ammonium acetate, and desalted on a C-18 Sep-Pack column (Waters).[56]

Experimental Measurement of KIE

Methods for N-Ribosyl Hydrolases and Transferases

Partial conversion of the isotopically labeled substrates to products generates the mixture of isotopically labeled products that is used to establish the KIE. Quantitative analysis of this ratio requires accurate analysis of (1) the isotopic ratio in the substrate, (2) the fraction of substrates converted to product, (3) separation of the unreacted substrates from the products without altering the isotopic ratio of the products, and (4) analysis of the isotopic ratio in the products. An example of the experimental approach for a KIE study for the nucleoside hydrolase reaction is provided in Fig. 8. The isotopic ratio in the substrate (labeled inosines) is measured

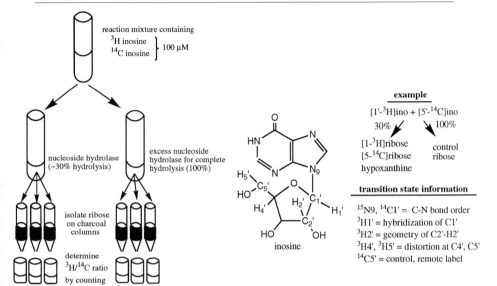

FIG. 8. The charcoal column method to measure KIE and transition-state information from specific isotopic labels [D. W. Parkin, *in* "Enzyme Mechanism from Isotope Effects" (P. F. Cook, ed.), p. 269. CRC Press, Boca Raton, FL, 1991; D. W. Parkin and V. L. Schramm, *Biochemistry* **26,** 913 (1987); D. W. Parkin, H. B. Leung, and V. L. Schramm, *J. Biol. Chem.* **259,** 9411 (1984)]. The reaction mixture (upper tube) is typically 1.5 ml of buffered reaction mixture containing 0.1–1 mM substrate and the desired ^{3}H and ^{14}C substrates to provide equal counts in both isotopes. The mixture is divided into portions to provide equal counts in the product ribose after the reactions have occurred [1.0 ml for the partial reaction (left tube) and 0.4 ml for the complete reaction (right tube)]. Following the enzymatic conversions to product, the reactions are quenched with acid (or EDTA for reactions dependent on divalent cations), and equal aliquots are applied to three or four charcoal–cellulose columns (three are shown) equilibrated previously with the reaction product to be collected. For nucleoside hydrolases, the columns are equilibrated with 50 or 100 mM ribose. For AMP nucleosidase (product is ribose 5-phosphate), the columns were equilibrated with 10 mM ribose 5'-phosphate. A control to establish that all substrate is retained on the column uses an additional column and the reaction mixture prior to reaction with the enzyme. The columns are eluted with the equilibration solutions, and 1.0-ml fractions are collected into scintillation vials, made to 20 ml with scintillation fluid, and counted to quantitate ^{3}H and ^{14}C at ~ 30 and 100% conversion to products. The observed ^{3}KIE = (^{3}H/^{14}C for 100%)/(^{3}H/^{14}C for 30%). Preparation of multiple samples from a single reaction mixture provides statistical analysis of the procedures for chromatography, sample preparation, and counting. Samples are counted for 10-min cycles until sufficient counts are accumulated to provide small counting errors, usually 10 to 15 cycles per sample set.

following complete conversion of inosine to ribose with excess enzyme. It is also possible to determine the ratio of $^3H/^{14}C$ in substrate inosines directly; however, analysis of the $^3H/^{14}C$ ratio in the products eliminates the possibilities that quenching is different when counting inosine and ribose or that the substrate contains a small fraction of impurity as ^3H or ^{14}C that is not converted to product and thus gives an inaccurate $^3H/^{14}C$ ratio in inosine. Chromatographic or other physical means of complete separation of unreacted substrate from product is essential. In the case of nucleoside hydrolase (Fig. 8), the product ribose is resolved efficently from inosine by chromatography on small columns of acid-washed, powdered charcoal–cellulose in a mixture to give good flow rates and is packed in disposable Pasteur pipettes. Ratios of dry charcoal to cellulose between 1:1 and 1:4 are satisfactory to provide the desired flow rate of approximately 0.5 ml/min. Elution with buffered, unlabeled ribose provides complete yields of product ribose and the complete retention of inosine. In our experience, the resolution of substrates from products on charcoal columns is superior to other chromatographic mediums and should be used when substrates and products separate by this method. In some cases, it is possible to quantitatively convert the products of a reaction to compounds that can be analyzed conveniently by charcoal columns.

KIE in Enzymes of RNA and DNA Processing

The conversion of RNA and DNA oligonucleotides to products with a single adenine removed presents a challenge in the quantitative separation and recovery of the oligonucleotide substrate from the product oligonucleotide. This problem is solved by the conversion of both substrate and product RNA and DNA to the corresponding nucleosides with methods that generate free ribose or 2-deoxyribose only at the site of depurination. The substrate sites where depurination has not occurred generate adenosine from this position. The ultimate products of the reactions yield a mixture of nucleosides from all sites not depurinated and ribose or 2-deoxyribose from all sites that have been depurinated (Fig. 9). These products are resolved conveniently on charcoal columns and analyzed for total counts and the $^3H/^{14}C$ ratio. Collection and counting of the product $^3H/^{14}C$ ratio in the partial reaction compared to that of the samples from the complete conversion gives the observed KIE.

KIE in Enzymes of Protein Covalent Modification

The action of bacterial exotoxins on GTP-binding proteins (G-proteins) has been developed as the first example in which the transition-state structure for protein covalent modification has been attempted using KIE

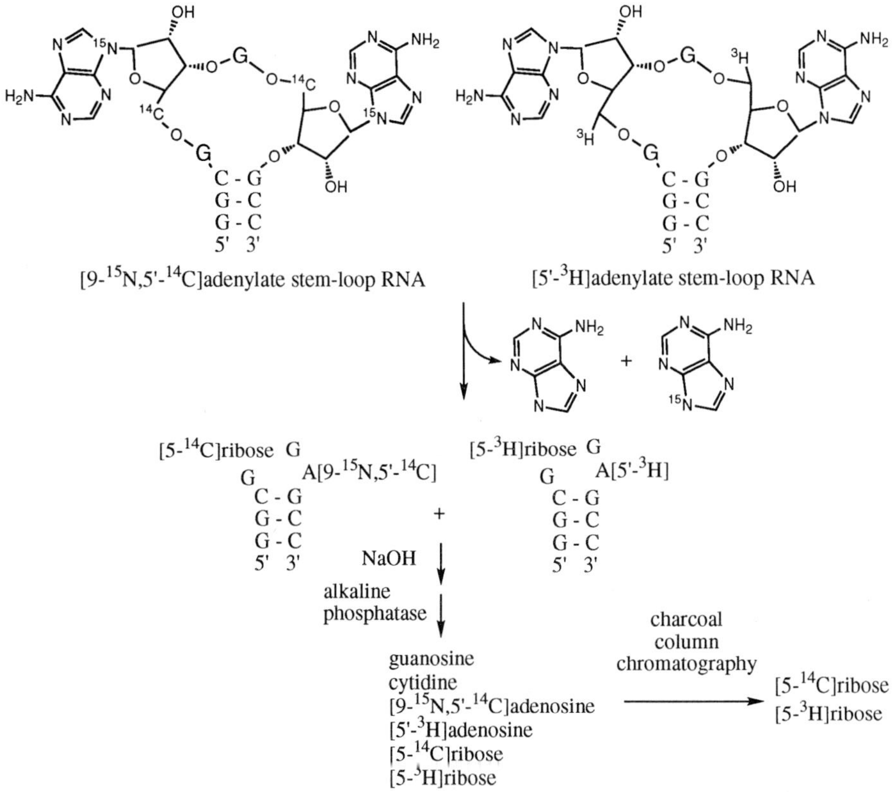

FIG. 9. Method to measure the ^{15}N KIE for and RNA depurination reaction (X.-Y. Chen and V. L. Schramm, unpublished observations, 1998). Stem-loop RNA with the labels in the site of depurination (left stem-loop RNA) or in a remote position (right stem-loop RNA) are incubated in companion reaction mixtures with ricin A-chain to provide ~30 or 100% hydrolysis of the susceptible bond, using the method described in Fig. 8. Adenine or [9-^{15}N]adenine is released in the ratio reflecting the ^{15}N KIE, leaving [5-^{14}C]ribose and [5-^{3}H]ribose at the depurination sites. Treatment with 0.3 M NaOH at 37° overnight followed by neutralization to pH 8 and alkaline phosphatase yields nucleosides for every base that has not been depurinated and ribose for every depurination site. The samples are then processed as in Fig. 8 to determine the KIE.

methods. The ADP-ribosylation of G-proteins by NAD$^+$ occurs with bacterial exotoxins, including cholera, diphtheria, and pertussis exotoxins. Covalent modification of other mammalian proteins by ADP-ribose is also common, but is not as well understood. The reactions catalyzed by bacterial exotoxins are

$$\text{NAD}^+ + \text{G-protein} \rightarrow \text{nicotinamide} + \text{ADP-ribosylated-G-protein}$$

TABLE IV
Isotopically Labeled NAD$^+$ Mixtures for Kinetic Isotope Effects

Labeled NAD$^+$ substrate	Kinetic isotope effect
[1'-^{3}H]NAD$^+$ + [5'-^{14}C]NAD$^+$	α secondary ^{3}H
[2'-^{3}H]NAD$^+$ + [5'-^{14}C]NAD$^+$	β secondary ^{3}H
[4'-^{3}H]NAD$^+$ + [5'-^{14}C]NAD$^+$	Remote ^{3}H
[5'-^{3}H]NAD$^+$ + [5'-^{14}C]NAD$^+$	Remote ^{3}H
[1'-^{14}C]NAD$^+$ + [4'-^{3}H]NAD$^+$	Primary ^{14}C
[4'-^{18}O]NAD$^+$	Direct α secondary ^{18}O
[1-^{15}N, 5'-^{14}C]NAD$^+$ + [4'-^{3}H]NAD$^+$	Primary ^{15}N
[1-^{15}N, 1'-^{14}C]NAD$^+$ + [4'-^{3}H]NAD$^+$	Combined primary ^{14}C + ^{15}N
[4'-^{18}O, 8-^{14}C]NAD$^+$ + [4'-^{3}H]NAD$^+$	α secondary ^{18}O

where the ADP-ribosylation site is variable. Reactions catalyzed by cholera, diphtheria, and pertussis toxins ADP-ribosylate a specific arginine, diphthamide (a modified histidine) or cysteine residues. The reaction of pertussis toxin involves ADP-ribosylation of the G$_{i\alpha 1}$ subunit of the trimeric GTP-binding complex. The G$_{i\alpha 1}$ subunit has been overexpressed in *Escherichia coli* and was purified for use at substrate levels to measure the KIE. Substrate trapping studies with labeled NAD$^+$ and a variable G$_{i\alpha 1}$ subunit demonstrated a negligible forward commitment for the reaction. Reactions to measure KIE were conducted with excess isotopically labeled NAD$^+$ (Table IV) and a limiting G$_{i\alpha 1}$ subunit. This procedure permits isotopic fractionation of the labeled NAD$^+$ substrate until the G$_{i\alpha 1}$ subunit is largely ADP-ribosylated. The resolution of the ADP-ribosyl-G$_{i\alpha 1}$ subunit from the labeled NAD$^+$ that had not reacted is accomplished by precipitation of the protein with 15% trichloroacetic acid. The pellet is washed with TCA to remove all unincorporated NAD$^+$. The pellets were dissolved in sodium dodecyl sulfate (SDS) and counted by scintillation counting to determine the KIE. The KIE measured for the ADP-ribosylation of the G$_{i\alpha 1}$ subunit by pertussis toxin are intrinsic and were used to solve the first transition state for the covalent modification of a protein.[57]

Correction of Observed KIE for Isotopic Depletion

The ^{3}H/^{14}C ratios obtained by the procedures just described are those for the total amount of labeled substrate that has been converted to products. The experimental KIE is dependent on the fractional conversion to product, as the ^{3}H/^{14}C ratio in the substrate changes as a function of the extent of the reaction. Comparison of the total product ribose counts in

[57] J. Scheuring, P. J. Berti, and V. L. Schramm, *Biochemistry* **37**, 2748 (1998).

the 30–40% conversion to the 100% conversion gives an accurate value of the fractional conversion to product. $KIE_{experimental}$ values are converted to KIE_{actual} values that would occur at 0% of substrate depletion using the expression[38]

$$KIE_{actual} = \ln[1 - (KIE_{experimental}) \times (\text{fraction converted})]/\ln[1 - \text{fraction converted}]$$

Details of the $^{3}H/^{14}C$ scintillation counting procedures to minimize errors and to accuracies of 0.2–0.3% have been described.[46,47] The value of KIE_{actual} can vary from unity (no KIE) to the intrinsic KIE, defined as the KIE for the TS with no contribution from steps other than the chemical structure of the TS. The most common problem faced in the determination of intrinsic KIE for enzyme-catalyzed reactions is forward and reverse commitments.[58–60]

Measurement of Commitment Factors and Binding Isotope Effects

With the potential for KIE to provide direct information of the enzymatic TS, why has the application been limited to a relatively few enzymatic reactions? Enzymes achieve catalytic power by multiple steps in the reaction coordinate, often of comparable energetic barriers, so that the chemical step which gives the intrinsic KIE may be buried between enzymatic steps that include forward and reverse commitment, rate-limiting product release, or rate-limiting enzymatic conformations (Fig. 10). These problems were recognized and quantitated by Northrop at the inception of the application of KIE analysis for enzymatic reactions.[58–60] Because most of the first KIE measurements were accomplished with $^{1}H/^{2}H$, methods for correcting commitment factors were proposed primarily for dehydrogenases. Hydride transfer represents the most optimal case in KIE measurements, as isotope effects from ^{2}H or ^{3}H transfer as hydride ions give large isotope effects related by the Swain–Schadd relationship ($^{3}H\text{-}KIE = {}^{2}H\text{-}KIE^{1.44}$) in classical mechanics of hydride transfer. Deviations from this relationship or from the temperature dependence for KIE in hydride transfer reactions provide indications of commitment factors or quantum mechanical tunneling.[59,61] When tunneling can be eliminated, the deviation from the Swain–Schadd

[58] D. B. Northrop, *Biochemistry* **14**, 2644 (1975).
[59] D. B. Northrop, *Annu. Rev. Biochem.* **50**, 103 (1981).
[60] D. B. Northrop, *in* "Isotope Effects on Enzyme-Catalyzed Reactions" (W. W. Cleland, M. H. O'Leary, and D. B. Northrop, eds.), p. 122. University Park Press, Baltimore, MD, 1977.
[61] B. J. Bahnson and J. P. Klinman, *Methods. Enzymol.* **249**, 374 (1995).

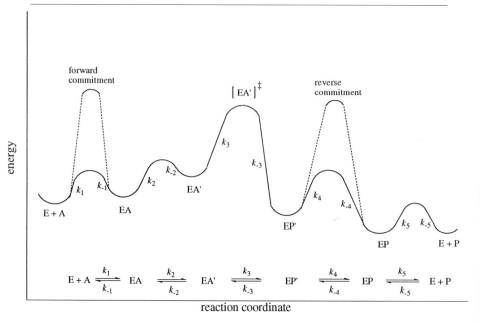

FIG. 10. Reaction coordinate diagram demonstrating forward and reverse commitments. The reaction coordinate diagram formed by the solid line demonstrates a reaction with chemistry as the highest energetic barrier for the catalytic cycle. This pattern yields intrinsic KIE. Forward commitment is indicated by the dashed line, where substrate molecules (A), which cross the binding barrier k_1, are more likely to react (k_3) than to return to E + A. Reverse commitment is indicated by the energetic barriers for EP′, which is more likely to be converted back to EA′ than to EP. Reprinted with permission, from the *Annual Review of Biochemistry*, Volume 67, © 1998, by Annual Reviews www.annualreviews.org.[14]

relationship can be used to establish the intrinsic KIE for hydride transfers. However, the decision that establishes or excludes tunneling is difficult. In the case of quantum mechanical tunneling, experimental KIE for hydride ion transfer can be large or small and may deviate significantly from the Swain–Schadd relationship.[62] Large KIE are not required in hydride tunneling, and it has been proposed that in cases of coupled motion at the TS, the KIE for hydride transfer may appear to follow classical limits of bond vibrational isotope effects, but that α secondary effects, coupled to the tunneling atom, may show deviations from the Swain–Schadd relationship.[61,63] Tunneling complications for hydride transfer reactions have caused the KIE that are the easiest to measure to become the most difficult to interpret. Quantum mechanical tunneling has been discussed in several

[62] D. Antoniou and S. D. Schwartz, *Proc. Nat. Acad. Sci. U.S.A.* **94,** 12360 (1997).
[63] W. P. Huskey and R. L. Schowen, *J. Am. Chem. Soc.* **105,** 5704 (1983).

reviews and reports and will not be discussed further here. Reactions without hydride transfer are not subject to tunneling effects and are interpreted more readily in the framework of classic bond vibrational theory.

Apart from the complications in the special case of quantum mechanical tunneling, the commitment factors can be measured by the following experimental methods. These can then be used to establish the intrinsic KIE using the relationship of Northrop[59]

$$^xV/K = (^xk + C_f + C_r \cdot {}^xK_{eq})/(1 + C_f + C_r) \tag{1}$$

where ^{x} is the isotopic label (^{3}H, ^{14}C, etc), $^xV/K$ is the experimental competitive KIE, which is KIE on the V/K kinetic constants when competing labels are used, ^xk is the intrinsic KIE on the chemical step k, C_f and C_r are forward and reverse commitments, and $^xK_{eq}$ is the isotope effect on the equilibrium constant for the chemical equilibrium. Analysis of ^xk is required for TS analysis based on KIE. Therefore, C_f and C_r must be known to solve for ^xk.

Forward Commitment and KIE

Forward commitment arises from exceptions to the case that a thermodynamic equilibrium is established between the pool of free substrate molecules and those that populate the TS.[59] The procedure to detect forward commitment depends on measuring the fraction of substrate bound in the Michaelis complex, which is converted to product and that which is released unchanged from the catalytic site for each catalytic event. Forward commitment is the probability function for the fate of each substrate molecule in the Michaelis complex. A reaction coordinate diagram illustrates the forward commitment problem (Fig. 10). The substrate trapping procedure established by Rose[64] provides a convenient solution to this problem. The enzyme is saturated with labeled substrate, followed by mixing with a large excess of the unlabeled substrate followed by several catalytic turnovers and quenching of catalytic activity. Analysis of the fraction of bound, labeled substrate converted to products gives the commitment factor.

The experimental protocol and the results of a substrate trapping experiment for the determination of forward commitment are shown schematically in Fig. 11. The fraction of bound substrate [bound substrate = free substrate concentration/(K_d + free substrate concentration)] is converted to product following five or more catalytic cycles (to permit equilibration of bound substrate with the new pool of diluted substrate and conversion of bound substrate to product) and provides the commitment factor. En-

---

[64] I. A. Rose, *Methods Enzymol.* **249**, 315 (1995).

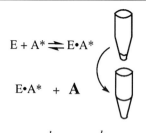

$E + A^* \rightleftharpoons E{\cdot}A^*$ step 1. form Michaelis complex
in small volume (e.g., 20 µl)
using labeled substrate (A*)

$E{\cdot}A^* + A$ step 2. dilute with >1000 fold
excess unlabeled A

$E + A^* \xleftarrow{k_{off}} E{\cdot}A^* \xrightarrow{k_{cat}} E + P^*$ step 3. several catalytic turnovers
to permit bound A* to react

$E + P \longleftarrow E{\cdot}A$ step 4. stop reaction and
measure A* and P* to
establish k_{cat}/k_{off}

purine nucleoside phosphorylase: commitment of bound inosine

29.7 µM
$E{\cdot}[8\text{-}^{14}C]\text{inosine}$

excess inosine
and variable PO_4

$E{\cdot}PO_4 \xleftarrow[9.3\ \mu M]{k_{off}} E{\cdot}[8\text{-}^{14}C]\text{inosine}{\cdot}PO_4 \xrightarrow[20.4\ \mu M]{k_{cat}} [8\text{-}^{14}C]\text{hypoxanthine}$

forward commitment = $C_f = k_{cat}/k_{off} = 2.2$

FIG. 11. Experimental measurement of substrate commitment factor [I. A. Rose, *Methods Enzymol.* **249**, 315 (1995)] for purine nucleoside phosphorylase (PNP) [P. C. Kline and V. L. Schramm, *Biochemistry* **32**, 13212 (1993)]. In step 1, 30 µM PNP was incubated for 10 sec with 400 µM [8-^{14}C]inosine in 50 mM triethanolamine, pH 7.5, in a volume of 20 µl. In step 2, 1 ml was added containing 5 mM inosine and variable concentrations of phosphate (100, 125, 200, 2000, and 4000 µM). Following 15 sec, the reactions were quenched with 100 µl 1 N HCl and analyzed by HPLC (C-18 reverse phase eluted with 5% methanol in H$_2$O) to establish the fraction of bound substrate converted to product (ribose 1-phosphate). The commitment factor is determined at saturating phosphate concentration.

zymes with two or more substrates, bound randomly, are analyzed easily by this approach, as the binary enzyme–substrate complex forms but does not react until the second substrate is added together with the large excess of unlabeled first substrate. Purine nucleoside phosphorylase demonstrated large forward commitment with bound inosine when phosphate was added, and the KIE for the phosphorolysis reaction were too small to be measured accurately (Fig. 11).[65,66]

Single substrate enzymes begin reaction as soon as the enzyme and labeled substrate are mixed, and the label/dilution steps must therefore occur on the time scale of the catalytic turnover rate. Measurement of forward commitment under these conditions requires rapid mix, chemical quench approaches. These methods have been used to determine forward commitments for NAD^+ in CD-38 and for ATP in S-adenosylmethionine synthetase.[55,67]

Forward commitment is the most common factor to obscure intrinsic kinetic isotope effects in reactions with an irreversible chemical step, e.g., in hydrolytic reactions with large negative free energy values. In enzymes where the conversion of enzyme-bound products to bound substrates does not occur, Eq. (1) reduces to $^xV/K = (^xk + C_f)/(1 + C_f)$. The observed isotope effects vary between intrinsic (^xk) and unity as forward commitment increases from negligable to dominant. Measurement of the chemical isotope effect and the forward commitment is sufficient to establish the nature of the intrinsic KIE unless the experimental isotope effect is so near unity that corrections are numerically unsound. In KIE measurements, the assumption is made that binding isotope effects are negligible. Binding isotope effects are discussed in more detail later.

Reverse Commitment

Reactions that are energetically reversible often exhibit reverse commitment [C_r in Eq. (1)] in which the enzyme-bound products react to reform enzyme-bound substrate before product release. Reverse commitment can also occur in reactions with irreversible overall reactions provided that the equilibrium between enzyme-bound substrates and products permits reversal. The equilibrium constant for enzyme-bound substrates and products is often nearer unity than the chemical reaction, thus reverse commitment must be considered for all reactions. Reverse commitment alone obscures the intrinsic KIE by the expression

$$^xV/K = (^xk + C_r \cdot {}^xK_{eq})/(1 + C_r)$$

where ^xk is the intrinsic isotope effect. The value of the observed KIE, $^xV/K$, can vary between the intrinsic KIE and the $^xK_{eq}$. Reverse commitments for chemically reversible reactions can be measured with the same tech-

[65] P. C. Kline and V. L. Schramm, *Biochemistry* **32,** 13212 (1993).
[66] P. C. Kline and V. L. Schramm, *Biochemistry* **34,** 1153 (1995).
[67] G. D. Markham, D. W. Parkin, F. Mentch, and V. L. Schramm, *J. Biol. Chem.* **262,** 5609 (1987).

niques as forward commitment, using the substrates for the reverse reaction. If present, the $^x K_{eq}$ must also be established to interpret the intrinsic KIE. Methods to measure $^x K_{eq}$ are simple and have been outlined previously in this series.[1] With experiments to measure both forward and reverse commitments, it has become experimentally feasible to measure intrinsic isotope effects for many enzymes.

Enzymes with Large Commitment Factors

Substrates in the Michaelis complex of enzymes with large forward and large reverse commitment factors are converted efficiently to enzyme-bound products. Bound products react to form bound substrates with a high probability relative to the release as free products. Large commitment factors are independent of the chemical thermodynamics of the reaction because the equilibrium constants for bound substrates may vary considerably from that in solution. This phenomenon has been well established by the NMR measurement of the equilibrium concentrations of substrates and products bound at catalytic sites.[68] Thus reverse commitment can occur in enzymes with freely reversible and in experimentally irreversible reactions. If the experimental KIE are near unity, even accurate measurement of the commitment factors does not permit an accurate analysis of the intrinsic isotope effect, as the correction factor is the product of the observed isotope effect and the correction factors [Eq. (1)]. An experimental solution for large commitments is to measure pre-steady-state kinetic isotope effects on partial reactions.

Overcoming Large Commitment Factors

A highly committed reaction coordinate diagram is shown in Fig. 12 and will be used to illustrate pre-steady-state reaction methods to overcome large commitments, even when the observed KIE are unity in the steady state. The reaction is initiated from a complex of enzyme with appropriately labeled substrates (E*A). In a two substrate reaction, this complex is nonreactive, and the reaction is initiated by rapid mixing with sufficient B to form the E*AB complex. The reaction is permitted to proceed until approximately 30% of the E*AB complex is converted to the E*PQ complex. This mixture is quenched chemically or physically before a significant fraction of E*PQ reverses to E*AB or dissociates to products. Under these conditions the KIE arises from isotopic fractionation of the E*AB complex, which contains a mixture of *A with the appropriate isotopic labels. Both forward and reverse commitments are avoided in this method, which uses

---

[68] M. Cohn and G. H. Reed, *Annu. Rev. Biochem.* **51**, 365 (1982).

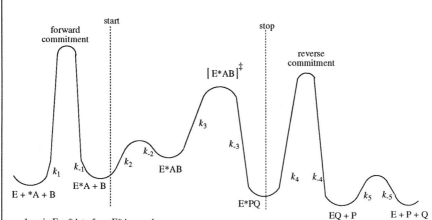

1. mix E + *A to form E*A complex
2. add B in large excess
3. quench reaction when ~30% E*AB is converted to E*PQ and before EQ + P forms
4. KIE is measured from isotopic ratio released from the E*PQ complex relative to *A
   following 100% conversion to *P

reaction coordinate

FIG. 12. Pre-steady-state analysis to circumvent commitment factors. The reaction coordinate diagram illustrates the unfavorable case of strong forward and reverse commitments in a bisubstrate enzymatic reaction. KIE for the chemical step can be measured by forming the E*A complex (where A indicates the mix of isotopically labeled substrates). Assuming no binding isotope effect on k_1 (usually diffusion controlled), the E*AB complex is formed by mixing with B. After a small fraction (~20%) is converted to E*PQ, the reaction is quenched. The product is analyzed in the same way as for normal KIE. Analysis of KIE for a series of reactions with conversions to E*PQ from 10 to 40% of E*AB can be used to extrapolate to the KIE at 0% conversion, the intrinsic KIE for the chemical step.

the enzyme as a stoichiometric reactant rather than a steady-state catalyst. This method measures specific steps in the reaction coordinate and contrasts to steady-state measurements where many steps in the cycle contribute to the reaction rate and therefore to the V/K KIE. The pre-steady-state approach to enzymatic TS analysis is certain to be widely applicable in complex enzymatic reactions, but only a few examples have been reported. The approach requires relatively large amounts of enzyme, knowledge of the reaction coordinate barriers, and sufficient knowledge of the reaction rates to stop the reaction when the desired fraction of labeled substrate is converted to enzyme-bound product. This approach has been used to determine the TS structure for the hydrolysis of inosine by purine nucleoside phosphorylase, where reverse commitment and rate-limiting product release obscured intrinsic KIE.[69] Single substrate enzymes can therefore be analyzed by partial catalytic turnover techniques to avoid the problem of reverse commitment. Transient KIE have been measured most often with dehydrogenases, where deuterium isotope effects and the easily measured spectral changes make the approach most accessible. The relatively limited number of studies available at that time were summarized in 1991.[70]

Binding Isotope Effects

It has been assumed that binding equilibrium isotope effects are small relative to the magnitude of KIE and have been disregarded in most studies of enzymatic KIE.[71] However, it is apparent from inspection of the reaction coordinate diagram that steady-state KIE studies are influenced by binding isotope effects arising from the interactions of substrate and enzyme in the Michaelis complex (e.g., Fig. 12; competitive KIE report on the isotopic responses for every rate constant included in the V/K expression for the specific mechanism). Because kinetic constants in this term include every step between free substrate and the first irreversible step, V/K KIE always include substrate binding effects. Equilibrium binding isotope effects between free and bound substrate (E + *A ↔ E*A) occur if the enzyme-induced distortion of the substrate is sufficient to cause bond-vibration differences at the atomic positions labeled with isotopes for the KIE experiments. Binding isotope effects are well documented in a small number of enzymatic reactions. For example, the binding isotope effect for [4-^{3}H]NAD$^+$ has been measured to be approximately 1.10 (10%).[72] Although an isotope effect of 10% is measured easily, the intrinsic KIE for hydride transfer reactions relative to tritium ion transfer in dehydrogenase reactions are much larger, with an upper limit near 13 (1300%), and are commonly found to be 3–6 (300–600%). For these reactions, the assumption that binding isotope effects are negligible is valid.[71] Reactions with C–C, C–N, C–S, and C–O bond scission at the TS typically cause small isotope effects of 1–13% for the primary ^{14}C atom and α and β secondary ^{3}H effects from 1 to 25%, depending on the nature of the TS. In these cases, a ^{3}H-binding isotope effect of 10% would be a major correction required in the calculation of the intrinsic KIE.

Experimental binding isotope effects can be measured by binding equilibrium experiments. Ultrafiltration or equilibrium dialysis followed by the

[69] P. C. Kline and V. L. Schramm, *Biochemistry* **34**, 1153 (1995).
[70] J. T. McFarland, *in* "Enzyme Mechanism from Isotope Effects" (P. F. Cook, ed.), p. 151. CRC Press, Boca Raton, FL, 1991.
[71] D. Northrop, *J. Chem. Ed.* **75**, 1153 (1998).
[72] R. LaReau, W. Wah, and V. Anderson, *Biochemistry* **28**, 3619 (1989).

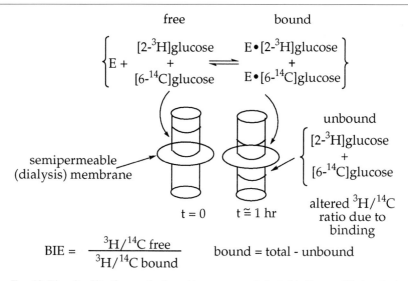

FIG. 13. The ultrafiltration method used to measure substrate-binding equilibrium isotope effects (B. Lewis and V. L. Schramm, unpublished observations, 1998). A solution (100 μl is typical) of buffered enzyme containing the appropriate ratio of labeled substrates (glucose in the example) is placed above an ultrafiltration membrane that is freely permeable to ligand but retains the enzyme. The conditions are selected to permit approximately 50% of the total ligand to bind. The upper chamber is placed under N_2 pressure, approximately 50% of the volume is forced through the membrane, and the bottom chamber is sampled for unbound ^3H and ^{14}C substrate. This is compared to the isotopic label in the original mixture. In the example shown here, the ^{14}C is the control as no equilibrium-binding isotope effect is expected with the carbon. Control experiments move the ^{14}C to other positions, keeping the ^3H at the 2 position to demonstrate that the binding isotope effect is independent of the placement for the remote (^{14}C) label.

analysis of bound and free isotopically labeled substrates provides a simple and direct measurement of isotope effects (Fig. 13). In ordered kinetic reaction mechanisms, trapping of the first substrate by extrapolation to a saturating concentration of the second has also been used to analyze binding isotope effects.[73] This method does not provide binding equilibrium effects because return over the substrate binding barrier (k_{-1} of Fig. 12) is prevented by saturation with the second substrate. The method therefore provides the KIE for the bimolecular formation (k_1 of Fig. 12) of the Michaelis complex (the transition state for substrate binding). Binding isotope effects are not subject to commitment factors or any other obscuring effects. Recent experiments have tested the hypothesis that substrate bind-

[73] E. Gawlita, W. S. Caldwell, M. H. O'Leary, P. Paneth, and V. E. Anderson, *Biochemistry* **34**, 2577 (1995).

ing to hexokinase causes binding isotope effects in the Michaelis complex of enzyme-glucose. Binding of specific [^{3}H]glucose substrates indicates that the remote ^{3}H-binding isotope effects can be as large as 9%.[74] These early results suggest that binding isotope effects may need to be considered in obtaining accurate intrinsic KIE in many enzymatic reactions. The results of isotope-edited difference Raman spectroscopy (Chapter 8 of this volume) lead to the same conclusion.[75] The Raman shifts observed in several Michaelis complexes indicate altered bond vibrational frequencies that correspond to significant binding isotope effects. It is expected that data from Raman or infrared-red spectroscopy and binding isotope effects will be complementary, as the physical basis of all three effects is the change in bond vibrational modes for the atoms of interest when free and when bound to the enzyme. In those cases where binding isotope effects are demonstrated, additional terms can be incorporated into the intrinsic KIE expression [Eq. (1)] to correct for the binding isotope effects.

Equation (2) indicates that substrate binding isotope effects will influence the calculation of the intrinsic binding isotope effects both by the equilibrium effect and by isotopic discrimination at the transition state for binding (the effect on k_1 discussed earlier)[74]:

$$^xV/K = (^xk^xK_a + {}^xk_1C_f + C_r{}^xK_{eq}{}^xK_a)/(1 + C_f + C_r) \tag{2}$$

In addition to the terms described in Eq. (1), ^xK_a is the substrate binding equilibrium isotope effect and $^xK_1$ is the effect on the initial binding of the substrate. Each of the terms in Eq. (2) can be evaluated experimentally. The binding expressions for a variety of mechanisms and the experimental impact of binding isotope effects require further evaluation.

Summary of Commitment Factors and Their Influence on KIE and Transition-State Analysis

The commitment and binding problems discussed earlier can be overcome experimentally in most enzymes. Enzymatic reactions with multiple chemical steps where the energies are approximately the same require pre-steady-state KIE approaches. Similar to other technological advances, there has been a period of experimental development to establish the techniques that permit the measurement of intrinsic KIE for enzymes of ever-increasing mechanistic complexity. Experimental and computational methods are now available to overcome most of the commitment factor barriers imposed by slow enzymatic steps and binding isotope effects for a broad range of important and interesting enzymatic reactions.

[74] B. Lewis and V. Schramm, unpublished results (1998).
[75] R. Callender and H. Deng, *Annu. Rev. Biophys. Biomol. Struct.* **23,** 215 (1994).

Matching of Intrinsic KIE to Transition-State Structures

The relative atomic vibrational environments in the reactant and TS molecules are the major factors that determine the magnitude of the intrinsic KIE. Conversely, the intrinsic KIE for an enzymatic reaction permits reconstruction of the atomic vibrational environment of the TS if sufficient isotope effects have been measured. This section provides brief examples of three experimentally determined intrinsic KIE and the steps that lead to a construction of a TS structure for the enzyme-stabilized TS. The steps outlined here were developed with computationally modest resources and are semiquantitative estimates of the TS structures. This approach systematically compares many possible TS structures in reaction coordinate space and assumes a linear relationship among bond orders, angles, and changes in atomic rehybridization.[67,76] Despite the semiquantitative approach, the TS methods have captured the major features that are altered between substrate and TS and give rise to the intrinsic KIE. Article [14] provides the most recent evolution of the systematic computational approach. A systematic quantum chemical computational approach matches intrinsic KIE to atomic structures of the TS. This new approach incorporates nonlinear features of bond order and geometric changes at the TS. Transition-state imbalance is a relatively recently recognized property of enzymatic TS, and future applications of TS determination will incorporate this feature of chemical bond changes. Coupling these methods to a systematic search of reaction coordinate space improves selection of the TS that is most consistent with intrinsic KIE.[7]

AMP Deaminase

Determination of the TS structure for AMP deaminase had the purpose of comparing the nature of the experimentally determined TS with the structure of coformycin 5'-phosphate. Coformycin is a natural product TS inhibitor that binds 10^7 more tightly than substrate to adenosine deaminase, and the 5'-phosphate shares an equal affinity for AMP deaminase. Deamination of AMP occurs by the enzymatic activation of a water nucleophile that attacks C-6 of the purine ring (Fig. 14).[77] The metalloenzyme contains a tightly bound Zn^{2+} that chelates the water and lowers the pK_a to create a catalytic site hydroxide ion. The catalytic mechanism is common to adenosine deaminase and AMP deaminase, as all amino acids in contact with the Zn^{2+} and purine ring are conserved in both amino acid sequences.[78] The

[76] F. Mentch, D. W. Parkin, and V. L. Schramm, *Biochemistry* **26,** 921 (1987).
[77] D. J. Merkler, P. C. Kline, P. Weiss, and V. L. Schramm, *Biochemistry* **32,** 12993 (1993).
[78] D. J. Merkler and V. L. Schramm, *Biochemistry* **32,** 5792 (1993).

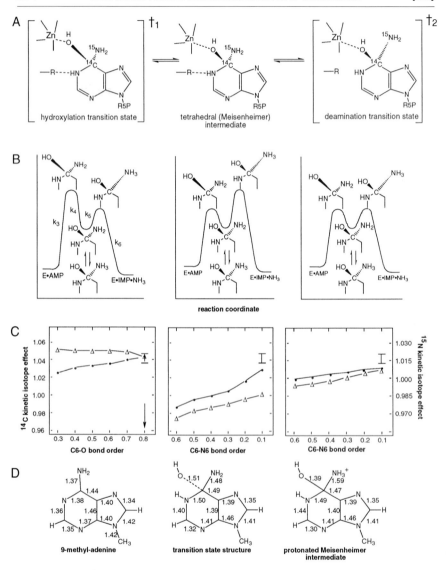

FIG. 14. Reaction coordinate, KIE, and transition-state structure for yeast AMP deaminase. (A) The three steps in the aromatic nucleophilic substitution to deaminate NH$_3$ from AMP. (B) Possible reaction coordinate diagrams with rate-limiting OH attack (left), rate-limiting NH$_3$ departure (middle), and equivalent rates for OH attack and NH$_3$ departure (right). (C) The KIE expected for each case is plotted as a function of attacking OH bond order (left, C-6–C bond order) or departing NH$_3$ bond order (middle and right, C-6–N-6 bond order). Superimposed on the predicted KIE as a function of bond order are the actual KIE values for ^{14}C and ^{15}N. Ordinate and abscissa scales have been adjusted so that both ^{14}C and ^{15}N

sequence homology outside the catalytic site is limited, and AMP deaminase is more than double the size of adenosine deaminase (352 and 810 amino acids, respectively) because of the presence of allosteric activation and inhibition regulatory domains.[79] The hydroxide oxygen reacts and displaces the amino-leaving group to convert the adenine ring to hypoxanthine. Nucleophilic substitutions at an aromatic ring are well known from chemical precedent and involve the formation of an unstable intermediate (the Meisenheimer complex) with the carbon under nucleophilic attack being converted from sp^2 to sp^3 hybridization, followed by protonation of the leaving group and its elimination, regenerating the sp^2 center.[80] The natural product inhibitor, coformycin, resembles this intermediate, thus it was proposed that the TS complex must be closely related to the Meisenheimer complex.[81]

Intrinsic KIE were established from the observations that (1) the KIE were the same for AMP over a wide range of V/K values with allosteric activator, allosteric inhibitor, and alternative slow substrates and (2) the [^{14}C]KIE was near the theoretical limit for any possible degree of bond formation to the attacking oxygen nucleophile.[77] The TS structure was determined from the KIE summarized in Fig. 14. In the first model of the TS, the extent of oxygen nucleophile to C-6 bond order was varied systematically and fixed at values of 0.3 to 0.8 with a full bond retained to the NH_2, which was fixed as the NH_2 and not permitted to become NH_3. The hybridization of C-6 was converted incrementally from sp^2 to sp^3 as the C-6–oxygen bond was formed. With fixed values for each C-6–O bond distance, the remainder of the molecule was minimized using MOPAC 6.0. The optimized structure from MOPAC 6.0 was used to establish the KIE that each TS structure would be expected to generate using BEBOVIB-IV. This method introduces only one variable, the distance to the nucleophile, with the remainder of the structure assigned by computational optimization. Two additional TS models were generated by fixing C-6 as an sp^3

[79] S. L. Meyer, K. L. Kvalnes-Krick, and V. L. Schramm, *Biochemistry* **28,** 8734 (1989).
[80] V. L. Schramm and C. Bagdassarian, *in* "Comprehensive Natural Product Chemistry" (C. D. Poulter, ed.), Vol. 5, pp. 71–100. Pergamon Press, Elsevier Science, New York, 1999.
[81] P. C. Kline and V. L. Schramm, *Biochemistry* **34,** 1153 (1994).

KIE are represented by the same value, shown with their standard error bars. The only reaction coordinate diagram consistent with the intrinsic KIE is rate-limiting OH attack followed by rapid protonation and departure of the NH_3. Agreement between experimental and calculated KIE was best at a C-6–O bond order of 0.8, indicated by the arrow. (D) 9-Methyl truncations of the substrate, transition state, and protonated Meisenheimer complexes. The bond lengths are shown for the optimized structures.

center with fully bonded oxygen and varying the bond order to the leaving group NH_3. Finally, a mechanism was tested in which two equal TS, one for hydroxide oxygen attack and one for NH_3 departure, contributed equally in two distinct TS separated by the Meisenheimer intermediate. The isotope effects calculated for each of these TS models were compared to the experimental isotope effects for [14]C-6 and for [15]N-6 as the leaving group. Only the TS with the attacking hydroxide ion and fully bonded NH_2 provided agreement with the experimental KIE and agreement was consistent with a bond order of 0.8 to the attacking oxygen. This TS is late in the nucleophilic attack portion of an $A_N + D_N$ mechanism and occurs just before formation of the Meisenheimer intermediate. Bond lengths for the MOPAC-6 optimized structures of the reactant, TS, and Meisenheimer complex for AMP deaminase are shown in Fig. 14.

Nucleoside Hydrolase

Nucleoside hydrolases are enzymes found in protozoan parasites, which contribute to the essential pathway of purine salvage in these purine auxotrophs.[82] The enzymes produce purine or pyrimidine bases and ribose and are considered targets for novel inhibitor design as mammals have no nucleoside hydrolases. The nonspecific nucleoside hydrolase from *Crithidia fasciculata* was used to meaure the KIE for labeled inosine substrates synthesized as outlined in Figs. 2 and 3. In addition to extensive kinetic studies, the KIE for each isotopically labeled position indicated that the enzyme expressed intrinsic KIE (Fig. 15).[33] A match between experimentally measured intrinsic KIE and a TS structure was obtained by the systematic variation of the C-1'–N-9 bond order with varying degrees of nucleophilic participation by the attacking water nucleophile. The reactant-state structure for inosine was taken from the crystal structure with the bond lengths for C–H bonds obtained by AMPAC calculations using the AM1 parameter set.[83] The initial structure for the TS ribosyl was the X-ray crystal structure coordinates for ribonolactone. An imidazole was used as the leaving group, and the bond changes, which characterize the leaving group, were estimated by AMPAC calculations as the C-1'–N-9 bond was broken. The reaction coordinate was generated by coupling the stretching motion of the breaking C-1'–N-9 bond with the C-1'–oxygen nucleophile bond. Inversion of configuration at C-1' was incorporated into the TS model with

[82] D. W. Parkin, B. A. Horenstein, D. R. Abdulah, B. Estupiñán, and V. L. Schramm, *J. Biol. Chem.* **266**, 20658 (1991).
[83] M. J. S. Dewar, E. F. Zoebisch, J. J. P. Healy, and J. Stewart, *J. Am. Chem. Soc.* **107**, 3902 (1985).

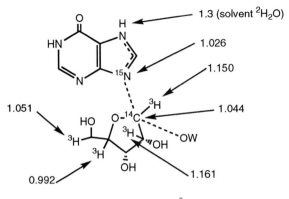

FIG. 15. Intrinsic KIE and transition-state bond lengths for the nucleoside hydrolase reaction. Arrows indicate the positions at which the intrinsic KIE were measured. Isotopic labels are indicated in the molecule, however; in practice each KIE was measured with the isotopic pairs as indicated in Table V. The table indicates bond lengths in substrate and transition state.

	bond lengths in Å	
bond	inosine	transition state
C1'-N9	1.477	1.97 ± 0.14
C1'-O4'	1.417	1.30 ± 0.05
C1'-C2'	1.530	1.502 ± 0.001
C2'-H2'	1.128	1.160 ± 0.001
N9-C8	1.372	1.332 ± 0.006
C8-N7	1.308	1.346 ± 0.006
C1'-WO	---	3.0 ± 0.4

an interaction force constant of ±0.05 between the stretching modes for the oxygen nucleophile–C-1' and C1'–N-9 with the angle bending modes for all bond angles with C1' at the center (six bond angles of the type OW–C-1'–H-1' and N-9–C-1'–O-4'). Reaction coordinates for the TS were compatible with the experimental KIE only in TS structures that had low bond order in the C-1'–N-9 bond and low but significant bond order to OW, the attacking nucleophile. The match between hypothetical TS structures and experimental KIE employed the BEBOVIB program restricted to 25 atoms for both reactant and TS models. Truncations used in the calculations included every atom within two bonds of the C-1'–N-9 bond, which defines the reaction coordinate. It is generally assumed that KIE (perturbations in bond vibrational environments) will not be significant at a distance of more than two atoms away from the site of bond scission at the TS.[10]

Calculated KIE for atoms of the labeled substrate as a function of the nature of the TS are shown in Fig. 16. The experimental KIE are compared to calculated KIE for dissociated (S_N1) TS, nucleophilic (S_N2) TS, and a mixed TS with weak participation of both the attacking nucleophile and the significant bond order to the leaving group. Matches of the calculated KIE and the experimental values indicate agreement of the mixed TS. However, the 5.1% KIE at the 5'-^{3}H position was unpredicted because it is four bonds away from the reaction center, and any reasonable chemical model of the TS would not perturb this bond vibrational environment. The substrate specificity of nucleoside hydrolase indicated that the 5'-hydroxyl

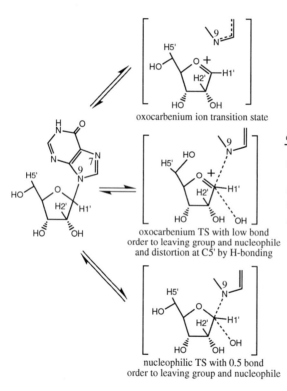

expected KIE

^{15}N9	1.04
^{3}H1'	1.35
^{14}C1'	1.00
^{3}H2'	1.09*
^{3}H5'	1.00

*dependent on dihedral angle N9-C1'-C2'-H2'

oxocarbenium ion transition state

expected KIE		observed KIE	
^{15}N9	1.03	^{15}N9	1.04
^{3}H1'	1.15	^{3}H1'	1.15
^{14}C1'	1.04	^{14}C1'	1.04
^{3}H2'	1.09	^{3}H2'	1.16
^{3}H5'	1.00*	^{3}H5'	1.05*

*KIE unexpected at this position unless enzyme-specific sp^3 distortion occurs here

oxocarbenium TS with low bond order to leaving group and nucleophile and distortion at C5' by H-bonding

expected KIE

^{15}N9	1.01
^{3}H1'	1.00
^{14}C1'	1.12*
^{3}H2'	1.00
^{3}H5'	1.00

*KIE dominated by C1' reaction coordinate motion

nucleophilic TS with 0.5 bond order to leaving group and nucleophile

FIG. 16. Calculated KIE for three potential reaction mechanisms compared to an intrinsic KIE data set. The upper and lower TS structures represent the extremes of dissociative and associative mechanisms calculated for the gas phase. The expected KIE were calculated from BEBOVIB calculations [B. A. Horenstein, D. W. Parkin, B. Estupiñán, and V. L. Schramm, *Biochemistry* **30**, 10788 (1991)]. The intrinsic KIE ruled out the limiting mechanisms and are in good agreement with the structure of the intermediate TS. Exact matches of the ^{3}H2' and ^{3}H5' are accommodated by the hyperconjugation and H-bonding effects specific to the enzymatic catalytic site.

is essential for catalytic activity, but not for substrate binding.[82] Together, the steady-state kinetic analysis and KIE results indicated that formation and stabilization of the TS require enzymatic interaction with the 5'-hydroxymethyl group in such a way that the sp^3 hybridization is distorted to give rise to a 5.1% normal KIE at this position. Normal isotope effects indicate increased vibrational freedom for the isotopic atom in the TS relative to the reference state. The most likely TS structure that could account for the remote KIE was obtained by rotation of the 5'-hydroxyl group to form a dihedral angle between O-5'–C-5'–C-4'–O-4' that locates the 5'-O near the ring oxygen, O-4'. In this conformation, the lone pair of O-5' is directed toward the ring oxygen, the site of ribooxocarbenium cation formation at the TS. A normal hydrogen bond to the hydrogen of the 5'-hydroxyl group in this geometry can easily account for the 5.1% KIE measured experimentally. This unexpected and unprecedented KIE predicted the unusual nucleoside TS configuration that was later established to have a dihedral angle of 336° when the X-ray crystal structure was solved with a transition-state inhibitor.[84] This geometry is unknown in nucleoside and nucleotide chemistry, where the average dihedral angle at this position is 233°.[85] The TS structure, which is in agreement with all of the experimental KIE, is indicated in Figs. 15 and 16. The C-1'–N-9 ribosidic bond is nearly broken at 1.97 ± 0.14 Å, and the attacking water nucleophile oxygen is at 3.0 ± 0.4 Å, in van der Waals contact at the TS. Loss of the bonding electrons without compensation from the attacking nucleophile leaves the ribosyl group with an electron deficiency and nearly a full positive charge. This introduces double bond character into the ribosyl ring, primarily between C-1' and O-4', the site of the ribooxocarbenium cation. Departure of the bonding electrons into the leaving group gives it a negative charge, and neutralization of the charge is accomplished by the protonation of N-7 at the TS. Because the N-ribosidic bond is not completely broken at the TS, it is conceivable that the protonation of N-7 is in progress at the TS. To test this hypothesis, the solvent deuterium KIE was measured to establish the role N-7 protonation plays in leaving group activation. The solvent deuterium KIE is small, 1.3 for k_{cat} and 0.99 for k_{cat}/K_m, whereas a TS that demonstrates proton transfer at the TS gives a KIE >2.0.[86] It was concluded that proton transfer to N-7 is complete at the TS, as the [15]N-9 KIE required imidazole bonding consistent with N-7 protonation.

Atomic structures for the truncated molecules used in the BEBOVIB

[84] M. Degano, S. C. Almo, J. C. Sacchettini, and V. L. Schramm, *Biochemistry* **37**, 6277 (1998).
[85] A. Gelbin, B. Schneider, L. Clowney, S. H. Hsieh, W. K. Olson, and H. M. Berman, *J. Am. Chem. Soc.* **118**, 519 (1996).
[86] D. M. Quinn and L. D. Sutton, *in* "Enzyme Mechanism from Isotope Effects" (P. F. Cook, ed.), p. 72. CRC Press, Boca Raton, FL, 1991.

analysis were converted into the full atomic structures. Atoms for the substrate inosine were taken from the X-ray crystal structure with the hydrogens positioned using Gaussian 92 with the STO-3G basis set.[87] The TS state structure was computed with the 3'-exo configuration of the ribose, as required by the β secondary ^{3}H KIE from 2'-^{3}H. All bonds defined in the final BEBOVIB-TS model (the structure that recreates the experimentally observed intrinsic KIE) were fixed, and the positions of the remaining bonds were minimized energetically with Gaussian 92 and the STO-3G basis set. The enzyme-bound product α-D-ribose was modeled as the 3-exo conformation based on its relationship to the structure of the TS and least-motion principles to achieve the product. Likewise, the enzyme-bound hypoxanthine product was modeled using the same computational parameters, with the site of ring protonation at N-7 corresponding to that of the TS rather than the preferred protonation at N-9 as occurs in solution. These parameters were selected to preserve the features of the TS imprinted onto the products in the next few bond vibrations that occur following TS separation and before release from the catalytic site. This approach permitted analysis of substrate, TS, and products to provide an electronic model of the reaction coordinate pathway.[88]

Purine Nucleoside Phosphorylase

The phosphorolysis of inosine catalyzed by purine nucleoside phosphorylase (PNP) plays a unique role in the development of the human immune system, as infants with a genetic deficiency of the enzyme develop a specific and ultimately fatal T-cell deficiency within 1 to 2 years of birth.[89] This physiological role has made PNP the target of inhibitor discovery efforts for agents to modulate T-cell disorders and malignancies of T-cell origin.[90,91] The PNP reaction was selected for TS analysis in an effort to apply TS practice to the design of TS inhibitors.

[87] M. J. Frisch, G. W. Trucks, M. Hend-Gordon, P. M. W. Gill, M. W. Wong, J. B. Foresman, B. G. Johnson, H. B. Schlegel, M. A. Robb, E. S. Replogle, R. Gomperts, J. L. Andres, K. Raghavachari, J. S. Binkley, C. Gonzalez, R. L. Martin, D. J. Fox, D. J. DeFrees, J. Baker, J. J. P. Stewart, and J. A. Pople, "Gaussian 92 User's Guide." Gaussian Inc., Pittsburgh, 1992.

[88] B. A. Horenstein and V. L. Schramm, *Biochemistry* **32,** 7089 (1993).

[89] M. S. Hershfield and B. S. Mitchell, *in* "The Metabolic Basis of Inherited Disease" (C. R. Scriber, A. L. Beaudet, W. S. Sly, and D. Valle, eds.), 7th ed., p. 1725. McGraw-Hill, New York, 1995.

[90] S. Niwas, P. Chand, V. P. Pathak, and J. A. Montgomery, *J. Med. Chem.* **37,** 2477 (1994).

[91] G. M. Walsh, N. S. Reddy, S. Bantia, Y. S. Babu, and J. A. Montgomery, *Hematol. Rev.* **8,** 87 (1994).

Kinetic isotope effects for the reaction using the same labeled inosines as for nucleoside hydrolase were near unity and could not be corrected to reliable values because of large forward and reverse commitment factors, the most unfavorable of the commitment cases described earlier (Table V). The problem of reverse commitment was solved by replacing phosphate with its analog arsenate, which reacts at a rate similar to phosphate in the PNP reaction, but forms α-D-ribose-1-arsenate as the chemically unstable product.[65] This product hydrolyzes to form ribose and arsenate with a rate constant greater than the reverse reaction and thus prevents the reverse reaction from occurring. The forward commitment factor was determined experimentally using the method of inosine substrate trapping described in Fig. 11. The commitment factor permits the calculation of intrinsic KIE from the experimental KIE measured in the arsenolysis reaction. The experimentially measured and the intrinsic KIE for the arsenolysis reaction (Table V) were used to generate a truncated structure of the TS with the bond lengths around the N-9–C-1′ bonds defined (Fig. 17). A comparison of the bond lengths for the reactant inosine and the TS indicates that the C-1′–N-9 bond retains approximately 0.3 bond order at the TS whereas the attacking arsenate anion is 3.01 Å away, in weak van der Waals contact. Although the arsenate is not bonded, the presence of arsenate at this distance is required to match the TS structure to the intrinsic KIE. The significance of these low bond orders in TS stabilization and on the values of the intrinsic KIE is discussed in more detail in article [14].[7] The residual 0.3 bond order to C-1′ indicates that the net charge on the ribooxocarbenium ion is near 0.7 and is apparent in the shortened C-1′–O-4′ bond from 1.42 to 1.29 Å. The value of the ^{15}N-9 KIE is consistent with the protonation of N-7 at the TS as would be expected for the charge separation leading to TS formation. These features are informative for TS inhibitor design, e.g., the weak interaction with the attacking anion nucleophile establishes

TABLE V
KIE FOR ARSENOLYSIS BY PNP

Substrate	Type KIE	Experimental KIE	Intrinsic KIE
[1′-³H]- + [5′-¹⁴C]inosine	α secondary	1.118 ± 0.003	1.141 ± 0.004
[2′-³H]- + [5′-¹⁴C]inosine	β secondary	1.128 ± 0.003	1.152 ± 0.003
[1′-¹⁴C]- + [5′-³H]inosine	Primary	1.022 ± 0.005	1.026 ± 0.006
[1′-¹⁴C]- + [4′-³H]inosine	Primary	1.020 ± 0.006	1.024 ± 0.007
[9-¹⁵N, 5′-¹⁴C]- + [5′-³H]inosine	Primary	1.009 ± 0.004	1.010 ± 0.005
[4′-³H]- + [5′-¹⁴C]inosine	γ secondary	1.007 ± 0.003	1.008 ± 0.004
[5′-³H]- + [5′-¹⁴C]inosine	δ secondary	1.028 ± 0.004	1.033 ± 0.005

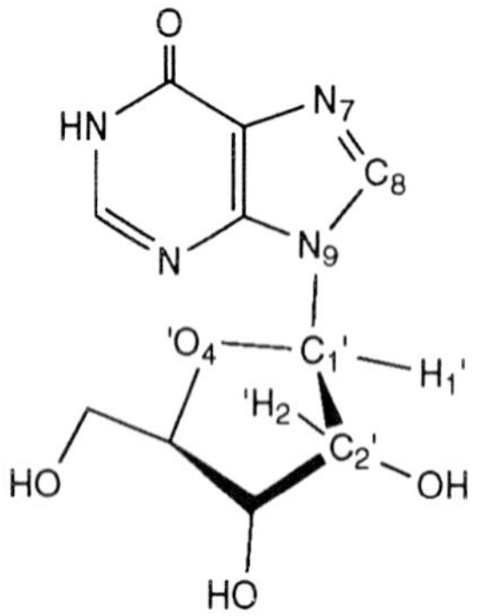

bond	Å bond length, inosine	bonds arsenolysis transition state	bonds hydrolysis transition state
C1'-N9	1.47	1.77 ± 0.04	1.90 ± 0.12
C1'-O4'	1.42	1.29 ± 0.02	1.32 ± 0.06
C1'-H1'	1.13	1.12 ± 0.01	1.12 ± 0.001
C1'-C2'	1.53	1.50 ± 0.02	1.50 ± 0.003
C2'-H2'	1.12	1.16 ± 0.01	1.16 ± 0.002
C1'-OAs	---	3.01 ± 0.02	2.96 ± 0.4
C8-N9	1.37	1.32 ± 0.02	1.33 ± 0.005
C4-N9	1.37	1.33 ± 0.02	1.39 ± 0.004
C8-N7	1.31	1.31 ± 0.02	1.34 ± 0.003

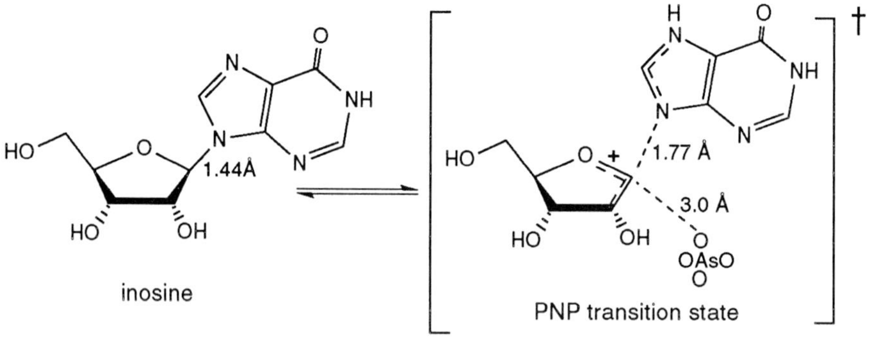

Fig. 17. Transition-state structures for arsenolysis and hydrolysis transition states of purine nucleoside phosphorylase. (Top) Bond lengths at transition states for the reactions. Errors in bond length relate to the error limits of the KIE used to establish the bond length. The transition state for hydrolysis is more advanced in the reaction coordinate than that for arsenolysis, as demonstrated by the increased C-1'–N-9 bond length for hydrolysis. The position of the arsenate or water nucleophile (indicate as C-1'–OAs) is the same for both transition states. (Bottom) Features that distinguish the substrate from the transition state.

that this group is not required as a covalent group in the design of a TS inhibitor (see later).

Summary of KIE and TS Relationships

The information content from KIE is greatest for large isotope effects, where the error is a small fraction of the experimental isotope effect. In this case, there is less ambiguity as to whether a small experimental KIE is a consequence of small bond differences at the TS or an unfavorable commitment factor. The magnitude of intrinsic KIE for three reactions for substitution at carbon are shown in Fig. 16. Kinetic isotope effects at the

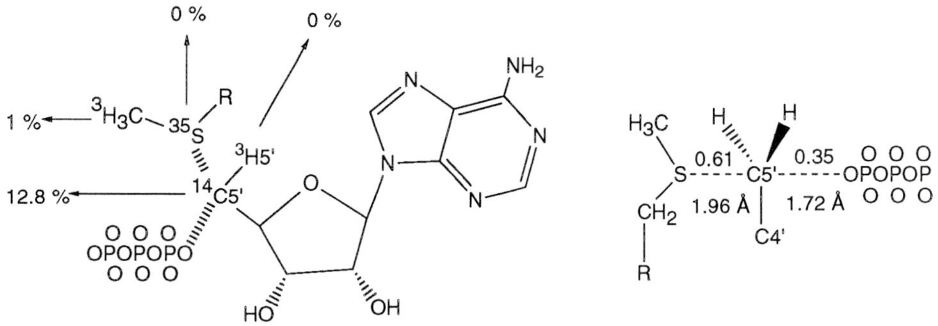

FIG. 18. Kinetic isotope effects and transition-state structure in the nucleophilic displacement catalyzed by S-adenosylmethionine synthetase. Individual intrinsic KIE are demonstrated in the structure on the left, which indicates the transition-state attack of the S of methionine on C-5′ of ATP to displace the tripolyphosphate. Isotope effects were measured individually using labeled ATP pairs, as indicated in Table I, or with labeled methionine pairs. Note the dominant isotope effect from ^{14}C-5′ at the reaction center. The bond orders (above) and bond lengths (below) at the TS are shown in the structure on the right. The mass difference between S and O causes the S–C-5′ and C-5′–O bonds to be vibrationally isoenergetic at the bond orders indicated in the figure, a characteristic of S_N2 reaction chemistry.

central atom of the reaction are dominated by reaction coordinate motion at the TS: the motion of the central atom away from the leaving group and toward the acceptor. An example of this is in nucleophilic (S_N2) substitutions at carbon, where at the TS the carbon atom is in a coordinated motion away from the leaving group and toward the attacking nucleophile. The displacment of the tripolyphosphate of ATP by a sulfur nucleophile in the S-adenosylmethionine synthetase mechanism results in a [5′-^{14}C]ATP kinetic isotope effect of 12.8% (Fig. 18).[92] This KIE is near the theoretical limit for the relative masses of ^{12}C/^{14}C and provides unambiguous proof that the TS involves symmetric nucleophilic participation of the attacking group and departure of the leaving group and that the commitment factors for the reaction are small. In this most favorable of cases, the single KIE provides nearly complete information about the bonding nature of the TS.

The reaction examples of Fig. 16 demonstrate that reactions involving hybridization changes at the TS provide strong signals in the measurement of kinetic isotope effects. The lack of isotope effects in reactions where chemistry is known to dominate the reaction coordinate can also provide useful information. At least four distinct types of TS information are available from intrinsic KIE.

[92] G. D. Markham, D. W. Parkin, F. Mentch, and V. L. Schramm, J. Biol. Chem. 262, 5609 (1987).

1. The primary ^{14}C isotope effect gives reaction coordinate motion at the TS, is large for nucleophilic displacements, is smaller in dissociative mechanism where some bond order remains to the leaving group at the TS, and is negligible in dissociative mechanisms where the TS is a fully developed carbenium ion. In this case, the reaction coordinate motion for carbon has ceased at the TS.

2. α Secondary KIE are dominated by the out of plane bending mode for the C–H bond. A change of hybridization from sp^3 to sp^2 results in increased vibrational freedom for the hydrogen atom and a resulting large normal isotope effect. The magnitude for this out of plane vibrational mode ^{3}H KIE can be calculated (in the gas phase) to be about 40% (1.40) as a limiting value; however, the experimental values are smaller, with α secondary kinetic isotope effects of 1.20 being common for dissociative mechanisms, but rarely larger than 1.3. The largest of these are observed where rehybridization to sp^2 is nearly complete at the TS, but where there is no significant participation of the attacking nucleophile. The out-of-plane C–H bending mode is sensitive to crowding by nearby heteroatoms so that even the weak participation of nucleophiles (e.g., at 3.0 Å) causes a significant change in the KIE.[76] In solution, there is chemical evidence that solvent atoms organize around a charged TS to dampen the out-of-plane motion, thereby decreasing the isotope effect relative to the gas phase.[44,93] Heavy atom α secondary isotope effects are useful in establishing bond changes at the TS. In the case of glucoside or riboside bond scission with dissociative and oxocarbenium-like TS, atoms that are neighbors to the anomeric carbon accumulate double bond character as the TS is reached. In sugars, the ring oxygen in the reactant molecule has a net bond order of 2, and in a fully developed sugar oxocarbenium ion, the oxygen is increasingly bonded to the anomeric carbon to give a net bond order approaching 3. This stiffening of the bond vibrational environment for the oxygen results in an inverse isotope effect, with the ^{18}O-labeled substrate reacting more rapidly than the ^{16}O substrate. Hydrolysis of the N-ribosidic bond to the nicotinamide of NAD$^+$ by acid-catalyzed solvolysis or by diphtheria toxin gives an inverse ^{18}O-4' KIE of 1.2%, whereas that of methyl β-glucoside is inverse 0.9%. Calculated ^{18}O isotope effects for this position can be inverse 2%.[44,45,94] Information from the ring ^{18}O KIE complements that from the primary ^{14}C and ^{15}N KIE in establishing the extent of rehybridization of the anomeric carbon at the TS. It should be apparent from this analysis that the α secondary heavy atom KIE will be negligible in nucleophilic displacements, as is demonstrated in the reaction of Fig. 18.

[93] G. W. Burton, L. B. Sims, J. C. Wilson, and A. Fry, *J. Am. Chem. Soc.* **99**, 3371 (1977).
[94] A. J. Bennet and M. L. Sinnott, *J. Am. Chem. Soc.* **108**, 7287 (1986).

3. The β secondary kinetic isotope effect provides both geometric and bonding information for the TS. Kinetic isotope effects in geometrically constrained reactants and analysis by computational chemisry have established that the dihedral angle between hydrogen isotopes β to the vacant p orbital of the leaving group determines the magnitude of the β isotope effect.[76,80,95] Hyperconjugative interations favor dihedral angles near $0°$ and $180°$ as these angles provide maximal stabilization of the electron-deficient TS of the type shown in Fig. 16 (middle panel). Of particular interest are cases in which the β kinetic isotope effect can be measured for both pro-R and pro-S stereochemistries. In this case, the relative magnitude of the KIE can be used to solve the equation

$$\ln {}^{x}H(KIE_{exp}) = \cos^2 \theta \ln[{}^{x}H(KIE_{max})] + {}^{x}H(KIE_{ind})$$

which describes the angular dependence of the β kinetic isotope effect in terms of the dihedral angle θ, the experimental KIE, the maximum KIE for a dihedral of $0°$ or $180°$, and the inductive isotope effect, which is 0.96 for ${}^{2}H$.[95] The geometric dependence of this β isotope effect is one of the most powerful tools for probing TS geometry, but has been used in a relatively few cases in enzymology.

4. Kinetic isotope effects decrease rapidly from the site of covalent bond modification at the TS. The general rule for solution or gas-phase chemistry is that the isotope effects are insignificant beyond the β position in molecules where the atoms are connected by single bonds. More remote isotope effects occur in conjugated ring systems where changes in conjugation alter bond order to groups. For example, the ionization of p-nitrophenol alcohol is influenced by the presence of ${}^{15}N$ in the nitro group, as electronic changes between alcohol and NO_2 are transmitted through the conjugated ring.[2] Remote kinetic isotope effects are rarely observed in solution chemistry in singly bonded reactants. In enzyme-catalyzed reactions, remote ${}^{3}H$ KIE are becoming common. In KIE studies with both nucleoside hydrolase and PNP, the [5'-${}^{3}H$]inosine label was designed as a remote position, because it is four bonds away from the reactive center. Nevertheless, it was discovered that this remote position causes a 5.1% (1.051) KIE in nucleoside hydrolase and a 3.3% KIE in PNP, whereas the same isotopically labeled molecule hydrolyzed by acid in solution gave small or insignificant KIE.[33,65,96] These observations forced a reexamination of remote KIE in enzyme-catalyzed reactions. It is apparent from these studies that hydrogen bonds to groups at sp^3 centers cause sufficient geometric distortion to change the out of plane modes of adjacent hydrogens. A normal energy

[95] D. E. Sunko, I. Szele, and W. J. Hehre, *J. Am. Chem. Soc.* **99,** 5000 (1977).
[96] D. W. Parkin, H. B. Leung, and V. L. Schramm, *J. Biol. Chem.* **259,** 9411 (1984).

hydrogen bond is sufficient to cause angular distortion of the tetrahedral geometry at an anchored sp^3 center and give rise to a kinetic isotope effect of 5.1% (1.051).[88] Based solely on the kinetic isotope effects at this remote center, an unprecedented TS geometry was predicted for the nucleoside hydrolase reaction. Remote $5'$-^{3}H KIE have now been documented in the hydrolysis of inosine by nucleoside hydrolases (two different enzymes), the arsenolysis and hydrolysis reactions catalyzed by PNP, and the hydrolysis of NAD^+ by cholera, diphtheria, and pertussis toxins.[14] In every case, the nonenzymatic solvolysis of the same labeled molecules has demonstrated smaller KIE at the remote position. These enzymes enforce distortion at remote positions from the chemistry of the TS.

Molecular Electrostatic Potential Surfaces of Substrates and Transition State

The molecular electrostatic potential surface of an enzymatic TS obtained from KIE was first reported in 1993 for nucleoside hydrolase.[88] As outlined in that report, the utility of the molecular electrostatic potential (MEP) permits comparison of the electron potential at all points in space for the TS and the substrate. The van der Waals surfaces of the molecules are of the most interest, as this distance from nuclei contains the electron interactions involved in H-bonding and ionic interactions, the major forces in protein–substrate interaction and TS stabilization. The catalytic potential of enzymes has been attributed to binding the TS complex more tightly than the substrate by the factor of the catalytic rate enhancement. Therefore the difference between MEP surfaces of the substrate and TS contain the answer to the tight binding of the TS. Analysis of the MEP surfaces of enzymatic TS has only been possible since enzymatic TS structures have become available from KIE investigations. The MEP has been widely used to describe the properties of stable molecules.[97–99] It represents the energy imposed on a point of unit charge positioned at the site of interrogation, a defined distance from the nuclei.[100] The procedure to obtain the MEP uses crystallographic data for the substrate and KIE data for the TS. Computational refinement is used to locate H atoms prior to calculation of the MEP for both substrate and TS.[88] The wave function for the molecule is determined in a point calculation with the bond angles and lengths fixed

[97] P. Politzer and G. Truhlar, eds., in "Chemical Applications of Atomic and Molecular Electrostatic Potentials." Plenum Press, New York, 1981.
[98] C. A. Venanzi, C. Plant, and T. J. Vananzi, *J. Med. Chem.* **35**, 1643 (1992).
[99] P. Sjoberg and P. Politzer, *J. Phys. Chem.* **94**, 3959 (1990).
[100] S. Srebrenik, H. Weinstein, and R. Pauncz, *Chem. Phys. Lett.* **20**, 419 (1973).

at the values determined from the KIE analysis. These calculations can be done at modest basis sets, e.g., STO-3G, as calculations using a range of basis sets have demonstrated that the MEP at the van der Waals surface is not strongly dependent on the basis set. For the few TS structures currently available, this level of MEP is adequate for visual representation and for the prediction of hydrogen bonding sites that differ between substrate and the TS. The MEP is determined from the wave function for the molecule using the CUBE function of Gaussian. For purposes of calculation and visualization, the van der Waals surfaces are considered to be at an electron density of 0.002 e/a_0^3.[88] Results of the CUBE function analysis were visualized with the AVS-Chemistry viewer (Advanced Visual Systems Inc. and Molecular Simulations Inc.). Since the original report of 1993, a variety of alternative molecular visualization programs have become available.

The MEP for substrate and inhibitor molecules can be compared to those of the TS following MEP analysis of the molecules. Substrates and inhibitor molecules are adjusted to be in conformations similar to that of the TS, and a single point calculation is performed to determine the wave function of each molecule using Gaussian. The MEP for all molecules is determined from the CUBE function as described earlier for the TS. These results provide quantitative data for visual and numerical comparison of the electrostatic features for substrate, inhibitors, and the TS. This relationship is now being used as a means to predict the ability of unknown molecules to bind as analogs of the TS.[101,102] Article [15] discusses the use of substrate, TS, and inhibitor MEP to train neural networks for the analysis of inhibitor binding energy.

An example of the relationship between the MEP of the TS and an established TS inhibitor is provided by the AMP deaminase study.[102] The TS for AMP deamination was characterized by the KIE and BEBOVIB analysis in Fig. 14. Substrate and TS were compared to the TS inhibitor coformycin 5'-phosphate, which is known to bind with an equilibrium dissociation constant of 10 pM relative to the substrate K_d of 0.5 mM under the same assay conditions. Thus, the inhibitor binds 5×10^7 more tightly than the substrate.[81] For purposes of MEP analysis, the ribose 5-phosphate group was truncated to a methyl group, as the sugar phosphate is common to substrate, TS, and inhibitor, but is chemically inert in the reaction. AMP deamination is a relatively simple and convenient reaction for TS analysis. The rigid purine ring system fixes neighboring atoms in the TS, and the attack of the hydroxide ion from the pro-R face of the purine ring defines

[101] C. K. Bagdassarian, B. B. Brauheim, V. L. Schramm, and S. D. Schwartz, *Int. J. Quant. Chem.* **60,** 73 (1996).
[102] C. K. Bagdassarian, V. L. Schramm, and S. D. Schwartz, *J. Am. Chem. Soc.* **118,** 8825 (1996).

the reaction coordinate quite closely. The results demonstrate a striking similarity between the MEP surfaces of the TS and the TS inhibitor (Fig. 19, upper panels; see color insert). Both the TS and the TS inhibitor are different from the MEP of the substrate. This difference defines the essential property of the TS that accounts for the 5×10^7 tighter binding of the TS analog than the substrate and the TS binding energy that leads to the facile deamination of AMP at a rate of $1200 \, sec^{-1}$.[103] This result provided evidence that electrostatic features of the transition state, when incorporated into a stable analog, result in powerful inhibition of the target enzyme.

The similarity among transition state, substrates, and transition-state analogs can be evaluated by comparing the properties of the MEP surfaces.[102] The molecules are first oriented to align the geometries of the van der Waals surfaces as closely as possible. The MEP surfaces are then compared for electrostatic similarity using an electrostatic grid at the van der Waals surfaces with typically 17 points per atom at the van der Waals surface. At each point on the TS, the corresponding point on the test molecule (substrate or inhibitor analog) is evaluated for distance and similarity of MEP. Identical molecules will score equivalent properties of geometry and MEP at every point. Close relatives of the TS will be similar for the sum of these measures, and distant relatives of the TS will be dissimilar at each point. The sum of these comparisons provides a similarity index that is scaled from a value of 1.0 for exact similarity to a value of zero for molecules with little similarity. The inhibition of AMP deaminase by TS inhibitors correlates to this measure of similarity to the TS, as do inhibitors for adenosine deaminase and AMP nucleosidase.[102] The correlation of binding energy with similarity to the TS electrostatics emphasizes that the complete molecular description is contained in the MEP at the van der Waals surface and the geometry of the molecules. These are the features used to design analogs as TS inhibitors. The TS state structure and its similarity to a TS inhibitor for purine nucleoside phosphorylase are discussed in the next section.

Designing Transition-State Inhibitors

The MEP surfaces of the TS most important for tight-binding interactions involve the formation of new H-bond donor and acceptor sites as substrate is converted to the TS. A comparison of AMP at the Michaelis complex of AMP deaminase and at the TS permits the identification of several new H-bond donor or acceptor sites available at the TS that were

[103] P. A. M. Michels, A. Polisczak, K. A. Osinga, O Misset, J. Van Beeumen, R. K. Wierenga, P. Borst, and F. R. Opperdoes, *EMBO J.* **5,** 1049 (1986).

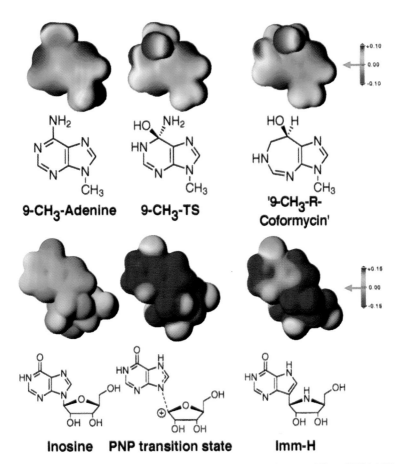

FIG. 19. Molecular electrostatic potential surfaces for the substrates, TS, and TS inhibitors for AMP deaminase and PNP. (Top) The electrostatic potential for N-9–CH$_3$ truncations of AMP, the AMP TS, and the TS inhibitor coformycin 5'-phosphate. Convential structures are shown below in the same geometry. The calibration colors are shown on the right and indicate the relative charge a unit point charge experiences at the molecular electrostatic potential surfaces corresponding to the van der Waals distance from the nuclei. The NH$_2$ group of adenine is planar with the conjugated ring. Conjugation is broken at the transition state with the attacking OH and the formation of sp^3 carbon at C-6. The NH$_2$ has not yet accepted a proton at the transition state, but now protrudes behind the ring and is rehybridized to expose the lone pair electrons. Loss of the conjugated ring results in the protonation of N-1, which changes the electrostatic potential. The TS inhibitor for AMP deaminase binds approximately 10^7 more tightly than the substrate. The electrostatic similarity between the TS and the TS-state inhibitor is apparent, as is their common difference from the substrate. The computational analysis is described in the text. (Bottom) A similar analysis for the reaction catalyzed by PNP. The ribooxocarbenium ion formed at the transition state is the site of attack by the phosphate anion. At the transition state, the phosphate oxygen is 3 Å away and is therefore not bonded. The structures are shown below without the phosphate anion. Note the similarity between the transition state and the transition-state inhibitor. This inhibitor binds approximately 10^6 more tightly than substrate.

not present in the substrate or are relocated to a new geometry. The catalytic site amino acids reposition from the Michaelis complex to form optimal contacts to the TS. Even six modest H bonds of 3 kcal/mol provide 18 kcal/mol energy for TS stabilization. Relocation of existing H-bond donor/ acceptors from the substrate to the TS can be exploited in the design of TS inhibitors. The relocation of H-bond sites in the conversion of AMP to the Meisenheimer-like TS complex of AMP is illustrated in Fig. 20. The example of AMP deaminase does not qualify as inhibitor design, as the natural product coformycin was designed by natural selection. However, it provides an example of the tight binding that occurs when the electrostatics of the inhibitor approach those of the TS. We can apply the same logic to TS solved by the methods of KIE analysis.

Purine Nucleoside Phosphorylase

   Transition-state analysis of the arsenolysis reaction catalyzed by PNP established a TS with ribooxocarbenium character and protonation of the leaving group (Fig. 17). The attacking phosphate anion lags far behind bond

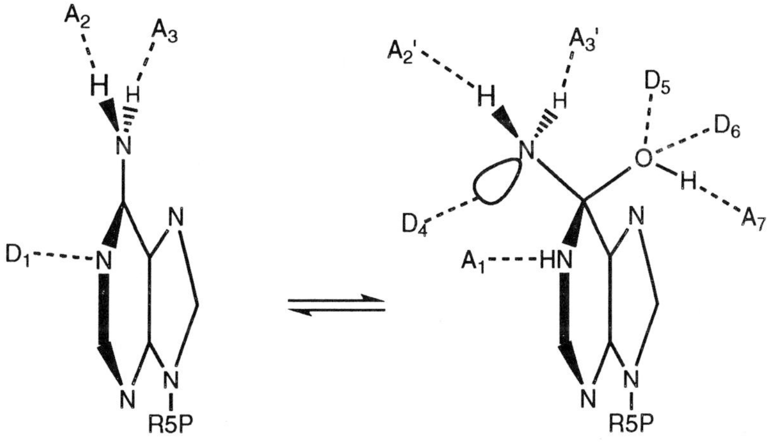

FIG. 20. An example of the new H bonds that can be formed at reacting atoms of the TS for AMP deaminase. Bonds altered in position and new sites for H-bond formation are key sites for transition-state interactions. The H-bond acceptors A_2 and A_3 are altered in geometry and H-bond donor D_1 is altered to an acceptor (A_1) at the transition state. New H-bond donor/acceptor sites are indicated by D_4-D_6 and by acceptor A_7. The sum of the ΔG for these new interactions is proposed to be the major contribution to TS formation. The crystal structure of adenosine deaminase has not yet been solved with an amino group in the position of the leaving group amino, therefore these contacts are hypothetical.

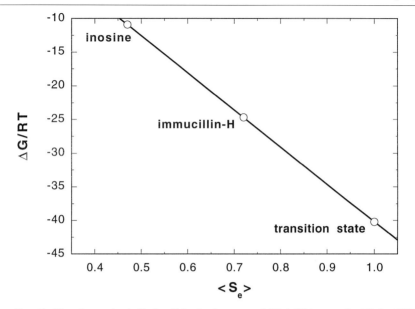

FIG. 21. The electronic similarity $\langle S_e \rangle$ of substrate and TS inhibitor to the TS for PNP [R. W. Miles, P. C. Tyler, R. H. Furneaux, C. K. Bagdassarian, and V. L. Schramm, *Biochemistry* **37,** 8615 (1998)]. The dimensionless binding free energy ($\Delta G/RT$) is determined experimentally for substrate inosine and TS inhibitor and is estimated for the transition state using the Wolfenden thermodynamic box principle for transition-state binding affinity [A. Radzicka and R. Wolfenden, *Methods Enzymol.* **249,** 284 (1995)]. The calculations for $\langle S_e \rangle$ are based on a combination of geometric and molecular electrostatic potential surface comparisons as indicated in the text. Reprinted with permission from the American Chemical Society.

breaking to the purine ring. Two major features of the purine nucleoside at the TS are the positive charge of the ribooxyocarbenium ion and the protonation of the leaving group. KIE analysis implicated N-7 as the site of protonation, although the X-ray crystal complex indicates the presence of an asparagine near N-7.[104,105] The low bond order to the attacking arsenate (and phosphate) nucleophile indicated that the TS analog need not incorporate the phosphate group, as the phosphate anion site can readily fill with free phosphate and accomplish the bond order of the transition state through van der Waals and ionic interactions. The substrate specificity of PNP includes hypoxanthine and guanine purine bases, and these features were included in the synthesis of TS analogs. The positive charge of the

[104] M. D. Erion, J. D. Stoeckler, W. C. Guida, R. L. Walter, and S. E. Ealick, *Biochemistry* **36,** 11735 (1997).

[105] M. D. Erion, K. Takabayashi, H. B. Smith, J. Kessi, S. Wagner, S. Honger, S. L. Shames, and S. E. Ealick, *Biochemistry* **36,** 11725 (1997).

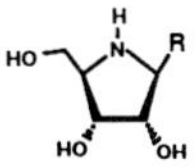

R	IU-nucleoside hydrolase[a] K_i	K_m/K_i	IAG-nucleoside hydrolase[b] K_i	K_m/K_i	$\dfrac{(K_m/K_i\ \text{IU-NH})}{(K_m/K_i\ \text{IAG-NH})}$
(I) H	$4.5 \pm 0.4\ \mu M$	25	$44 \pm 4\ \mu M$	0.41	60
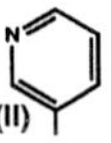 (II) pyridyl	$7.9 \pm 0.6\ \mu M$	14	$>240^c\ \mu M$	<0.075	>187
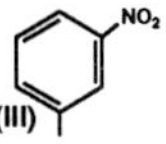 (III) NO_2-phenyl	$7.5 \pm 0.5\ \mu M$	15	$>360^c\ \mu M$	<0.05	>296
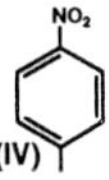 (IV) NO_2-phenyl	$1.1 \pm 0.1\ \mu M$	101	$>360^c\ \mu M$	<0.05	>2,000
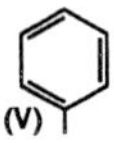 (V) phenyl	300 ± 27 nM	370	$180 \pm 15\ \mu M$	0.1	3,700
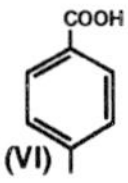 (VI) COOH-phenyl	96 ± 7 nM	1,160	$>480^c\ \mu M$	<0.04	>30,800
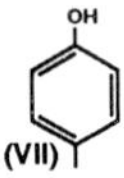 (VII) OH-phenyl	75 ± 4 nM	1,480	$35 \pm 2\ \mu M$	0.51	2,900
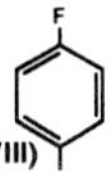 (VIII) F-phenyl	57 ± 5 nM	1,950	$205 \pm 14\ \mu M$	0.088	22,180
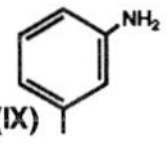 (IX) NH_2-phenyl	51 ± 4 nM	2,180	$38 \pm 4\ \mu M$	0.47	4,590
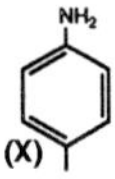 (X) NH_2-phenyl	30 ± 2 nM	3,700	$12 \pm 1\ \mu M$	1.5	2,470
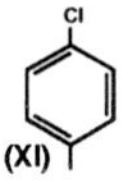 (XI) Cl-phenyl	30 ± 1.4 nM	3,700	$190 \pm 8\ \mu M$	0.095	39,000
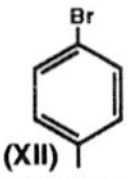 (XII) Br-phenyl	28 ± 4 nM	3,960	$113 \pm 6\ \mu M$	0.16	24,890

[a]The K_m for inosine was $111 \pm 17\ \mu M$ under these assay conditions. K_m/K_i values are for inosine as substrate. [b]The K_m for inosine was $18 \pm 1\ \mu M$ under these assay conditions. K_m/K_i values are for inosine as substrate. [c]No inhibiton was observed at 80, 120, 120, and 160 μM II, III, IV, and VI, respectively, when assayed at 75 μM inosine. The indicated inhibitor constants are lower limits of the constants based on the sensitivity of the assay.

FIG. 22. Isozyme specificity of TS inhibitors for IU-nucleoside hydrolase and IAG-nucleoside hydrolase. The dissociation constants for inhibitor (K_i) and the factor by which the inhibitor binds relative to substrate (K_m/K_i) are compared for the nonspecific IU-nucleoside hydrolase and the purine specific IAG-nucleoside hydrolase. From D. W. Parkin, G. Limberg, P. C. Tyler, R. H. Furneaux, X.-Y. Chen, and V. L. Schramm, *Biochemistry* **36**, 3528 (1997). Reprinted with permission from the American Chemical Society.

inhibitor	IU-NH	IAG-NH
	K_I (μM)	
I	0.04	0.02
		nonspecific
II	0.007	0.0008
III	0.4	>100
		IU-specific
IV	0.2	>100
V	>100	0.7
		IAG-specific
VI	>100	0.7

ribooxocarbenium was incorporated with a 4′-imino group replacing the 4′-oxygen. This group has a pK_a of 6.5 in the related phenyliminoribitol.[106] Chemical stability was introduced with a carbon bond to the purine base using 9-deazahypoxanthine and 9-deazaguanine. Carbon substitution at the 9 position also changes the pK_a of the N at N-7. Inosine and guanosine have pK_a values of approximately 2 at this position. As the TS is formed, the pK_a increases at N-7 and is proposed to be protonated or in a low-barrier hydrogen bond as the TS is achieved.[69] The N-7 position of immucillin-H (Imm-H) and immucillin-G has a pK_a >8.5, resembling the elevated pK_a of the TS (Fig. 20, lower panel).[107]

Both immucillin-H and immucillin-G are slow-onset, tight-binding inhibitors of PNP from calf spleen or from human erythrocytes. Equilibrium dissociation constants (K_i^*) are in the range of 23 to 72 pM with this combination of inhibitors and enzymes. Complete inhibition of PNP occurs with a single catalytic site of the PNP homotrimer occupied by inhibitor, and binding at one site prevents equivalent binding of inhibitor molecules at the remaining two sites. Inhibitor binding stoichiometry indicates that the enzyme works in an obligate sequential catalytic fashion. Dissociation of the enzyme–inhibitor complex is slow, e.g., $t_{1/2}$ = 4.8 hr for the complex of bovine PNP and immucillin-H. The geometries and molecular electrostatic potential surfaces of substrate, TS, and TS inhibitor were compared. The relationship between the binding energy $(\Delta G/RT)$ and the molecular electrostatic potential surface similarity $\langle S_e \rangle$ were correlated in the same manner that was used for the interactions of TS inhibitors with AMP deaminase, AMP nucleosidase, and adenosine deaminase (Fig. 21).[101,102] In this comparison, the geometric and electrostatic similarity of inosine to the transition state is 0.483, whereas that of immucillin-H is 0.723. The transition-state inhibitor binds 7×10^5 more tightly than substrate.[107] The inhibitors have potential in the treatment of T-cell proliferative disorders and provide a practical example of the application of TS structure to TS inhibitor design.

[106] B. A. Horenstein and V. L. Schramm, *Biochemistry* **32,** 9917 (1993).
[107] R. W. Miles, P. C. Tyler, R. H. Furneaux, C. K. Bagdassarian, and V. L. Schramm, *Biochemistry* **37,** 8615 (1998).

FIG. 23. Design of TS inhibitors with isozyme specificity for one or both of the nucleoside hydrolase isozymes (R. W. Miles, P. C. Tyler, R. H. Furneaux, G. Evans, and V. L. Schramm, unpublished data, 1998). Inhibitors I and II contain purine leaving-group contacts and the ribooxocarbenium ions mimic and bind tightly to both enzymes. Inhibitors III and IV are unable to be leaving-group activated, a major energetic interaction for IAG-NH, and are IU specific. Inhibitors V and VI lack the ribosyl-specific interactions required for TS formation in IU-NH by containing all purine activation groups and thus are IAG-NH specific.

Nucleoside Hydrolase

One of the protozoan nucleoside hydrolases (IU-NH) is nonspecific for the aglycone and accepts both purine and pyrimidine nucleosides as substrate. KIE were used to establish a ribooxocarbenium ion character at the TS with protonation of the purine leaving group as a secondary interaction for TS stabilization (see Fig. 16). The major factor in TS stabilization is ribooxocarbenium ion interaction and was demonstrated by the inhibition with iminoribitol, which binds 38-fold more tightly than substrate, even without any elements of the leaving group (Fig. 22). The addition of hydrogen bond acceptor or hydrophobic groups at the 1-position of iminoribitol provided inhibitors binding 120 and 12,000 times more tightly than substrate.[108] The nonspecific nature of the leaving-group interactions for TS formation are demonstrated by the requirement for a ribooxycarbenium mimic in tight-binding analogs. In contrast, a variety of substituent groups can be used to replace the protonated purine of the transition state while retaining nM binding ability. A large number of inhibitors for IU-NH have been used to train neural networks to recognize features important to the tight-binding interactions of the transition state (see article [15]).

The specificity of transition-state analogs is illustrated by the differing specificities with two isozymes of nucleoside hydrolases. The IU-NH has low-leaving group specificity, high ribose specificity, and is inhibited by most iminoribitol analogs. Inhibitor binding is improved by hydrophobic groups or by hydrogen bond acceptors as purine or pyrimidine analogs (Fig. 23). In contrast, the purine-specific IAG-NH catalyzes the same reaction, but half of the transition-state binding energy is involved in purine contacts to activate the purine-leaving group.[107]

Transition-state inhibitors designed specifically for the nonspecific-leaving group of the IU-NH bind poorly to the IAG-NH. By incorporating features of leaving groups that are accepted by both of the nucleoside hydrolase isozymes, TS inhibitors have been designed and synthesized that bind tightly to both nucleoside hydrolase isozymes (I and II of Fig. 23). The hydrophobic and nonspecific-leaving group properties of IU-NH permit tight binding of III and IV (Fig. 23), which are inert for IAG-NH. Conversely, precise leaving-group interactions with lower specificity for the ribooxycarbenium ion of the TS permit V and VI of Fig. 23 to bind tightly to IAG-NH while being inert for IU-NH (Miles, Tyler, Evans, Furneaux, and Schramm, unpublished results, 1998). Transition-state information permits the design of isozyme-specific inhibitors. These inhibitors have poten-

[108] D. W. Parkin, G. Limberg, P. C. Tyler, R. H. Furneaux, X.-Y. Chen, and V. L. Schramm, *Biochemistry* **36**, 3528 (1997).

tial for species-specific interaction with target enzymes, an ultimate goal of logically designed transition-state inhibitors.

Conclusions

Experimental methods are now available for obtaining transition-state information for most enzymatic reactions. Intrinsic KIE are established and used to calculate TS bonding geometry. The bonding structure provides information for calculation of the molecular electrostatic potential surface. Differences between the molecular electrostatic potential surfaces of the substrate and the TS indicate the hydrogen bonding and charge patterns to incorporate into synthetic TS analogs. This approach has now been used for the design of several powerful transition-state inhibitors. The combination of quantum computational chemistry, enzymatic mechanisms, and KIE provides a new level of information for enzymatic catalysis. The reliable success of the approach provides confidence that enzymatic TS structures from KIE analysis provide a sufficiently accurate structure of the TS for the practical purpose of TS inhibitor design.

[14] Determining Transition States from Kinetic Isotope Effects

By Paul J. Berti

Introduction

Kinetic isotope effects (KIEs[1]) have long been used for qualitative analyses of enzyme mechanisms. Qualitative analysis of KIEs has been a cornerstone of enzymology in showing, for example, the orderedness of reactions, the identification of rate-limiting steps, and identifying which bonds are being formed or broken at the TS. It is only since the mid-1980s that it has become feasible to routinely use precise measurements of KIEs to quantitatively determine the transition states (TSs) of enzymatic reac-

---

[1] Abbreviations: BEBOVA, bond energy/bond order vibrational analysis; EIE, equilibrium isotope effect; EXC, contribution from vibrationally excited states to an IE; GVFF, general valence force field; IE, isotope effect; KIE, kinetic isotope effect; LSC, liquid scintillation counting; MMI, contribution from the change in mass and moments of inertia to an IE; NAD$^+$, nicotinamide adenine dinucleotide; PDB, Protein Databank file format; RHF, restricted Hartree-Fock; SVFF, simple valence force field; TS, transition state; TSI, transition-state imbalance; ZPE, contribution from zero point energy to an IE.

Copyright © 1999 by Academic Press
All rights of reproduction in any form reserved.
0076-6879/99 $30.00

tions. The quantitative analysis of KIEs has become possible due to a number of technical advances that have made it practical to routinely synthesize complex molecules with isotopic labels at any one of many specific positions, to measure KIEs accurately and precisely, and to analyze the experimental KIEs via the structure interpolation approach to bond energy/bond order vibrational analysis (BEBOVA) to yield precise, quantitative determinations of experimental TSs. This contribution will describe the process of proceeding from observed KIEs to the experimental TS using structure interpolation and BEBOVA.

The biggest change in TS determination since Sims and Lewis's comprehensive description of BEBOVA[2] has been the huge increase in computational power available to the average biochemist. At one time, the molecular models used in BEBOVA analyses were truncated as far as possible to make them computationally tractable. Today, desktop computers can routinely handle molecules of 100 atoms. Another advance has been the introduction of the structure interpolation technique of generating many test TS structures automatically. This approach allows many more structures to be tested with greater accuracy than was possible using manually adjusted TS models. In the past, many simplifying assumptions were made in the modeling process. Given the greater accuracy of TS determination that is now routinely possible, these assumptions can be reexamined. The theoretical and practical complexities in determining a TS are still formidable, but the invaluable increase in our knowledge of enzyme mechanisms from each experimental TS, and the impossibility of achieving these results by any other technique, fully justify the effort.

The method for deriving an experimental TS from experimental KIEs will be described, as far as possible, in terms of general principles, although reference will often be made to the specific case where the approach is most developed: nucleophilic substitutions on NAD^+ (Fig. 1), including hydrolysis[3-6] and ADP-ribosylation reactions of peptide substrates.[7,8] Our interest in nucleophilic substitutions on NAD^+ stems from its role as a substrate in bacterial toxin-catalyzed ADP-ribosylation reactions. All the hydrolytic TSs follow bimolecular, A_ND_N[9] (S_N2) mechanisms with highly

[2] L. B. Sims and D. E. Lewis, in "Isotope Effects: Recent Developments in Theory and Experiment" (E. Buncel and C. C. Lee, eds.), Vol. 6, p. 161. Elsevier, New York, 1984.
[3] P. J. Berti and V. L. Schramm, J. Am. Chem. Soc. **119**, 12069 (1997).
[4] P. J. Berti, S. R. Blanke, and V. L. Schramm, J. Am. Chem. Soc. **119**, 12079 (1997).
[5] K. A. Rising and V. L. Schramm, J. Am. Chem. Soc. **119**, 27 (1997).
[6] J. Scheuring and V. L. Schramm, Biochemistry **36**, 4526 (1997).
[7] J. Scheuring, P. J. Berti, and V. L. Schramm, Biochemistry **37**, 2748 (1998).
[8] J. Scheuring and V. L. Schramm, Biochemistry **36**, 8215 (1997).
[9] R. D. Guthrie and W. P. Jencks, Acc. Chem. Res. **22**, 343 (1989).

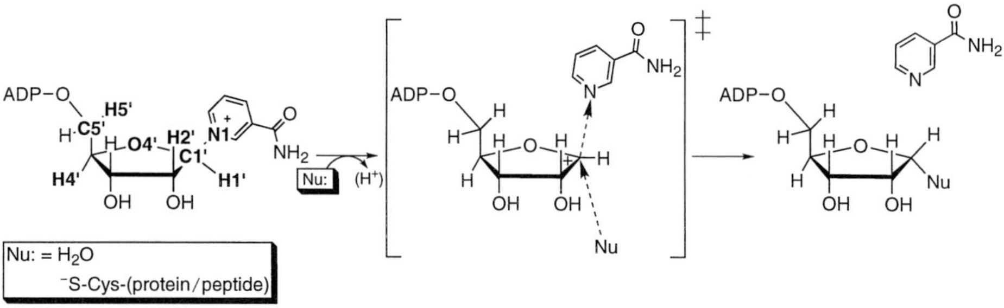

FIG. 1. Nucleophilic substitutions on NAD⁺. (Left) Sites of isotopic labels (^{3}H, ^{14}C, ^{15}N, ^{18}O) are indicated in the reactant. (Middle) The reactions pass through an A_ND_N TS before forming the products (right).

dissociative TSs, similar to the TSs for other N-glycoside hydrolysis reactions examined in this laboratory.[10–16] In these TSs, the N-glycosidic bond is almost completely broken, with low but significant bonding to the incoming water nucleophile. TSs for the ADP-ribosylation reactions catalyzed by pertussis toxin are more concerted than the hydrolytic reactions.

Theory

Transition State and Its Structure

In any reaction proceeding between two stable species, the reacting molecules must pass through higher energy states between reactants and products. One of these states is of particular interest, the transition state. The TS is the highest energy point on the lowest energy path between reactants and products. The significance of the TS for enzymologists was noted by Pauling,[17] who pointed out that enzymes catalyze reactions by binding to and stabilizing TSs in preference to either substrates or products. Furthermore, the extent to which an enzyme increases the rate over the uncatalyzed reaction is equal to the extent to which it binds the TS more

---

[10] P. C. Kline and V. L. Schramm, *Biochemistry* **34**, 1153 (1995).
[11] D. J. Merkler, P. C. Kline, P. Weiss, and V. L. Schramm, *Biochemistry* **32**, 12993 (1993).
[12] P. C. Kline and V. L. Schramm, *Biochemistry* **32**, 13212 (1993).
[13] B. A. Horenstein, D. W. Parkin, B. Estupinan, and V. L. Schramm, *Biochemistry* **30**, 10788 (1991).
[14] D. W. Parkin, F. Mentch, G. A. Banks, B. A. Horenstein, and V. L. Schramm, *Biochemistry* **30**, 4586 (1991).
[15] D. W. Parkin and V. L. Schramm, *Biochemistry* **26**, 913 (1987).
[16] F. Mentch, D. W. Parkin, and V. L. Schramm, *Biochemistry* **26**, 921 (1987).
[17] L. Pauling, *Chem. Eng. News* **24**, 1375 (1946).

tightly than the substrates or products. There are now several enzymes that have been shown to catalyze reactions with rate enhancements of ca. 10^{17}-fold,[18–21] including orotate decarboxylase. This rate enhancement, taken with the equilibrium dissociation constants for the substrate, implies that the equilibrium dissociation constant for the enzyme·TS complex would be 10^{-24} M.[18] If it were possible to capture a small fraction of that binding energy in a TS mimic—a stable molecule that reproduces geometric and charge features of the TS—a potent inhibitor would result.

The potential energy surface of a reaction is often likened to a mountain pass between two valleys. Both the hiker and the molecule seek the lowest energy route between the valleys (reactants and products), passing over the lowest barrier (i.e., the TS), rather than over a mountain peak. Topologically, the TS is at a saddle point on the potential energy surface, with the energy minimized in every dimension except the reaction coordinate, where it is a maximum. The TS is a stationary state, i.e., the structure is at equilibrium, with no net forces acting on any atom. In a regular bond, any distortion results in a force tending to restore the structure back to a minimum energy structure. In contrast, any distortion in the reaction coordinate direction causes forces *away* from the TS that increase with increasing displacement from the stationary point. The definition used here of the reaction coordinate is that it is a normal vibrational mode that has an imaginary frequency.[22] Displacement of atoms along this normal mode results in a decrease in energy and movement of the molecule either forward toward products, or back toward reactants.

Because of the emphasis on understanding enzyme·substrate interactions that promote catalysis, and on designing TS mimics as inhibitors, analysis of TSs tends to focus on their structure. It is intrinsic to the process of determining TS structures, however, to also characterize the reaction coordinate. For example, in the highly dissociative A_ND_N TSs of glycoside nucleophilic substitutions, the residual bond order to the leaving group and the incipient bond order to the incoming nucleophile (both <0.05) are so low that it seems they should have little or no observable effect on the KIEs. The question then arises whether these groups are present in the TS at all. It is the contribution of the leaving group and the nucleophile to the reaction coordinate frequency (and its associated contribution to the primary KIEs) that makes it possible to unambiguously judge whether they

[18] A. Radzicka and R. Wolfenden, *Science* **267,** 90 (1995).
[19] B. G. Miller, T. W. Traut, and R. Wolfenden, *J. Am. Chem. Soc.* **120,** 2666 (1998).
[20] R. Wolfenden, C. Ridgway, and G. Young, *J. Am. Chem. Soc.* **120,** 833 (1998).
[21] R. Wolfenden, X. Lu, and G. Young, *J. Am. Chem. Soc.* **120,** 6814 (1998).
[22] W. E. Buddenbaum and V. J. Shiner, Jr., *Can. J. Chem.* **54,** 1146 (1976).

are present at the TS, i.e., that the reaction has a bimolecular, A_ND_N mechanism, not a stepwise, $D_N + A_N$ (S_N1) mechanism.

Balls-and-Springs Vibrational Models and Geometry

The physical origin and significance of KIEs have been very well described in several reviews[23–25] and will not be covered in detail here. Isotope effects are primarily a vibrational phenomenon, in the same sense as infrared (IR) and Raman spectroscopies. KIEs reflect a *change* in the vibrational environment of a given atom between the reactant and the TS of a reaction. If there is a change in molecular structure in the vicinity of an isotopic label on going from the reactant to the TS, there will be changes in its vibrational environment and therefore in the frequencies of normal vibrational modes. These frequency changes lead to an IE. There is also a (usually small, but not always) contribution from the isotopic label to a change in mass and moments of inertia of the molecule.

Isotope effects are an inherently quantum phenomenon; they arise as a consequence of the application of the Heisenberg uncertainty principle to vibrational frequencies. An atom in a molecule cannot be at rest, even at 0 K, because that would violate the principle that the position and momentum of a particle cannot be perfectly known. Thus, molecules possess a vibrational energy even at 0 K, the zero point energy. The zero point energy is a function of the vibrational frequency, which is in turn a function of the mass of the vibrating atoms and the force constant (Fig. 2). There is also a small (near room temperature) contribution from normal modes that are vibrationally excited (not in the vibrational ground state).

Vibrational energy (E_v) is a simple function of frequency:

$$E_v = (v + 1/2)h\omega \tag{1}$$

where v is the vibrational quantum number, h is Planck's constant, and ω is the angular frequency ($= 2\pi\nu$). Thus, the lowest vibrational energy an oscillator may have is the zero point energy, when $v = 0$ and $E_v = h\omega/2$. Changes in zero point energies between molecules are generally the largest contributor to IEs. The other factors that contribute to IEs are contributions from excited-state energies, the change in mass and moments of

[23] W. P. Huskey, *in* "Enzyme Mechanism from Isotope Effects" (P. F. Cook, ed.), p. 37. CRC Press, Boca Raton, FL, 1991.

[24] J. Suhnel and R. L. Schowen, *in* "Enzyme Mechanism from Isotope Effects" (P. F. Cook, ed.), p. 3. CRC Press, Boca Raton, FL, 1991.

[25] W. E. Buddenbaum and V. J. Shiner, Jr., *in* "Isotope Effects on Enzyme-Catalyzed Reactions" (W. W. Cleland, M. H. O'Leary, and D. B. Northrop, eds.), p. 1. University Park Press, Baltimore, MD, 1977.

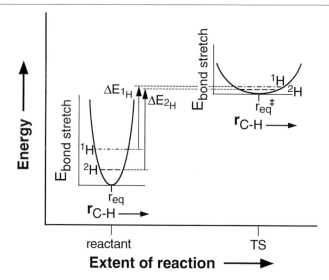

FIG. 2. Each inset shows the energy of changing the length of a C–H bond from its equilibrium length in the reactant (r_{eq}) and the TS ($r_{eq}^{\ddagger}$). Zero point energies for ^1H and ^2H (dashed horizontal lines) depend on the mass and the bond-stretching force constant. For a normal IE, the zero point energy decreases *more* for the light isotope (^1H) than the heavy isotope (^2H) when the bond-stretching force constant decreases at the TS. This leads to a difference in activation energies ($\Delta\Delta E^{\ddagger} = \Delta E_{2H} - \Delta E_{1H}$), which is the source of the ZPE contribution to the IE. This illustration is for a secondary IE; for a primary KIE, the reaction coordinate normal mode at the TS has no zero point energy because it is an imaginary frequency and therefore has no restoring force. For a primary KIE, $\Delta\Delta E^{\ddagger} = E_{2H,\text{reactant}} - E_{1H,\text{reactant}}$.

inertia, transmission effects, and tunneling.[26,27] The contribution from transmission coefficient effects is in the realm of variational TS theory, where barrier recrossing can lead to transmission coefficients <1.[28–30] In general, barrier recrossing becomes significant for reactions in the gas phase, at high temperature, where small atoms are transferred between much larger groups. There is no evidence in the literature indicating transmission coefficients <1 for enzymatic reactions. Tunneling through an energetic barrier, rather than passing over it, has been observed only in reactions involving electron, hydron, or hydride transfer, reactions that will not be considered here.

[26] A. Kohen and J. P. Klinman, *Acc. Chem. Res.* **31**, 397 (1998).
[27] Y. Cha, C. J. Murray, and J. P. Klinman, *Science* **243**, 1325 (1989).
[28] D. G. Truhlar, W. L. Hase, and J. T. Hynes, *J. Phys. Chem.* **87**, 2664 (1983).
[29] S. C. Tucker, D. G. Truhlar, B. C. Garrett, and A. D. Isaacson, *J. Chem. Phys.* **82**, 4102 (1985).
[30] D. G. Truhlar and B. C. Garrett, *Acc. Chem. Res.* **13**, 440 (1980).

The expression for calculating equilibrium isotope effects (EIEs) from vibrational frequencies was formulated by Bigeleisen and Goeppert Mayer,[31] with additions by Bigeleisen and Wolfsberg[32] to calculate KIEs. The contributions to the IEs are multiplicative:

$$KIE = MMI \cdot ZPE \cdot EXC \qquad (2)$$

where each factor is the contribution to the KIE from the change in mass and moments of inertia (MMI), zero point energy (ZPE), and excited-state energy (EXC).

In a stable nonlinear molecule, there are 3N-6 normal modes. In the TS, however, one of these modes becomes the reaction coordinate with an imaginary frequency. The imaginary frequency is sometimes expressed as a negative number, although it is correct to write a frequency of $100i$ cm^{-1}, rather than -100 cm^{-1}. The reaction coordinate normal mode has no restoring force; it is aperiodic. It therefore makes no contribution to ZPE or EXC. Thus, for a KIE, ZPE and EXC are calculated using 3N-7 normal modes for the TS molecule.

The expression for each term is

$$MMI = \frac{\left(\left(\frac{^{light}M}{^{heavy}M}\right)^{3/2} \frac{^{light}I_x \, ^{light}I_y \, ^{light}I_z}{^{heavy}I_x \, ^{heavy}I_y \, ^{heavy}I_z}\right)_{TS}}{\left(\left(\frac{^{light}M}{^{heavy}M}\right)^{3/2} \frac{^{light}I_x \, ^{light}I_y \, ^{light}I_z}{^{heavy}I_x \, ^{heavy}I_y \, ^{heavy}I_z}\right)_{reactant}} \qquad (3a)$$

where $^{light}M$ is the molecular mass with the light isotope and $^{light}I_x$ is the moment of rotational inertia in the x-axis.

$$ZPE = \frac{\left[\prod_i^{3N-7\ddagger} e^{-(^{light}u_i - ^{heavy}u_i)/2}\right]_{TS}}{\left[\prod_i^{3N-6} e^{-(^{light}u_i - ^{heavy}u_i)/2}\right]_{reactant}} \qquad (3b)$$

where $^{light}u_i$ is the ith normal mode, u_i is $h\nu_i/k_B T$ (h is Planck's constant, ν_i is frequency, k_B is the Boltzman constant, and T is absolute temperature) and 3N-7‡ is the number of normal modes in a nonlinear TS, where N is the number of atoms. For linear molecules, use 3N-6‡ normal modes.

[31] J. Bigeleisen and M. Goeppert Mayer, *J. Chem. Phys.* **15**, 261 (1947).
[32] J. Bigeleisen and M. Wolfsberg, *Adv. Chem. Phys.* **1**, 15 (1958).

3N-6 is the number of normal modes in a nonlinear stable molecule. For linear molecules, use 3N-5.

$$\text{EXC} = \frac{\left[\displaystyle\prod_{i}^{3N-7\ddagger} \frac{1 - e^{(-^{\text{heavy}}u_i)}}{1 - e^{(-^{\text{light}}u_i)}} \right]_{\text{TS}}}{\left[\displaystyle\prod_{i}^{3N-6} \frac{1 - e^{(-^{\text{heavy}}u_i)}}{1 - e^{(-^{\text{light}}u_i)}} \right]_{\text{reactant}}} \tag{3c}$$

MMI can be calculated directly, as in Eq.(3a). For EIEs, it can also be calculated from the vibrational product (VP) by application of the Teller–Redlich rule:

$$\text{MMI} = \text{VP} = \frac{\left(\displaystyle\prod_{i}^{3N-6} \frac{^{\text{light}}\nu_i}{^{\text{heavy}}\nu_i} \right)_{\text{final}}}{\left(\displaystyle\prod_{i}^{3N-6} \frac{^{\text{light}}\nu_i}{^{\text{heavy}}\nu_i} \right)_{\text{initial}}} \tag{4}$$

For KIEs, the reaction coordinate mode contributes to the MMI, which can also be expressed as:

$$\text{MMI} = \frac{^{\text{light}}\nu^*}{^{\text{heavy}}\nu^*} \times \text{VP} \tag{5}$$

where $^{\text{light}}\nu^*$ is the reaction coordinate frequency for the light isotope and VP is the vibrational product calculated using 3N-6 normal modes for the reactant state and 3N-7‡ for the TS.

Even though IEs are a quantum phenomenon, ZPE and EXC factors are simple functions of vibrational frequencies. These frequencies can be calculated by using purely mechanical models of the molecules. To a very good approximation, the vibrational frequencies of a molecule can be described in terms of a balls-and-springs harmonic oscillator model. Atoms are treated as balls of defined mass, and bonds are treated as ideal massless springs with defined strengths. At its equilibrium length, the spring exerts no force. If a displacement of the atoms changes the length of the spring, it will exert a restoring force proportional to the displacement, following Hook's law:

$$F = -k \cdot \Delta x, \tag{6}$$

where F is the restoring force, k is the spring force constant (in units of force per unit displacement), and Δx is the displacement from equilibrium spring length.

The simplest vibrational model is a diatomic molecule, with two balls connected by a spring. With this model, the harmonic bond stretching frequency can be calculated using:

$$\nu = \frac{1}{2\pi}\sqrt{k/\mu} \tag{7}$$

where ν is the vibrational frequency, k is the spring force constant, μ is the reduced mass of the oscillator, $m_1 m_2/(m_1 + m_2)$, where m_1, m_2 = mass of the spheres.

A diatomic molecule has only one vibrational motion, bond stretching. In polyatomic molecules, molecular vibrations do not occur as isolated vibrations of individual bonds. Rather, individual bond vibrations become coupled, depending on symmetry and frequency, into a discrete number of complex motions, the normal modes of vibration.[33,34] Normal modes correspond to the observable vibrational frequencies detected by IR or Raman spectroscopies (although not all normal modes are IR or Raman active, depending on symmetry). There are 3N-6 normal modes in nonlinear molecules and 3N-5 in linear molecules. In normal coordinate analysis, the frequencies and motions of individual atoms for each normal mode are determined. Vibrational frequency is synonymous with vibrational energy; changes in vibrational frequency caused by an isotopic label translate into changes in the relative energies of the reactant and TS, i.e., a change in activation energy, and therefore a change in reaction rate. Thus, the change in mass associated with an isotopic label leads to the change in relative reaction rates: the KIE.

Thus, it is possible to calculate KIEs based on balls-and-springs vibrational models of molecules by applying Eq. (3) to those frequencies. The next step is to make the connection between molecular structures and vibrational frequencies.

The relationship between structure and frequency arises from the relationship between bond length and the force required to distort a bond. The relation between bond length and stretching force constants is described by the Pauling bond order and Badger's rule. Pauling bond orders relate bond length to bond order:

$$n_{ij} = e^{(r_1 - r_{ij})/0.3} \tag{8}$$

where n_{ij} is the Pauling bond order between atoms i and j, r_1 is the length of a single bond of that type, and r_{ij} is the length of the bond between atoms i and j.

[33] E. B. Wilson, Jr., J. C. Decius, and P. C. Cross, "Molecular Vibrations: The Theory of Infrared and Raman Vibrational Spectroscopy." McGraw-Hill, New York, 1955.
[34] D. C. Harris and M. D. Bertolucci, "Symmetry and Spectroscopy: An Introduction to Vibrational and Electronic Spectroscopy." Dover Publications, New York, 1978.

Badger's rule describes the relationship between bond order and bond strength, i.e., the force constant for that bond:

$$F_{ij} = F_1 \cdot n_{ij} \tag{9}$$

where F_{ij} is the force constant for the bond between atoms i and j and F_1 is the force constant for a single bond between those atom types.

Force constants are expressed in units of force per unit displacement, the same units as k in Hook's law, generally mdyne/Å. Evidence supporting the use of the Pauling and Badger rules in BEBOVA analysis have been reviewed.[2,35,36] This establishes the connection between geometry and vibrational frequencies through the variation in bond stretching force constants with bond length. Thus, the chain of causality is changes in bond lengths between reactant and TS lead to changes in bond-stretching force constants, which leads to changes in vibrational frequencies, which leads to IEs.

Structural changes also affect bond bending force constants, although the relationship between bond angles and bending force constants is not as clear cut. Sims and Lewis[2] developed a series of functions to describe the variation in bending force constants as a function of bond angle (see later).

Competitive KIEs

Large isotope effects ($>$ca. 1.5) can be determined with acceptable precision from the steady-state kinetic constants for isotopically pure species. This is the direct, or noncompetitive, method. For smaller IEs, such as secondary hydrogen or heavy atom IEs, competitive, or isotope discrimination, methods are used to obtain measurements precise enough to be useful for quantitative TS analysis. In competitive IE measurements, a mixture of isotopically labeled and unlabeled substrates is allowed to react with the enzyme. The isotopologs compete with each other as substrates, which means that their relative rates of reaction are determined by the relative values of k_{cat}/K_M, the "specificity constant."[37] Competitive KIEs measure IEs on k_{cat}/K_M. This is true regardless of the concentration of each isotopic label, i.e., whether the isotopic labels are trace, radioactive isotopes, or nonradioactive labels present as a large molar fraction of the total substrate concentration. This is contrary to occasional statements in the literature that this is true only for trace (i.e., radioactive) labels.[38] There

[35] H. S. Johnston, "Gas Phase Reaction Rate Theory." Ronald Press, New York, 1966.
[36] G. W. Burton, L. B. Sims, J. C. Wilson, and A. Fry, *J. Am. Chem. Soc.* **99**, 3371 (1977).
[37] A. R. Fersht, "Enzyme Structure and Mechanism." Freeman, New York, 1985.
[38] D. W. Parkin, *in* "Enzyme Mechanism from Isotope Effects" (P. F. Cook, ed.), p. 269. CRC Press, Boca Raton, FL, 1991.

are many techniques for measuring competitive IEs, including isotope ratio mass spectrometry,[39,40] whole molecule mass spectrometry,[4,41,42] liquid scintillation counting of radioisotopes,[38] and nuclear magnetic resonance (NMR).[43]

Irreversible Step(s) and Commitment to Catalysis

In the simplest case, the kinetic constant k_{cat}/K_M (or V/K) reflects partitioning between the free reactant in solution and the TS of the *first irreversible step* of the enzymatic reaction. In this case, KIEs on k_{cat}/K_M reflect the IEs between these same species, the reactant in solution, and the TS for the first irreversible step. The meaning of k_{cat}/K_M, and therefore the meaning of KIEs, may change if there are intermediates in the chemical mechanism; i.e., if there are chemical species with finite lifetimes that can partition either forward or backward in the reaction pathway. If these intermediates partition mostly back to the substrate and are therefore in equilibrium with it, then the meaning of the KIEs is unchanged. If there is significant partitioning of these intermediates both forward and backward, then the TS leading to each intermediate will contribute to the KIEs. The case of one intermediate is discussed further later.

There is no reason why the first irreversible step must be a chemical step. If every time a substrate molecule binds to the enzyme it reacts to give products, i.e., if there is full commitment to catalysis (C_f), then the first irreversible step is association of the substrate and enzyme to form the Michaelis complex. In this case, the observed KIEs would be for diffusion of the enzyme and substrate together, which would carry no information on the chemical mechanism. Ideally one seeks to find reaction conditions where the commitment to catalysis is zero, i.e., where $k_2 \gg k_3$ in Eq. (10).

$$E + S \underset{k_2}{\overset{k_1}{\rightleftharpoons}} E \cdot S \xrightarrow{k_3} E \cdot P \xrightarrow{k_5} E + P \qquad (10)$$

where E is the enzyme, S is the substrate, E·S is the Michaelis complex, and k_n is the intrinsic rate constant on step n. This can be accomplished by using alternate substrates, adjusting reaction conditions, or using a mutant enzyme with increased k_2/k_3. When commitment is zero, the observable KIEs are those of the chemical step(s).

[39] M. H. O'Leary, *Methods Enzymol* **64,** 83 (1980).
[40] P. M. Weiss, *in* "Enzyme Mechanism from Isotope Effects" (P. F. Cook, ed.), p. 291. CRC Press, Boca Raton, FL, 1991.
[41] B. J. Bahnson and V. E. Anderson, *Biochemistry* **30,** 5894 (1991).
[42] E. Gawlita, P. Paneth, and V. E. Anderson, *Biochemistry* **34,** 6050 (1995).
[43] D. A. Singleton and A. A. Thomas, *J. Am. Chem. Soc.* **117,** 9357 (1995).

If conditions of zero commitment cannot be found, it is possible to measure the commitment and derive the intrinsic KIEs of the chemical steps from the observed KIEs using Eq. (11)[44,45]:

$$KIE_{obs} = \frac{KIE_{intrinsic} + C_f + EIE \cdot C_r}{1 + C_f + C_r} \qquad (11)$$

where KIE_{obs} is the observed KIE, $KIE_{intrinsic}$ is the intrinsic KIE on the chemical step, C_f is the commitment to catalysis in the forward direction, C_r is the commitment to catalysis in the reverse direction, and EIE is the EIE between reactants and products. The reverse commitment to catalysis, C_r, may be neglected if the reaction is run under conditions that are effectively irreversible.

If a nonchemical step, such as a protein conformational change or product release, is the irreversible step, then once again the observable KIEs will not include a contribution from the chemical step. If this situation is encountered, then a change in reaction conditions may increase the rate of the nonchemical step or slow the chemical step, thereby allowing the KIEs on the chemical step to be expressed.[7,10]

So far, we have only considered the case where one step is cleanly the irreversible step. In the case where one intermediate is formed, formation and/or breakdown of the intermediate will contribute to the observable KIEs, depending on the rate constants. Let us consider a kinetic mechanism with one intermediate:

$$E + S \underset{k_2}{\overset{k_1}{\rightleftarrows}} E \cdot S \underset{k_4}{\overset{k_3}{\rightleftarrows}} [E \cdot S'] \overset{k_5}{\longrightarrow} E \cdot P \overset{k_7}{\longrightarrow} E + P \qquad (12)$$

where $E \cdot S'$ is the enzyme-bound intermediate with a finite lifetime.

The k_{cat}/K_M on this reaction will be

$$\frac{k_{cat}}{K_M} = \frac{k_1 k_3 k_5 k_7}{k_2 k_4 k_6 \left(1 + \dfrac{k_7}{k_6}\left(1 + \dfrac{k_5}{k_4}\left(1 + \dfrac{k_3}{k_2}\right)\right)\right)} \qquad (13)$$

The observable competitive KIEs will be

$$KIE = \frac{\dfrac{\alpha_1 \alpha_3}{\alpha_2}\left(\dfrac{\alpha_5}{\alpha_4} + \dfrac{k_5}{k_4}\right)}{1 + \dfrac{k_5}{k_4}} \qquad (14)$$

where α_n is the intrinsic isotope effect on step n.

[44] D. B. Northrop, *Annu. Rev. Biochem.* **50**, 103 (1981).
[45] I. W. Rose, *Methods Enzymol.* **64**, 47 (1980).

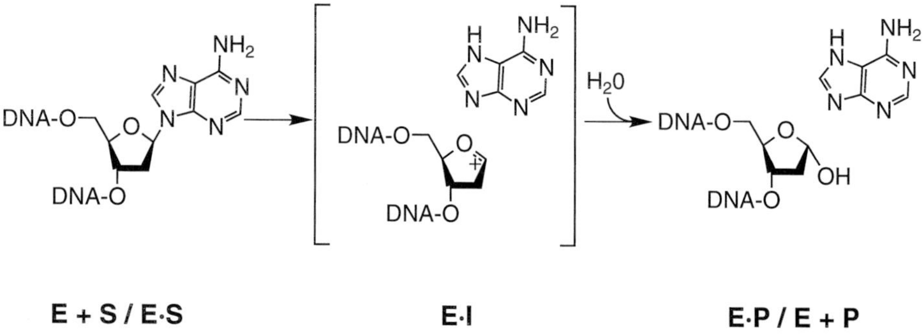

FIG. 3. Ricin-catalyzed DNA hydrolysis. This reaction follows the kinetic scheme in Eq. (12). Ricin depurinates 28S RNA *in vivo*.

The observable isotope effect depends on k_5/k_4, the partitioning of the enzyme-bound intermediate, $E \cdot S'$. If $k_5/k_4 \gg 1$, then k_3 is the first irreversible step, and KIE $= (\alpha_1\alpha_3)/\alpha_2$. The KIE will be that between the reactant in solution and the TS of the first chemical step. If $k_5/k_4 \approx 0$, then k_5 will be the first irreversible step, and KIE $= (\alpha_1\alpha_3\alpha_5)/(\alpha_2\alpha_4)$. This is the KIE between the reactant and the TS for the second chemical step. Alternately, and equivalently, it is the product of the EIE for $E \cdot S'$ formation $(\alpha_1\alpha_3)/(\alpha_2\alpha_4)$ and the KIE of $E \cdot S'$ breakdown, α_5. At intermediate values of k_5/k_4, the observable KIE varies monotonically between those for the extreme values of k_5/k_4. Whether it is possible to determine k_5/k_4 by BEBOVA will depend on the particular system.

In the case of ricin-catalyzed depurination of DNA[46] (Fig. 3), experimental KIEs showed that the TS did not involve a bimolecular nucleophilic $A_N D_N$ displacement conforming to Eq. (10). Rather, they showed that an oxocarbenium ion · enzyme complex with a finite lifetime (more than several bond vibrations, ca. 10^{-12} sec) was being formed, i.e., the reaction was proceeding through a $D_N + A_N$ mechanism. The experimental primary $1'$-^{14}C KIE was 1.015 ± 0.001, compared with the calculated KIEs for $k_5/k_4 \gg 1$ (KIE $= 1.018$) and $k_5/k_4 \approx 0$ (1.010). At the same time, the primary 1-^{15}N KIE was 1.023 ± 0.004, compared with the calculated KIEs of 1.028 for $k_5/k_4 \gg 1$, and 1.024 for $k_5/k_4 \approx 0$. In this case, the calculated KIEs were in good agreement with the experimental KIEs, but were not sufficiently different from each other to make it possible to estimate the value of k_5/k_4 from the experimental KIEs. The ricin example is a reminder that although competitive KIEs always measure the IEs up to the first

[46] X. Y. Chen, P. J. Berti, and V. L. Schramm, manuscript in preparation.

irreversible step, one does not know beforehand (and sometimes afterward) which step that is.

Route from KIEs to Transition-State Structures

Overall Strategy

It is not possible to calculate a TS directly from experimental KIEs. Rather, it is necessary to create a test TS, then calculate what the KIEs would be for that TS. The calculated KIEs are then compared with the experimental ones. KIEs for the test TS structure are calculated using the BEBOVA approach. Bond energy/bond order vibrational analysis attempts to recreate the vibrational frequencies of the reactant and test TS based on simple rules relating geometry and force constants (see later). Test TS structures for BEBOVA have been created using the structure interpolation method, an improvement on the old, *ad hoc* approach.

In the old, *ad hoc* method used in this laboratory, one searches for the TS by adjusting the structure of the TS manually. Structural parameters that are expected to contribute to the KIEs are adjusted and the KIEs calculated. The structure is then readjusted iteratively to arrive at a structure that gives calculated KIEs in agreement with experimental KIEs. The limitations of this approach include the fact that it is possible to create structures that are subtly incorrect or to fail to make important adjustments to the structure because they are expected (incorrectly) not to affect the KIEs. A second limitation is that only one TS is determined, without any indication whether there is another structure, or group of structures, that would match the experimental KIEs. Third, the process is very time-consuming. Finally, because one does not know what the KIEs should be for a given reaction, it is not possible to recognize enzyme-caused or other unusual effects. The transition state imbalance effects observed for the pertussis toxin-catalyzed ADP-ribosylation of protein $G_{i\alpha1}$[7] were recognized as such only because the experimental KIEs were different from those expected based on the NAD^+ hydrolysis reactions.[3,4]

For these reasons, the structure interpolation approach was developed.[3] Rather than trying to reach one possible TS, an algorithm is used to generate many test TS structures throughout reaction space (Fig. 4), and KIEs are calculated for all the test TSs. Plots are made showing the areas in reaction space where calculated KIEs match experimental KIEs for each isotopic label (Fig. 5). If the modeling process has been successful, there will be a point in reaction space where calculated KIEs match experimental KIEs for all of the isotopic labels. This is the experimentally determined TS.

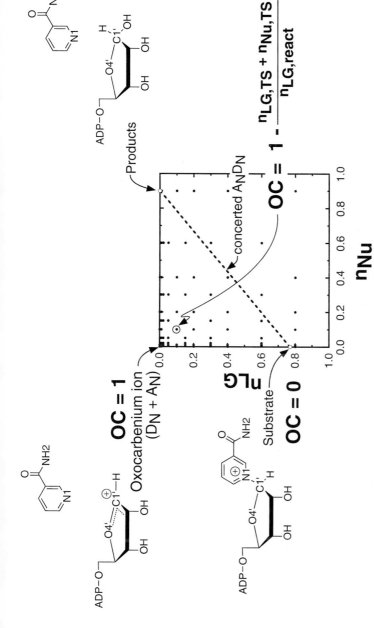

Fig. 4. Reaction space for hydrolysis of NAD^+ to ADP-ribose and nicotinamide. The main processes of the reaction are breaking the bond to the leaving group ($n_{LG} \equiv n_{C-1'-N1}$) and forming the bond to the nucleophile ($n_{Nu} \equiv n_{C-1'-Nu}$). A classical $D_N + A_N$ mechanism involves departure of the leaving group ($n_{LG} = 0$) to form an oxocarbenium ion intermediate plus the first product, nicotinamide. In the second step, the nucleophile approaches to form the C-1'–Nu bond. In a fully concerted $A_N D_N$ mechanism, the formation of n_{Nu} exactly matches the loss of n_{LG}, as shown by the diagonal dashed line. The dots in reaction space represent the 85 test TS structures generated by structure interpolation. OC is the oxocarbenium ion character described in the text.

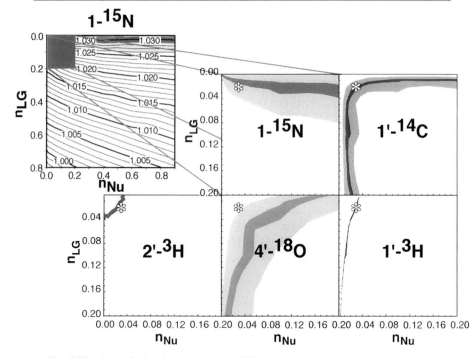

F𝅘𝅥𝅮G. 5. Match of calculated to experimental KIEs in reaction space for the diphtheria toxin-catalyzed hydrolysis of NAD^{+4}. (Top left) Contour plot of calculated $1\text{-}^{15}N$ KIEs throughout reaction space. (Top middle) Expansion of the dissociative section of reaction space for $1\text{-}^{15}N$ KIEs. The shading shows the areas in reaction space where the calculated KIEs match the experimental KIEs exactly (dark gray) or within the 95% confidence interval (light gray). The asterisk is the experimentally determined TS. Isotopic labels as shown in Fig. 1.

If more than one TS is consistent with experimental KIEs, that will be obvious also.

The structure interpolation approach is similar to systematic searches of reaction space that have been described previously,[47,48] with several changes. The two reference structures in structure interpolation are either X-ray crystal structures or high-level quantum mechanically optimized structures, rather than structures based on ideal bond lengths and angles. All the internal coordinates of the test TS structures are varied between the values of the first and second reference structures, rather than relying on simplified models of structure variation throughout reaction space, and the structure interpolation algorithm was based on the results of quantum

[47] J. Rodgers, D. A. Femec, and R. L. Schowen, *J. Am. Chem. Soc.* **104**, 3263 (1982).
[48] D. E. Lewis, L. B. Sims, H. Yamataka, and J. McKenna, *J. Am. Chem. Soc.* **102**, 7411 (1980).

mechanical optimizations of intermediate structures. This allowed accurate modeling of the structures throughout reaction space.

The BEBOVA approach achieves its greatest power if a *unified model* of a given reaction type can be created. Nucleophilic substitutions on NAD$^+$ are a case in point. Transition states of nonenzymatic[3] and three enzyme-catalyzed hydrolysis reactions[4] (and unpublished results), plus the two ADP-ribosylation TSs,[7,8] were determined using a unified model of NAD$^+$ nucleophilic substitution. A unified model differs from any other TS determination only in that the same calculation system has been used successfully to determine several different TSs of that type, lending support to the correctness of each step in the process. Thus, the same reference structures, same interpolation algorithm, same force constants, and same vibrational model were used for all these TS determinations.

A unified model of a reaction yields several advantages. (a) Speed: Once the calculations are complete, the TS for any reaction of that type can be determined simply by matching the already calculated KIEs to the new experimental ones. (b) Reliability: If many TSs can be determined from a single unified model, this is a strong indication of the validity of that model. (c) Sensitivity: BEBOVA detects differences between TSs with much greater precision than the absolute values of bond orders for a single TS. This characteristic is especially useful in understanding enzyme mechanisms because the difference between TSs for the enzymatic and nonenzymatic reactions reflects the interaction between the inherent reaction pathway of the substrate and the enzyme's use of binding energy to lower the energetic barrier to the TS.

The description of the process will be in two parts, the first covering BEBOVA and the second covering structure interpolation.

Bond Energy/Bond Order Vibrational Analysis

The program used for all the following BEBOVA work was BEBOVIB-IV,[49] written by Sims, Burton, and Lewis. It is available through the Quantum Chemistry Program Exchange and is the primary program for BEBOVA studies. There is one other program for doing bond vibrational analysis, VIBIE.[50,51] VIBIE does not automatically vary force constants as a function of structure, as BEBOVIB does, so they must be included in the input directly.

[49] L. B. Sims, G. W. Burton, and D. E. Lewis, "BEBOVIB-IV, QCPE No. 337," Quantum Chemistry Program Exchange, Department of Chemistry, University of Indiana, Bloomington, IN, 1977.
[50] W. P. Huskey, *J. Am. Chem. Soc.* **118,** 1663 (1996).
[51] T. E. Casamassina, and W. P. Huskey, *J. Am. Chem. Soc.* **115,** 14 (1993).

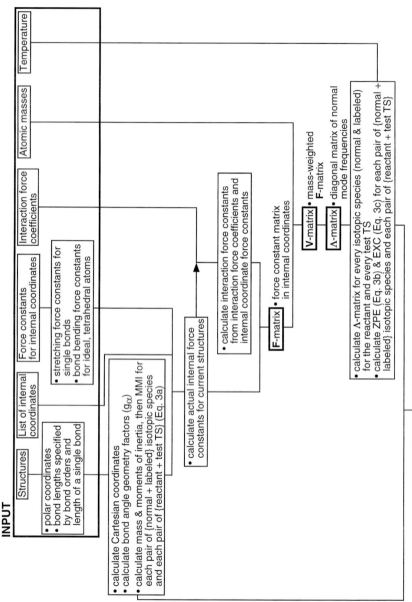

BEBOVA versus Quantum Mechanical KIEs. As an alternative to BEBOVA, it is sometimes possible to use an entirely quantum mechanical approach to TS determination. Increases in computational power have made it possible to perform quantum mechanical calculations at usefully high levels of theory on molecules of biologically relevant size, although only in a vacuum or a continuum dielectric reaction field. Using electronic structure programs, the reactant and TS structures for a reaction may be determined. Calculating the vibrational frequencies of these structures allows KIEs to be calculated and compared with experimental ones. If the calculated KIEs match the experimental ones, that will, in effect, confirm the computational result and make BEBOVA calculations unnecessary. It continues to be the general case, however, that the computed and experimental KIEs (and therefore the TSs) do not match, despite sometimes heroic efforts. At this point, if one is limited to using only a quantum mechanical approach, one is left knowing what the TS is not, which is not generally a useful result. Quantum mechanical calculations are a crucial component in solving TSs via the structure interpolation approach, but are unlikely on their own to yield the correct TS. Quantum mechanical calculations are still constrained by the limits of computer power and of theory. These limitations are likely to remain significant factors into the foreseeable future.

Using BEBOVIB. The principles of BEBOVA discussed by Sims and Lewis[2] in their excellent description of the program BEBOVIB have not changed, but certain details of the process have. BEBOVA works by attempting to recreate the vibrational frequencies of the molecules involved. Once the frequencies have been calculated, the vibrational (ZPE and EXC) contributions to the KIE can be calculated, as well as the contribution from the change in mass and moments of inertia (MMI). The functioning of BEBOVIB is illustrated schematically in Fig. 6.

The input required for BEBOVA analysis are (a) structures of the initial and final states, (b) list of internal coordinates, (c) force constants for each internal coordinate, (d) (optionally) interaction or "off-diagonal" force coefficients or constants, (e) list of atomic masses, (f) temperature, and (g) the reaction coordinate [implicit to (c) and (f)].

FIG. 6. Schematic description of BEBOVIB operation. The route from input, at the top, to the calculated KIE, at the bottom, is illustrated. Procedures are in light outline and objects are in dark outline. This scheme assumes that the internal coordinate force constants for each structure are being calculated from the force constants for single bonds and ideal, tetrahedral bond angles in the input. It also assumes that interaction force coefficients are supplied. It is possible to provide internal coordinate force constants and/or interaction force constants directly without BEBOVIB recalculating them.

a. STRUCTURES. Structures of the initial and final states (for KIE modeling, this is the reactant and the TS) are required. Because IEs report on the *change* in vibrational environment, structures of the reactant and the TS are equally important and have equal effects on calculated KIEs. The creation of models of the reactant and test TS structures will be discussed under the heading "Structure Interpolation."

It has been a general practice in BEBOVA analyses to use cutoff models; this includes the NAD^+ substitution reactions discussed here. In a cutoff model, atoms more than a couple of bonds away from an isotopic label are deleted from the model. The justification for using cutoff models is the assumption that IEs are local effects, that the vibrational environment of any atom is not affected significantly by atoms more than a few bonds away. In the past, cutoff models were used because computers were not powerful enough and did not have enough memory to handle large molecules. In 1999, even desktop computers can easily handle molecules with 100 atoms, so this limitation is no longer important and the use of cutoff models is less necessary.

Stern and Wolfsberg[52] (see also Ref. 2) described the conditions under which cutoff models could be made. Models consisting of all atoms within three bonds of any site of isotopic substitution will give acceptable cutoff models by the criteria of Stern and Wolfsberg. If certain conditions are met, cutoff models consisting of all atoms within two bonds of any isotopic label, or even one bond, are possible.

The use of cutoff models can add an artifact to calculated KIEs, however, in the form of inaccurate MMI factors. For heavy atom KIEs, the contribution of the MMI factor can be large, up to one-half of the total IE.[23] If the size of the model molecules is significantly smaller than the real molecules, there is the risk of introducing an artifact in the MMI factor. The use of cutoff models should be evaluated for each reaction studied.

The format for structure input in BEBOVIB is a polar coordinate system. Bond lengths are not given directly, rather the bond order between two connected atoms is defined, along with the length of a single bond of that type. Structure input in VIBIE is in Cartesian coordinates or in a Z-matrix[53] of internal coordinates.

[52] M. J. Stern, and M. Wolfsberg, *J. Chem. Phys.* **45,** 4105 (1966).

[53] M. J. Frisch, G. W. Trucks, H. B. Schlegel, P. M. W. Gill, B. G. Johnson, M. A. Robb, J. R. Cheeseman, T. Keith, G. A. Petersson, J. A. Montgomery, K. Raghavachari, M. A. Al-Laham, V. G. Zakrzewski, J. V. Ortiz, J. B. Foresman, J. Cioslowski, B. B. Stefanov, A. Nanayakkara, M. Challacombe, C. Y. Peng, P. Y. Ayala, W. Chen, M. W. Wong, J. L. Andres, E. S. Replogle, R. Gomperts, R. L. Martin, D. J. Fox, J. S. Binkley, D. J. Defrees, J. Baker, J. P. Stewart, M. Hend-Gordon, C. Gonzalez, and J. A. Pople, "Gaussian 94, Revision C.2, D.4." Gaussian, Pittsburgh, PA, 1995.

b. LIST OF INTERNAL COORDINATES. Once the structures have been defined, it is necessary to define the connectivities between atoms and which types of vibrational motions may occur. There are five types of internal coordinates: bond stretch, bond angle bend, out-of-plane bend, linear bend, and torsion (Fig. 7). Out-of-plane bends are used in any structure containing three atoms bonded to a central atom where all four atoms are coplanar. Only one out-of-plane internal coordinate needs to be defined for a set of four coplanar atoms because displacing any of the terminal atoms out of the plane is equivalent to any other (because a plane can be defined with the central atom and any two of the three terminal atoms). Linear bends are defined for each set of three colinear atoms. For any three colinear atoms, two linear bends are defined by two virtual atoms connected to the central atom with bonds that are perpendicular to the linear bonds and perpendicular to each other. Acetylene contains two overlapping sets of three colinear atoms and requires two sets of two linear bends to be defined. The torsional force constants are for librations of the torsional angles, not free rotation.

Force constants may be supplied to the program as either force constants to be used directly or force constants for single bonds (or "ideal" bonds for bends), which BEBOVIB then uses to calculate the actual force constants for a given structure. The treatment to be used for the force constants is included in the list of internal coordinates, although the values of the

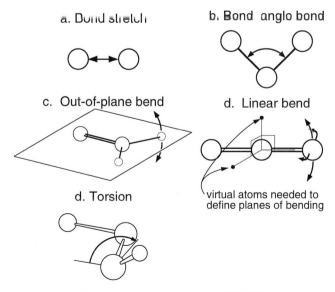

a. Bond stretch
b. Bond angle bond
c. Out-of-plane bend
d. Linear bend
virtual atoms needed to
define planes of bending
d. Torsion

FIG. 7. Internal coordinates used in BEBOVA.

force constants are listed separately. The stretching force constant is a simple function of the bond order, $F_{ij} = F_1 \cdot n_{ij}$ [Eq. (9)].

Several treatments for the force constants for bond angle bends are possible. Sims and Lewis[2,36] recommend:

$$F_{ijk} = (n_{ij}n_{jk})^x \cdot g_\alpha \cdot F_1 \tag{15}$$

where F_{ijk} is the bending force constant for the angle, α, formed by bonds i–j and j–k; x is 0.5 or 1.0 (see later); and g_α is the hybridization factor, $g_\alpha = 1.39 + 1.17 \cdot \cos \alpha$.

The value of x should usually be 0.5, except in the case of large changes in n_{ij} or n_{jk}, where 1.0 is preferable. The hybridization factor, g_α, attempts to reproduce observed trends in bond-bending force constants in certain hydrocarbons and other types of compounds (see references in 2, 36). For an ideal sp^3 carbon atom, with $\alpha = 109.5°$, $g_\alpha = 1.0$; it decreases as α increases, to 0.81 at 120° and 0.22 at 180°.

Only 3N-6 nonredundant internal coordinates are required to specify the normal modes of a molecule. A significant disadvantage of using nonredundant coordinates is that identical bond angle bending coordinates are treated unequally. For an sp^3 atom, e.g., methane, there are six possible bond angle bends, but one of those is redundant (Fig. 8). If five bending modes for methane are used, the normal modes must be described as linear combinations of the five (unequal) bending force constants, which would

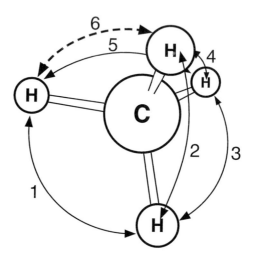

FIG. 8. Redundant bond-bending coordinates in methane. With five bond-bending internal coordinates defined, the sixth (dashed line) is redundant for specifying atom positions and vibrational frequencies, but is still useful. The order of numbering is arbitrary.

be represented in BEBOVIB as off-diagonal elements of the **F** matrix (see later). In BEBOVIB calculations, it is simpler to use six bending modes for methane, including one redundant one, all with equal force constants. BEBOVIB can calculate vibrational frequencies from an arbitrarily large number of internal force constants (i.e., >3N-6). It should be remembered when using force constants from the literature that they may have been derived from experimental vibrational frequencies using nonredundant sets of internal coordinates.

c. FORCE CONSTANTS FOR EACH INTERNAL COORDINATE. Internally, BEBOVIB builds an **F** matrix that contains, as its diagonal elements, the force constants for each internal coordinate (see later), i.e., **F** is a square matrix of dimension n, where n is the number of internal coordinates, with F_{ii} the force constant for internal coordinate i. The off-diagonal elements will be discussed later. An **F** matrix containing only diagonal elements is referred to as a simple valence force field (SVFF, as distinct from a general valence force field, GVFF, which includes off-diagonal elements). Generally, SVFFs are used in BEBOVA analyses, with the off-diagonal forces used to generate the reaction coordinate if necessary.

Values of force constants designed specifically for use in vibrational analyses can be found in several sources in the literature.[2,33,36,49] For NAD$^+$ TS determinations, the force constants for bond stretches and bond angle bends were taken from the AMBER force field.[54,55] This force field was derived by reference to the normal modes of model compounds and contains force constants for a large variety of atom types, making it a good source of force constants for use with BEBOVIB.

d. INTERACTION FORCE COEFFICIENTS OR CONSTANTS (OPTIONAL). Force constants for the internal coordinates described in the previous section are the diagonal components of the **F** matrix. They represent the restoring force exerted upon distortion of a bond. Interaction force constants are the off-diagonal components of the **F** matrix. They represent the interactions between internal coordinates. If, for example, stretching a given bond (internal coordinate i) reduces the force constant for a bond angle bend (internal coordinate j), then there is an interaction between those internal coordinates. This interaction constant is given by the value of matrix element F_{ij} (or F_{ji}; the matrix is symmetrical). A *positive* interaction force constant couples an *increase* in the value of one internal coordinate with a *decrease*

[54] W. D. Cornell, P. Cieplak, C. I. Bayly, I. R. Gould, K. M. Merz, Jr., D. M. Ferguson, D. C. Spellmeyer, T. Fox, J. W. Caldwell, and P. A. Kollman, *J. Am. Chem. Soc.* **117**, 5179 (1995).
[55] S. J. Weiner, P. A. Kollman, D. A. Case, U. C. Singh, C. Ghio, G. Alagona, S. Profeta, Jr., and P. Weiner, *J. Am. Chem. Soc.* **106**, 765 (1984).

in the other. A negative interaction force constant has the opposite effect. Coupling a bond stretch with a bond angle bend with a positive interaction force constant means that stretching the bond is coupled with closing the bond angle, and vice versa. Coupling two bond stretches with a positive interaction force constant means that the stretching of one bond is coupled with compression of the other. Reaction coordinate normal modes for bimolecular substitutions are generated using interaction force constants (see later).

Interaction force constants may be defined directly as a force constant, although, more commonly, they are defined as an interaction force coefficient (a_{ij}). This allows the interaction force constant to vary along with the internal coordinate force constants as they are adjusted as a function of the structure. The interaction force coefficient (a_{ij}) gives an interaction force constant based on the equation:

$$F_{ij} = a_{ij} \cdot \sqrt{F_{ii}F_{jj}} \tag{16}$$

In general, it has been found that acceptable accuracy in IE calculations can be achieved using SVFFs for stable molecules.[2,36] Off-diagonal elements can be used to generate the reaction coordinate mode, as described later.

e. List of Atomic Masses. Each isotopically labeled species is defined by listing the atomic mass of each atom in the molecule for initial and final structures. Many isotopologs may be defined in a single file so IEs at every labeled position can be calculated in one run.

f. Temperature. Vibrational frequencies are independent of temperature, but IEs are temperature dependent because the proportion of vibrationally excited molecules increases with temperature, which changes the contribution from the EXC term.[31,32,56,57] IEs at up to four temperatures can be calculated in a given run with BEBOVIB.

g. Reaction Coordinate. As discussed earlier, the reaction coordinate is the one normal mode in the TS that has an imaginary frequency. The reaction coordinate is generated through appropriate values of the internal coordinate force constants and/or interaction force constants discussed in c and d, respectively, but is treated separately here.

The reaction coordinate can be broken down conceptually into the main processes, and coupled processes. The main processes are the bond-making and bond-breaking events in the reaction. For a nucleophilic substitution, the main processes depend on the mechanism. For an A_ND_N mechanism, they are the concerted departure of the leaving group and approach of the nucleophile. In a $D_N + A_N$ mechanism, it is either the unimolecular depar-

[56] M. J. Stern, W. Spindel, and E. U. Monse, *J. Chem. Phys.* **52**, 2022 (1970).
[57] M. J. Stern, W. Spindel, and E. U. Monse, *J. Chem. Phys.* **48**, 2908 (1968).

ture of the leaving group or the nucleophilic attack on the electrophilic intermediate, depending on which is the first irreversible step. One coupled process for nucleophilic substitutions is the inversion of stereochemistry of the electrophilic carbon in an umbrella-like motion called the Walden inversion. In the NAD^+ example, this was achieved in part by introduction of a positive interaction force coefficient between the N-1–C-1' bond stretch and the N-1–C-1'–X bond bends, which meant that stretching the N-1–C-1' bond made decreasing the N-1–C-1'–X angles more favorable, facilitating Walden inversion. The inversion of stereochemistry was coupled with the main processes by using interaction force coefficients between bond bending and bond stretching forces (see later).

The first decision in modeling a reaction coordinate is to identify the main processes and decide which forces should be coupled to those main processes. For an A_ND_N mechanism, the main process is the concerted movement of the leaving group and nucleophile. The Walden inversion is coupled to this. Should there be other motions included in the reaction coordinate? Let us consider the nucleophilic attack of water on NAD^+ as an example.

As illustrated in Fig. 9a, a number of structural changes occur to NAD^+ on forming an oxocarbenium ion-like A_ND_N TS. A number of bonds, including C-1'–O-4' and C-1'–C-2', become stronger in compensation for the

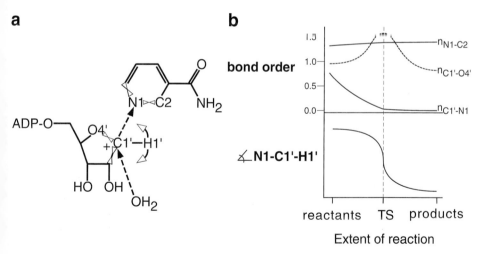

FIG. 9. Selection of structural changes to be coupled to the main processes in forming the reaction coordinate. (a) The main processes of this A_ND_N reaction are shown with dark arrowheads. Associated structural changes are shown with open arrowheads. (b) Changes in selected bond orders (top) or the ∠N-1–C-1'–H-1' bond angle through the course of the reaction (bottom). Curve shapes are approximate, based on electronic structure calculations.[68]

positive charge on the ribosyl ring. In the nicotinamide ring, the N-1–C-2 and N-1–C-6 bonds become stronger to compensate for the loss of the N-1–C-1′ bond. Because these structural changes occur as the reaction proceeds from the reactants to the TS, should they not be included in the reaction coordinate motion? Let us consider three bond lengths in turn (Fig. 9b, top). Electronic structure calculations show that $n_{C-1'-O-4'}$ increases from 0.96 to 1.59 as the ribosyl ring becomes more oxocarbenium ion like.[4] Because the TS is bimolecular, the bond order from the electrophilic carbon to the nucleophile water, $n_{C-1'-Nu}$ (or n_{Nu}), is increasing as the bond order to the leaving group, $n_{C-1'-N-1}$ (or n_{LG}), decreases. At some point, which we expect to be at or near the TS, the increase in $n_{C-1'-Nu}$ balances the decrease in $n_{C-1'-N-1}$. At this point the ribosyl ring achieves its maximal oxocarbenium ion character, i.e., the C-1′–O-4′ bond is at its shortest length and will start to lengthen as $n_{C-1'-Nu}$ continues to increase. Thus, at or near the TS, the length of the C-1′–O-4′ bond stops changing and is constant as the reaction coordinate proceeds through the TS toward products. Therefore, the C-1′–O-4′ bond stretch should not be coupled to the reaction coordinate normal mode. The bond order $n_{N-1-C-2}$ (and $n_{N-1-C-6}$) in the nicotinamide ring continues to increase as the TS is passed and a nicotinamide molecule is formed. In this case, however, the relative change in bond order[58] is small, going from $n_{N-1-C-2} = 1.32$ in the reactant to 1.39 at the TS and 1.40 in nicotinamide, so its contribution to the reaction coordinate will be small and is therefore neglected. The Walden inversion, as followed by the N-1–C-1′–H-1′ angle, proceeds monotonically from reactants, through the TS, to the products (Fig. 9b, bottom). It is included in the reaction coordinate motion. In summary, the reaction coordinate motion for an $A_N D_N$ mechanism will include the main processes, the bonds being made and broken, plus the coupled process, Walden inversion of the electrophilic carbon.

Main process reaction coordinate motions are generated using appropriate values of the internal coordinate force constants and/or the interaction force constants. The simplest reaction coordinate motion would be for a unimolecular dissociation. In this case the imaginary frequency of the reaction coordinate is generated by giving the internal coordinate for the breaking bond a zero or small negative force constant. Nucleophilic attack on the intermediate of a $D_N + A_N$ reaction is equivalent to a dissociation (in the reverse direction) and is treated in the same way.

---

[58] As discussed in Ref. 3, IEs are a nonlinear function of bond order, being more closely correlated with the relative change in bond order. Thus the IEs associated with a 1.06-fold increase in the N-1–C-2 bond order will be very much smaller than the 40-fold decrease in the C-1′–N-1 bond order.

For an $A_N D_N$ mechanism, the main process involves a concerted motion of the leaving group and the nucleophile. The reaction coordinate mode with an imaginary frequency is generated by coupling the bond stretches from the electrophilic carbon to the leaving group and the nucleophile. In NAD^+ hydrolysis, the C-1'–N-1 and C-1'–Nu bond stretches were coupled with a positive interaction force coefficient, so a stretch of one bond made compression of the other more favorable. By making the interaction force coefficient greater than 1 (1.1 in this case), the asymmetric Nu $\rightarrow$ C-1' $\rightarrow$ N-1 stretch acquired an imaginary frequency. An interaction force coefficient, rather than a interaction force constant, is used in these situations so that the proper reaction coordinate motion is generated automatically for different test TS structures with different values of $n_{C-1'-Nu}$ and $n_{C-1'-N-1}$. One advantage of generating a reaction coordinate through interaction forces, as opposed to simply assigning negative force constants to the Nu–C-1' and C-1'–N-1 bond stretches, is that the symmetric Nu $\rightarrow$ C-1' $\leftarrow$ N-1 bond stretch remains a harmonic vibration, as it should. In addition, by using an interaction force coefficient, the interaction force constant for a given test TS is calculated depending on the internal coordinate force constants for leaving group and nucleophile bonds. Because of this, the reaction coordinate imaginary frequency will have the desired characteristic of depending on $n_{C-1'-N-1}$ and $n_{C-1'-Nu}$, being higher for concerted and lower for less concerted (more dissociative) TSs.[22] A general method for generating a preselected imaginary frequency in a preselected normal vibrational motion has been described.[25,59]

The Walden inversion motion is generated by coupling the C-1'–Nu and C-1'–N-1 bond stretches to the bond angle bends centered on C-1', i.e., X–C-1'–Y, where X = N-1 and Nu, and Y = H-1', C-2', and O-4' (Fig. 1). An interaction force coefficient of 0.05 for coupling the C-1'–N-1 bond stretch to bends with X = N-1 and −0.05 to bends with X = Nu were used,[3,4] based on the work of Horenstein et al.[13] and Markham et al.[60] (Fig. 10). Similarly, interaction force coefficients were used to couple the C-1'–Nu bond to bends where X = Nu (0.05) and X = N-1 (−0.05). A general method for generating interaction force constants has been described.[25,59]

In an $A_{xh}D_H D_N$ (E_2) elimination mechanism, where two groups depart in concert, a negative interaction force coefficient would be used so that the concerted departure of both groups would have an imaginary frequency, whereas an asymmetric stretch would be a harmonic motion.

[59] W. E. Buddenbaum and P. E. Yankwich, *J. Phys. Chem.* **71**, 3136 (1967).
[60] G. D. Markham, D. W. Parkin, F. Mentch, and V. L. Schramm, *J. Biol. Chem.* **262**, 5609 (1987).

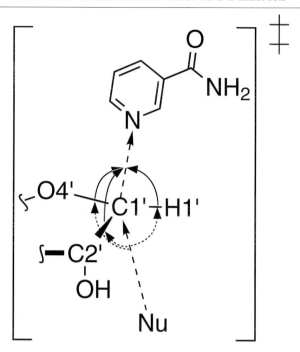

Fig. 10. Coupling of bond bending forces to the C-1'–N-1 bond stretch to create the Walden inversion motion. (Solid curves) Bond bends N-1–C-1'–Y (Y = H-1', O-4', C-2') are coupled with positive interaction force coefficients. (Dashed curves) Bond bends Nu–C-1'–Y are coupled with negative interaction force coefficients to the C-1'–N-1 bond stretch. These same bending internal coordinates are coupled through interaction force coefficients to the C-1'–Nu stretch, but with the opposite sign.

BEBOVIB calculates KIEs as KIE = MMI · ZPE · EXC, where MMI is calculated directly [Eq. (3a)]. The contribution to the KIE from the reaction coordinate is obtained by calculating MMI and VP [Eq. (4)] directly, then using $^{light}\nu*/^{heavy}\nu* = $ MMI/VP [see Eq. (5)]. Individual reaction coordinate frequencies are also calculated for each isotopolog from the eigenvalues and reported. In principle, the ratio of frequencies from eigenvalues should be the same as the ratio calculated from MMI/VP. In practice, rounding and other arithmetic errors mean that the ratios differ slightly.

Structure Interpolation

The structure interpolation method involves generating many test TS structures for which KIEs are calculated and compared with the experimental KIEs. It works by interpolating test TS structures throughout reaction space based on two reference structures. One of the reference structures will be the reactant or product. The other will be a high energy (possibly hypothetical) intermediate that is close in structure to the expected TS structure, such as an oxocarbenium ion for glycoside hydrolyses or a tetrahedral intermediate for amide hydrolysis.

CTBI (Cartesian Coordinates to Transition-State Structures to BEBOVIB Format Using Internal Coordinates). Structure interpolation is performed using the program CTBI: **C**artesian coordinates to **t**ransition-state structures to **BEBOVIB** format, using **i**nternal coordinates (written in QBasic, available from the author). The input includes the two reference structures and a list of values for the independent variables (n_{LG} and n_{Nu} for a nucleophilic substitution). The algorithm for determining interpolated structures is written into the program and would have to be changed for other reactions.

CTBI performs the following procedures: (a) Read in reference structures in Cartesian coordinates, including atom connectivities (PDB[61] format). (b) Parse structures to (i) find which atoms belong to the leaving group, the nucleophile, and the rest of the molecule and (ii) find cyclic fragments. (c) Convert the molecular definitions from Cartesian to internal coordinates. (d) Using an interpolation algorithm, create test TS structures throughout reaction space. (e) Convert the resulting test TS structures back to Cartesian coordinates and write the structures in the polar coordinate format for BEBOVIB.

The reference structures (step a), the parsing process (step b), and the interpolation process (step d) are described in some detail. Steps c and e are straightforward matrix and arithmetic calculations and will not be discussed further.

Reaction Space and Interpolation Philosophy. Reaction space (Fig. 4) is defined by the two independent variables: the internal coordinates that describe the "main processes" of the reaction, i.e., the bonds that are being made or broken. In a nucleophilic substitution, the main processes are breaking the bond to the leaving group and forming the bond to the nucleophile. Thus, the independent variables in reaction space are n_{LG} and n_{Nu}. In an elimination reaction, the main processes would be breaking the two leaving group bonds. All other structural changes that occur in the reaction are calculated as a function of the independent variables. For example, in NAD^+ hydrolysis, the reaction space was defined as a function of n_{LG} and n_{Nu}, and the reorganization of the ribosyl ring structure in oxocarbenium

[61] F. C. Bernstein, T. F. Koetzle, G. J. B. Williams, E. F. Meyer, Jr., M. D. Brice, J. R. Rodgers, O. Kennard, T. Shimanouchi, and M. Tasumi, *J. Mol. Biol.* **112,** 535 (1977).

ion formation was a dependent variable, governed by the values of n_{LG} and n_{Nu}.

The structure interpolation method attempts to create a reasonable test TS structure for any values of n_{LG} and n_{Nu}, whether or not those values are themselves reasonable. For example, it is clearly unreasonable to have a pentavalent TS, with $n_{LG} = n_{Nu} = 1$, but CTBI would extrapolate from known structures in an attempt to create a "reasonable" pentavalent structure. In essence, structure interpolation answers the question: If the TS occurred at this arbitrary point in reaction space (i.e., these values of n_{LG} and n_{Nu}), what would the structure be? The question of which values of n_{LG} and n_{Nu} constitute the correct TS is then answered by calculating KIEs for each isotopic label for each test TS structure. There is no attempt to limit the independent variables to reasonable values because what is reasonable is not clearly known until after the TS has been determined. There is no information on the reaction coordinate in the structure interpolation-derived structures; this is a function of the internal coordinate and/ or interaction forces in BEBOVIB.

Advantages of the structure interpolation approach over the old, *ad hoc* approach have already been discussed. It also has the advantages of speed and accuracy over one possible alternative, that of using electronic structure optimizations at fixed values of n_{LG} and n_{Nu}. The speed advantage arises because structure interpolation is simply an arithmetic manipulation of structures. The accuracy advantage arises from the fact that interpolated structures are derived from high accuracy reference structures to which empirical adjustments can be made to account for experimentally known phenomena, such as hyperconjugation, that are not modeled adequately in electronic structure calculations (see later).

Structure Interpolation in Internal Coordinates. Structure interpolations are performed in internal coordinates. Molecular structures are defined in the same Z-matrix internal coordinates (Fig. 7) as used by the electronic structure program GAUSSIAN.[53] These internal coordinates are of the same type as used for BEBOVA, but without the implicit motion. Thus a bond stretching in BEBOVA becomes a bond length when the structure is being defined. Similarly, bond angles and torsions are used. Out-of-plane bends and linear bends are not used in defining molecular structure.

The reason for using internal coordinates for interpolations is illustrated in Fig. 11. Interpolation between the atom at position 4 and at 4' is accomplished by increasing the torsional angle d_{3214}. No other internal coordinates are changed. If the structures were interpolated in Cartesian coordinates, the atom 4 would move in a straight line (shown by the dashed line and shaded circle), an unrealistic motion that would result in a decrease in r_{14} at intermediate values. Reference structure input is in Cartesian coordi-

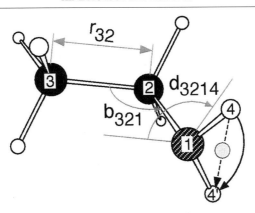

Fig. 11. Internal coordinates used in defining molecular structures and internally by CTBI during structure interpolation. The consequences of performing structure interpolations in Cartesian coordinates are illustrated for atom 4 (see text). With internal coordinates, this movement is accomplished by rotation about the C-2–O-1 bond by varying the dihedral angle d_{3214}.

nates, and the BEBOVIB formated output is in polar (i.e., a variation on Cartesian) coordinates, so CTBI converts the structures to and then back from internal coordinates.

The Interpolation Algorithm Applied to NAD^+ Hydrolysis. The structure interpolation technique is best explained in the context of a concrete example: NAD^+ hydrolysis reactions. The details of interpolation will vary with each reaction type, so this can only be considered a template, or archetypal, interpolation.

Based on other glycoside hydrolytic reactions studied in this and other laboratories,[10,14,16,62–65] NAD^+ hydrolysis was expected to proceed through an asynchronous, oxocarbenium ion-like A_ND_N TS. The emphasis, therefore, was on modeling the oxocarbenium ion-like section of reaction space accurately (the top left in Fig. 4). The first reference structure was the reactant, NAD^+. The second reference structure was the hypothetical oxocarbenium ion, plus free nicotinamide. These would be the intermediates in a $D_N + A_N$ mechanism (Fig. 4). The ribosyl ring, leaving group, and nucleophile were each treated differently.

[62] L. J. Mazzella, D. W. Parkin, P. C. Tyler, R. H. Furneaux, and V. L. Schramm, *J. Am. Chem. Soc.* **118**, 2111 (1996).

[63] V. L. Schramm, *in* "Enzyme Mechanism from Isotope Effects" (P. F. Cook, ed.), p. 367. CRC Press, Boca Raton, FL, 1991.

[64] D. W. Parkin, B. A. Horenstein, D. R. Abdulah, B. Estupinan, and V. L. Schramm, *J. Biol. Chem.* **266**, 20658 (1991).

[65] M. L. Sinnott, *Chem. Rev.* **90**, 1171 (1990).

For the ribosyl ring, the simplifying assumption was made that the structure of the ribosyl ring is a function of one parameter, the oxocarbenium ion character (OC), i.e., the similarity between the ribose ring in a given structure and the pure oxocarbenium ion. By definition, OC = 0 in the reactant NAD^+ molecule, and OC = 1 in the oxocarbenium ion. OC is defined as a simple function of n_{LG} and n_{Nu} in the test TS structure ($n_{LG,TS}$ and $n_{Nu,TS}$) and in the reactant ($n_{LG,reactant}$):

$$OC = 1 - (n_{Nu,TS} + n_{LG,TS})/n_{LG,reactant} \qquad (17)$$

In the simplest model, each internal coordinate in the ribose ring would be the linear combination of the first and second reference structures. When transition state imbalance[66,67] effects in the ribosyl ring were recognized and elucidated further by electronic structure calculations,[68] the ribosyl ring structure was made an exponential function of OC, i.e., $OC^{1.71}$. For example, for a bond length in the ribosyl ring ($r_{ribosyl,TS}$):

$$r_{ribosyl,TS} = r_{ribosyl,reactant} + OC^{1.71} \cdot (r_{ribosyl,oxocarbenium} - r_{ribosyl,reactant}) \qquad (18)$$

Internal coordinates in the nicotinamide leaving group (r_{nic} for bond lengths) were a function of n_{LG} only:

$$r_{nic,TS} = r_{nic,reactant} + (1 - n_{LG,TS}/n_{LG,reactant})* (r_{nic,nic} - r_{nic,reactant}) \qquad (19)$$

where $r_{nic,nic}$ is the corresponding bond length in the crystal structure of nicotinamide.[69]

To a first approximation, the internal structure of the nicotinamide ring should not depend on n_{Nu}, and the internal structure of the nucleophile should not depend on n_{LG}. In NAD^+ hydrolysis reactions, the nucleophile water was a single atom of mass = 18 amu, with no internal structure (i.e., O–H bonds) to adjust.

The Interpolation Algorithm: The General Case. The NAD^+ hydrolysis case was relatively straightforward because it was known beforehand that the TS would occur somewhere in the top left corner of reaction space (Fig. 4). In other cases, it may be necessary to use more than one pair of reference structures for different parts of reaction space. For example, in a phosphorolysis reaction, it is possible, *a priori,* that an associative mechanism is followed, with nucleophile approach leading leaving group departure. In this case, a pentacoordinate phosphorus structure may be needed as a reference structure. In elimination reactions, where $D_N + A_{xh}D_H$ (E1),

[66] C. F. Bernasconi, *Acc. Chem. Res.* **25,** 9 (1992).
[67] C. F. Bernasconi, *Adv. Phys. Org. Chem.* **27,** 119 (1992).
[68] P. J. Berti and V. L. Schramm, manuscript in preparation.
[69] W. B. Wright and G. S. D. King, *Acta Crystallogr.* **7,** 283 (1954).

$A_{xh}D_HD_N$ (E2), and $A_{xh}D_H + D_N$ (E1$_{cB}$) mechanisms are all possible, several pairs of reference structures may be needed.

Reference Structures. Two reference structures are required for a structure interpolation. The first reference structure is the reactant molecule. The preferred source for the reactant molecule structure would be a crystallographic structure. As the reactant is a stable molecule, it is possible that there will be structures of that molecule, or a closely related one, available through one of the crystallographic databases.[61,70] Because the majority of crystal structures do not include the coordinates of hydrogen atoms, these have to be added and their positions optimized using electronic structure calculations.

If several sources of structural information are available, all those structures should be considered in creating the first reference structure. For instance, there were several crystallographic structures of NAD^+ available from both small molecule and protein crystal structures. All of the crystal structures fell into two categories, having either 3'-*endo* or 2'-*endo* ribose ring puckers.[3] Unexpectedly, and unlike other (e.g., adenylyl) nucleotides,[71,72] the conformation of the ribose ring attached to the nicotinamide has a significant effect on bond orders. Therefore, there was a significant difference in KIEs calculated using the different conformations, and it was not possible to find a TS that matched the experimental KIEs using the 2'-*endo* conformer, or a combination of 3'-*endo* and 2'-*endo* conformers, as the reactant model. Only the 3'-*endo* reactant gave calculated KIEs that matched the measured KIEs. This was unexpected because, in general, ribose rings in nucleotides have weak conformational preferences, they interchange conformations readily, and there is little effect of ribose ring conformation on the overall structure.[71,73,74] Coupling constants from solution NMR offered a resolution to this apparent dilemma by showing that the distribution of ring conformers in solution is skewed toward 3'-*exo* and 2'-*exo*,[75,76] conformers that have not been observed in the solid state. These conformers are expected to have, like the 3'-*endo* conformer, longer

[70] F. H. Allen, J. E. Davies, J. J. Galloy, O. Johnson, O. Kennard, C. F. Macrae, E. Mitchell, G. F. Mitchell, J. M. Smith, and D. G. Watson, *J. Chem. Inf. Comput. Sci.* **31,** 187 (1991).

[71] S. L. Moodie and J. M. Thornton, *Nucleic Acids Res.* **21,** 1369 (1993).

[72] A. Gelbin, B. Schneider, L. Clowney, S. H. Hsieh, W. K. Olson, and H. M. Berman, *J. Am. Chem. Soc.* **118,** 519 (1996).

[73] J. M. Thornton and P. M. Bayley, *Biopolymers* **16,** 1971 (1977).

[74] C. Altona and M. Sundaralingam, *J. Am. Chem. Soc.* **95,** 2333 (1973).

[75] N. J. Oppenheimer, *in* "The Pyridine Nucleotide Coenzymes" (J. Everse, B. Anderson, and K.-S. You, eds.), p. 51. Academic Press, New York, 1982.

[76] N. J. Oppenheimer, *in* "Pyridine Nucleotide Coenzymes" (D. Dolphin, R. Poulson, and O. Avramovic, eds.), p. 185. Wiley-Interscience, New York, 1987.

C-1′–N-1 and C-4′–O-4′ bonds than the 2′-*endo* conformer, due to the anomeric effect.[3,77,78] Thus, it was necessary to consider structural information from a variety of sources to arrive at an appropriate first reference structure for NAD$^+$ hydrolysis. Because KIEs reflect a *change* in vibrational environment, the reactant molecule is as important in determining KIEs as the TS.

The second reference structure will be a computationally derived TS or a high energy (or hypothetical) intermediate structure for which electronic structure calculations are needed. Crystal structures of similar molecules may be useful. For example, earlier models[13] of the oxocarbenium ion of the ribose ring were based in part on the structure of ribonolactone, a molecule that, like the oxocarbenium ion, contains a planar, ribose-like ring structure.[79]

For the NAD$^+$ hydrolysis studies, the second reference structure consisted of two parts, the oxocarbenium ion and the nicotinamide molecule. The nicotinamide model was from a crystal structure,[69] with hydrogens added computationally. The oxocarbenium ion was an *ab initio* structure, optimized at the RHF/6-31G** level of theory. The large magnitude of 2′-^{3}H KIEs of NAD$^+$ hydrolysis are due to hyperconjugation.[3,13,80,81] The RHF-optimized structure produced KIEs much lower than the experimental KIEs because Hartree–Fock optimizations are inherently unable to reproduce electron correlation effects, including hyperconjugation. Empirical adjustments to the oxocarbenium ion were made to reproduce the experimental 2′-^{3}H KIEs, as discussed later.

REFERENCE STRUCTURE INPUT IN CTBI. Reference structure input is in Cartesian coordinates, specifically in Protein Databank[61] (PDB) format. The PDB format is used because all the bonds in a molecule are listed in the "CONECT" section of the file. A full connectivity list of all bonded atoms is needed for parsing and pruning the molecule. No explicit bonding information is included in the Z-matrix internal coordinate format as atoms may be defined with respect to atoms to which they are not bonded. Also, there is no ring closure information (see later). Finally, Z-matrix structures are not unique in that the same structure may be defined with different lists of internal coordinates. CTBI requires that the atom numbering and connectivities be identical in the first and second reference structures.

Both reference structures must contain all the atoms that will be present

---

[77] C. L. Perrin, K. B. Armstrong, and M. A. Fabian, *J. Am. Chem. Soc.* **116**, 715 (1994).

[78] A. J. Briggs, R. Glenn, P. G. Jones, A. J. Kirby, and P. Ramaswamy, *J. Am. Chem. Soc.* **106**, 6200 (1984).

[79] Y. Kinoshita, J. R. Ruble, and G. A. Jeffrey, *Carbohydr. Res.* **92**, 1 (1981).

[80] M. Ashwell, X. Guo, and M. L. Sinnott, *J. Am. Chem. Soc.* **114**, 10158 (1992).

[81] D. E. Sunko, I. Szele, and W. J. Hehre, *J. Am. Chem. Soc.* **99**, 5000 (1977).

in the test TS structures. For instance, the first reference structure includes the nucleophile. The bond length is unimportant as this will be adjusted according to the list of n_{Nu}'s provided, but its orientation relative to the ribosyl ring must be correct.

In internal coordinate definitions of molecular structures, and the polar coordinate format of BEBOVIB, each successive atom in the structure is defined by reference to a previously defined atom (the first atom is the origin; its position is not defined). Strictly speaking, these methods of defining structures give only atomic positions, with no information on bonding, so atoms can, in principle, be defined in any order. However, it is common in these methods, and it is required by CTBI, that each successive atom in a structure be specified by reference to a lower numbered atom to which it is bonded (Fig. 12).

The order of atom numbering is important in CTBI for two reasons. First, in the BEBOVIB structure definition, each atom must be bonded to an already defined atom. Thus, atom 2 must be bonded to 1. Atom 3 may be bonded to 1 or 2, and so on. Second, this numbering scheme makes possible the pruning procedure for finding cyclic structures described later.

EMPIRICAL ADJUSTMENTS TO REFERENCE STRUCTURES. The example of hyperconjugation cited earlier is a specific case of a general problem: the limited accuracy of electronic structure calculations in representing real molecules. One of the major strengths of the structure interpolation approach is its ability to include alternate sources of information in creating the reference structures.

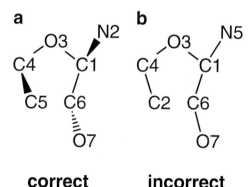

correct   **incorrect**

FIG. 12. Atom serial numbers in a fragment of an N-riboside structure for CTBI input. (a) Correct numbering, with each successive atom specified in reference to existing atoms. The C-5–C-6 bond is implicit because atom C-5 is specified by atoms C-4, O-3, and C-1, and atom C-6 is specified by atoms C-1, O-3 and C-4 (or C-1, N-2, and O-3). (b) Incorrect numbering, with C-2 specified in reference to C-1, to which it is not bonded.

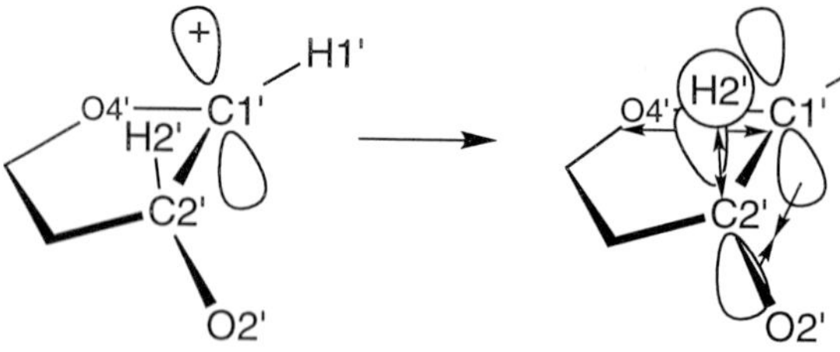

FIG. 13. Hyperconjugation of the C-2'–H-2' bond (see text).

Evidence from several sources was incorporated in modeling hyperconjugation: theoretical considerations of the angular dependence,[81] molecular orbital theory to explain the source and effects of hyperconjugation,[82] experimental KIEs from previous studies,[5,13,80] and density functional theory calculations that take some electron correlation effects into account.[83] Hyperconjugation arises from the interaction of the occupied π-symmetry orbital of the β-carbon, C-2', with the developing vacant p-orbital of the anomeric carbon as the leaving group departs[82] (Fig. 13). This leads to a lengthening of the C-2'–H-2' bond, which in turn causes the large 2'-^{3}H KIE. Because the cause and many of the characteristics of KIEs arising from hyperconjugation are understood, it was possible to model this effect by adjusting the structure of the oxocarbenium ion in the second reference structure. This was done by decreasing $n_{C-2'-H-2'}$ in the oxocarbenium ion, with an equal increase in $n_{C-1'-C-2'}$ and decrease in $n_{C-1'-O-4'}$ to conserve the total bond order to C-1'. The adjustment was increased until the calculated 2'-^{3}H KIE at the TS matched the experimental KIE. The fact that the match of calculated to experimental 2'-^{3}H KIEs is excellent in all the NAD$^+$ hydrolysis and ADP-ribosylation reactions supports this approach.

Structure Parsing. Because the interpolation algorithm treats the leaving group, ribosyl ring, and nucleophile differently, CTBI parses the structures to find which atoms belong to each fragment of the molecule. The program assumes that atom 1 is the electrophilic carbon (C-1') and that atom 2 is the leaving group nitrogen (N-1). Atoms other than 1 that are bound to atom 2 are part of the leaving group; all others are part of the ribosyl moiety. At present, the nucleophile is the last atom defined so it must be only one atom.

Structure Pruning to Find Rings. The significance of cyclic structures is that even though all the bond connectivities must be included to create a reasonable vibrational model of the molecule, one of the bonds in a cyclic

[82] W. J. Hehre, *Acc. Chem. Res.* **8,** 369 (1975).
[83] Unpublished results.

structure is not defined when the structure is converted from Cartesian to internal coordinates. When a molecular structure is defined in terms of internal coordinates, one of the bonds in the ring is implicit; it is not needed to define the position of any atom (see Fig. 12). Which bond is the "ring closure" bond is somewhat arbitrary, being governed by the atoms' serial numbers, but it is important because (a) the ring closure bond must be added to the list of internal coordinates to generate a complete vibrational model, (b) BEBOVIB requires that ring closure bonds be included in the structure definition, and (c) beacause of the possibility of distortion of nonplanar rings (see later).

THE RING CLOSURE BOND. A *ring closure* bond is defined as a bond that is not required for specifying atomic positions (i.e., it is redundant for defining the molecular structure), but that is specified in order to close a cyclic structure. This bond is required to generate a complete list of internal coordinates (i.e., stretching, different types of bending, and torsions) for the vibrational model of the molecule.

CTBI uses a pruning algorithm to recognize the presence of a ring (or rings) and to decide which is the ring closure bond. Because a list of all bonds is provided along with the Cartesian coordinates in the input file, it is simply a matter of finding which bond is implicit when the molecular structure is defined in internal coordinates. CTBI recognizes rings by "pruning" from the terminal atoms toward the middle. Terminal atoms (with no other atoms bound to them) are deleted iteratively until there are no more terminal atoms. At this point, the molecule contains only ring atoms, or internal atoms connecting rings. Because of the way that atoms in a molecule are numbered, any atom connected *only* to lower numbered atoms (a) is a ring atom and (b) is one atom in the ring closure bond. The next highest numbered atom is the other atom in the ring closure bond. In a fused ring system (e.g., purines), two or more ring closure bonds must be defined. This is done automatically by CTBI.

DISTORTION OF NONPLANAR RINGS. Some distortion is inevitable at the site of ring closure in nonplanar rings. The problem is illustrated more easily than explained. Consider a ribosyl ring undergoing a conformational change where its ring pucker flips from 3-*endo* (Fig. 14) to 3-*exo*. Given the serial numbers of the atoms as shown in Fig. 14, the bond connecting C-5–C-6 is the ring closure bond. If a structure is generated that is halfway between the two stable conformers, then atom C5 will be moved to position C-5′ by changing the dihedral angle $\angle$C-1–O-3–C-4–C-5 to 0°. The C-4–C-5 bond length remains unchanged, but the implicit bond C-5–C-6 is shortened from 1.53 to 1.35 Å, giving an increase in the Pauling bond order from 0.98 to 1.80. Because bond-stretching force constants are directly proportional to bond order, in this case the bond stretch force constant for

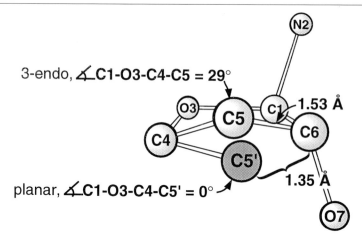

FIG. 14. Distortion of a nonplanar ring structure in structure interpolation (see text).

the C-5–C-6 bond will be almost doubled whereas there will be no change in the C-4–C-5 bond stretch. This will affect the vibrational frequencies involving that bond, and therefore the calculated isotope effects. This is an extreme example, but any structure interpolation of nonplanar rings will exhibit this type of distortion to some extent.

To avoid this problem altogether would require performing a full geometric optimization of each test TS structure, which would be too expensive computationally. Alternatively, the problem can be alleviated to a large extent by using redundant atoms. Two C-5 atoms are defined in each reference model, both with the same Cartesian coordinates. One, C-5a, is bonded to C-4, but not C-6, and its internal coordinates are specified relative to C-4. The other, C-5b, is bonded to C-6 (not to C-4 or C-5a), with its internal coordinates specified in relation to C-6. Force constants and vibrational frequencies involving C-4 are calculated using C-5a, similarly with C-6 and C-5b. Because neither C-4 nor C-6 is subject to the types of distortion seen in the example in Fig. 14, secondary ^{3}H KIEs can be calculated with better accuracy for hydrogens bound to C-4 and C-6.

Matching Test Transition-State Structures to Experimental KIEs

We can now use the structure interpolation method to generate many test TS structures and BEBOVA analysis to calculate the KIE for each isotopic label for each test TS structure. The last step is to interpret the calculated KIEs, i.e., to find which test TS structure gives calculated KIEs that most closely match the measured, experimental KIEs.

CONTOUR DIAGRAMS. Finding the experimental TS is accomplished most conveniently by plotting contour diagrams of the calculated KIEs (Fig. 5). The top left panel of Fig. 5 shows a plot of the calculated 1-^{15}N KIE as a function of position in reaction space. The 1-^{15}N KIE depends most strongly on n_{LG} and less on n_{Nu}. The top middle panel of Fig. 5 is an expansion of the top left corner of reaction space, in the region of highly dissociative TSs. The gray shading in Fig. 5 represents the areas in reaction space where the calculated KIEs match the experimental values exactly (dark gray) or within the 95% confidence interval (light gray). Similar plots are shown for the other isotopically labeled positions. If both the KIE measurements and the modeling process were accurate, there should be (at least) one point in reaction space where the calculated and experimental KIEs match for all the labeled positions. This was the case for the diphtheria toxin-catalyzed hydrolysis of NAD$^+$ (Fig. 5). The position of the experimental TS in reaction space is marked by an asterisk in Fig. 5. The calculated KIEs for the TS were within experimental error for all the isotopic positions, except for the 4'-^{18}O label, where it fell just outside of experimental error. Having found a TS that matches the experimental KIEs, it remains to (a) assess the accuracy of the structure, (b) add atoms back to the cutoff model, if necessary, and (c) analyze the structure with respect to geometry, charge distribution, and so on with an eye to explaining enzyme mechanisms.

Transition-State Accuracy and Sources of Error

Broadly speaking, there are two sources of uncertainty in assessing the accuracy of the experimental TS: errors in measuring KIEs and errors in the modeling process.

Many of the issues concerning the accuracy and precision of KIE measurements are those of analytical chemistry in general, whereas some are specific to IEs, such as whether greater accuracy can be achieved by measuring isotope ratios in the products or residual substrates,[84] or the quenching of counts. Many of these issues have been discussed by Parkin.[38]

Assuming that the experimental KIEs are correct, the second source of uncertainty is in the modeling process. There is no theory to allow a systematic evaluation of the accuracy of TSs determined from experimental KIEs; one must consider each case individually to find evidence to corroborate the KIE-based TS. The accuracy of TSs derived from KIEs is difficult to assess because KIEs provide data on TSs that is not available from other sources. Approaches such as linear-free energy relationships, inhibition constants for TS analogs or crystallographic structures, do provide informa-

[84] R. G. Duggleby and D. B. Northrop, *Bioorg. Chem.* **17,** 177 (1989).

tion on the TS, but only on a qualitative level that will support or contradict the general nature of the TS determined by KIEs, but not address the more precise, quantitative conclusions that can be achieved with KIEs.

Based on TSs solved in this laboratory, several strong enzyme inhibitors have been designed, lending support to KIE-derived TSs. These include formycin monophosphate with AMP nucleosidase (K_d = 43 nM),[85] a p-nitrophenylamidrazone compound with nucleoside hydrolase IU (K_d = 2 nM),[86] and immucilin-H with purine nucleoside phosphorylase (K_d = 23 pM).[87] The structure of nucleoside hydrolase IU cocrystallized with a TS mimic inhibitor[88] strongly corroborated the TS determined from KIEs.[13,89,90] These studies substantiated the conclusion that all these enzymes catalyze reactions passing through oxocarbenium ion-*like* TSs, but were unable to provide evidence to differentiate between A_ND_N and $D_N + A_N$ TSs; this is the unique domain of KIEs. These examples all represent *post facto* support for KIE-derived TSs, and none address the finer, quantitative distinctions of which KIEs are capable.

In some studies, estimates of the range of possible TSs that were consistent with the experimental KIEs were made by fitting TSs to the upper and lower limits of the error range of the experimental KIEs.[11,13,14,16,60] This approach provides an indication of how sensitive the experimental TS is to variation in the KIEs. This approach is useful because it gives some indication of the range of TSs that are possible, but it should be remembered that it assumes that the only source of error is in the KIE measurements and that the calculated KIEs for a given TS are exactly correct. The error range is also used in the structure interpolation approach in that the contour plots contain information on the exact measured KIEs (dark gray, Fig. 5) and the confidence interval (light gray). The significance of the confidence interval for a given single experimental KIE is not as clear, however, when the TS is defined as the intersection of the contour bands for many KIEs.

One promising approach that has not yet been used would take advantage of the high accuracy of calculated frequencies possible with many *ab initio* methods, particularly density functional theory methods,[91–93] to act

[85] W. E. DeWolf, Jr., F. A. Fullin, and V. L. Schramm, *J. Biol. Chem.* **254,** 10868 (1979).
[86] M. Boutellier, B. A. Horenstein, A. Semenyaka, V. L. Schramm, and B. Ganem, *Biochemistry* **33,** 3994 (1994).
[87] R. W. Miles, P. C. Tyler, R. H. Furneaux, C. K. Bagdassarian, and V. L. Schramm, *Biochemistry* **37,** 8615 (1998).
[88] M. Degano, S. C. Almo, J. C. Sacchettini, and V. L. Schramm, *Biochemistry* **37,** 6277 (1998).
[89] B. A. Horenstein and V. L. Schramm, *Biochemistry* **32,** 9917 (1993).
[90] B. A. Horenstein and V. L. Schramm, *Biochemistry* **32,** 7089 (1993).
[91] A. P. Scott and L. Radom, *J. Phys. Chem.* **100,** 16502 (1996).
[92] J. A. Pople, A. P. Scott, M. W. Wong, and L. Radom, *Isr. J. Chem.* **33,** 345 (1993).
[93] M. W. Wong, *Chem. Phys. Lett.* **256,** 391 (1996).

as a reference for the BEBOVIB vibrational model. *Ab initio* optimized reactant and TS structures would not necessarily be expected to be correct, but the vibrational frequencies calculated for those structures are likely to be highly accurate. If the BEBOVIB vibrational model could duplicate the *ab initio*-based KIEs for the *ab initio* optimized reactant and TS structures, that would support the accuracy of the BEBOVIB KIE calculations. If the vibrational model used with BEBOVIB is known to be accurate, that would be evidence for the accuracy of the (different) TS determined by matching BEBOVIB-based KIEs to the experimental KIEs.

Other evidence that may be considered includes the consistency of the TS with existing chemical reactivity results and the consistency of results between related TSs in a unified model. With the TS for solvolytic hydrolysis of NAD$^+$,[3] even though the experimental KIEs indicated an extremely dissociative TS with very little nucleophile bond order, it was possible to be confident of the mechanism because a large body of reactivity data on NAD$^+$ and related ribosides supported an A_ND_N mechanism rather than $D_N + A_N$.[3] The combination of KIE and reactivity data on their own gave good confidence for the TS of the solvolytic hydrolysis reaction. The unified model used to determine the TSs of five other reactions[94] with no changes to the reference structures, interpolation algorithm, or vibrational model gave a strong indication that all are essentially correct.

Another source of confidence in the results is if the experimental KIEs force one to reach previously unexpected conclusions about the TS that can be shown to be true, or at least reasonable. Three examples are given. The first has already been discussed, the effect of 3'-*endo* versus 2'-*endo* ring conformation in the reactant on calculated KIEs. A second example is in the cholera[5] and pertussis[6] toxin-catalyzed hydrolyses of NAD$^+$, for which TSs were determined before the introduction of structure interpolation, using the *ad hoc* method, without using quantum chemically derived reference structures for guidance. In those cases, it was possible to match calculated to experimental KIEs only when the total bond order to the anomeric carbon, C-1', *increased* between the reactant and the highly dissociative TSs. The same result was observed with the TSs solved by the structure interpolation method, namely those for diphtheria toxin-catalyzed[4] and solvolytic hydrolysis.[3] A variety of electronic structure calculations on the oxocarbenium ion support the conclusion that the π-bonding interactions formed from C-1' to O-4' and C-2' at the TS are greater in

---

[94] The TSs for diphtheria toxin catalyzed hydrolysis[4] and ADP-ribosylation of protein $G_{i\alpha 1}$[7] have been published. Reanalysis of the KIEs hydrolyses catalyzed by cholera[5] and pertussis[6] toxins gave TSs very similar to that for the diphtheria toxin reaction, and reanalysis of the KIEs for pertussis toxin-catalyzed ADP-ribosylation of a 20-mer peptide[8] gave a TS identical to that for ADP-ribosylation of $G_{i\alpha 1}$.

Experimental data

Bond energy/bond
order vibrational
analysis

Structure
interpolation

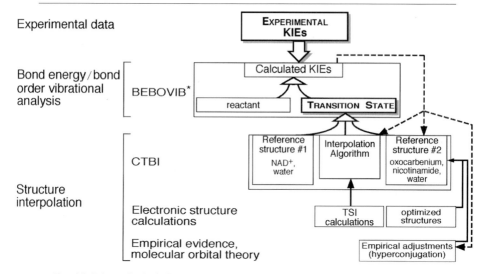

FIG. 15. Schematic depiction of the process of determining a TS. The goal of the process is to find a TS giving calculated KIEs that match the experimental KIEs. The arrows represent, roughly, the flow of information during the process. Dashed lines represent "feedback" from calculated KIEs to earlier stages in the modeling process. The left column shows the three separate parts of the process. The middle column shows the computer program or other source of information. *The function of BEBOVIB is illustrated in Fig. 6.

sum than the loss of the N-glycosidic bond order[3] (and unpublished results). The third example is the observation of transition state imbalance (TSI) in the ADP-ribosylation of protein $G_{i\alpha 1}$.[7] TSI is relatively well known in some contexts, such as proton transfer,[66,67,95,96] but was not expected in this case. It was impossible to match the experimental KIEs using the structure interpolation method without accounting for TSI effects in the interpolation algorithm, even though the structural differences were very small. The differences in bond lengths in the TS if TSI was neglected in the interpolation algorithm were $\Delta r_{C\text{-}1'\text{-}O\text{-}4'} = -0.028$ Å, $\Delta r_{C\text{-}1'\text{-}C\text{-}2'} = -0.012$ Å, $\Delta r_{C\text{-}1'\text{-}H\text{-}1'} = 0.006$ Å. The smallness of these changes illustrates an important property of the unified model: that changes in TS structure can be detected more sensitively than the absolute structure. In other words, the TS structures are not accurate to within 0.01 Å. They are based on reference structures derived from crystal structures and calculations, both of which have >0.01 Å errors in atom position. Whatever the accuracy may be of

[95] C. F. Bernasconi and P. J. Wenzel, *J. Am. Chem. Soc.* **116**, 5405 (1994).
[96] C. F. Bernasconi, *Tetrahedron* **41**, 3219 (1985).

any one of the TS structures, deviation by one structure from the unified model is detected with great precision. The TSI effects were unexpected, but electronic structure calculations and a consideration of the changes in molecular orbital interactions on TS formation confirm their existence. The fact that such small structural changes could be detected highlights the accuracy and precision possible in TS determinations.

Summary

BEBOVA-based TS determination has been very successful in elucidating enzyme mechanisms at a level of detail that would be otherwise inaccessible. The resulting TS structures have been used successfully as the basis for designing TS mimics as enzyme inhibitors with dissociation constants to 10^{-11} M. The structure interpolation approach has systematized the process of finding a TS, increasing both the speed and the accuracy of TS determination.

The combination of information from several TSs into a unified model increases the accuracy of the process significantly and results in an extremely sensitive probe of changes in TS with varying reaction conditions (i.e., enzymatic vs nonenzymatic reactions, different enzymes, or different nucleophiles). The TS determination process is summarized in Fig. 15.

Acknowledgments

The author thanks Vern Schramm for support and advice throughout this work and for the freedom to explore new directions in TS analysis, as well as other members of the Schramm laboratory whose work is discussed here: Drs. Xiangyang Chen, Robert Miles, Kathy Rising, and Johannes Scheuring. The author also thanks Drs. Phil Huskey of Rutgers University and Alvan Hengge of Utah State University for their critical reading and thoughtful comments on the manuscript. This work was supported by NIH Research Grant AI34342 to Vern L. Schramm and by a postdoctoral fellowship from the Natural Sciences and Engineering Research Council (Canada) to P.J.B.

[15] Computational Methods for Transition State and Inhibitor Recognition

By BENJAMIN B. BRAUNHEIM and STEVEN D. SCHWARTZ

Introduction

The discovery of enzymatic inhibitors for therapeutic use is often accomplished through random searches. Hundreds of thousands of molecules are tested with refined enzyme for signs of inhibition. This process is expensive and time-consuming, costing a great deal of money and taking many years. The development of computational and theoretical methods for prediction of the binding constant of a putative inhibitor, prior to synthesis and testing, would facilitate the discovery of novel inhibitors.

The first attempts to develop computational methods to predict the potency of an inhibitor prior to synthesis are described broadly as quantitative structure activity relationship (QSAR) studies. QSARs are polynomial equations with n terms, where n corresponds to the number of ways regions of a molecule can be altered.[1] The value of these n coefficients are varied depending on how changes at those regions affect binding.[1] It is believed that once enough enzyme inhibition experiments have been done and the polynomial has been adjusted properly, it could be used to predict the inhibition constant of molecules that have not been tested experimentally.[1] The problems with these techniques lie in their need for the user to define a functional relationship between molecular structure and molecular action. In the QSAR approach, or any approach where a person is charged with adjusting a mathematical model, the investigator must use variations in the structure of a molecule as the motivation for changing the value of coefficients in the model. It is not possible to predict *a priori all* the effects a change to a substrate molecule will have on enzymatic action. An example of this is the fact that variations in double bond conjugation can have only slight effects on the size and electrostatic potential of a group, but it can have a large effect on the electrostatic potential of the molecule as a whole, particularly if the enzyme polarizes the substrate on formation of the transition state. Polarizable groups of molecules can have a large effect on the electrostatic potential of a molecule, even if the only field acting on the molecule is that of the molecule itself. The following statements were

---

[1] E. J. Ariens, "QSAR: Quantitative Structure-Activity Relationships in Drug Design," p. 3. A. R. Liss, 1989.

Copyright © 1999 by Academic Press
All rights of reproduction in any form reserved.
0076-6879/99 $30.00

made about the limitations of traditional QSAR: "With drugs one usually isn't even sure of the atom from which to base steric effects. Second, in obscurity are electronic effects. Again, with drugs part of the problem is to decide which is the key atom from which substituent effects should be measured.... To summarize, a big problem of traditional QSAR is describing molecules to the computer. This is the most time-consuming and ambiguous part of a QSAR analysis."[2]

This article describes an approach that is different from that of a QSAR. We do not use classically derived quantities such as volume, hydrophobicity, or number of specific groups to describe molecules. We use *ab initio* quantum mechanics to describe molecules. In addition, we describe molecules as coincidentally oriented surfaces that vary in geometry and electrostatic potential. QSARs describe molecules as a collection of features that are independent of orientation. As a result of this simplification, libraries can be searched more easily with the QSAR approach, but subtle information about the bonding capability of a molecule is lost. The energy of noncovalent interactions such as ionic interactions and hydrogen bonds drops off with $1/r$. Van der Waals interactions drop off with $1/r^{12}$. As a result of these distance dependencies, the relative geometric position information of groups is vital to the task of simulating molecular recognition.

The structure of this article is as follows: it begins with a brief description of transition-state structures and enzyme stabilization. The concept of a transition-state analog will be central to our discovery process. We then describe how quantum mechanics and the electrostatic potential are used to discover putative inhibitors. We next describe the use of mathematical similarity measures for quantum systems in the prediction of binding strength. Application to three enzyme systems will be used to demonstrate the strength and weaknesses of the method. We then discuss the application of neural computation networks to the recognition problem. The method will be demonstrated through a study of an enzyme system not amenable to similarity measure analysis. The article concludes with future directions.

Enzyme-Stabilized Transition-State Structures

Along the reaction coordinate from reactant to product, the reactant reaches its most unstable configuration at the transition state. Transition-state stabilization theory says that enzymes increase reaction rates by stabilizing the transition-state configuration. The theory suggests that enzymes bind the transition-state structure with high energy contacts. The atomic

[2] Y. Martin, K. Kim, and T. C. Lin, "Advances in Quantitative Structure-Property Relationships," p. 3. JAI Press, Greenwich, CT, 1996.

and electronic structure of the transition state gives information about the enzyme active site when the enzyme is in the transition-state stabilizing structure. Molecules in the transition-state structure are used in our approach because they represent the structure that the enzyme evolved to bind most tightly. With this information one can design stable molecules that mimic the geometry and electrostatic potential of the enzyme-stabilized transition state. If one were to find a stable molecule that could bind to the enzyme in the same way as the transition state, that molecule would bind strongly to the enzyme through slow onset inhibition and destroy enzymatic action.

The most basic description of a molecule is that of the quantum mechanical wave function. The quantum properties of an inhibitor are in fact what an enzyme-active site will recognize; we therefore use quantum mechanics as the descriptor of molecules. Quantum mechanics is important for simulating molecular recognition because the molecular interactions that define recognition are sensitive to subtle variations caused by intra- and intermolecular polarizations. Polarizations across conjugated bonds can have a large effect on binding energy. Polarizations of large atoms such as Br or I can have profound effects on binding. We will argue that to be effective as a recognition algorithm, any attempt to simplify the description of a molecule before presentation to a comparison algorithm must not remove information concerning the electrostatic potential on the surface of the entire molecule and the relative position of points on the surface of the molecule.

We create quantum descriptions of molecules in the following way: First, the molecular structures are energy minimized using semiempirical methods. Molecules with many degrees of freedom are configured such that they all have their flexible regions in the same relative position. Then the wave function for the molecule is calculated with the program Gaussian 94.[3] From the wave function, the electrostatic potential is calculated at all points around and within the molecule. The electron density, the square of the wave function, is also calculated at all points around and within the molecule. With these two pieces of information the electrostatic potential at the van der Waals surface can be generated. Such information sheds light on the kinds of interactions a given molecule can have with the active site.[4] Regions with electrostatic potentials close to zero are likely to be capable of van der Waals interactions, regions with a partial positive or negative charge can serve as hydrogen bond donor or acceptor sites, and regions with even greater positive or negative potentials may be involved in

[3] "Gaussian 94, Revision C.2." Gaussian, Pittsburgh, PA, 1995.
[4] B. A. Horenstein and V. L. Schramm, *Biochemistry* **32,** 7089 (1993).

coulombic interactions. The electrostatic potential also conveys information concerning the likelihood that a particular region can undergo electrophilic or nucleophilic attack.[5] Since molecules described by quantum mechanics have a finite electron density in all space, a reasonable cutoff is required to define a molecular geometry. One choice is the van der Waals surface, within which 95% of the electron density is found. One can approximate the van der Waals surface closely by finding all points around a molecule where the electron density is close to $0.002 \pm \delta$ electrons/bohr.[3] δ is the acceptance tolerance. δ is adjusted so that about 15 points per atom are accepted, creating a fairly uniform molecular surface, as shown previously.[6] The information about a given molecular surface is thus described by a matrix with dimensions of $4 \times n$, where n is the number of points for the molecule and the row vector of length 4 contains the x, y, z coordinates of a given point and the electrostatic potential at that point.

Similarity Measures

Having defined the necessary inputs to our recognition algorithms, the quantum similarity measure will now be described. Our work with similarity measures is based on the principle that stable molecules that are similar in structure to the transition state make good transition-state inhibitors. The structure of a molecule is defined for this application as the electrostatic potential at the van der Waals surface of a molecule.[6,7] The molecules are oriented for maximum geometric coincidence and the degree of similarity (both in electrostatic potential and in geometry) of the surfaces is used to generate an output. Two molecules that are quite similar will produce an output close to one, whereas two different molecules will generate an output close to zero.[6,7] This method is most useful when searching for transition-state mimics. If the transition-state structure is known from heavy atom kinetic isotope effect experiments,[8–10] the similarity measure can be used to compare ground-state molecules to the transition state. Ground-state molecules that are similar to the enzyme-induced transition state are potent inhibitors.[6] A strong inhibitor can be effective even if the substrate is more concentrated in the solution by a factor of 10^6.

[5] B. E. Evans, G. N. Mitchell, and R. Wolfenden, *Biochemistry* **14,** 621 (1975).
[6] C. K. Bagdassarian, V. L. Schramm, and S. D. Schwartz, *J. Am. Chem. Soc.* **118,** 8825 (1996).
[7] C. K. Bagdassarian, B. B. Braunheim, V. L. Schramm, and S. D. Schwartz, *Int. J. Quant. Chem. Quant. Biol. Symp.* **23,** 73 (1996).
[8] D. W. Parkin, F. Mentch, G. A. Banks, B. A. Horenstein, and V. L. Schramm, *Biochemistry* **30,** 4586 (1991).
[9] D. J. Merkler, P. C. Kline, P. Weiss, and V. L. Schramm, *Biochemistry* **32,** 12993 (1993).
[10] F. Mentch, D. W. Parkin, and V. L. Schramm, *Biochemistry* **26,** 921 (1987).

Molecular similarity measures derive from the idea that different molecules can be compared and the degree to which they both share the same qualities can be measured.[11] We have defined a similarity measure as

$$S = \frac{\int \varepsilon_A(r)\varepsilon_B(r)\,\mathrm{d}r}{\sqrt{\int \varepsilon_A(r)\,\mathrm{d}r}\,\sqrt{\int \varepsilon_B(r)\varepsilon_B(r)\,\mathrm{d}r}} \tag{1}$$

Let ε_A and ε_B be some quality of interest that the two molecules posses, such as electron density or electrostatic potential; r is the position vector. Regions within the van der Waals surface of a molecule are not important in noncovalent interactions at biological temperatures, thus this integral is computationally expensive and compares regions of the molecules that are irrelevant from a biological point of view.

We thus define a discretized similarity measure that compares points at the van der Waals surfaces of two molecules:

$$S = \frac{\sum\limits_{i=1}^{nA}\sum\limits_{j=1}^{nB} \varepsilon_i^A \varepsilon_j^B \exp(-\alpha r_{ij}^2)}{\sqrt{\sum\limits_{i=1}^{nA}\sum\limits_{j=1}^{nA} \varepsilon_i^A \varepsilon_j^A \exp(-\alpha r_{ij}^2)}\,\sqrt{\sum\limits_{i=1}^{nB}\sum\limits_{j=1}^{nB} \varepsilon_i^B \varepsilon_j^B \exp(-\alpha r_{ij}^2)}} \tag{2}$$

Here ε_i^A and ε_j^B are the electrostatic potential at point i on molecule A and point j on molecule B; r_{ij} is the distance between two points. Unlike the similarity measure that uses a point-by-point integration over all space, this similarity measure compares points on surfaces that are not coincident in space. Thus each comparison of n points must be weighted. This is accomplished by an exponential decay. The term in the exponent is negative so all surface point comparisons that occur over large distances have less effect than those of small distances. The term is squared to reduce computation time. α acts to modulate the effects of the distance term r_{ij}; when α is small, points separated by large distances have little effect on the similarity output whereas the opposite is true for large values of α. Figure 1 shows how the similarity measure output of a molecular surface point comparison varies with α.[7] This graph shows the similarity output for five inhibitors of two enzymes. (R)- and (S)-coformycin are inhibitors of AMP deaminase; they along with AMP are compared to the transition state of this reaction. Formycin and AMP are compared to the transition state of AMP nucleosidase. The important point of this graph is that the lines for any two molecules' S never cross. For any physically reasonable value of α the similarity output for two molecules will remain in the same relative order.

[11] P. G. Mezey, "Shape in Chemistry: An Introduction to Molecular Shape Topology." VCH, New York, 1993.

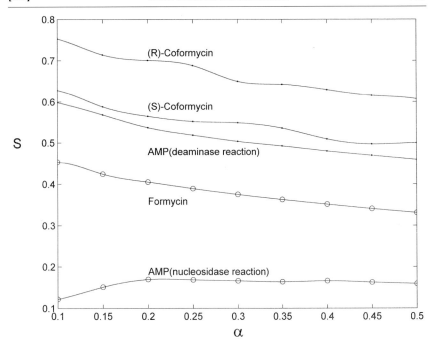

FIG. 1. Graph of similarity (S) as a function of α.

Physically reasonable values of α range from 0.1 to 0.5 bohr^{-2}. With $\alpha =$ 0.1, 0.3, 0.5 bohr^{-2} the exponential decays to e^{-1} for distance of 1.67, 0.97, and 0.75 Å, respectively.[6] Use of similarity measures requires that the molecules be oriented for maximum geometric coincidence. This can be accomplished in a variety of different ways. In our original work,[6,7] we defined a geometric similarity measure in which ε was replaced by unity. We demonstrated that maximization of this measure maximized geometric coincidence. To orient two molecules with a geometric similarity measure one assigns a unit mass to all points nA and nB and translates both molecule centers of mass to the origin. Then one assigns unit charge to all points and allows one of the molecules to undergo center of mass rotations until the similarity is maximized. One hundred thousand random reorientations were sufficient to locate the optimal orientation. This approach is computationally impractical for large molecules and does not work well for molecules that are dissimilar in structure. In such cases one can orient molecules by choosing three common atoms in the backbone of the molecule and label them a, b, and c. a is translated to the origin and this translation is performed on b and c and all the surface points. The basis set is rotated

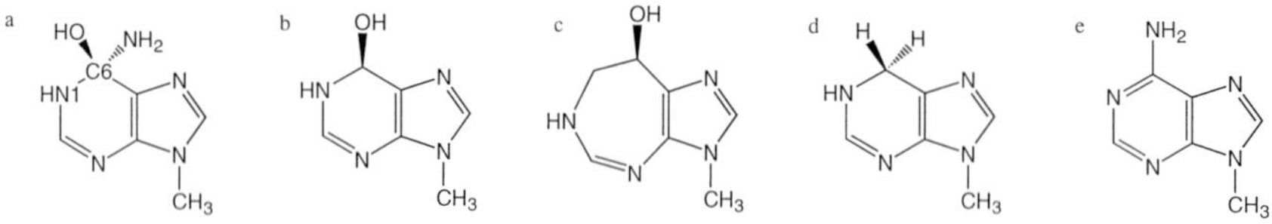

FIG. 2. Adenosine deaminase transition state (a), hydrated purine ribonucleoside (b), (R)-coformycin (c), 1,6-dihydropurine ribonucleoside (d), and adenosine (e).

such that b is on the positive z axis. Then the basis set is rotated so that c is in the positive x, z plane.

We now report on the application of the molecular surface point similarity measure to three different enzyme systems: adenosine deaminase, AMP nucleosidase, and cytidine deaminase.

Adenosine Deaminase

This enzyme facilitates the chemical hydrolysis of an amine group from adenosine converting it to inosine. The three inhibitors used here are hydrated purine ribonucleoside, 1,6-dihydropurine ribonucleoside, and (R)-coformycin. Stick figures of methyl derivatives of adenosine, the transition state structure, and inhibitors are shown in Fig. 2. Enzymatic hydroxylation at C-6 and protonation at N-1 of adenosine represent the highest energetic barrier to catalysis and yield the transition-state structure. Because AMP deaminase, for which the transition-state structure is known from kinetic isotope effects,[12] has similar chemistry and catalytic site structure to adenosine deaminase, that transition state is constructed through analogy with the AMP deaminase transition state. All molecules in Fig. 2 have methyl groups in place of the ribose 5'-PO_4 (at the N-9 positions of adenosine and the transition state, at the N-3 position of (R)-coformycin, and at the analogous loci in the remaining two inhibitors). Those residues are constant and are not involved in the chemistry of the reaction and are assumed to affect the electrostatic potentials and geometries of the five molecules equally. Equilibrium constants for the binding of the molecules to adenosine deaminase are found in Radzicka and Wolfenden,[13] who review the original work. $K_{TX} = 1.5 \times 10^{-17}$ M for the transition state; $K_i = 3.0 \times 10^{-13}$ M for the hydrate of purine ribonucleoside; (R)-coformycin binds with an inhibition constant of $K_i = 1 \times 10^{-11}$ M; and 1,6-dihydropurine ribonucleoside is the weakest binding inhibitor with $K_i = 5.4 \times 10^{-6}$ M. Finally, $K_m = 3.0 \times 10^{-5}$ M for the substrate adenosine. For a visual representation

[12] P. C. Kline and V. L. Schramm, *J. Biol. Chem.* **269**, 22385 (1994).
[13] A. Radzicka and R. Wolfenden, *Methods Enzymol.* **249**, 284 (1995).

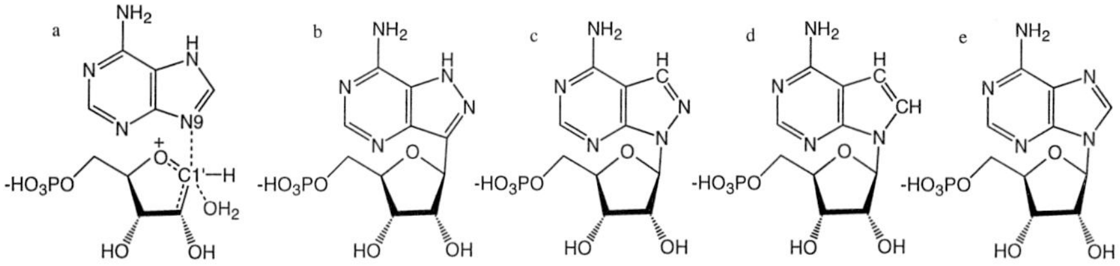

FIG. 3. AMP nucleosidase transition state (a), formycin 5′-phosphate (b), aminopyrazolo pyrimidine ribonucleoside (c), tubercidin 5′-phosphate (d), and AMP (e).

of the molecular electrostatic potential surfaces for the structures shown in Fig. 2, see Bagdassarian et al.[6] and Kline and Schramm.[12]

AMP Nucleosidase

Figure 3 shows AMP, the transition state for the nucleosidase reaction, and three inhibitors, formycin 5′-PO$_4$, 4-aminopyrazolo-(3,4-*d*)pyrimidine-1-ribonucleotide, and tubercidin 5′-PO$_4$. The chemistry of this reaction proceeds with a protonation at the N-7 position of AMP and the inclusion of an attacking hydroxyl nucleophile at the C-1′ carbon to give the transition state. The C-1′ to N-9 bond is partially broken with a bond order of 0.2 at the transition state.[14] Orientation of the ribose with respect to the purine across the breaking glycosidic bond is important for this reaction, so both the ribose and the phosphate group are included in the calculations. The phosphate groups are neutralized by ionic interactions in the bound form of these molecules, and quantum mechanical calculations to give the electrostatic potentials are performed with protonated phosphates.[6] The equilibrium constants are[15] $K_{TX} = 2 \times 10^{-17}$ M for the transition state of AMP; $K_i = 4.3 \times 10^{-8}$ M for formycin 5′-PO$_4$; $K_i = 1.0 \times 10^{-5}$ M for aminopyrazolo pyrimidine ribonucleotide; $K_i = 5.1 \times 10^{-5}$ for tubercidin 5′-PO$_4$; and $K_m = 1.2 \times 10^{-4}$ M for the substrate AMP. More detailed biochemistry and the electrostatic potential surfaces can be found in Bagdassarian et al.[6] and Ehrlich and Schramm.[14]

Cytidine Deaminase

This enzyme catalyzes the hydrolysis of the amine group on cytidine to give uridine and ammonia as products. For this system, more inhibitors have been studied. Twelve molecules are used here to analyze binding to this enzyme, and the usefulness of the method is explored more fully. The

[14] J. L. Ehrlich and V. L. Schramm, *Biochemistry* **33**, 8890 (1994).
[15] W. E. DeWolf, F. A. Fullin, and V. L. Schramm, *J. Biol. Chem.* **254**, 10868 (1979).

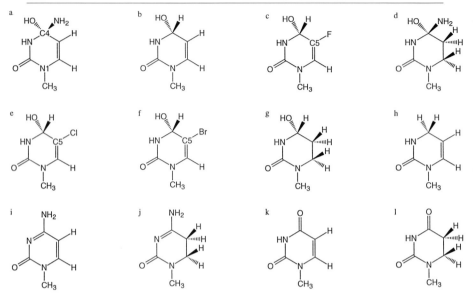

FIG. 4. Cytidine deaminase series: transition state for cytidine (a), hydrated pyrimidine-2-one ribonucleoside (b), hydrated 5-fluoropyrimidine-2-one ribonucleoside (c), transition state for 5,6-dihydrocytidine (d), hydrated 5-chloropyrimidine-2-one ribonucleoside (e), hydrated 5-bromopyrimidine-2-one ribonucleoside (f), 3,4,5,6-tetrahydrouridine (g), 3,4-dihydrozebularine (h), cytidine (i), 5,6-dihydrocytidine (j), uridine (k), and 5,6-dihydrouridine (l).

detailed structure of the transition state for this enzyme has yet to be determined by isotope effect methods; nonetheless, the crystal structure of the enzyme complexed with the transition-state analog 5-fluoropyrimidin-2-one ribonucleoside is known and serves as a starting point for the construction of a model transition-state structure.[16] Figure 4a shows a schematic of this transition-state model for cytidine. The reaction mechanism is assumed to be similar to that for adenosine deaminase. The C-4 to O (of the attacking $-OH$) bond distance is constrained at 1.67 Å, corresponding to that found in the crystal structure of the enzyme–inhibitor complex. The angles to the attacking $-OH$ and to the leaving group $-NH_2$ are modeled by analogy to the AMP (or adenosine) deaminase reaction where a detailed transition-state structure is available. The rest of the molecule is energy minimized as described later. Methyl derivatives are used again for the transition states, substrates, and inhibitors of this system. The 12 molecules in Fig. 4 are in order of decreasing binding strength to cytidine deaminase. The

[16] L. Betts, S. Xiang, S. A. Short, R. Wolfenden, and C. W. Carter, Jr., *J. Mol. Biol.* **235**, 635 (1994).

transition state for cytidine has a binding constant of[17] $K_{TX} = 4 \times 10^{-16}$ M. Cytidine, the substrate, has a binding constant of[17] $K_m = 5 \times 10^{-5}$ M. 5,6-Dihydrocytidine is another substrate with[18] $K_m = 1.1 \times 10^{-4}$ M, and the transition state resulting from it with $K_{TX} = 7.9 \times 10^{-10}$ M (calculated from information given in Frick et al.[17]). The products corresponding to these substrates are uridine and 5,6-dihydrouridine with $K_i = 2.5$ and 3.4×10^{-3} M, respectively.[17] The remaining inhibitors are pyrimidine-2-one ribonucleoside, which binds as the hydrate[16] with a K_i estimated to be 2×10^{-13} M.[17] The bound form (hydrate) of 5-fluoropyrimidine-2-one ribonucleoside has an equilibrium constant of 3.9×10^{-11} M.[19] We applied the same water addition constant offered in Carlow et al.[19] to calculate the inhibition constants of 5-chloropyrimidine-2-one ribonucleoside, $K_i = 7.1 \times 10^{-9}$ M,[20] and 5-bromopyrimidine-2-one ribonucleoside,[20] $K_i = 1.8 \times 10^{-8}$ M. Finally, 3,4,5,6-tetrahydrouridine has $K_i = 1.8 \times 10^{-17}$ M (the average of the two quantities given in Frick et al.[17]) and 3,4-dihydro-zebularine has[18] $K_i = 3.0 \times 10^{-5}$ M. For a comprehensive review of cytidine deaminase and the related adenosine deaminase, see Radzicka and Wolfenden.[13]

Constrained energy minimizations for all inhibitors for the three enzyme systems were required. For both AMP nucleosidase and AMP deaminase, on which adenosine deaminase is modeled, structures for the transition states were derived from heavy atom kinetic isotope effects.[8-10] All inhibitor and substrate structures for these systems are energy minimized to accommodate the bonds defined by kinetic isotope methods and are influenced by the transition-state geometry. For example, the $-OH$ group of (R)-coformycin is positioned and constrained to mimic the experimentally derived orientation of the transition-state $-OH$ group. The ab initio calculations are done with the 3-21G basis set at the Hartee–Fock level of theory. The transition-state structure for cytidine bound to cytidine deaminase is deduced from the closely related transition state of adenosine deaminase. The remaining 11 molecules in this series are energy minimized except for substituents, which are analogous to that defined by the transition-state geometry.

Results of the application of Eq. (2) are shown in Tables I and II. The names of the compounds are listed on the far left; the experimentally determined binding free energies are listed in the second column from the

[17] L. Frick, C. Yang, V. E. Marquez, and R. Wolfenden, *Biochemistry* **28**, 9423 (1989).
[18] S. Xiang, S. A. Short, R. Wolfenden, and C. W. Carter, Jr., *Biochemistry* **34**, 4516 (1995).
[19] D. C. Carlow, S. A. Short, and R. Wolfenden, *Biochemistry* **35**, 948 (1996).
[20] J. J. McCormack, V. E. Marquez, P. S. Liu, D. T. Vistica, and J. S. Driscoll, *Biochem. Pharmacol.* **29**, 830 (1980).

TABLE I
PREDICTION OF LIGAND-BINDING ENERGY BY A MOLECULAR SURFACE POINT
SIMILARITY MEASURE

Enzyme/molecule	$\Delta G/RT$ (experimental)	S	$\Delta G/RT$ (S)
AMP nucleosidase			
Transition state	−39	1.000	−39
Formycin	−17	0.434	−18
Aminopyrazolopyrimidine ribonucleotide	−12	0.310	−14
Tubercidin	−9.9	0.298	−13
AMP	−9.0	0.173	−9.0
		Average deviation	1.2
Adenosine deaminase			
Transition state	−39	1.000	−39
Hydrated purine ribonucleoside	−29	0.765	−27
(R)-Cofomycin	−25	0.604	−19
1,6-Dihydropurine ribonucleoside	−12	0.677	−23
Adenosine	−10	0.428	−10
		Average deviation	3.8

TABLE II
PREDICTION OF LIGAND-BINDING ENERGY BY A WELL-TRAINED NEURAL NETWORK
FOR CYTIDINE DEAMINASE

Enzyme/molecule	$\Delta G/RT$ (experimental)	S	$\Delta G/RT$ (S)
Cytidine deaminase			
Transition state for cytidine	−36	1.00	−36
Hydrated pyrimidine-2-one ribonucleoside	−27	0.87	−30
Hydrated 5-fluoropyrimidine-2-one ribonucleoside	−24	0.78	−26
Transition state for 5,6-dihydrocytidine	−21	0.88	−30
Hydrated 5-chloropyrimidine-2-one ribonucleoside	−19	0.70	−22
Hydrated 5-bromopyrimidine-2-one ribonucleoside	−18	0.68	−21
3,4,5,6-Tetrahydrouridine	−16	0.76	−25
3,4-Dihydrozebularine	−10	0.68	−22
Cytidine	−9.9	0.39	−9.9
5,6-Dihydrocytidine	−9.1	0.28	−5.3
Uridine	−6.0	0.58	−18
5,6-Dihydrouridine	−5.7	0.45	−12
		Average deviation	5.3

left. These values are divided by the gas constant and absolute temperature, making them dimensionless. The middle column shows the output of the similarity measure. The second column from the right shows the binding energy as predicted by the similarity measure. For both AMP nucleosidase and adenosine deaminase, the binding free energies of the transition states and of the substrates are defined by experimental values. Predicted values of $\Delta G/RT$ from S are made by linear extrapolation between these values to give values for the inhibitors.[6] In cases where there are few inhibitors with known affinities for the enzyme, the similarity measure was extremely effective.

Similarity Measure Results and Discussion

For AMP nucleosidase, the errors in $\Delta G/RT$ as predicted by the similarity measure are 0.0 for the transition state (by construction), 1.0 for formycin, 2.0 for aminopyrazolo pyrimidine ribonucleotide, 3.1 for tubercidin, and 0.0 for AMP (again, by construction). The average deviation for the similarity measure when used in the study of these AMP nucleosidase inhibitors is 1.2. The average deviation for the adenosine deaminase inhibitors for the similarity measure is 3.8.

The similarity measure predictions for the enzyme/inhibitor system of cytidine deaminase are reported in column 3 of Table II. It is unable to give accurate predictions. Many of the molecules, especially the 5,6-hydrogenated ones, when compared with the transition state (the one for cytidine), give large errors in the predicted binding energy. The products uridine and 5,6-dihydrouridine are scored with 12 and 6.3 $\Delta G/RT$ deviation, respectively, by the similarity measure.

The study of cytidine deaminase has shown that it is crucial that the comparison algorithm be able to learn the binding rules of the system because binding energy is not always a linear function of similarity to the transitions state. The product uridine, a poor inhibitor, was predicted to be a good inhibitor. This is an example of the fact that sometimes products look like the transition state, but enzymes have evolved to bind products weakly to minimize back reactions. The similarity measure is not capable of determining complex sets of rules that govern binding energy. The study of cytidine deaminase led our group to the development of methods that could determine such rules. These learning approaches are broadly called computational neural networks and are described in the following section.

Neural Networks

A computational neural network is a computer algorithm that, during its training process, can learn features of input patterns and associate these with an output. Neural networks learn to approximate the function defined by the input/output pairs. The function is never specified by the user. After the learning phase, a well-trained network should be able to predict an output for a pattern not in the training set. In the context of the present work, the neural net is trained with a set of molecules—the transition state, substrate, and inhibitors for a given enzyme—until it can associate with every quantum mechanical description of the molecules in this set, a free energy of binding (which is the output). Then the network is used to *predict* the free energy of binding for unknown molecules.

The use of neural networks for the prediction of binding free energies becomes crucial in the case where the geometric and electrostatic similarity are poor predictors of binding energy. Enzymes interact with specific regions of the molecules they bind. Similarity measures that compare inhibitors to the transition state usually weigh all regions of the inhibitor as being equally important to the binding process. A difference between the inhibitor and the transition state distant to any region involved in binding will affect the similarity output adversely, leading to a lowered measure, but this difference will not affect binding. The neural network is conditioned to *ignore* such regions. It is also desirable to use a neural network when the enzyme might be modifying the inhibitors chemically in the first stages of binding (e.g., protonation). If this is the case, the molecular features of the inhibitor that bind to the enzyme might be altered from the entity used in the calculation of similarity. In the study of cytidine deaminase the similarity measure was ineffective because the products looked like the transition state. A neural network could learn to identify product-like features in the same way that the enzyme has and to predict poor binding.

Neural networks have been used previously for the task of simulating biological molecular recognition. Gasteiger *et al.*[21] have used Kohonen self-organizing networks to preserve the maximum topological information of a molecule when mapping its three-dimensional surface onto a plane. Wagener *et al.*[22] have used autocorrelation vectors to describe different molecules. In that work,[22] the molecular electrostatic potential at the molecular surface was collapsed onto 12 autocorrelation coefficients. Neural networks were used by Weinstein *et al.*[23] to predict the mode of action of different chemotherapeutic agents. The effectiveness of the drugs on different malignant tissues served as descriptors, and the output target for the

[21] J. Gasteiger, X. Li, C. Rudolph, J. Sadowski, and J. Zupan, *J. Am. Chem. Soc.* **116,** 4608 (1994).
[22] M. Wagener, J. Sadowski, and J. Gasteiger, *J. Am. Chem. Soc.* **117,** 7769 (1995).
[23] J. N. Weinstein, K. W. Kohn, M. R. Grever, V. N. Viswanadhan, L. V. Rubinstein, A. P. Monks, D. A. Scudiero, L. Welch, A. D. Koutsoukos, A. J. Chiausa, and K. D. Paull, *Science* **258,** 447 (1992).

network was the mode of action of the drug (e.g., alkylating agent, topoisomerase I inhibitor). Tetko et al.[24] used an approach similar to the autocorrelation vectors. So and Richards[25] used networks to learn and to predict biological activity from QSAR descriptions of molecular structure. Neural networks were used by Thompson et al.[26] to predict the amino acid sequence that the HIV-1 protease would bind most tightly, and this information was used to design HIV protease inhibitors.

The present work is a departure from all previous work because the quantum mechanical electrostatic potential at the van der Waals surface of a molecule is used as the physicochemical descriptor. The *entire* surface for each molecule, represented by a discrete collection of points, serves as the input to the neural network. To preserve the geometric and electrostatic integrity of the training molecules, a collapse onto a lower dimensional surface is avoided. After alignment of the inhibitor molecule for maximal geometrical overlap with the transition-state structure, the electrostatic potentials on the inhibitor surface are mapped onto a reference surface. Input patterns for a neural network are presented in the form of a vector with entries $(l_1, l_2, \ldots, l_n)$. Because the molecules are represented by a $4 \times n$ matrix, a method is needed to discard the x, y, z coordinates but to maintain spatial information. This is accomplished by mapping the surface points of every molecule onto a reference surface. Mapping ensures that similar regions on different molecules enter the same part of the neural network. To minimize the amount of geometric information loss, we augment the network with geometric information. Geometrical information is generated by a reference surface that was larger, in all regions, than the surfaces used in the study (a sphere of diameter larger than the largest molecule). During the nearest neighbor mapping, the shortest distance between each point on the reference and each point on the inhibitor is calculated. The input to the neural network includes the electrostatic potential at many points and the distance of that point from the nearest point on the reference surface. Using a reference surface larger than the other surfaces permits a uniform outward mapping. In the limit, with an infinite number of points, all mappings are normal to the surface of the inhibitor, and the mapping distances will be as small as possible. To approach this limit, a 10-fold excess of points was selected to describe the molecules. The surfaces of the molecules are described by 150 points per atom. The reference sphere that the points are mapped onto is described by a smaller

[24] I. V. Tetko, V. Y. Tanchuk, N. P. Chentsova, S. V. Antonenko, G. I. Poda, V. P. Kukhar, and A. I. Luik, *J. Med. Chem.* **37**, 2520 (1994).
[25] S.-S. So and W. G. Richards, *J. Med. Chem.* **35**, 3201 (1992).
[26] T. B. Thompson, K.-C. Chou, and C. Zheng, *J. Theor. Biol.* **177**, 369 (1995).

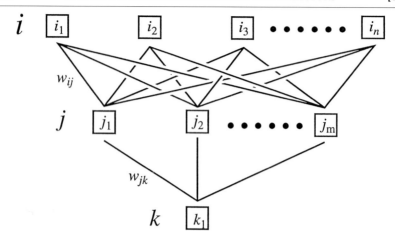

Fig. 5. Schematic of a three-layer neural network with an n dimensional input layer and a single output layer neuron.

number of points, 15 times the average number of atoms in the molecules of the study. As a result of mapping to the reference sphere, all molecules are described by the smaller number of points.

Computational neural networks are composed of many simple units operating in parallel. These units and the aspects of their interaction are inspired by biological nervous systems. The network function is determined largely by the interactions between units. Networks learn by adjusting the values of the connections between elements.[27] The neural network employed in this study is a feed forward with back propagation of error network that learns with momentum. The basic construction of a back propagation neural network has three layers: input, hidden, and output. The input layer is where input data are transferred. The link between the layers of the network is one of multiplication by a weight matrix, where every entry in the input vector is multiplied by a weight and sent to every hidden layer neuron so that the hidden layer weight matrix has the dimensions n by m, where n is the length of the input vector and m is the number of hidden layer neurons. A bias is added to the hidden and output layer neurons; it scales all the arguments before they are input into the transfer function.[27]

Referring to the schematic in Fig. 5, the input layer is represented by the squares at the top of the diagram. The weights are represented by the lines connecting the layers: w_{ij} is the weight between the ith neuron of the

[27] L. Fausett, "Fundamentals of Neural Networks." Prentice Hall, New Jersey, 1994.

input layer and the jth neuron of the hidden layer, and w_{jk} is the weight between the jth neuron of the hidden layer and the kth neuron of the output layer. In this diagram the output layer has only one neuron because the target pattern is a single number $\Delta G/RT$. The hidden layer input from pattern number 1 for neuron j, $h_j^I(1)$, is calculated,

$$h_j^I(1) = b_j + \sum_{i=1}^{n} x_i^o(1) \times w_{ij} \tag{3}$$

where x_i^o (1) is the output from the ith input neuron, w_{ij} is the element of the weight matrix connecting input from neuron i with hidden layer neuron j, and b_j is the bias on the hidden layer neuron j. This vector h_j^I is sent through a transfer function, f. This function is nonlinear and usually sigmoidal, taking any value and returning a number between -1 and 1.[27] A typical example is

$$f(h_j^I) = \frac{1}{1 + e^{-h_j^I}} - 1 \equiv h_j^o \tag{4}$$

The hidden layer output, h_j^o, is then sent to the output layer. The output layer input o_k^I is calculated for the kth output neuron

$$o_k^I = b_k + \sum_{j=1}^{m} h_j^o w_{jk} \tag{5}$$

where w_{jk} is the weight matrix element connecting hidden layer neuron j with output layer neuron k. The output layer output, o_k^o, is calculated with the same transfer function given earlier. The calculation of an output concludes the feed forward phase of training. Back propagation of error is used in conjunction with learning rules to increase the accuracy of predictions. The difference between $o_k^o(1)$ and the target value for input pattern number 1, $t_k(1)$, determines the size and sign of the corrections to the weights and biases during back propagation. The relative change in weights and biases is proportional to a quantity δ_k:

$$\delta_k = (t_k - o_k^o) \times f'(o_k^I) \tag{6}$$

where f' is the first derivative of Eq. (4). Corrections to the weights and biases are calculated:

$$\Delta w_{jk} = \alpha \delta_k h_j^o \tag{7}$$
$$\Delta b_k = \alpha \delta_k \tag{8}$$

The size corrections are moderated by α, the learning rate; this number ranges from zero to one exclusive of the end points. The same learning rule is applied to the hidden layer weight matrix and biases. In most adaptive systems, learning is facilitated with the introduction of noise. In neural networks this procedure is called learning with momentum. The correction to the weights of the output layer at iteration number τ is a function of the correction of the previous iteration, $\tau - 1$, and μ, the momentum constant:

$$\Delta w_{jk}(\tau) = \alpha \delta_k h_j^o + \mu \Delta w_{jk}(\tau - 1) \qquad (9)$$
$$\Delta b_k(\tau) = \alpha \delta_k + \mu \Delta b_k(\tau - 1) \qquad (10)$$

The same procedure is applied to hidden layer weights and biases. The correction terms are added to the weights and biases, concluding the back propagation phase of the iteration. The network can train for hundreds to millions of iterations, depending on the complexity of the function defined by the input/output pairs. This type of back propagation is a generalization of the Widrow–Hoff learning rule applied to multiple-layer networks and nonlinear differentiable transfer functions.[28]

Input vectors and the corresponding output vectors are used to train until the network can approximate a function.[28] The strength of a back propagation neural network is its ability to form internal representations through the use of a hidden layer of neurons. For example, the "exclusive or" problem demonstrates the ability of neural networks, with hidden layers, to form internal representations and to solve complex problems. Suppose we have four input patterns [(0, 1) (0, 0) (1, 0) (1, 1)] with output targets [1, 0, 1, 0], respectively. A perceptron or other single layer system would be unable to simulate the function described by these four input/output pairs. The only way to solve this problem is to learn that the two types of inputs work together to affect the output. In this case the least similar inputs cause the same output, and the more similar inputs have different outputs.[28] The ability required to solve this problem is not unlike that required to find the best inhibitor when it does not share all the same quantum features of the transition state. It is this inherent ability of neural networks to solve complex puzzles that makes them well conditioned for the task of simulating biological molecular recognition. As a result of the mapping during input preparation, each input neuron is given information (electrostatic potential or geometry) about the nearest point on the surface of the inhibitor. It is as though each input neuron is at a fixed point on the sphere around the inhibitors, judging each inhibitor in the same way the enzyme active site would.

The left-most columns of Table III show the results of the neural network approach in the study of cytidine deaminase. The (11) column shows the results of a neural network that was trained with 11 of the 12 molecules

[28] D. E. Rumelhart, G. E. Hinton, and R. J. Williams, "Parallel Distributed Processing," Vol. 1. MIT Press, Massachusetts, 1986.

TABLE III
PREDICTION OF LIGAND-BINDING ENERGY BY A WELL-TRAINED NEURAL NETWORK
FOR CYTIDINE DEAMINASE

Enzyme/molecule	$\Delta G/RT$ (experimental)	S	$\Delta G/RT$ (S)	$\Delta G/RT$ (neural net) (11)	(7)
Cytidine deaminase					
Transition state for cytidine	−36	1.00	−36	−30	
Hydrated pyrimidine-2-one ribonucleoside	−27	0.87	−30	−27	−26
Hydrated 5-fluoropyrimidine-2-one ribonucleoside	−24	0.78	−26	−19	
Transition state for 5,6-dihydrocytidine	−21	0.88	−30	−23	
Hydrated 5-chloropyrimidine-2-one ribonucleoside	−19	0.70	−22	−19	−18
Hydrated 5-bromopyrimidine-2-one ribonucleoside	−18	0.68	−21	−17	
3,4,5,6-Tetrahydrouridine	−16	0.76	−25	−15	
3,4-Dihydrozebularine	−10	0.68	−22	−12	−13
Cytidine	−9.9	0.39	−9.9	−8.3	
5,6-Dihydrocytidine	−9.1	0.28	−5.3	−7.5	
Uridine	−6.0	0.58	−18	−6.1	−9.9
5,6-Dihydrouridine	−5.7	0.45	−12	−6.2	
	Average deviation		5.3	1.7	2.2

used in the study. The neural network predicted the binding energy of only 1 of the molecules and its value is reported in this column. The (7) column reports the results of neural network predictions when the network was trained with 7 molecules. Molecules left out of the training set were chosen at random such that they represented a large range in binding energy and that they were spaced evenly throughout this range. A situation similar to the "exclusive or" problem was encountered in the study of cytidine deaminase. The products uridine and 5,6-dihydrouridine are similar structurally to the transition state and found to be so by the similarity measure. This makes the similarity measure predictions error prone because binding energy is not a linear function of similarity to the transition state. As with the "exclusive or" problem, the network was able to learn that two similar inputs can cause very different outputs.

An even better example of the use of the neural network recognition method is found in a study of the inosine–uridine-preferring nucleoside hydrolase (IU-NH). Protozoan parasites lack *de novo* purine biosynthetic pathways and rely on the ability to salvage nucleosides from the blood of their host for RNA and DNA synthesis.[29] The IU-NH from *Crithidia fasciculata* is unique and has not been found in mammals.[30] This enzyme catalyzes the *N*-ribosyl hydrolysis of all naturally occurring RNA purines and pyrimidines.[30] The active site of the enzyme has two binding regions: one binding ribose and the other binding the base. The inosine transition state requires $\Delta\Delta G = 17.7$ kcal/mol activation energy: 13.1 kcal/mol are used in activation of the ribosyl group and only 4.6 kcal/mol are used for activation of the hypoxanthine-leaving group.[31] Analogs that resemble the inosine transition state have proven to be powerful competitive inhibitors of this enzyme both geometrically and electronically and could be used as antitrypanosomal drugs.[30]

The transition state for these reactions feature an oxocarbenium ion achieved by the polarization of the C-4' oxygen C-1' carbon bond of ribose. The C-4' oxygen is in proximity to a negatively charged carboxyl group from Glu-166 during transition-state stabilization.[30] This creates a partial double bond between the C-4' oxygen and the C-1' carbon, causing the oxygen to have a partial positive charge and the carbon to have a partial negative charge. Nucleoside analogs with iminoribitol groups have a secondary amine in place of the C-4' oxygen of ribose and have proven to be effective inhibitors of IU-NH.

IU-NH acts on all naturally occurring nucleosides (with C-2' hydroxyl groups). The lack of specificity for the leaving groups results from the small number of amino acids in this region to form specific interactions: Tyr-229, His-82, and His-241.[30] The only crystal structure data available concerning the configuration of bound inhibitors were generated from a study of the enzyme bound to *p*-aminophenyliminoribitol (pAPIR) (Fig. 6, t4). Tyr-229 relocates during binding and moves above the phenyl ring of pAPIR. The side chain hydroxyl group of Tyr-229 is directed toward the cavity that would contain the six-member ring of a purine, were it bound.[30] His-82 is 3.6 Å from the phenyl ring of pAPIR and in the proper position for positive charge-π interactions to occur.[30] His-241 has been shown to be involved in leaving-group activation in the hydrolysis of inosine, presumably as the proton donor in the creation of hypoxanthine.[30]

A similarity measure does not work well in the prediction of binding energy of inhibitors of IU-NH because the enzyme has many substrates. The chemistry involved in catalyzing the scission of the *N*-ribosidic bonds

[29] D. J. Hammond and W. E. Gutteridge, *Mol. Biochem. Parasitol.* **13**, 243 (1984).
[30] M. Degano, S. C. Almo, J. C. Sacchettini, and V. L. Schramm, *Biochemistry* **37**, 6277 (1998).
[31] D. W. Parkin, G. Limberg, P. C. Tyler, R. H. Furneau, X. Y. Chen, and V. L. Schramm, *Biochemistry* **36**(12), 3528 (1997).

of these substrates is quite different. Therefore the transition states for the various substrates are quite different. Because there is only one known transition-state structure, that of inosine, it would be unwise to presume that the binding energy of all inhibitors would be a linear function of similarity only in the inosine transition state. The neural network has the capacity to learn the binding rules of this enzyme if enough information is present in the training set.

To study this enzyme, inhibitors with different base and ribose analogs were used. The crystal structure of the enzyme bound to p-aminophenylimi-noribitol[30] shows the relative position of the phenyl group with the iminori-bitol group. This compound is identical on rotation of 180°. This being the case, the structure offers little information as to how molecules that are ambiguous on rotation should be oriented. Molecule t1, the transition state, probably binds to the enzyme in a particular configuration with the hypoxan-thine group extended either over the C-2′ carbon of the ribose group or in the opposite direction, but there is no way to tell which of these is correct from existing data. Subtle orientation discrepancies are unlikely to introduce sufficient misinformation (if the orientation is conserved for all of the molecules) for this to be a problem for the neural network. Both orientations are used to demonstrate that the neural network is not depen-dent on the selection of a "correct" orientation. The neural network was trained with the patterns of each molecule in both orientations. The network was trained twice with the patterns of molecules that are identical on rotation of 180°; this prevents the network from being biased to the other patterns.

In addition to using both configurations about the ribosidic bond, there are other degrees of freedom that have to be consistent for all molecules. We assume that the enzyme will bind all molecules in a similar low energy conformation. The neural network need not know the configuration as long as the conformation of all molecules we present to the neural network is consistent. The known crystal structure of p-aminophenyliminoribitol bound to IU-nucleoside hydrolase is used as the model conformation.

The inosine transition-state structure is stabilized by a negatively charged carboxyl group within the active site 3.6 Å from the C-4′ oxygen.[32] In order to simulate this aspect of the active site, we included a negatively charged fluoride ion (at the same relative position of the nearest oxygen of the carboxyl group) in the calculations of the electrostatic potential at the van der Waals surface.

[32] B. A. Horenstein, D. W. Parkin, B. Estupinan, and V. L. Schramm, *Biochemistry* **30,** 10788 (1991).

In an attempt to simulate a real-world application of the use of our method in drug discovery, the structures of 41 molecules were used but the binding energy of only 22 of the molecules was known at the time of study. The binding energies of the other 19 molecules were yet to be determined. The 22 molecules in the training set included eight substrates and substrate analogs, the transition state of inosine, 11 inhibitors from a previous publication, and 2 inhibitors with known but unpublished binding energies. The neural network was trained with 22 molecules as inputs. The output (training target) is the binding free energy of the molecule with the enzyme.

Training a neural network requires variation of four adjustable parameters: number of hidden layer neurons, learning rate, momentum constant, and number of training iterations. The only way to tell that a network is well trained is to minimize the training set prediction error. This can be calculated by taking the difference between the target$_i$ value for a molecule (experimentally determined binding energy) and the number the neural network predicted for that pattern$_i$ and summing the absolute value of this number for all the molecules in the training set. As training progresses, the training set prediction error will decrease. Minimizing training set error is not without negative consequence; overtraining occurs when the network trains for too many iterations, has too many hidden layer neurons, has too large of a learning rate, or has too small of a momentum constant. The only way to tell that a neural network has not been overtrained is to have it make a prediction for a pattern not in the training set, i.e., see if the network can generalize from the information contained in the input/output pairs of the training set and apply that information to a molecule it has not trained with. In accommodation of this fact we employ 1 of the 22 training set molecules as an adjuster molecule. This molecule is left out of the training set during training and is used to check if the neural network was overtrained. The procedure is to train the neural network until the prediction set error has decreased into a plateau. We then end training and test the neural network with the adjuster molecule. If the neural network predicts the binding energy of the adjuster molecule within 5%, then the construction of that neural network is saved; if the prediction is more than 5% off, a new construction is chosen. This procedure is repeated until a construction is found that allows the neural network to predict the binding energy of the adjuster molecule within 5%. This is done for all 22 molecules in the training set, and these 22 best neural network constructions are used to predict binding energies for the 19 molecules in the predicting set. The final reported result is the average of the predictions for each of the network constructions.

IU-NH Neural Network Results

Figure 6 shows the structure of the 22 molecules in the training set with their experimental binding energies with IU-NH. Figure 7 shows the 19 molecules in the predicting set, with their experimentally determined binding energies (E) and their predicted binding energies (P). Figure 8 shows a graphic representation of data: prediction versus experimentally determined binding energy. It is clear that the method is effective over the entire range of binding energies. The method is able to distinguish between inhibitors that bind with a little, medium, and large amount of energy. The inhibitors in the predicting set were tested only experimentally in ranges below $-9.81 \Delta G/RT$ binding energy, so any predictions above this number are considered to have zero deviation if the experimental binding energy is greater than $-9.81 \Delta G/RT$. The experimental dimensionless binding energy range is $31.6 \Delta G/RT$, the average deviation is $0.90 \Delta G/RT$, and the largest deviation is $2.69 \Delta G/RT$.

The accuracy of the results leads us to believe that the method was able to create a mathematic model that describes the function defined by the input output pairs. There are specific successes that are the product of synthetic reasoning on the part of the neural network that warrant discussion. The prediction for prediction set molecule number one (p1 of Fig. 7) was an example of such synthetic reasoning. Training set molecule number eleven (t11 of Fig. 6) and t13 have nitro groups at the *para* and *meta* positions, respectively. Nitro groups, such as fluoro groups (like that of p1), are capable of electron withdrawing and little electron donating. The neural network could have learned about the negative effects electron-withdrawing substitutions to pyrimidine analogs have on binding from nitro-substituted inhibitors in the training set. The algorithm also learned of the negative effects of *meta* versus *para* substitutions from studying the nitro inhibitors. Because *p*-flourophenyliminoribitol (t6) was in the training set, the network was able to combine these two pieces of information and subtract about 2 $\Delta G/RT$ from the binding energy of t6 when generating a prediction of *m*-fluorophenyliminoribitol (p1). The prediction of p2 is slightly more complicated, as there are no guanosine analogs in the training set, but the neural network could have learned about the negative effects of electron-donating groups to purine analogs from the implied structure activity relationship present in the training set: t1 > t15 > t16. Prediction set molecule number 2 has an electron-donating amine group at C-4 of the purine six member ring. This group is the only difference between p2 and p3; notice the difference between the predictions. The prediction of p3 was also generated by synthetic reasoning. p3 has three features that are conducive to

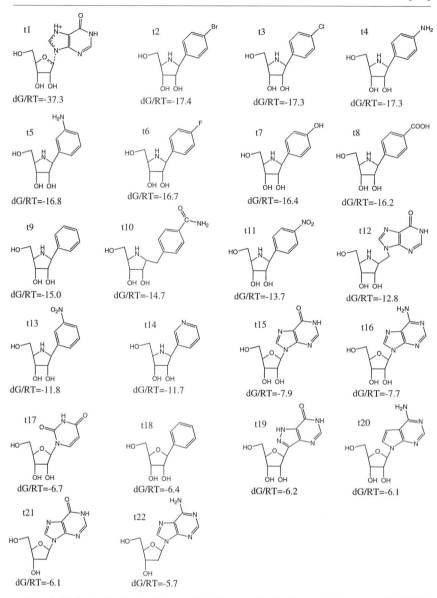

Fig. 6. Molecules in the training set, binding energies in dimensionless units of $\Delta G/RT$. Most of these values are in Ref. 37, the others have yet to be published.

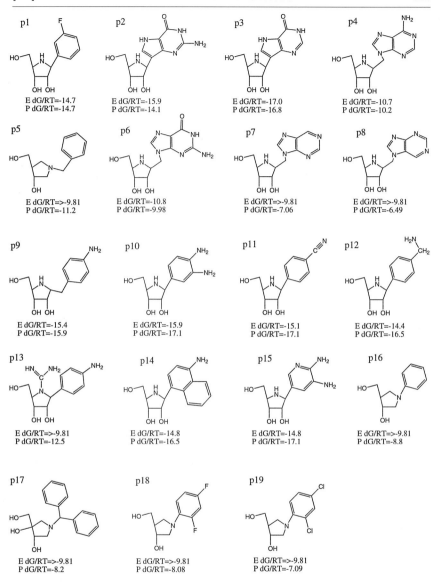

FIG. 7. Molecules in the predicting set. Experimentally determined binding affinities (E) and prediction (P).

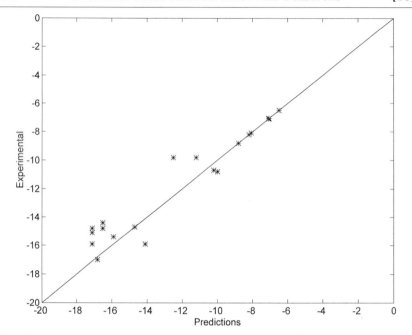

FIG. 8. Neural network predictions versus experimental data for IU-nucleoside hydrolase.

tight binding: an electron-withdrawing group at C-6, a partially positive charged hydrogen at N-7, and an iminoribitol group. The neural network combined these pieces of information and predicted a large binding energy. The prediction of binding energy for p4 also required a combination of facts. Prediction set molecule 4 has two qualities that make it a poor inhibitor of IU-NH: (1) p4 has a substituted C-1'–N-9 glycocidic bond, the neural network could have learned about the negative effects of this feature from t10 and t12, and (2) p4 has an electron-donating group at C-6. As already mentioned, these groups decrease the binding of purine analogs. Prediction set molecule 4 does have an iminoribitol group that is a tight binding group, but overall the molecule is a poor inhibitor and is identified as such by the neural network.

These results show that this quantum neural network technology can make completely blind predictions of enzyme/inhibitor-binding constants in a complex biochemical system. To highlight the nature of this accomplishment, we examined the different nature of transition-state structures for the two different kinds of substrates: purines and pyrimidines. The transition state of inosine is the only one for which there is a determined structure, which is shown in Fig. 6 (t1). The transition-state structure for inosine is created by polarization of the ribosyl group across the C-4' oxygen C-1' bond and protonation of N-7 of the purine group. This protonation would be impossible when dealing with the pyrimidine uridine, t17, as there is no place for this group to receive a proton (the electrons of N-3 are involved in the ring conjugation). Therefore it is clear that these two types of substrates have quite different transition-state structures and that the rules of binding pyrimidine analogs are quite different from those of binding purines. For pyrimidine analogs, the binding energy tends to decrease with increasingly electron-withdrawing substitutions. The opposite trend is seen with purine analogs. From the accuracy of the results, it is clear that the neural network was able to distinguish between purine and pyrimidine analogs and to apply the correct rules when evaluating the two families in the predicting set.

Molecular Importance and Hidden Layer Magnitudes

We have devised a way to probe trained neural networks and get information about what is most important about the molecules and what gives them their efficacy. This method can aid in the design of new inhibitors. This technique probes part of a trained neural network directly and reveals which parts of the molecules the neural network found most important.

This technique relies on the fact that the weights in the hidden layer associated with regions of the input important in binding have large absolute values. The network is presented with an input pattern and an output; to minimize the error the network must recognize regions that change and affect the binding energy relative to those regions that change and do not affect binding energy. This recognition occurs when weights of the neural network are adjusted so that important regions are multiplied by large weights and unimportant regions are multiplied by small weights. Documentation of this behavior is made by inspection of the absolute values of every number in the hidden-layer weight matrix of a trained network. The matrix is collapsed into a vector V_i by summing on j where $j = 1, \ldots, m$ and m is the number of hidden layer neurons:

$$V_i = \sum_{j=1}^{m} |w_{ij}| \tag{11}$$

and where i refers to the input surface points. Large values for V_i represent regions found to be important to the neural network, whereas small values represent regions found to be unimportant. Figure 9 (see color insert) shows a colored graphic representation of the hidden-layer weight matrix of a neural network trained with the molecules of the cytidine deaminase series.

All the patterns were generated by mapping the electrostatic potential of different molecules onto a 5-bromo transition-state surface (instead of a sphere as in the IU-NH study). The common 5-bromo transition-state geometry is used to identify those regions on the molecules found as most important by the neural net. This is represented by coloring points on a van der Waals surface with large V_i values yellow and regions with small values blue. Regions on molecular surfaces with intermediate weights have a mixture of the two colors.

Figure 9a (see color insert) shows the first half of the hidden-layer weight matrix of the neural network associated with the electrostatic potential used in the study of cytidine deaminase. Figure 9b (see color insert) shows the bottom half of that weight matrix associated with the geometry of the van der Waals surface. On inspection of Fig. 9a, the attacking hydroxyl nucleophile and protonated N-3 were both characterized by large weights and therefore were deemed important to the neural network. It should be noted that C-6 has large weights, which is likely caused by the fact that the bond between C-6 and C-5 is saturated and unsaturated in various inhibitors and is responsible for large effects on binding energy. Figure 9b shows that the geometry of the C-5 region is the only region that affects binding greatly. This is consistent with crystallographic data, as there is a nonbinding region of the enzyme that interferes sterically with inhibitors that have large substitutions at that region. Inhibitors with increasingly bulky substitutions at C-5 have decreasing binding energies.

Conclusions

The very accurate predictions of inhibitor potency of IU-nucleoside hydrolase could not have been done by any method that did not use quantum mechanically derived features. In a QSAR study the IU-NH inhibitors would have been described to the computer in terms of their volumes, hydrophobicity, H-bonding donor and acceptor sites, and other experimentally or classically derived features. These descriptions would not have provided the information necessary to model the binding preferences of IU-NH. This is demonstrated by inspection of the weaker inhibitors in the training set. 3-Pyridine-iminoribitol (t14) is the weakest binding substituted iminoribitol: it binds with less energy than iminoribitol alone (iminoribitol $\Delta G/RT = -12.3$, 3-pyridineiminoribitol $\Delta G/RT = -11.7$). It is not likely that the pyridine group is interacting with the leaving-group binding region with a positive $\Delta\Delta G$. It is more likely that the quantum mechanical nature of pyridine is affecting the electrostatic potential of the iminoribitol group and decreasing its ability to bind the enzyme. Because subtle quantum mechanical differences between the molecules dictate the binding prefer-

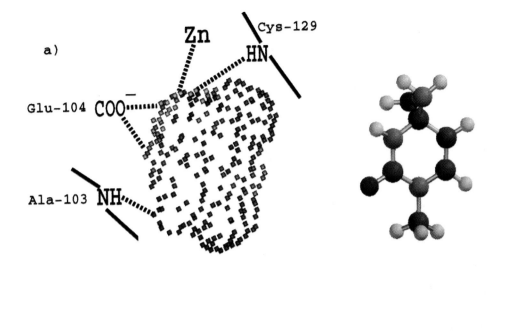

FIG. 9. Representation of the hidden-layer weight matrix of a trained neural network. It provides evidence that the neural network found the regions of the inhibitors, which are important in binding, to be important in the calculation. The range of these values is from 0.118 to 15.70.

ences of IU-NH, an approach that uses quantum mechanically-derived features is required.

The molecular surface point similarity measure is most effective when there are few inhibitors with known inhibition constants for the enzyme. The small training sizes of adenosine deaminase and AMP nucleosidase made it difficult to train the neural network, and neural network predictions for these systems were less accurate than those generated by the similarity measure. The IU-nucleoside hydrolase enzyme system and its inhibitors were studied by our group with the molecular surface point similarity measure. The similarity measure worked when purine analogs were compared to the inosine transition state. The binding strengths of pyrimidine analogs were not predicted accurately by the similarity measure when the analogs were compared to the inosine transition state. The reason is that the binding rules for this enzyme are different for purines and pyrimidines.

The most important result is that even though we do not know *a priori* what important molecular quantum features are needed to bind IU-NH, the information necessary to know this was presented adequately to the neural network during training, and this information was used by the neural network to make accurate predictions. Our method uses quantum mechanical descriptions of molecules, and with this information the neural network functions as the ultimate structure/activity relationship machine. It is able to determine what quantum structural features are of importance and form a highly accurate association of these with binding energy.

This is not a surprise from a mathematic sense, as the most common application for neural networks in engineering is in pattern classification and the creation of priority trees. The neural network first identifies that there are two or more groups of molecules that have one mutually exclusive characteristic that has an effect on the output. For our purposes, this characteristic could be the fact that purines have a double ring structure and pyrimidines have a single ring. Once two distinct groups of molecules have been identified, rules for evaluating the two can be learned, and the rules of one group can be the same or different from the rules of the other group. Examples of this from the IU-nucleoside hydrolase study are methyl substitutions that lengthen the glycosidic bond, decrease binding energy for purines and pyrimidines, but extremely electronegative groups (such as protons and double-bonded oxygens) increase the binding energy for purines ($\Delta G/RT$ for t1 > t19 > t20) but electron-withdrawing groups decrease the binding energy of pyrimidines ($\Delta G/RT$ for t2 < t6 < t11). As shown earlier with the "exclusive or" problem, the ability of a neural network to solve a problem when it appears to contain contradictory information is not a new finding. What is new is that this ability is being employed to solve a real-world problem in biochemical engineering.

The ability of the neural network to model the system defined by the input/output pairs is potentially useful for the search of chemical libraries. It might also be useful to aid in the design of new drugs. In the cytidine deaminase system, the weight matrix of the network was adjusted such that regions important in binding were owned by large weights. This was found to be true for the two kinds of information in the input. This technique could be helpful to researchers, giving them the ability to know what regions are to be avoided sterically and what regions are important in terms of electrostatic potential.

Author Index

Numbers in parentheses are footnote reference numbers and indicate that an author's work is referred to although the name is not cited in the text.

A

Abdulah, D. R., 336, 339(82), 385
Abe, H., 105, 109(22)
Abeles, R. H., 222, 226, 228(24)
Abeygunawardana, C., 22, 23, 23(25), 219, 222, 223, 224(29), 226, 226(28–30), 229(28–30), 231(29, 30), 232(28–30), 235(28, 29), 236(28, 29), 238, 239(28–30), 240(30), 241, 243(29), 244(28, 29), 251(35–37), 253, 304
Abraham, M. H., 256, 275(51)
Abu-Dari, K., 221
Adalsteinsson, H., 221
Admiraal, S. J., 72, 73, 84(7, 13)
Agnew, B. J., 105, 109(22)
Aguilar, V., 118
Ahmadian, M. R., 92
Ahmed, S. A., 116, 122, 124, 124(37, 49), 125, 125(49), 126(49, 53)
Akasaka, K., 224
Akrories, K., 313
Alagona, G., 377
Al-Awar, O., 94
Alber, T., 125
Alberts, B., 30, 36(5), 94
Alberty, R. A., 62, 253
Albery, W. J., 3, 11, 12, 162, 202, 276
Albrecht, G., 260
Aleshin, A. E., 304
Al-Laham, M. A., 374
Al-Laham, V. G., 302
Allen, F. H., 387
Almo, S. C., 304, 339, 393, 416, 417(30)
Altona, C., 227, 387
Amano, T., 92
Amatore, C., 276(18), 277
Ambulos, N. P., 285
Ampe, C., 109

Andersen, B. N., 54
Anderson, K. S., 111, 117, 118(41, 69), 119(41), 120, 121, 121(41), 122(41), 123, 123(46), 124, 125(53a), 126(41, 53a), 127(53a), 128(53a), 129, 129(41), 132(69), 137, 137(41), 139(85, 86)
Anderson, P., 90
Anderson, V. E., 188, 330, 331, 365
Anderson, W. F., 77
Andres, J. L., 302, 340, 374
Angelastro, J. M., 94, 97(8), 106(3)
Antonenko, S. V., 411
Antoniou, D., 324
Antonny, B., 82
Arabshahi, A., 38, 53
Ariens, E. J., 398
Armon, T., 29
Armstrong, K. B., 388
Armstrong, R. N., 269
Asakura, S., 97
Ash, E. L., 229
Ashwell, M., 388, 390(80)
Atha, D. H., 54
Atkins, P. W., 253
Atkinson, S. J., 109
Atkinson, T., 191(77), 200
Aull, J., 137
Austin, C. A., 63
Axelsen, P. H., 134
Ayala, P. Y., 302, 374

B

Babu, Y. S., 340
Bachovchin, W. W., 229
Bacon, D., 77

Baenziger, J. E., 178(36), 179
Bagdassarian, C. K., 177, 177(29), 179, 183(29), 201(29), 302, 305, 335, 347, 348(102), 350, 353, 353(101, 102), 354(107), 393, 401, 402(7), 403(6, 7), 405(6), 409(6)
Bahnson, B. J., 323, 324(61), 365
Bajdor, K., 177
Baker, E. N., 246
Baker, J., 302, 340, 374
Baldwin, R. L., 261
Balogh-Nair, V., 177, 177(25), 181(16)
Bamburg, J. R., 105, 109(22)
Banik, U., 122, 126(48)
Banks, G. A., 357, 385(14), 394(14), 401, 407(8)
Bannister, C. E., 276(17), 277
Bantia, S., 340
Bao, D., 234
Barbacid, M., 85
Bares, J. E., 262
Barrow, G. M., 253
Barshop, B. A., 169
Barstow, D. A., 191(77), 200
Bartels, K., 271
Bartness, J. E., 262
Bash, P. A., 226, 232
Bates, A. D., 65, 66
Batke, J., 112, 113
Batterham, T. J., 309
Bauer, R. W., 64
Bauerle, R., 125
Baugher, B. W., 113
Bax, A., 229, 231
Bayley, P. M., 387
Bayly, C. I., 377
Beauchamp, J. L., 247
Becker, E. D., 227
Begeleisen, J., 305
Behravan, G., 13, 14(14)
Beissner, R. S., 113
Bell, J. B., 11
Bell, R. P., 277, 280
Benedict, H., 221
Benkovic, S. J., 136, 196, 303
Bennet, A. J., 344
Benson, S. W., 248
Berche, P., 106
Berchtold, H., 56, 80

Berg, D. E., 62, 63(46)
Berghuis, A. M., 73, 74(10, 22, 39), 75(10, 22, 39), 76, 76(10), 78(10), 81(10), 83(10), 85, 86(22, 39), 88(22)
Berman, D. M., 92
Berman, H. M., 339, 387
Bernasconi, C. F., 279, 282(27–30), 386, 396, 396(66, 67)
Bernhard, S. A., 112
Bernstein, F. C., 383, 387(61), 388(61)
Bernstein, R. B., 303
Berti, P. J., 307, 322, 331, 333(7), 344(44, 45), 355, 356, 365(4), 366(7), 367, 368(3, 4, 7), 371(3, 4), 374(3), 376(3), 379(68), 380(3, 4), 381(3, 4), 386, 387(3), 388(3), 395(3, 4, 7), 396(3, 7)
Bertolucci, M. D., 363
Bertunik, H. D., 304
Betts, G. F., 112
Betts, L., 406
Beusen, D., 128
Bhriain, N. N., 67
Bigay, J., 82
Bigeleisen, J., 361, 378(31, 32)
Binkley, J. S., 302, 340, 374
Blakley, R. L., 193, 196, 309
Blanchard, J. S., 313
Blanke, S. R., 307, 344(45), 356, 365(4), 368(4), 371(4), 380(4), 381(4), 395(4)
Blow, D. M., 208, 271
Blum, J. J., 38
Bode, W., 269
Boggs, J. E., 183
Bolduc, J. M., 203
Bolin, J. T., 196
Bommerman, J. C., 14(30), 24
Borasio, G. D., 87
Bordwell, F. G., 262, 262(62), 263, 282
Borenstein, B. A., 311
Borst, P., 348
Botera, R., 29
Bourne, H. R., 71, 77(2), 78(2), 80(2)
Boutellier, M., 393
Boyle, W. J., Jr., 282
Brady, K., 226
Braiman, M. S., 177
Brand, S. J., 54
Bratt, G. T., 313
Brauer, P. R., 54

Brauman, J., 267
Braunheim, B. B., 302, 347, 353(101), 398, 401, 402(7), 403(7)
Brayer, G. D., 226
Breton, J., 177
Brice, M. D., 383, 387(61), 388(61)
Briggs, A. J., 388
Brikun, I., 62, 63(46)
Britton, H. G., 12
Brønsted, J. N., 280
Brothers, P., 229
Brown, I. D., 186, 198(47)
Brown, P. O., 66
Brownson, C., 309
Bruice, T. C., 221
Brünger, A., 76
Bryant, F. R., 29
Brzovic, P., 117, 118, 122, 125
Bubb, M. R., 107, 108(34)
Buddenbaum, W. E., 358, 359, 381, 381(22, 25)
Bukau, B., 29
Bulow, L., 142
Bunting, J. W., 284
Burgess, S. M., 32
Burgner, I. J. W., 186, 198(48)
Burgner, J. W., 177, 177(19–24, 26, 28), 179, 181(26), 186, 188(24), 189(19, 21), 190(28), 191(24), 196, 198(26, 49, 67)
Burgner II, J. W., 189, 190(54), 191(54–56)
Burton, D. E., 302
Burton, G. W., 344, 364, 371, 376(36), 377(36, 49)
Bystroff, C., 196

C

Cai, S. J., 143
Calcaterra, L. T., 280
Caldwell, J. W., 377
Caldwell, W. S., 331
Callender, R., 176, 177, 177(18–29), 179, 179(10), 181(16, 18, 26), 183, 183(29), 186, 188(24), 189(19, 21), 190(28), 191(24), 193, 196, 198(26, 48), 201(29), 332
Campbell, I. D., 290

Campbell-Crawford, A. N., 276
Cannon, W. R., 303
Caplow, M., 105
Caput, D., 132
Carey, F. A., 254
Carey, P. R., 177, 178, 185, 192, 193(59)
Carlier, M. F., 94, 97(6), 103, 105, 106(2), 109(21)
Carlsson, M., 286
Caron, M. G., 73
Carroll, L. J., 152, 155(7), 156, 159, 160, 161, 163(9), 165
Carter, C. W., Jr., 406
Carter, P., 244
Carty, R. P., 211
Casamassina, T. E., 371
Case, D. A., 377
Casey, M. L., 281
Cassidy, C. S., 24, 222, 223, 230, 244(19), 251(32, 34), 252, 253
Cha, S.-S., 226
Cha, Y., 360
Chabre, M., 82, 105
Chakraborty, T., 106
Challacombe, M., 302, 374
Chan, A. Y., 177, 177(29), 183(29), 201(29)
Chand, P., 340
Chang, S., 142
Chapman, E., 269
Cheeseman, J. R., 302, 374
Chen, C.-X., 177, 177(25), 181(16)
Chen, D., 177
Chen, H. L., 276(13), 277, 284(13), 285(13), 293(13)
Chen, W., 302, 374
Chen, X., 288, 289
Chen, X.-Y., 302, 316, 318(56), 351, 353, 367, 416
Chen, Y.-Q., 177, 177(18), 181(18), 193, 196
Chen-Howard, M. C., 77
Chentsova, N. P., 411
Chia, W. N., 190, 191(77), 200
Chiang, Y., 276(15, 16), 277
Chiausa, A. J., 410
Cho, H.-S., 226
Cho, M.-J., 226
Choay, J., 54
Chock, B. P., 112
Chock, P. B., 122, 126(48)

Choi, K. Y., 226
Chothia, C. H., 271
Chou, K.-C., 411
Christianson, D. W., 287
Chua, C. Y., 260
Chua, N. H., 103, 109(21)
Chwang, W. K., 276(14), 277, 296(12)
Cieplak, P., 377
Cioslowski, J., 302, 374
Ciscanik, L., 152, 163(6)
Clapper, D. L., 313
Clark, B. F. C., 76, 78(20), 89(20)
Clarke, A., 177, 177(28), 190, 190(28), 191(77), 200
Cleland, W. W., 3, 17(19), 19, 20, 22, 23(22), 24(22), 33, 149, 221, 222, 245(2), 251, 301, 304, 328(1)
Closs, G. L., 280
Clowney, L., 339, 387
Cohen, A. O., 276
Cohen, G. H., 209, 210(22)
Cohn, M., 5, 6(5), 328
Coleman, D. E., 70, 73, 74(10, 22, 47), 75(10, 22, 47), 76, 76(10), 78(10), 81(10), 83(10), 86(22), 88, 88(22)
Coleman, J. E., 290
Condeelis, J., 94, 106(5)
Connick, R. E., 240
Cook, A. G., 151
Cook, P. F., 301
Cool, R. H., 77
Copeland, R. A., 185(76), 200
Cornell, W. D., 377
Cornforth, F. J., 262
Cosloy, S., 177, 181(16)
Cossart, P., 106
Cotecchia, S., 73
Cothup, N. B., 185, 186(46)
Cottam, P. F., 128
Cozzarelli, N. R., 63, 64(50), 65, 65(50), 66
Craik, C. S., 244
Crans, D. C., 186, 198(49)
Crawford, I. P., 121, 125
Creighton, T. E., 116
Cross, P. C., 363, 377(33)
Cruickshank, F. R., 248
Cullis, P. M., 29, 66
Curran, J. S., 276

D

Dabiri, G. A., 106
Dahlquist, F. W., 307
Daimay, L.-V., 185, 186(46)
Danenberg, K., 135
Danenberg, P., 135
Daniel, F., 253
Daniel, K. W., 73
Dartzenberg, M. D., 54
Davenport, R. C., 125, 226, 232
Davies, D. R., 116, 122, 124, 124(37, 49), 125(49), 126(49, 53), 209, 210(22)
Davies, G. J., 66
Davies, J. E., 387
Davis, L., 230, 251(40), 253
Decius, J. C., 363, 377(33)
De Fabo, E. C., 229
Defrees, D. J., 302, 340, 374
Degano, M., 304, 339, 393, 416, 417(30)
Delaria, K., 177, 181(16)
Delbaere, L. T., 226
De Meio, R. M., 53
Demerec, M., 62, 63(45)
Demoss, J. A., 116
Deng, H., 176, 177, 177(18–29), 179, 179(10), 181(16, 18, 26), 183, 183(29), 186, 188(24), 189(19, 21), 190(28), 191(24), 198(26, 48, 49), 201(29), 332
Denisov, G. S., 221
Deterre, P., 82
Deuter, R., 77
deVos, A. M., 73, 76, 80(9), 83(9), 85(9), 87
Dewar, M. J. S., 336
DeWolf, W. E., Jr., 394, 405
Didry, D., 103, 105, 109(21)
Dijkstra, B. W., 203
Dittmann, J., 143
Dix, D. B., 68(77, 79), 69
Dixon, D. A., 47
Dixon, M. M., 208, 210(17)
Dodson, E. J., 66
Dodson, G., 66
Dogonadze, R. R., 277
Dolejsi, M. K., 49
Domann, E., 106
Donald, R., 135, 137(82)
Donato, J., 137
Donohue, J., 221

Dorman, C. J., 67
Drabble, K., 178(33, 35), 179
Drago, R. S., 185(74), 200
Drewe, W. F., 118, 119(43), 129, 130(73)
Driscoll, J. S., 407
Drucker, G. E., 262
Drueckhammer, D. G., 251, 251(27, 28), 252
Duce, P. P., 256, 275(51)
Duggleby, R. G., 393
Dunaway-Mariano, D., 149, 152, 155, 155(7),
    156, 157, 158, 159, 160, 161, 163(6, 9),
    164(8), 165, 167, 168(8), 169, 169(8), 170,
    171, 173(8), 174, 174(8), 175
Dunlap, R., 137
Dunn, M. F., 117, 118, 119(43), 122, 125, 126,
    129, 130(73)
Dwivedi, P. C., 184, 185(44)
Dyer, D. H., 203

E

Eads, J. C., 42, 310
Ealick, S. E., 350
Easterby, J. S., 112, 113(8)
Ebel, F., 106
Ebrahemian, S., 229, 243(44)
Eccleston, J. F., 68(79), 69, 72
Eckert, M., 251(31), 252
Eggleston, A. K., 47
Ehrlich, J., 129, 405
Eigen, M., 242
Elcock, A. H., 134, 135(80), 143
Eliason, R., 277, 296(12)
Elliston, K. O., 143
Emsley, J., 22, 246, 251(6)
Eriksson, A. E., 286, 287, 287(59), 288(65),
    293(65)
Erion, M. D., 350
Estupinan, B., 177, 177(29), 183(29), 199,
    201(29), 304, 311, 336, 336(33), 338,
    339(82), 345(33), 357, 381(13), 385,
    388(13), 390(13), 394(13), 417
Etter, M., 246(11), 247
Evans, B. E., 401
Evans, C., 142
Evans, G., 353
Evans, J. A., 94, 106(2)

Evans, M. G., 277
Evans, S. V., 79
Exner, O., 254

F

Fabian, M. A., 388
Faerman, C. H., 134
Farber, G. K., 125, 201, 203, 203(2), 204,
    208(9), 210(9, 10), 211(9), 212(10)
Farifield, F. R., 49
Farrell, K. W., 101
Fasella, P., 111, 112(1)
Fateley, W. G., 185, 186(46)
Fausett, L., 412, 413(27)
Femec, D. A., 370
Ferguson, D. M., 377
Ferguson, K. M., 87
Fersht, A. R., 91, 269, 364
Feuerstein, J., 72
Field, C. M., 94
Fierke, C. A., 115, 136, 150, 287
Filman, D. J., 196
Fink, A. L., 203
First, E. A., 269
Fischer, A. M., 54
Fisher, L. M., 11, 12, 63
Fisk, C. L., 227
Fitzpatrick, P. A., 204
Flanagan, J., 29
Flanagan, P. W. K., 282
Fogarasi, G., 183
Foresman, J. B., 302, 340, 374
Forsman, C., 286, 289(58)
Foster, J. W., 37, 67
Fothergill, L. A., 44
Fox, D. J., 302, 340, 374
Fox, T., 377
Frainier, L., 282
Frey, P. A., 23, 24, 38, 53, 222, 223, 230,
    244(19), 251(32–34), 252, 253
Freyberg, D. P., 221
Frick, L., 407
Frieden, C., 169
Frisch, M. J., 302, 340, 374
Fritsch, J., 260
Frolow, F., 208, 210(18)

Fromm, H. J., 304
Fry, A., 344, 364, 376(36), 377(36)
Fukushima, T., 112
Fukuyama, M., 282
Fullin, F. A., 394, 405
Funderburk, L. H., 255
Furneaux, R. H., 302, 350, 351, 353, 354(107), 385, 393, 416
Furukawa, K., 196

G

Galloy, J. J., 387
Galunsky, B., 208
Gandour, R. D., 301
Ganem, B., 177, 177(29), 183(29), 201(29), 393
Ganoth, D., 29
Garay, R. P., 54
Garcia-Viloca, M., 305
Gardner, J. D., 54
Garrett, B. C., 360
Garvey, E. P., 132, 134(76)
Gassman, P. G., 22, 23(22a), 222, 251, 285
Gasteiger, J., 410
Gatev, G. G., 267
Gawlita, E., 331, 365
Gelabert, R., 305
Gelbin, A., 339, 387
Gellert, M., 49, 65, 66, 67
Gerhardt, E., 124, 125(53a), 126(53a), 127(53a), 128(53a)
Gerlt, J. A., 22, 23(22a), 222, 251, 269, 285, 304
German, E. D., 277
Gerstel, B., 106
Geuron, M., 229
Geyer, M., 80
Ghambeer, R. K., 309
Ghannam, A. F., 297
Ghio, C., 377
Gholson, R. K., 38, 40, 44(26)
Giesemann, W., 129
Gill, P. M. W., 302, 340, 374
Gill, S. C., 39
Gillespie, D., 62, 63(45)
Gilman, A. G., 72, 73, 74(10, 22, 39), 75(10, 22, 39), 76, 76(10), 78(10), 80, 81, 81(10), 82(32), 83(10), 84(3), 85, 86(22, 39), 87, 88, 88(22), 92, 103
Gilson, M. K., 261, 264(57), 272(57)

Gindin, V. A., 222
Girvin, M. E., 302
Gittis, A. G., 23, 223, 224(29), 226, 226(29), 229(29), 231(29), 232(29), 235(29), 236(29), 239(29), 243(29), 244(29), 251(37), 253
Glaid, A. J., 188
Glasfeld, A., 203
Glenn, R., 388
Goeppert Mayer, M., 361, 378(31)
Goldberg, J. M., 177, 177(18), 181(18)
Golden, D. M., 248
Goldsmith, E. J., 92
Goldstein, E. K., 67
Golubev, N. S., 221, 222
Gomperts, R., 302, 340, 374
Gonzalez, C., 302, 340, 374
Gonzalez-Lafont, A., 305
Goody, A. G., 87
Goody, R. S., 72, 76, 77, 80(19), 84(19)
Gottesman, S., 29
Gouin, E., 106
Gould, I. R., 377
Grasselli, J. G., 185, 186(46)
Green, N. J., 280
Greene, D., 135, 137(82)
Grever, M. R., 410
Groesbeek, M., 183
Gross, J. W., 28, 39, 40, 41(27), 44(22)
Grubmeyer, C., 28, 38, 39, 39(16), 40, 40(16), 41(27), 42, 43(16), 44(22), 46, 46(16), 49, 302, 310
Grumont, R., 132
Grunwald, E., 272, 282
Guajardo, R., 49
Guida, W. C., 350
Gullion, T., 128
Guo, X., 388, 390(80)
Gurka, D., 260
Gutfreund, H., 112
Guthrie, C., 32
Guthrie, J. P., 282
Guthrie, R. D., 356
Gutteridge, W. E., 416

H

Ha, N.-C., 226
Hackney, D. D., 49

Hagglund, C., 142
Håkansson, K., 286
Hames, G. G., 150
Hamlin, R. C., 196
Hamm, H. E., 73, 76, 76(11), 77(21), 80(21), 81(21), 82, 83(36), 84(36), 85(21), 87(11)
Hammes, G., 3, 115
Hammond, D. J., 416
Hammond, G. S., 277
Handler, P., 36, 37, 38(14), 40(10a)
Hanna, L. S., 40
Hardcastle, F. D., 176, 186(1), 191(1)
Harel, M., 208, 210(18)
Harkins, R. N., 44
Harris, D. C., 363
Harris, T. K., 219, 222, 223, 226, 226(30), 229(30), 231(30), 232(30), 238, 239(30), 240(30), 241, 251(35), 253, 304
Hartl, F. U., 29
Hartman, F. C., 125
Hase, W. L., 360
Hassid, A. I., 276
Hasson, T., 49
Haugen, G. R., 248
Hautala, J. A., 282
Hawkinson, D. C., 285
Hayaishi, O., 37, 40(15), 43(15), 313
Hayashi, H., 222, 230, 231, 232(46), 251(39), 253
Hayes, R. N., 313
Healy, J. J. P., 336
Hehre, W. J., 345, 388, 390, 390(81)
Hemsley, J., 219, 224(1), 225(1)
Henderson, R. J., 271
Hend-Gordon, M., 302, 340, 374
Hepler, J. R., 73
Herschlag, D., 22, 72, 73, 84(7, 13), 91, 198, 221, 246, 251(2, 3), 260(1, 3), 261(1), 263(1, 2), 264(1), 265, 268, 269(1, 2), 270(2), 271(1, 3), 272, 272(1), 304
Hershfield, M. S., 340
Hershko, A., 29
Hess, S. L., 40
Heuser, J. A., 109
Hibbert, F., 22, 219, 224(1), 225(1), 246, 251(6)
Higacki, T., 231, 232(45)
Higashijima, T., 87
Higgins, C. F., 67
Higgins, N. P., 66

Hildebrandt, P. G., 185(76), 200
Hilgenfeld, R., 56, 80
Hine, J., 254, 257, 257(45), 274(45, 53)
Hing, A. W., 128
Hinton, G. E., 414
Hinz, H.-J., 188
Hirotsu, K., 222, 230, 231, 232(46), 251(39), 253
Ho, C., 128
Hofsteenge, J., 54
Holbrook, J. J., 177, 177(28), 189, 190, 190(28), 191(77), 200
Holden, H., 90, 141
Holl, S., 128
Holland, D., 269
Holmgren, A., 62, 63(47)
Homandberg, G. A., 208
Hong, Y., 103, 109(21)
Honger, S., 350
Honig, B., 261, 264(56–58), 272(56–58)
Honjo, T., 37, 40(15), 43(15)
Honzatko, R. B., 304
Hopewell, R., 63
Hopfield, J. J., 32, 67
Hopkins, N. H., 221
Hore, P. J., 228
Horenstein, B. A., 199, 200, 302, 303(18), 304, 336, 336(33), 338, 339(82), 340, 345(33), 346(88), 347(88), 353, 357, 381(13), 385, 385(14), 388(13), 390(13), 393, 394(13, 14), 400, 401, 407(8), 417
Horn, V., 125
Horvath, G., 176, 193(2)
Horwich, A. L., 29
Horwitz, S. B., 101, 105(17)
Hostomska, Z., 133, 135(77)
Houben, K. F., 118
Howard, J. B., 49
Howland, E., 133, 135(77)
Hsieh, S. H., 339, 387
Huang, L., 183
Hubbard, R. E., 246
Huber, R., 269, 271
Hude, C. C., 125
Hughes, J., 54
Huskey, W. P., 226, 234, 302, 306, 324, 359, 371, 374(23)
Huttner, W. B., 54
Hvidt, A., 238
Hwang, Y.-W., 77

Hyde, C. C., 116, 122, 124(37, 49), 125, 125(49), 126(49)
Hynes, J. T., 360

I

Ibanez, I. B., 222, 228(23)
Illenyi, J., 176, 193(2)
Imbert, T., 54
Imsande, J., 37, 38(14)
Inesi, G., 36
Iñiguez-Lluhi, J. A., 73, 85, 88
Isaacson, A. D., 360
Itikawa, H., 62, 63(45)
Ito, J., 121
Ivanetich, K. M., 134, 135(81), 137
Iwamatsu, A., 109

J

Jackman, J. E., 287
Jacks, T., 69
Jaenicke, R., 188
James, M. N. G., 226
Janin, J., 271
Janson, C. A., 133, 135(77)
Jansson, B.-H., 14(32), 24
Jeffrey, G. A., 246, 251(5), 388
Jencks, W. P., 28, 29, 29(3), 30(4), 31(3, 4), 49, 72, 90, 91, 190, 198, 246, 249, 249(4), 250, 251(7, 19), 255, 255(19), 259(19), 260(19), 271(4), 272(4, 18), 274(19), 275(19, 50), 277, 356
Jensen, P. R., 64, 67(54)
Jensen, R. T., 54
Jewell, D. A., 287, 289
Jockusch, B. M., 106
Joesten, M., 184, 185(41)
John, J., 87
Johnson, B. G., 302, 340, 374
Johnson, C. D., 281
Johnson, K. A., 115, 117, 118(41), 119(41), 120, 121, 121(41), 122(41), 123, 126(41), 129(41), 135(28), 136, 137, 137(28, 41), 150
Johnson, M. J., 10
Johnson, O., 387

Johnston, H. S., 364
Jones, M. E., 11
Jones, P. G., 388
Jones, T. A., 286, 287(59)
Jonsson, B.-H., 13, 14(14), 286, 289(58), 297
Joo, S., 226
Jordan, F., 234
Joris, L., 260
Joshi, V. C., 112
Jurnak, F., 177, 181(16)

K

Kabsch, W., 76, 80(19), 84(19), 92
Kagamiyama, H., 222, 230, 231, 232(46), 251(39), 253
Kahn, V., 38
Kalitzer, H. R., 80
Kalk, K. H., 203
Kan, C.-C., 133, 135(77)
Kang, F., 106, 107, 108(34)
Karem, K., 67
Karim, A. M., 68(77), 69
Karplus, M., 226, 232
Karpusas, M., 269
Karr, T. L., 101
Kasche, V., 208
Kasende, O., 260
Kato, Y., 251(30), 252
Katunin, V. I., 49
Kawasaki, H., 125
Kayastha, A., 122
Kebarle, P., 267
Keillor, J. W., 49
Keith, T., 302, 374
Keleti, T., 111, 112, 112(1), 113
Keller, J. M., 54
Keller, K. A., 54
Kemp, D. S., 281
Kennard, O., 246(10), 247, 383, 387, 387(61), 388(61)
Kent, S. B. H., 269
Kenyon, G. L., 269
Kessi, J., 350
Kettner, C., 234
Khalifah, R. G., 286
Khalil, M., 312
Kharkats, Y. I., 277
Khundkar, L. R., 303

Kicska, G., 302
Kiefer, W., 179
Kim, A. Y., 121, 123, 123(46)
Kim, J. H., 304
Kim, J.-S., 226
Kim, K., 399
Kim, K. K., 226
Kim, M., 177
Kim, S.-H., 73, 76, 80(9), 83(9), 85(9), 87
Kim, S. W., 226
Kim, Y., 296
King, G., 261, 264(55), 272(55), 386, 388(388)
Kingma, J., 39
Kinoshita, Y., 388
Kinosita, K., 49
Kintanar, A., 222, 229(26), 230, 231(26), 232(26), 244(26), 251(39), 253
Kirby, A. J., 388
Kirkgaard, K., 65
Kirkwood, J. G., 257, 274(52)
Kirsch, J. F., 177, 177(18), 181(18), 244
Kirschner, K., 117, 118, 119(44), 125, 129(44), 130(44)
Kirschner, M. W., 94, 105
Kisker, C., 49
Kistler, K. A., 293, 294
Kjeldgaard, M., 56, 76, 78(20), 80, 87, 89(20)
Kleinkauf, H., 143
Kleuss, C., 80
Klibanov, A. M., 204
Kline, A. D., 238
Kline, P. C., 302, 303(18), 326, 327, 330, 333, 335, 335(77), 341(65), 345(65), 347(81), 353(69), 357, 366(10), 385(10), 394(11), 401, 404, 405(12), 407(9)
Klinman, J. P., 323, 324(61), 360
Knighton, D. R., 133, 134, 135(77, 80)
Knowles, J. R., 3, 11, 12, 151, 162, 202, 229, 249, 269(17), 272, 272(17), 273(17, 84, 85)
Kobilka, B. K., 73
Kochi, J. K., 276(18), 277
Kociolek, K., 128
Kocks, C., 106
Koeppl, G. W., 277
Koetzle, T. F., 383, 387(61), 388(61)
Kogo, K., 231, 232(45)
Kohen, A., 360
Kohn, K. W., 410
Kollman, P. A., 377
Komives, E. A., 249, 269(17), 272(17), 273(17)

Konasewich, D. E., 276
Korn, E. D., 94, 106(2)
Kosaka, A., 40, 44(26)
Koshland, D. E., 94, 203
Koutsoukos, A. D., 410
Kowalczykowski, S. C., 47
Kozarich, J. W., 269
Kozasa, T., 92
Krahn, J. M., 304
Kraulis, P. J., 77, 209(24), 211
Kraut, J., 193, 196
Kreditch, N. M., 62, 63(48)
Kreevoy, M. M., 22, 23(22), 24(22), 219, 222, 233, 251, 276, 296
Krengel, U., 76, 80(19), 84(19)
Kresge, A. J., 266, 275(65), 276, 276(11, 13–16), 277, 281, 282, 282(36), 284, 284(13), 285(13), 290, 291, 291(11, 73), 292, 292(73), 293(13), 295(11), 296(12)
Krishnan, B. R., 62, 63(46)
Kristofferson, D., 95, 97(12)
Krupka, R. M., 49
Krupp, R. A., 47
Kukhar, V. P., 411
Kukla, D., 271
Kulakauskas, S., 62, 63(46)
Kuliopulos, A., 222, 243
Kuramitsu, S., 231, 232(46)
Kuznetsov, A. M., 277
Kvalnes-Krick, K. L., 335
Kvassman, J., 112

L

Labaune, J. P., 54
Lady, J. H., 185(75), 200
Laidler, K. L., 15
Laine, R. O., 106, 107, 108(34)
Laipis, P. J., 287, 288, 289, 290, 290(70), 291, 291(73), 292, 292(73), 293(70), 297(70)
Lambooy, P. K., 94, 106(2)
Lambright, D. G., 73, 76(11), 82, 83(36), 84(36), 87(11)
Lane, A., 117, 118, 119(44), 125, 129(44), 130(44)
Langen, R., 80, 83(31), 84(31)
Langer, T., 29
Lapetina, E. Y., 73

LaReau, R., 188, 330
Largman, C., 244
Larson, J. W., 221, 247, 267(12)
Larson, T. M., 21
Laskowski, M. J., 185(76), 200
Laskowski, M., Jr., 208
Latajka, Z., 184
Lauder, S. D., 47
Laurent, V., 103, 109(21)
Lautwein, A., 92
Lawen, A., 143
Leberman, R., 87
Lee, E., 73, 74(10, 22, 39), 75(10, 22, 39), 76, 76(10), 78(10), 80, 81(10), 83(10), 85, 86(22, 39), 88, 88(22)
Lee, F. S., 261, 264(55), 272(55)
Lee, H. C., 313, 315, 316, 327(55)
Lee, M., 118(69), 124, 125(53a), 126(53a), 127(53a), 128(53a), 129, 130, 131, 132(69, 70)
Leffler, J. E., 277
Lefkowitz, R. J., 73
Leitner, E., 143
Leja, C. A., 118
Lelander, L., 301, 306(6)
Lemp, G. F., 54
Lenzen, C., 77
Leplawy, M., 128
Leung, H. B., 307, 308(43), 311, 319, 345
Levich, V. G., 277
Lewis, B., 331, 332
Lewis, D. E., 302, 337(10), 370, 371, 377(49)
Lewis, E. S., 280
Leyete, A., 54
Leyh, T. S., 47, 48, 49, 53, 53(16), 54, 55, 55(37, 38), 56, 56(16, 38), 57(16, 38), 60(38), 62, 63(40, 46)
Li, C., 302
Li, H., 224
Li, R., 109
Li, X., 410
Li, Y.-K., 222
Liang, N., 280
Liang, P. H., 124, 125(53a), 126(53a), 127(53a), 128(53a), 137, 139(85, 86)
Liang, T.-C., 222, 228(24)
Liang, T. M., 219, 276
Liang, Z., 297
Liepinsh, E., 242
Ligai, S. S., 222

Liljas, A., 189, 286, 287, 287(59), 288(65), 293(65)
Limbach, H. H., 221
Limberg, G., 351, 353, 416
Lin, J., 24, 222, 223, 230, 244(19), 251(34), 253
Lin, T. C., 399
Lindbladh, C., 142
Linder, M. E., 73, 74(10), 75(10), 76(10), 78(10), 81(10), 83(10)
Lindskog, S., 13, 14(14, 32), 24, 286, 289(58), 290, 294(54), 297
Link, T. M., 302, 316, 318(56)
Lipmann, F., 53
Liu, C., 49, 53(16), 54, 55(37, 38), 56(16, 38), 57(16, 38), 60(38)
Liu, K., 249, 269(17), 272(17), 273(17), 303
Liu, L. F., 65
Liu, P. S., 407
Lluch, J. M., 305
Lodi, P. J., 229
LoGrasso, P. V., 287, 288, 289
Loh, S. N., 235, 236(52, 53), 237(52), 246, 251(3), 260(3), 271(3)
Lolis, E., 125
Lormeau, J. C., 54
Lu, X., 358
Luecke, H., 251(29), 252
Lugtenburg, J., 183
Luik, A. I., 411
Lumry, R., 4
Luo, P., 261
Lyon, J., 137

M

Maarouei, R. M., 54
Machesky, L. M., 108, 109
Macrae, C. F., 387
Maegley, K. A., 73, 84(13)
Mahroof-Tahir, M., 186, 198(49)
Makhatadze, G. I., 221
Makinen, M. W., 203
Mandelkow, E., 106
Mandelkow, E. M., 106
Mangel, W. F., 211

Mann, M., 109
Manor, D., 177, 177(25), 181(16)
Mäntele, W., 177
Marcus, R. A., 276, 280(1)
Mardiguian, J., 54
Margerrison, E. E. C., 63
Margerum, D. E., 276(17), 277
Margolin, Z., 262
Margolis, R. L., 99
Markham, G. D., 327, 333(67), 343, 381, 394(60)
Markham, G. H., 53
Markley, J. L., 222, 226, 228(23), 230, 235, 236(52, 53), 237(52), 251(38), 253
Marlier, J. F., 14(31), 24
Marquez, V. E., 407
Marshall, G. R., 128
Martin, C., 177
Martin, D. P., 251(27), 252
Martin, E., 54, 55(38), 56(38), 57(38), 60(38)
Martin, J., 29
Martin, R. B., 82
Martin, R. L., 302, 340, 374
Martin, Y., 399
Marx, A., 106
Matchett, W. M., 116
Mathews, C. K., 112
Mathews, F. S., 269
Matthews, B. W., 208, 210(17)
Matthews, D. A., 133, 134, 135(77, 80), 193, 196
Matthews, W. S., 262
Mattis, J. A., 208
Maulitz, A. H., 221
Maurizi, M. R., 29
Mautner, M., 267
Maxwell, A., 29, 66, 67
Mazzella, L. J., 385
McAllister, M. A., 221
McCallum, R. J., 262
McCammon, J. A., 134, 135(80), 143
McCollum, G. J., 262
McCormack, J. J., 407
McCormick, F., 71, 77(2), 78(2), 80(2)
McDermott, A., 224, 225
McDowell, L. M., 118(69), 129, 130, 131, 132(69, 70)
McFarland, J. T., 330
McHenry, C. S., 137

McKay, R., 118(69), 128, 129, 130, 131, 132(69, 70)
McKenna, J., 370
McMahon, T. B., 221, 247, 267(12)
Meek, T. D., 132, 134(76)
Meesters, H. A. R., 39
Mehl, A. F., 152, 155, 155(7), 156, 157, 158, 159, 160, 161, 163(9), 164(8), 165, 167, 168(8), 169, 169(8), 170, 171, 173, 173(8), 174, 174(8), 175
Meisheri, K. D., 54
Meister, A., 90
Melander, L., 284
Melki, R., 103, 105, 109(21)
Mentch, F., 327, 333, 333(67), 343, 344(76), 357, 381, 385(14, 16), 394(14, 16, 60), 401, 407(8, 10)
Menzel, R., 49
Meotner, M., 267
Merkler, D. J., 333, 335(77), 357, 394(11), 401, 407(9)
Mertens, K., 54
Merz, K. M., 287
Merz, K. M., Jr., 377
Mesangeau, D., 54
Metzler, C. M., 222, 229(26), 231, 231(26), 232(26, 45, 46), 244(26)
Metzler, D. E., 222, 223, 229(26), 230, 231, 231(26), 232(26, 45, 46), 244(26), 251(39), 253
Meyer, E. F., Jr., 383, 387(61), 388(61)
Meyer, S. L., 335
Mezey, P. G., 402
Miake-Lye, R., 30, 36(5)
Michaelis, G., 151, 208
Michels, P. A. M., 348
Migeon, J. C., 73
Mihalick, J. E., 267
Milburn, M. V., 73, 76, 80(9), 83(9), 85(9), 87
Mildvan, A. S., 22, 23, 23(25), 219, 222, 223, 224(29), 226, 226(28, 29, 30), 229(28–30), 231(29, 30), 232(28–30), 235(28, 29), 236(28, 29), 238, 239(28–30), 240(30), 241, 243, 243(29), 244, 244(28, 29), 251(35–37), 253, 304
Miles, E. W., 116, 117, 118(41), 119(41), 120, 121, 121(41), 122, 122(41), 123, 123(46), 124, 124(37, 49), 125, 125(49), 126(41, 48, 49, 53), 129(32, 41), 137(41)
Miles, R. W., 302, 350, 353, 354(107), 393

Miller, B. G., 358
Miller, D. L., 77
Miller, H. P., 101
Miller, J. R., 280
Miller, K. W., 178(36), 179
Miller, S. A., 282
Miller, W. G., 64
Miran, S., 142
Misset, O., 348
Mitchell, B. S., 340
Mitchell, E., 387
Mitchell, G. F., 387
Mitchell, G. N., 401
Mitchison, T. J., 94, 105, 109, 109(22)
Miwa, M., 313
Mixon, M. B., 74(22), 75(22), 76, 86(22), 88(22)
Miyahara, I., 222, 230, 231, 232(46), 251(39), 253
Mock, W. L., 260
Model, P., 62, 63(47)
Moinet, G., 54
Molenaar, D., 64, 67(54)
Mollova, E. T., 222, 230, 231, 232(45, 46), 251(39), 253
Monks, A. P., 410
Monse, E. U., 378
Montgomery, J. A., 302, 340, 374
Moodie, S. L., 387
Mook, W. G., 14(30), 24
Moore, W. T., 39, 44(22)
Mooseker, M. S., 49
Moreno, M., 305
More O'Ferrall, R. A., 280
Morino, Y., 231, 232(45)
Morley, J. S., 54
Morris, J. J., 256, 275(51)
Morrison, J. F., 196, 303
Mosbach, K., 142
Motamedi, H., 143
Muchmore, C. R., 304
Muench, K. A., 29
Muir, T. W., 269
Mullins, R. D., 109
Munshi, C., 315, 316, 327(55)
Murdoch, J. R., 277
Murphy, D. B., 106
Murray, C. J., 360
Musick, W. D. L., 37
Mylonakis, S. G., 276

N

Nabedryk, E., 177
Nageswara Rao, B. D., 5, 6(5)
Nakamura, S., 37, 40(15), 43(15)
Nanayakkara, A., 302, 374
Narlikar, G. J., 272
Nathanson, N. M., 73
Navia, M. A., 143
Nazaret, C., 54
Neer, E. J., 35
Neuhard, J., 62
Newald, A. F., 62, 63(46)
Ngo, K., 117
Nguyen, B. L., 54
Nguyen, T. T., 29
Nicholls, A., 261, 264(56), 272(56)
Niebuhr, K., 106
Niehrs, C., 54
Nielsen, S. O., 238
Nigata, S., 125
Nishimura, S., 73, 76, 80(9), 83(9), 85(9)
Nishizuka, Y., 37, 40(15), 43(15)
Nissen, P., 80
Niwas, S., 340
Noel, J. P., 73, 76, 76(11), 77(21), 80(21), 81(21), 82, 83(36), 84(36), 85(21), 87(11)
Noji, H., 49, 92
Nolan, R. D., 73
Northrop, D. B., 3, 7, 11, 13, 14(10), 16, 17(6, 17), 19(17), 301, 323, 325(59), 330, 366, 393
Novoa, W. B., 188
Nozari, M. S., 185(74), 200
Nyborg, J., 76, 78(20), 80, 87, 89(20)
Nygaard, P., 62

O

Oatley, S. J., 196
O'Brien, W. E., 151
O'Byrne, C. P., 67
O'Dea, M. H., 65, 67
Oh, B.-H., 226
Oh, S.-W., 276
Ohayon, H., 106
Oldenburg, N., 64, 67(54)
O'Leary, M. H., 11, 14(29, 31), 24, 301, 331, 365

Oleynek, J. J., 54
Olmstead, W. N., 262
Olson, W. K., 339, 387
O'Neal, H. E., 248
Onuffer, J. J., 244
Oosawa, F., 93, 97
Oppenheimer, N. J., 387
Opperdoes, F. R., 348
Oram, M., 63
Orbons, L. P. M., 227
Orosz, F., 113, 114(23)
Ortiz, J. V., 302, 374
Orville-Thomas, W. J., 184, 185(44)
Osinga, K. A., 348
Otlewski, J., 185(76), 200
Otting, G., 242
Ovadi, J., 111, 112, 112(1), 113, 114(23)
Owen, H., 177

P

Padlan, E. A., 116, 124(37)
Page, M. I., 90, 190, 249, 272(18)
Pai, E. F., 56, 76, 80(19), 84(19)
Pan, P., 126
Paneth, P., 11, 14(29), 24, 331, 365
Panson, G. S., 260
Pantaloni, D., 94, 103, 105, 106(?), 109(21)
Paranawithana, S. R., 287
Pardi, A., 224
Parkin, D. W., 199, 304, 307, 308, 308(43),
    309(47), 311, 319, 323(46, 47), 327, 333,
    333(67), 336, 336(33), 338, 339(82), 343,
    344(76), 345, 345(33), 351, 353, 357, 364,
    381, 381(13), 385, 385(14, 16), 388(13),
    390(13), 393(38), 394(13, 14, 16, 60), 401,
    407(8, 10), 416, 417
Parr, R. G., 302
Parris, K. D., 122, 124, 124(49), 125(49),
    126(49, 53)
Parson, W. W., 177
Parsons, J. F., 269
Pasqualini, J. R., 54
Patapoff, T. W., 177
Patel, S., 63
Pathak, V. P., 340
Pauling, L., 303, 357
Paull, K. D., 410
Pauncz, R., 346

Pawlak, Z., 260
Pedersen, K., 280
Peebles, C. L., 66
Penfield, K. W., 280
Penfound, T., 37
Peng, C. Y., 302, 374
Perez-Alverado, G. C., 229, 243(44)
Perrin, C. L., 388
Peterman, B. F., 15
Petersson, G. A., 302, 374
Peticolas, W. L., 177
Petitou, M., 54
Petrounia, I. P., 269
Petsko, G. A., 76, 80(19), 84(19), 125, 203,
    204, 212(14), 226, 232, 249, 269, 269(17),
    272(17), 273(17)
Petterson, G., 112
Pettijohn, D. E., 67
Pettit, F., 112
Pfister, C., 82
Pistor, S., 106
Plant, C., 346
Plateau, P., 229
Plummer, K., 63
Poda, G. I., 411
Podrasky, A. E., 101
Podtelejnikov, A. V., 109
Polanyi, J. C., 303
Polanyi, M., 277
Polgar, L., 207
Polisczak, A., 348
Politzer, P., 346
Pollack, R. M., 229, 269, 285
Pollard, T. D., 109
Ponelle, M., 143
Pongor, G., 183
Pople, J. A., 302, 340, 374, 393
Portnoy, D. A., 106, 109
Posner, B. A., 88
Post, C. B., 196, 198(67, 68)
Potter, M. J., 134, 135(80)
Poyner, R. R., 269
Preiss, J., 36, 40(10a)
Prendergast, F. G., 185(76), 200
Primrose, W. U., 178(33), 179
Privalov, P., 221
Privé, G. G., 73, 80(9), 83(9), 85(9)
Profeta, S., Jr., 377
Pruszynski, P., 276(15), 277
Puddington, L., 54

Pulay, P., 183
Purich, D. L., 93, 94, 95, 97(8, 12), 100, 101, 106, 106(3, 7), 107, 107(7), 108(34)
Pusztai, M., 178
Pusztay, L., 176, 193(2)

Q

Qiwen, W., 238
Quillen, J. M., 121, 123(46)
Quinn, D. M., 339
Quiocho, F. A., 251(29), 252

R

Radom, L., 393
Radzicka, A., 10, 303, 350, 358, 404, 407(13)
Rafferty, M. A., 307
Raghavachari, K., 302, 340, 374
Raiford, S. S., 227
Raines, R. T., 137
Rajavel, M., 28, 29, 39, 40, 41, 41(27), 43, 44(22)
Rakshys, J. W., 260
Ramaswamy, P., 388
Rand-Meier, T., 307
Rao, C. N. R., 184, 185(44)
Rao, N. N., 62
Ratajczak, H., 184, 185(44)
Rault, M., 142
Raushel, F., 90, 141, 142
Raw, A., 74(37, 39), 75(37, 39), 80, 85, 86(37, 39)
Ray, J. W. J., 186, 198(48)
Ray, W. J., 177, 177(26), 181(26), 186, 189, 190(54), 191(54–56), 198(26, 49)
Ray, W. J., Jr., 17
Ray, W. J. J., 196, 198(67, 68)
Raycheba, J. T., 276(17), 277
Rayment, I., 21, 90, 141, 269
Raymond, K. N., 221
Rebek, J., Jr., 251(30), 252
Rebholz, K. L., 11, 14(10)
Reddy, N. S., 340
Redlinski, A., 128
Reed, G. H., 21, 269, 328
Reed, L. J., 112

Reed, T., 129
Rees, D. C., 49
Regan, J. W., 73
Rehfield, J. F., 54
Rehrauer, W. M., 47
Reinhard, M., 106
Reinsch, J., 137
Reiser, C. O., 56, 80
Remington, S. J., 251, 251(27), 252, 269
Ren, X., 289, 290(70), 293(70), 297(70)
Rensland, H., 87
Replogle, E. S., 302, 340, 374
Reshetnikova, L., 56, 80
Reuille, R., 143
Rhee, K. W., 177
Rhee, S., 122, 124, 124(49), 125(49), 126(49, 53)
Richard, J. P., 255, 275(50)
Richards, W. G., 411
Ridenour, C. F., 224, 225
Ridgeway, C., 358
Ringe, D., 203, 204, 226, 232, 249, 269(17), 272(17), 273(17)
Ripoll, D. R., 134
Rising, K. A., 313, 314, 318(54), 356, 390(5), 395(5)
Rivett, A. J., 54
Robb, M. A., 302, 340, 374
Robbins, P. W., 53
Robert, G. C. K., 178(33), 179
Roberts, J. W., 221
Roberts, M. F., 62
Robillard, G., 222
Robinson, D., 178(33), 179
Robishaw, J. D., 85
Roca, J., 65, 66(57)
Roczniak, S., 244
Rodgers, J. R., 370, 383, 387(61), 388(61)
Rodnina, M. V., 49
Rogers, A. S., 248
Roos, D., 135, 137(82)
Rose, D., 125
Rose, I. A., 165, 325, 326
Rose, I. W., 366
Rose, M. C., 276, 284(7)
Rosenberg, R. D., 54
Rosenblatt, J., 94, 105, 109, 109(22)
Rosing, J., 28
Rossmann, M. G., 189
Roth, G. J., 94

Roth, J. R., 54
Rothemberg, M. E., 255, 275(50)
Rothschild, K. J., 177, 178(36), 179
Rousseau, D. L., 181
Rowe, T. C., 65
Rowlett, R. S., 286, 297
Roy, M., 118
Rozeboom, H. J., 203
Rubin, J., 260
Rubinstein, L. V., 410
Ruble, J. R., 388
Rudolph, C., 410
Rudolph, F. B., 113
Ruhlen, R., 105
Ruhlmann, A., 271
Rumelhart, D. E., 414
Russel, M., 62, 63(47)
Russell, S. T., 261, 264(54), 272(54)
Rutter, W. J., 244

S

Sacchettini, J. C., 42, 304, 310, 339, 393, 416, 417(30)
Sadowski, J., 410
Saenger, W., 219, 246, 251(5)
Sagatys, D. S., 276(13), 277, 284(13), 285(13), 292(13)
Salerno, C., 111, 112(1)
Salmon, E. D., 105
Sampson, N. S., 272, 273(84, 85)
Sander, C. A., 56
Sanders, D. A., 71, 77(2), 78(2), 80(2)
Sanger, J. M., 106
Santaballa, J. A., 276(16), 277
Santi, D. V., 132, 134, 134(76), 135, 135(81), 137, 137(82)
Santolini, J., 103, 109(21)
Sato, Y., 276
Saunders, W. J., Jr., 301, 306(6)
Sauve, A. S., 315, 316, 327(55)
Savelsbergh, A., 49
Sayers, E., 121, 123(46)
Schaad, L. J., 184, 185(41)
Schaefer, J., 118(69), 128, 129, 130, 131, 132(69, 70)
Schah-Mohammedi, P., 221
Schatz, B., 54
Schechter, H., 282

Scheffzek, K., 80, 92
Scheiner, S., 184, 295
Schepp, N. P., 276(15), 277
Scherer, N. F., 303
Scheuring, J., 356, 366(7), 368(7), 371(7, 8), 395(6–8), 396(7)
Schiff, P. B., 101, 105(17)
Schindelin, H., 49
Schirmer, N. K., 56, 80
Schlegel, H. B., 302, 340, 374
Schlesener, C. J., 276(18), 277
Schlessman, J. L., 49
Schlichting, I., 123, 124, 125(53a), 126(53a), 127(53a), 128(53a)
Schlicting, I., 87
Schmid, F., 188
Schmidt, G., 77
Schnackerz, K., 129
Schneider, B., 339, 387
Schneider, T., 124, 125(53a), 126(53a), 127(53a), 128(53a)
Schneider-Scherzer, E., 143
Schorgendorfer, K., 143
Schowen, K. B., 232, 303
Schowen, R. L., 232, 301, 306, 324, 359, 370
Schramel, A., 29
Schramm, V. L., 177, 177(29), 183(29), 199, 200, 201(29), 301, 302, 303(18), 304, 305, 307, 307(14), 308, 308(43), 309(47), 311, 312, 313, 314, 315, 316, 318(54, 56), 319, 322, 323(47), 326, 327, 327(55), 330, 331, 332, 333, 333(67), 335, 335(77), 336, 336(33), 338, 339(82), 340, 341(65), 343, 344(44, 45, 76), 345, 345(33, 65), 346(14, 88), 347, 347(81, 88), 348(102), 350, 351, 353, 353(69, 101, 102), 354(107), 356, 357, 365(4), 366(7, 10), 367, 368(3, 4, 7), 371(3, 4, 7, 8), 374(3), 376(3), 379(68), 380(3, 4), 381, 381(3, 4, 13), 385, 385(10, 14, 16), 386, 387(3), 388(3, 13), 390(13), 393, 394, 394(11, 13, 14, 16, 60), 395(3, 4, 6–8), 396(3, 7), 400, 401, 402(7), 403(6, 7), 404, 405, 405(6, 12), 407(8–10), 409(6), 416, 417, 417(30)
Schroder, H., 29
Schuering, J., 322
Schultz, P. G., 269
Schwager, P., 271
Schwartz, B., 251, 251(28), 252

Schwartz, S. D., 302, 305, 324, 347, 348(102), 353(101, 102), 398, 401, 402(7), 403(6, 7), 405(6), 409(6)
Schweins, T., 73, 80, 83(31), 84(31), 85(12)
Schwert, G. W., 188
Sconick, L. R., 287
Scott, A. P., 393
Scott, R. D., 222, 229(26), 231, 231(26), 232(26, 45, 46), 244(26)
Scott, W. G., 203
Scudiero, D. A., 410
Seaton, B. A., 226, 232
Segura, E., 39, 40, 41(27), 44(22)
Seljee, F., 203
Semenyaka, A., 393
Senkowski, B. Z., 260
Shafiee, A., 143
Shall, S., 313
Shames, S. L., 350
Shan, S., 22, 221, 246, 251(1–3), 260(1, 3), 261(1), 263(1, 2), 264(1), 265, 268, 269(1, 2), 270(2), 271(1, 3), 272(1), 304
Shanks, J., 105
Sharp, K. A., 261, 264(58), 272(58)
Shaw, R., 248
Shchepkin, D. N., 221
Shen, C. C., 280
Shimanouchi, T., 383, 387(61), 388(61)
Shiner, V. J., Jr., 358, 359, 381(22, 25)
Shirvanee, L., 125
Short, S. A., 406, 407
Shorter, J., 254
Shulman, R. G., 222
Sieck, L. W., 267
Siegel, L. M., 53
Sielecki, A. R., 226
Sigler, P. B., 73, 76, 76(11), 77(21), 80(21), 81(21), 82, 83(36), 84(36), 85(21), 87(11)
Sikorski, J. A., 137
Silman, I., 134, 208, 210(18)
Silverman, D. N., 276, 286, 287, 288, 289, 289(58), 290, 290(70), 291, 291(73), 292, 292(73), 293, 293(70), 294, 294(54), 297(70)
Silverton, E. W., 209, 210(22)
Simon, I., 77
Simon, K., 176, 193(2)
Simons, R. W., 64
Simonsson, I., 14(32), 24, 290

Simpson, F. B., 13
Sims, L. B., 302, 337(10), 344, 356, 364, 364(2), 370, 371, 376(36), 377(2, 36, 49)
Sinden, R. R., 67
Singer, P., 203
Singer, P. T., 211
Singh, U. C., 377
Singleton, D. A., 365
Singleton, R. J., 53
Singleton, S. F., 303
Sinha, N. K., 112
Sinnott, M. L., 344, 385, 388, 390(80)
Sirawaraporn, W., 132
Sjoberg, P., 346
Sklenár, V., 229, 231
Skoble, J., 109
Slaich, P. K., 178(33), 179
Slater, E. C., 28
Sloan, D., 40, 177, 177(20, 22, 23)
Smalås, A., 211
Small, W. C., 142
Smallcomb, S. H., 231, 232
Smallwood, C. J., 221
Smigel, M. D., 87
Smiley, J. A., 11
Smirnov, S. N., 221, 222
Smith, H. B., 350
Smith, J. L., 304
Smith, J. M., 387
Smith, L. D., 38
Smolen, P., 112
Smrcka, A. V., 85
Smulson, M., 313
Sng, J., 63
Snoep, J. L., 64, 67(54)
Snyder, K. B., 101
So, S.-S., 411
Sondek, J., 82, 83(36), 84(36)
Sousa, R., 49
Southwick, F. S., 93, 94, 106, 106(7), 107, 107(7), 108(34)
Spellmeyer, D. C., 377
Spindel, W., 378
Spiro, T., 177, 178, 185(76), 200
Spivey, H. O., 40, 44(26), 112
Sprang, S. R., 70, 73, 74(10, 22, 39), 75(10, 22, 39), 76, 76(10), 78(10), 80, 81(10), 83(10), 85, 86(22, 39), 88(22), 92
Sprinzl, M., 80
Srebrenik, S., 346

Sreedharan, S., 63
Srere, P. A., 111, 112(3), 142
Srivastava, D. K., 112
Stafford, W. F., 109
Stahl, N., 250, 251(19), 255(19), 259(19), 260(19), 274(19), 275(19)
Staverman, W. H., 14(30), 24
Steel, C., 272
Stefanidis, D., 284
Stefanov, B. B., 302, 374
Steindel, S. J., 189
Steiner, H., 286
Steiner, Th., 219
Steinmetz, A. C., 204
Steitz, J. A., 221
Stern, M. J., 374, 378
Sternberg, C., 54
Sternweis, P. C., 81, 82(32), 87
Stewart, J. P., 302, 336, 340, 374
Stitt, B. L., 29
Stoddard, B. L., 201, 203
Stoeckler, J. D., 350
Stone, S. R., 54, 196
Stoops, J. K., 112
Stossel, T. P., 94
Strasser, A., 125
Stroud, R. M., 124, 133(52), 134(52)
Stuehr, J., 276, 284(7)
Su, C.-T., 208, 210(18)
Sudmeier, J. L., 229
Sugimura, T., 313
Sugino, A., 65, 66
Sugio, S., 249, 269(17), 272(17), 273(17)
Suhnel, J., 306, 359
Sullivan, S. A., 247
Summers, M. F., 229
Sundaralingam, M., 387
Sundberg, R. J., 254
Sunko, D. E., 345, 388, 390(81)
Suo, Y., 49, 53(16), 54, 56, 56(16), 57(16), 63(40)
Sussman, J. L., 134, 208, 210(18)
Sutton, L. D., 339
Suziedelis, K., 62, 63(46)
Svensson, L. A., 286
Sweet, R. M., 203, 211, 271
Swift, T. J., 240
Sykes, B. D., 227
Szabo, A., 29
Szele, I., 345, 388, 390(81)

T

Tabouret, M., 106
Taft, R. W., 260
Taira, K., 196
Takabayashi, K., 350
Talalay, P., 22, 23(25), 222, 226(28), 229(28), 232(28), 235(28), 236(28), 239(28), 243, 244, 244(28), 251(36), 253
Tamamura, J. K., 66
Tamura, J. K., 66
Tan, R. C., 134
Tanabe, K., 196
Tanase, S., 231, 232(45)
Tanchuk, V. Y., 411
Tanford, C., 28
Tang, W.-J., 73
Tanhauser, S. M., 287, 290, 291, 291(73), 292, 292(73)
Taoka, S., 288, 293, 294
Tapon-Bretaudiere, J., 54
Tashian, R. E., 287
Tasumi, M., 383, 387(61), 388(61)
Taussig, R., 73
Taylor, J. T., 53
Taylor, P. J., 256, 275(51)
Taylor, R., 246(10), 247
Temm-Grove, C. J., 106
Teng, H., 39
Terpstra, P., 39
Terry, B. J., 100
Tesmer, J. J. G., 92
Tetko, I. V., 411
Tewey, K. M., 65
Thijs, R., 184, 185(42)
Thirup, S., 80
Thoden, J., 90, 141
Thomas, A. A., 365
Thomas, S. L., 73
Thompson, R. C., 67, 68(77, 79), 69, 69(75)
Thompson, T. B., 411
Thompson, W. C., 101
Thornberg, L. D., 229, 243(44)
Thornton, J. M., 387
Thorson, J. S., 269
Thrall, S. H., 159, 160, 161, 163(9), 165
Timasheff, S. N., 204, 212(14)
Tiraby, G., 203
Tjandra, N., 128
Tobin, J., 23, 222, 251(32, 33), 252, 253

Toledo, L. M., 251(30), 252
Tomcasanyi, T., 62, 63(46)
Tompa, P., 113
Tong, H., 230, 251(40), 253
Tong, L., 73, 76, 80(9), 83(9), 85(9), 87
Tonge, P. J., 178, 192, 193(59)
Torriani, A., 62
Traber, R., 143
Traut, T. W., 358
Trinczek, B., 106
Tritz, G. J., 37
Trucks, G. W., 302, 340, 374
Truhlar, D. G., 296, 360
Truhlar, G., 346
Trujillo, M., 135, 137(82)
Truong, T. N., 134
Tse, Y.-C., 65
Tsen, W., 297
Tskuada, H., 208
Tsuzuki, S., 196
Tu, C. K., 286, 287, 288, 289, 289(58), 290,
290(70), 291, 291(73), 292, 292(73), 293,
293(70), 294, 297(70)
Tucker, S. C., 360
Turner, D. L., 228
Tyler, P. C., 302, 350, 351, 353, 354(107), 385,
393, 416

U

Uchimaru, T., 196
Ueda, N., 85
Unsworth, C. D., 54
Usher, K. C., 251, 251(27), 252

V

Valencia, A., 56
Vananzi, T. J., 346
Van Beeumen, J., 348
van Beilen, J., 39
Vandekerckhove, J., 109
van der Marel, G. A., 227
van Dooren, S. J. M., 64, 67(54)
Vanier, N. R., 262
van Mourik, J. A., 54
van Schijndel, H. B., 54

van Workum, M., 64, 67(54)
Vargha, A., 183
Varin, C., 54
Varmus, H., 69
Venanzi, C. A., 346
Venderheyden, L., 185(73), 200
Venkatasubban, K. S., 287
Verbeet, M. P., 54
Verschueren, K. H., 203
Vertessy, B., 112, 113
Vetter, I., 87
Villafranca, J., 90, 152, 163(6)
Vinitsky, A., 38, 39, 39(16), 40(16), 43(16),
46, 46(16), 49
Vinogradov, S. N., 246
Vistica, D. T., 407
Viswandhan, V. N., 410
Vitullo, V. P., 276
Vogt, T. F., 56, 63(40)
vonder Saal, W., 152, 163(6)
von Hippel, P. H., 39
von Hipple, P. H., 49
von R. Schleyer, P., 260

W

Wachs, I. E., 176, 186(1), 191(1)
Wagener, M., 410
Wagner, G., 224
Wagner, S., 350
Wah, W., 330
Wakil, S. J., 112
Walker, R. A., 105
Wall, M. A., 88
Walseth, T. F., 313
Walsh, C. T., 303
Walsh, G. M., 340
Walsh, R., 248
Walter, R. L., 350
Walter, U., 106
Wan, W., 188
Wang, H.-C., 152, 163(6)
Wang, J. C., 49, 65, 66(57)
Wang, R., 54, 55(37)
Wang, Z., 251(29), 252
Wanner, B. L., 46
Ward, S., 178(35), 179
Ware, J., 94

Warshel, A., 73, 80, 83(31), 84(31), 85(12), 91, 261, 264(54, 55), 272(54, 55)
Washtien, W. L., 132
Watson, D. G., 387
Watson, H. C., 44
Watson, J. D., 221
Webb, M. R., 72, 94, 106(2)
Weber, G., 143
Weber, J. P., 112
Wedekind, J. E., 21, 269
Wegner, A., 99
Wehland, J., 106
Wei, A., 226
Weiner, A. M., 221
Weiner, D. P., 29, 66
Weiner, P., 377
Weiner, S. J., 377
Weinstein, H., 346
Weinstein, J. N., 410
Weiss, P., 306, 333, 335(77), 357, 365, 394(11), 401, 407(9)
Welch, G. R., 112, 113
Welch, L., 410
Welch, M. D., 109
Wells, J., 244
Welsh, K. M., 133, 135(77)
Weng, G., 177, 177(25), 181(16)
Wenger, R. M., 143
Wenzel, P. J., 396
Wesenberg, G., 90, 141
West, S. C., 29
Westerhoff, H. V., 64, 67, 67(54)
Westheimer, F. H., 257, 274(52), 284
Westler, W. M., 222, 226, 230, 251(38), 253
Wharton, C. W., 178, 178(33–35), 179
Whetsel, K. B., 185(75), 200
White, A. J., 178, 178(33–35), 179
White, I. A., 290
Whitt, S., 23, 222, 251(32, 33), 252, 253
Wierenga, R. K., 348
Wigley, D. B., 66, 190
Wilke, T. M., 92
Wilks, H. M., 191(77), 200
Williams, F. T., Jr., 282
Williams, G. J. B., 383, 387(61), 388(61)
Williams, R. J., 414
Wilson, E. B., Jr., 363, 377(33)
Wilson, J. C., 344, 364, 376(36), 377(36)
Wilson, K. J., 177
Wilson, L., 99, 101

Winer, A., 188
Winn, S. I., 44
Winter, D., 109
Wintermeyer, W., 49
Wirz, J., 276(15, 16), 277
Wishart, D. S., 227
Witholt, B., 39
Wittinghofer, A., 76, 77, 80, 80(19), 84(19), 87, 92
Woehl, E. U., 118, 126
Wolfenden, R., 10, 269, 303, 350, 358, 401, 404, 406, 407, 407(13)
Wolfsberg, M., 305, 361, 374, 378(32)
Wong, K. K., 313
Wong, L. F., 276(17), 277
Wong, M. L., 94
Wong, M. W., 302, 340, 374, 393
Wood, H. G., 151
Wrenn, R. F., 169
Wright, H. T., 271
Wright, W. B., 386, 388(388)
Wu, K. K., 186, 198(47)
Wu, R. W., 229, 243(44)
Wubbolts, M. G., 39
Wüthrich, K., 224, 238, 242

X

Xia, G. X., 103, 109(21)
Xia, Z.-X., 269
Xiang, S., 406, 407
Xu, Y., 42, 155, 157, 158, 159, 164(8), 167, 168(8), 169, 169(8), 170, 171, 173(8), 174, 174(8), 175, 302, 310
Xue, L., 244

Y

Yagi, M., 94
Yamada, H., 224
Yamaizumi, Z., 73, 76, 80(9), 83(9), 85(9)
Yamataka, H., 370
Yamdagni, R., 267
Yang, C., 407
Yang, M., 55
Yang, S., 303
Yang, W., 302

Yang, X. J., 121, 123(46), 125
Yang, X.-Y., 123
Yankwich, P. E., 381
Yano, T., 231, 232(46)
Yanofsky, C., 125
Yao, N., 251(29), 252
Yashphe, J., 62
Yasuda, R., 49
Yeaman, S., 112
Yee, K. C., 282
Yennawar, H. P., 204, 208(9), 210(9, 10), 211(9), 212(10)
Yennawar, N. H., 204, 208(9), 210(9), 211(9)
Yodh, J. G., 29
Yoshida, M., 49, 92
Young, G., 358
Yu, X., 36
Yue, K. T., 177
Yue, T. K., 177

Z

Zakrewski, V. G., 302, 374
Zalkin, H., 304

Zawrotney, M. E., 229, 243(44)
Zeegers-Huyskens, J. Th., 185(73), 200
Zeegers-Huyskens, T., 184, 185(42)
Zeegers-Huyskens, Th., 260
Zeile, W., 106
Zeng, C., 304
Zewail, A. H., 303
Zhang, Z., 249, 269(17), 272(17), 273(17)
Zhao, C., 22, 23, 23(25)
Zhao, Q., 222, 224(29), 226, 226(28, 29), 229(28, 29), 231(29), 232(28, 29), 235(28, 29), 236(28, 29), 239(28, 29), 243(29), 244(28, 29)
Zheng, C., 411
Zheng, J., 177, 177(18, 20, 22–24, 28), 181(18), 186, 188(24), 190(28), 191(24), 198(49)
Zhong, J. M., 77
Zhu, D. M., 122, 126(48)
Zieger, B., 94
Zigmond, S. H., 109
Zocher, R., 143
Zoebisch, E. F., 336
Zon, G., 125
Zundel, G., 251(31), 252, 260
Zupan, J., 410

Subject Index

A

Actin polymerization
  minimal mechanism, 93
  nucleotide exchange promotion by F, 93
  self-assembly models
    ATP role
      actin depolymerizing factor control, 103, 105
      bidirectional polymerization with hydrolysis at time of monomer addition, 99–101
      delayed hydrolysis after monomer addition, 101–102
      difficulty of study, 110–111
      overview, 94
      timing control of assembly and disassembly, 102–103, 110
      unidirectional polymerization with hydrolysis at time of monomer addition, 97–99
    condensation equilibrium model, 94–96
    elongation without nucleotide hydrolysis, 96–97
Activation energy
  energy for catalysis, 20–24
  rate constant relationship, 4–5
  rate acceleration calculation
    bimolecular enzymatic reactions, 9–10
    unimolecular enzymatic reactions, 6
  temperature effects, 13–15
Adenosine deaminase, molecular surface point similarity measures in binding energy analysis, 404–405, 407, 409
S-Adenosylmethionine synthetase, kinetic isotope effect modeling of transition state, 343
AMP, isotopic labeling for kinetic isotope effect studies, 309
AMP deaminase
  catalytic mechanism, 333, 335
  inhibitor design, 348–349, 394
  isotope effects and transition state modeling, 335–336

molecular electrostatic potential surface of transition state, 347–348
AMP nucleosidase, molecular surface point similarity measures in binding energy analysis, 405, 407, 409
APS, *see* ATP sulfurylase
ATP
  actin polymerization role, *see* Actin polymerization
  ATPase classification, 28
  free energy of hydrolysis, 28
  isotopic labeling for kinetic isotope effect studies, 308, 310, 312
  molecular discrimination in hydrolysis, *see also* ATP sulfurylase; DNA gyrase; Nicotinate phosphoribosyltransferase
    clock mechanism, 34–35
    conformational coupling, 28–29, 50–52
    conformer affinities for substrates and products, 31–34, 36
    examples of enzymes, 29, 49
    ion transporter energetics, 30–36
    leaks, origin, 52
    necessary imperfection, 49, 52, 70
    optimization of ATP use, 32–34
    RecA, 47–48
    thermodynamic modeling, 30–31, 35
ATP sulfurylase
  AMP formation, 57
  bond breaking, interdependence, 56
  central complex
    chemistry, 55
    formation, 54–55
  coupled versus uncoupled mechanisms, 54
  functions of activated sulfates, 53–54
  phosphate plugging of coupling efficiency leak, 60–63
  reaction cycle and subunits, 52–53
  stoichiometry and energy transfer, 57
  structural modeling of coupling step, 56–57
  transient product synthesis, 57–58, 60

B

Badger–Bauer relationship, *see* Raman spectroscopy
BEBOVA, *see* Bond energy/bond order vibrational analysis
Bond energy/bond order vibrational analysis, transition state structure
  ab initio optimized structures, 394–395
  AMP deaminase, 335
  BEBOVIB program
    advantages, 372
    atomic mass lists, 378
    force constants for each internal coordinate, 377
    inputs, overview, 373
    interaction force coefficients, 377–378
    internal coordinates, 375–377
    reaction coordinate, 378–382
    structures of initial and final states, 374
    temperature inputs, 378
  computational power, 356
  nucleoside hydrolase, 337, 339–340
  reliability of technique, 395–397
  software, 371, 373
  structure interpolation
    Cartesian coordinates to transition-state structures to BEBOVIB format using internal coordinates, 383, 388–389
    internal coordinates, 384–385
    nicotinamide adenine dinucleotide hydrolysis application, 385–388
    overview, 368, 370–371, 382–383
    reaction space and interpolation philosophy, 383–384
    reference structures
      high-energy intermediate, 388
      hyperconjugation, 390
      input in CTBI program, 388–389
      parsing, 390
      pruning to find rings, 390–392
      reactant, 387–388
      sources of structural information, 387–388
    test structures, isotope effect calculations, 368, 392–393
  transition state imbalance, 396–397
  unified models of reaction types, 371, 395
  Walden inversion motion, 379–381

Brønsted relation
  linear-free energy relationship analysis of hydrogen bonds, 254–261, 269, 274–276
  Marcus theory and proton transfer reactions, 280–282, 291–292

C

Carbamoyl phosphate synthase
  catalytic activities and subunits, 140–142
  substrate channeling
    kinetic studies, 142
    putative channel, 141
    rationale, 140
Carbonic anhydrase
  catalytic stages, 286
  free energy profile, 13
  His-64 in proton shuttling, 286–287
  Marcus parameters for proton transfer
    Brønsted plots, 291–292
    His-64 proton transfer, 291–293
    His-67 proton transfer, 293
    human isozyme III as model system, 287
    imidazole and lutidine buffers, proton transfer, 293–294
    intrinsic kinetic barrier, 295
    kinetic isotope effects, 292–293
    mutagenesis and disruption of pK_a values, 288–289
    pH titration, 290
    rate measurements, 289–290
    work function dominance, 290–291, 295–297
  rate-limiting steps, 285–286
Channeling, *see* Substrate channeling
Chymotrypsin
  mechanism, overview, 207–208
  organic solvent trapping of intermediates for crystallography
    acyl enzyme intermediate, 209–207
    enzyme–product complex, 211–212
    first tetrahedral intermediate structure, 208–209
    motions occuring between intermediates, 212–215
    native enzyme, 208

pre-second tetrahedral intermediate structure, 211
Citrate synthase, *see* Malate dehydrogenase–citrate synthase
Commitment, *see* Rate-limiting step
Conformational change
  ATP-driven molecular discrimination, *see also* ATP sulfurylase; DNA gyrase; Nicotinate phosphoribosyltransferase
    clock mechanism, 34–35
    conformational coupling, 28–29, 50–52
    conformer affinities for substrates and products, 31–34, 36
    examples of enzymes, 29, 49
    ion transporter energetics, 30–36
    leaks, origin, 52
    necessary imperfection, 49, 52, 70
    optimization of ATP use, 32–34
    RecA, 47–48
    thermodynamic modeling, 30–31, 35
  free energy profiling, 8–9, 11–12
  tryptophan synthase, hydrophobic tunnel substrate channeling, 126–127
CPS, *see* Carbamoyl phosphate synthase
Cyclosporin synthetase, substrate channeling, 143
Cytidine deaminase
  molecular surface point similarity measures in binding energy analysis, 405–407, 409
  neural network analysis of binding energies, 414–415, 424, 426

D

DHFR, *see* Dihydrofolate reductase
Dihydrofolate reductase, *see also* Thymidylate synthase–dihydrofolate reductase
  dihydrofolate pK_a determination at active site with difference Raman spectroscopy, 193–195
DNA, isotopic labeling for kinetic isotope effect studies, 316–318
DNA gyrase
  ATP coupling
    ATPase activation, 65–66
    efficiency, 66–67
    regulation in function, 67

stoichiometry, 66
DNA topoisomerase, classification, 63
  reaction overview, 63–65
  subunits, 65

E

EF-Tu, *see* Elongation factor-Tu
Electrostatic interaction, energy for catalysis, 21–22
Elongation factor-Tu
  GTP hydrolysis dynamics, *see* $G_{i\alpha1}$
  imperfect energetic coupling, *see* Kinetic proofreading
Enolase, energy for catalysis, 21–22
Equilibrium constant, derivation for enzymatic reactions, 3

F

FK506 synthetase, substrate channeling, 143–144
Fractionation factor, *see* Low-barrier hydrogen bond
Free energy
  activation, *see* Activation energy
  hydrogen bonds, linear free energy relationships
    Brønsted coefficients and slopes, 254–261, 269, 274–276
    intramolecular hydrogen bond strength, 261–263, 266–268
    substituted phenols, 254
  profiles
    activation energy, 4–5
    carbonic anhydrase, 13
    conformational change profiling, 8–9, 11–12
    enzymatic bimolecular reactions, 7–9
    enzymatic unimolecular reaction, 5–6
    nonenzymatic unimolecular reaction, 4
    orotidylate decarboxylase, 11
    partition ratios of intermediates, 9
    product concentration effects, 6, 11
    proline racemase, 11–13
    pyruvate phosphate dikinase, construction of profile calculations, 172–174

dissociation constant determinations, 164–165
equilibrium constant determinations, 167–168
kinetic model in simplification, 163
physiological conditions, modeling, 174, 176
rate constant determinations, 160, 162, 164–166
time courses for single turnover reactions/simulations, 168–172
rate-limiting step, 17, 19
software for construction, 24, 26–27
structure–function study applications, 163
substrate concentration effects, 6–7

G

$G_{i\alpha 1}$
ADP-ribosylation, transition state imbalance, 396
GTP hydrolysis
associative versus disociative mechanism, 72–73
magnesium ions in catalysis, 91
microscopic reversibility, 91–92
overview, 71–72
stages, 90
transition state, 72
X-ray crystallography with ligands
crystallization conditions, 74–75
data collection, 74, 76
data sets, 73–74
domains, 76
GDP complex, 86–88
GDP · P$_i$ ternary complex trapping, 85–86, 91
ground state enzyme–substrate complex, 78, 80–81, 90
hydrogen bonding interactions, 76–78
intermediates, forcing within crystals, 88–90
regulators of G-protein signaling, mechanism of action, 92
transition state as modeled by GDP · AlF$_4^-$ · Mg^{2+}, 81–85, 90–91

kinetic isotope effect analysis of transition state, 322
protein–protein interactions in signal transduction, 71
G-protein, see also $G_{i\alpha 1}$
kinetic isotope effect analysis of transition state, 320–322
regulators of G-protein signaling, mechanism of action, 92
timing of filament assembly reactions, 103
Ground state destabilization, enzyme catalysis, 21
GTP
hydrolysis by G-proteins, see $G_{i\alpha 1}$
imperfect energetic coupling, see ATP sulfurylase; Kinetic proofreading
tubulin polymerization role, see Tubulin polymerization

H

Haldane relationship, constraints on enzymatic reactions, 3, 15
Hine equation, rate enhancement calculation, 257, 274–276
Hydride transfer reaction
quantum mechanical tunneling complications in kinetic isotope effects, 324–325
Swain–Schadd relationship, 323–324
Hydrogen bond, see also Low-barrier hydrogen bond
acid–base catalysis, 24
Badger–Bauer relationships in Raman spectroscopy, 183–185
energy for catalysis, 22, 246, 248, 254–257, 261, 304–305
fluoride bonding with ethanol, energetics, 247
linear free energy relationships
Brønsted coefficients and slopes, 254–261, 269, 274–276
substituted phenols, 254
strength, 22, 219, 221
transition state binding, 304–305
Hypoxanthine–guanine phosphoribosyltransferase, low-barrier hydrogen bonds in catalysis, 304–305

I

Inosine, isotopic labeling for kinetic isotope effect studies, 308–309
Inosine–uridine nucleoside hydrolase
difference Raman spectroscopy, electrostatic interactions and inhibitor design, 200–201
neural network analysis of binding energies
accuracy, 419, 424
binding sites, 416
orientation of ligand binding, 417
substrate specificity, 416–417
training of network, 418
transition state structures and affinities, 418–419, 422–424
transition state structure, 199
Isotope effect, *see* Kinetic isotope effect

K

Ketosteroid isomerase, hydrogen bonds and energy for catalysis, 22–23
KIE, *see* Kinetic isotope effect
Kinetic isotope effect
Dadger's rule, relating bond order and bond strength, 383–384
contributions to isotope effects, 359–361
excited-state energy, calculation, 362
Heisenberg uncertainty principle, application to vibrational frequencies, 359
k_{cat}/K_M and irreversible step, 365
Marcus theory correlation of proton transfer reactions, 284–285, 292–293
moment of inertia, calculation, 361–362, 382
Pauling bond order, relating bond length and bond order, 363
rate-limiting step determination, 15–16, 19
ricin, irreversible step and isotope effect, 367–368
transition state structure analysis
S-adenosylmethionine synthetase, 343
AMP deaminase
catalytic mechanism, 333, 335
inhibitor design, 348–349, 394

isotope effects and transition state modeling, 335–336
molecular electrostatic potential surface of transition state, 347–348
binding isotope effects, 330–332
bond energy/bond order vibrational analysis of transition state structure
ab initio optimized structures, 394–395
AMP deaminase, 335
BEBOVIB inputs, 373–382
computational power, 356
nucleoside hydrolase, 337, 339–340
reliability of technique, 395–397
software, 371, 373
test structures, isotope effect calculations, 368, 392–393
transition state imbalance, 396–397
unified models of reaction types, 371, 395
Walden inversion motion, 379–381
commitment to catalyis
chemical versus nonchemical steps, 365–366
forward commitment, 325–327
large commitment factors, overcoming with transient kinetic analysis, 328–332
minimization for isotope effect studies, 365
reverse commitment, 327–328, 366
competitive radiolabeled technique, 306–307, 364–365
computational approaches to structure elucidation, overview, 333, 368
correction for isotopic depletion, 322–323
distance sensitivity of substituted isotope centers, 345–346
DNA and RNA processing enzymes, 320
error sources, 393–394
generalizations in interpretation, 305–306
hydride transfer
commitment factor measurement, 325
quantum mechanical tunneling complications, 324–325

Swain–Schadd relationship, 323–324
labeling of substrates
  nicotinamide adenine dinucleotide,
    313, 316, 356–357
  nucleosides and nucleotides, 308–
    310, 312
  oligonucleotides, 316–317
  purification, 317–318
magnitude of effects, 306
mass spectrometry accuracy, 307
molecular electrostatic potential surface of substrates and transition
  state, 346–348, 355
nucleoside hydrolase
  inhibitor design, 354–355, 394
  isotope effect measurement, experi-
    mental setup, 318, 320
  molecular electrostatic potential sur-
    face of transition state, 346
  transition state modeling, 336–340,
    342–343
out of plane C–H bending, 344
overview of transition state analysis
  and inhibitor design, 307
primary carbon-14 isotope effect, 344
protein-modifying enzymes, 320–322
purine nucleoside phosphorylase
  commitment factors, 341
  functions and inhibition rationale,
    340
  inhibitor design, 349–350, 353–354,
    394
  isotope effects for arsenolysis, 341
  transition state modeling, 341–
    342
quantum mechanical calculations,
  373
reaction coordinate motion at transition state, 344, 358, 378–382
α secondary isotope effect, 344
β secondary isotope effect, 345
structure interpolation
  Cartesian coordinates to transition
    state structures to BEBOVIB
    format using internal coordi-
    nates, 383
  internal coordinates, 384–385
  nicotinamide adenine dinucleotide
    hydrolysis application, 385–388,
    395–396

  overview, 368, 370–371, 382–383
  reaction space and interpolation phi-
    losophy, 383–384
  reference structures, 387–392
vibrational energy relationship to frequency, 359
vibrational modeling with balls and
  strings, 362–363
zero-point energy, calculation, 361
Kinetic proofreading
  imperfect energetic coupling, 68–69
  overview, 67–69
  stoichiometry of polypeptide elongation
    reaction, 69

L

Lactate dehydrogenase
  active site structure, 187–188
  difference Raman spectroscopy
    stereospecific hydride transfer, 188–
      189
    substrate binding energetics, 188
    transition state stabilization, 189–
      191
LBHB, see Low-barrier hydrogen bond
LDH, see Lactate dehydrogenase
Low-barrier hydrogen bond
  characteristics, 22, 219, 221
  energy for catalysis, 22–23, 246, 248,
    254–257, 261
  geometrical effects on energetics,
    270–271
  hypoxanthine–guanine phosphoribosyl-
    transferase, 304–305
  mutation analysis of enzymes, 273–274
  proton nuclear magnetic resonance
    bond lengths, 224–226
    criteria for identification, 222–223
    data acquisition, 225–229
    deshielding of resonance, 224
    fractionation factor
      bond length determination, 234–235
      deuterium exchange reaction,
        232–233
      factors less than unity, 233
      measurement, data acquisition and
        calculations, 235–237

quantum mechanical theory, 233–234

free energy of formation
mutation studies of enzymes, 243–244
pH titration in measurement, 244–245
overview of enzymes, 221–222, 230, 245
proton exchange rates
limiting mechanisms, 239
measurement, data acquisition and calculations, 240–243
mechanism representation in proteins, 238–239
protection factor, 237–239
resonance assignment
J coupling, 229
mutagenesis of enzyme, 229, 231
nuclear Overhauser effects, 232
temperature measurement and calibration, 227–228
water suppression, 228–229
serine protease mechanism, 23–24, 244
staphylococcal nuclease, 269
strength, 22, 219
triosephosphate isomerase
active site interactions with transition state, 272–273
catalytic problem description, 248–250
entropic barrier, lowering by enzyme, 272
strengthening of hydrogen bonds accompanying charge rearrangement
basis for effect, 251–254
catalytic contribution, quantitative estimation, 254–257, 261, 266–268
dielectric environment lowering by enzyme, 261, 264–266, 274
Hine equation and rate enhancement calculation, 257, 274–276
linear free energy analysis of intramolecular hydrogen bond strength, 261–263, 266–268
multiple hydrogen bond analysis, 269–270
overview, 250–251
Lysozyme, organic solvent transfer of crystals, 205

M

Malate dehydrogenase–citrate synthase, substrate channeling, 142–143
Marcus rate theory, see Proton transfer
MEP, see Molecular electrostatic potential
Microtubule, see Tubulin polymerization
Molecular electrostatic potential, surface of substrates and transition state, 346–348, 355
Molecular surface point similarity measures, binding energy analysis
accuracy, 409
adenosine deaminase, 404–405, 407, 409
AMP nucleosidase, 405, 407, 409
cytidine deaminase, 405–407, 409
indications for use, 425
overview, 401–404
quantum mechanical descriptions of molecules, 400–401
similarity as a function of α, 402–403

N

NAPRTase, see Nicotinate phosphoribosyltransferase
Neural network, binding energy calculation
advantages, 410, 425–426
architecture of network, 412–414
back propagation, 414
cytidine deaminase analysis, 414–415, 424, 426
hidden layer magnitudes and inhibitor designs, 423–424
inosine–uridine nucleoside hydrolase analysis
accuracy, 419, 424
binding sites, 416
orientation of ligand binding, 417
substrate specificity, 416–417
training of network, 418
transition state structures and affinities, 418–419, 422–424
inputs and mapping, 411–412
learning, 410, 414
output layer, 413
overview of biological applications, 410–411

pattern classification, 425
weights and biases, 413–414
Nicotinamide adenine dinucleotide, isotopic labeling for kinetic isotope effect studies, 313, 316, 318, 356–357
Nicotinamide guanine dinucleotide, isotopic labeling for kinetic isotope effect studies, 313
Nicotinate phosphoribosyltransferase
active site, 40–41
assay, 39–40
ATP dependence and independence, 37–39, 43
ATP-driven molecular discrimination energetics, 44–46
physiological significance, 46–47
RecA comparison, 47–48
cloning and overproduction of recombinant *Salmonella* enzyme, 39
conformational changes, 44
ligand affinities, 41
pH effects, 40–41
rapid quench studies, 42
reaction cycle, 36–37, 40–42
side reactions, 43
stoichiometry of reaction, 43–44
NMR, *see* Nuclear magnetic resonance
Nuclear magnetic resonance
low-barrier hydrogen bond studies by proton resonance
bond lengths, 224–226
criteria for identification, 222–223
data acquisition, 225–229
deshielding of resonance, 224
fractionation factor
bond length determination, 234–235
deuterium exchange reaction, 232–233
factors less than unity, 233
measurement, data acquisition and calculations, 235–237
quantum mechanical theory, 233–234
free energy of formation
mutation studies of enzymes, 243–244
pH titration in measurement, 244–245
overview of enzymes, 221–222, 230, 245
proton exchange rates

limiting mechanisms, 239
measurement, data acquisition and calculations, 240–243
mechanism representation in proteins, 238–239
protection factor, 237–239
resonance assignment
J coupling, 229
mutagenesis of enzyme, 229, 231
nuclear Overhauser effects, 232
temperature measurement and calibration, 227–228
water suppression, 228–229
tryptophan synthase, solid-state studies of substrate channeling
advantages over solution-state studies, 128
aminoacrylate intermediate, 128–132
serine binding site, 132
Nucleoside hydrolase, *see also* Inosine–uridine nucleoside hydrolase
inhibitor design, 354–355, 394
isotope effect measurement, experimental setup, 318, 320
molecular electrostatic potential surface of transition state, 346
transition state modeling, 336–340, 342–343

O

Oligonucleotide, isotopic labeling for kinetic isotope effect studies, 316–318
Orotidylate decarboxylase
free energy profile, 10–11
rate acceleration, 10, 358

P

PGM, *see* Phosphoglucomutase
Phosphoglucomutase
difference Raman spectroscopy
symmetric stretch frequency of substrate phosphate, 181, 198
vanadate spectroscopy, 198
steps in catalysis, 196, 198

5-Phosphoribosyl-1-pyrophosphate, isotopic labeling for kinetic isotope effect studies, 310

PNP, *see* Purine nucleoside phosphorylase

PPDK, *see* Pyruvate phosphate dikinase

Proline racemase, free energy profile, 11–13

Protection factor, *see* Low-barrier hydrogen bond

Proton transfer, *see also* Low-barrier hydrogen bond

  carbonic anhydrase

    His-64 in proton shuttling, 286–287

    Marcus parameters for proton transfer

      Brønsted plots, 291–292

      His-64 proton transfer, 291–293

      His-67 proton transfer, 293

      human isozyme III as model system, 287

      imidazole and lutidine buffers, proton transfer, 293–294

      intrinsic kinetic barrier, 295

      kinetic isotope effects, 292–293

      mutagenesis and disruption of pK_a values, 288–289

      pH titration, 290

      rate measurements, 289–290

      work function dominance, 290–291, 295–297

    rate-limiting steps, 285–286

    stages in catalysis, 286

  Marcus rate theory

    Brønsted relations, 280–282

    intrinsic barriers, 276–279

    inverted region, 280

    kinetic isotope effects, 284–285

    overview, 276–277

    variable transition state principle, 277–278

    work terms, 282–284

PRPP, *see* 5-Phosphoribosyl-1-pyrophosphate

Purine nucleoside phosphorylase

  commitment factors, 341

  functions and inhibition rationale, 340

  inhibitor design, 349–350, 353–354, 394

  isotope effects for arsenolysis, 341

  transition state modeling, 341–342

Pyruvate phosphate dikinase

  evolution and efficiency, 162

  free energy profile construction

    calculations, 172–174

    dissociation constant determinations, 164–165

    equilibrium constant determinations, 167–168

    kinetic model in simplification, 163

    physiological conditions, modeling, 174, 176

    rate constant determinations, 160, 162, 164–166

    time courses for single turnover reactions/simulations, 168–172

  kinetic mechanism, 151–152, 162–163

  rapid quench elucidation of chemical pathway

    data acquisition, 155

    instrumentation, 155

    proposed pathways, 151–152

    radiolabeled compound preparation, 154–155

    single turnover experiment design, 152–154

    time courses for single turnover reactions

      $[\beta\text{-}^{32}P]ATP$, phosphate and pyruvate, 156–157, 168

      $[\gamma\text{-}^{32}P]ATP$, divalent cation effects, 160

      $[\gamma\text{-}^{32}P]ATP$ with and without phosphate, 159–160

      $[\gamma\text{-}^{32}P]ATP$, phosphate and pyruvate, 157–159, 168

      magnesium versus cobalt-activated enzyme, 168

      phosphoenol pyruvate, 155–156

      phosphoenol pyruvate, AMP and pyrophosphate, 156, 168

      simulations, 169–172

Q

QSAR, *see* Quantitative structure activity relationship

Quantitative structure activity relationship

  limitations, 399

  overview of inhibitor design, 398

R

Raman spectroscopy
  difference spectroscopy
    advantages in protein analysis, 178–179
    dihydrofolate reductase, dihydrofolate
      pK_a determination at active site,
      193–195
    inosine–uridine nucleoside hydrolase
      electrostatic interactions and inhibi-
      tor design, 200–201
    transition state structure, 199
  instrumentation, 179, 181
  isotope editing, 178, 332
  lactate dehydrogenase analysis
    active site structure, 187–188
    stereospecific hydride transfer,
      188–189
    substrate binding energetics, 188
    transition state stabilization, 189–191
  ligand-bound proteins, 177–178, 181
  local oscillators, 183–185
  phosphoglucomutase
    symmetric stretch frequency of sub-
      strate phosphate, 181, 198
    vanadate spectroscopy, 198
  principle, 179
  sensitivity, 183
  serine protease, acyl intermediate anal-
    ysis, 192–193
  transition state analog binding,
    181–182
  interpretation tools
    Badger–Bauer relationships, 183–185
    entropy determination, 187
    ionization states, 185–186
    phosphate–vanadate spectroscopy, 186
  resolution, 176–177, 181
  spectral crowding in resonance spectros-
    copy, 177
Rapid quench
  advantages over steady-state measure-
    ments, 149–150
  dead time, 150
  free energy profile construction
    calculations, 172–174
    dissociation constant determinations,
      164–165
    equilibrium constant determinations,
      167–168

kinetic model in simplification, 163
physiological conditions, modeling,
  174, 176
rate constant determinations, 160, 162,
  164–166
time courses for single turnover reac-
  tions/simulations, 168–172
kinetic mechanism, 151–152, 162–163
nicotinate phosphoribosyltransferase studies, 42
pyruvate phosphate dikinase, elucidation
  of chemical pathway
  data acquisition, 155
  instrumentation, 155
  proposed pathways, 151–152
  radiolabeled compound preparation,
    154–155
  single turnover experiment design,
    152–154
  time courses for single turnover reac-
    tions
    [β-^{32}P]ATP, phosphate and pyruvate,
      156–157, 168
    [γ-^{32}P]ATP, divalent cation effects,
      160
    [γ-^{32}P]ATP with and without phos-
      phate, 159–160
    [γ-^{32}P]ATP, phosphate and pyruvate,
      157–159, 168
    magnesium versus cobalt-activated
      enzyme, 168
    phosphoenol pyruvate, 155–156
    phosphoenol pyruvate, AMP and py-
      rophosphate, 156, 168
  simulations, 169–172
substrate channeling analysis
  thymidylate synthase–dihydrofolate re-
    ductase, 135–137
  tryptophan synthase, 117–118
Rate acceleration
  bimolecular enzymatic reactions, 9–10
  orotidylate decarboxylase, 10, 358
  unimolecular enzymatic reactions, 6
Rate-limiting step
  commitments to catalysis, 15–16, 20
  energy barriers, 17
  free energy profiling, 17, 19
  isotope effects, 15–16, 19
  least conductive step, 17, 19
  stickiness of substrate dissociation, 20

RecA, ATP-driven molecular discrimination, 47–48
Ribonucleic acid, isotopic labeling for kinetic isotope effect studies, 316–318
Ribonucleotide triphosphate reductase, isotopic labeling of substrates for kinetic isotope effect studies, 309–310
Ricin, irreversible step and isotope effect, 367–368
RNA, see Ribonucleic acid

physiological relevance, 111–112
thymidylate synthase–dihydrofolate reductase, see Thymidylate synthase–dihydrofolate reductase, substrate channeling
transient time approximation, 113–114
tryptophan synthase, see Tryptophan synthase, substrate channeling
Subtilisin, organic solvent transfer of crystals, 205

S

Serine protease, see also Chymotrypsin
acyl intermediate analysis by difference Raman spectroscopy, 192–193
hydrogen bonds and energy for catalysis, 23–24
steps in catalysis, 191–192
Similarity measure, see Molecular surface point similarity measures
Stopped-flow
advantages over steady-state measurements, 149–150
dead time, 150
substrate channeling analysis
thymidylate synthase–dihydrofolate reductase, 137–139
tryptophan synthase, 117–118
Substrate channeling
carbamoyl phosphate synthase
catalytic activities and subunits, 140–142
kinetic studies, 142
putative channel, 141
rationale, 140
criteria for establishing
steady-state assays, 113–115
structural analysis, 116
transient kinetic assays, 115–116
cyclosporin synthetase, 143
diffusion comparison, 113
electrostatic channeling, 133–135
enzyme complexes, general features, 144–145
FK506 synthetase, 143–144
isotope dilution assays, 114–115
malate dehydrogenase–citrate synthase, 142–143

T

Temperature, effects on activation energy, 13–15
Thymidylate synthase–dihydrofolate reductase, substrate channeling
catalytic activities, 132–133
domains, 133
electrostatic channeling, 133–135
transient kinetic analysis
pulse chase, 139
rapid quench, 135–137
stopped-flow fluorescence, 137–139
TIM, see Triosephosphate isomerase
Transient time, approximation in substrate channeling, 113–114
Transition state
binding
affinity, 303–304, 357–358
energetics, 304–305, 358, 399–400
catalytic theory, 303, 399–400
difference Raman spectroscopy
inosine–uridine nucleoside hydrolase, transition state structure, 199
lactate dehydrogenase, transition state stabilization, 189–191
transition state analog binding analysis, 181–182
$G_{i\alpha1}$, GTP hydrolysis, 72, 81–85, 90–91
kinetic isotope effect analysis
S-adenosylmethionine synthetase, 343
AMP deaminase
catalytic mechanism, 333, 335
inhibitor design, 348–349, 394
isotope effects and transition state modeling, 335–336
molecular electrostatic potential surface of transition state, 347–348

binding isotope effects, 330–332
bond energy/bond order vibrational
    analysis of transition state
    structure
  ab initio optimized structures,
      394–395
  AMP deaminase, 335
  BEBOVIB inputs, 373–382
  computational power, 356
  nucleoside hydrolase, 337, 339–340
  reliability of technique, 395–397
  software, 371, 373
  test structures, isotope effect calcula-
      tions, 368, 392–393
  transition state imbalance, 396–397
  unified models of reaction types,
      371, 395
  Walden inversion motion, 379–381
commitment to catalyis
  chemical versus nonchemical steps,
      365–366
  forward commitment, 325–327
  large commitment factors, overcom-
      ing with transient kinetic analy-
      sis, 328–332
  minimization for isotope effect stud-
      ies, 365
  reverse commitment, 327–328, 366
competitive radiolabeled technique,
    306–307, 364–365
computational approaches to structure
    elucidation, overview, 333, 368
correction for isotopic depletion,
    322–323
distance sensitivity of substituted iso-
    tope centers, 345–346
DNA and RNA processing enzymes,
    320
error sources, 393–394
generalizations in interpretation,
    305–306
hydride transfer
  commitment factor measurement,
      325
  quantum mechanical tunneling com-
      plications, 324–325
  Swain–Schadd relationship, 323–
      324
labeling of substrates

nicotinamide adenine dinucleotide,
    313, 316, 356–357
nucleosides and nucleotides, 308–
    310, 312
oligonucleotides, 316–317
purification, 317–318
magnitude of effects, 306
mass spectrometry accuracy, 307
molecular electrostatic potential sur-
    face of substrates and transition
    state, 346–348, 355
nucleoside hydrolase
  inhibitor design, 354–355, 394
  isotope effect measurement, experi-
      mental setup, 318, 320
  molecular electrostatic potential sur-
      face of transition state, 346
  transition state modeling, 336–340,
      342–343
out-of-plane C–H bending, 344
overview of transition state analysis
    and inhibitor design, 307
primary carbon-14 isotope effect, 344
protein-modifying enzymes, 320–322
purine nucleoside phosphorylase
  commitment factors, 341
  functions and inhibition rationale,
      340
  inhibitor design, 349–350, 353–354,
      394
  isotope effects for arsenolysis, 341
  transition state modeling, 341–342
quantum mechanical calculations, 373
reaction coordinate motion at transi-
    tion state, 344, 358, 378–382
α secondary isotope effect, 344
β secondary isotope effect, 345
structure interpolation
  Cartesian coordinates to transition
      state structures to BEBOVIB
      format using internal coordi-
      nates, 383
  internal coordinates, 384–385
  nicotinamide adenine dinucleotide
      hydrolysis application, 385–388,
      395–396
  overview, 368, 370–371, 382–383
  reaction space and interpolation phi-
      losophy, 383–384

reference structures, 387–392
lactate dehydrogenase, transition state stabilization, 189–191
lifetime, 302–303
Marcus rate theory, variable transition state principle, 277–278
molecular surface point, similarity measures in binding energy modeling
accuracy, 409
adenosine deaminase, 404–405, 407, 409
AMP nucleosidase, 405, 407, 409
cytidine deaminase, 405–407, 409
indications for use, 425
overview, 401–404
quantum mechanical descriptions of molecules, 400–401
similarity as a function of α, 402–403
neural networks in binding energy calculation
advantages, 410, 425–426
architecture of network, 412–414
back propagation, 414
cytidine deaminase analysis, 414–415, 424, 426
hidden layer magnitudes and inhibitor designs, 423–424
inosine–uridine nucleoside hydrolase analysis
accuracy, 419, 424
binding sites, 416
orientation of ligand binding, 417
substrate specificity, 416–417
training of network, 418
transition state structures and affinities, 418–419, 422–424
inputs and mapping, 411–412
learning, 410, 414
output layer, 413
overview of biological applications, 410–411
pattern classification, 425
weights and biases, 413–414
triosephosphate isomerase, active site interactions with transition state, 272–273
Triosephosphate isomerase
active site interactions with transition state, 272–273

catalytic problem description, 248–250
entropic barrier, lowering by enzyme, 272
hydrogen bonds, strengthening accompanying charge rearrangement
basis for effect, 251–254
catalytic contribution, quantitative estimation, 254–257, 261, 266–268
dielectric environment lowering by enzyme, 261, 264–266, 274
Hine equation and rate enhancement calculation, 257, 274–276
linear free energy analysis of intramolecular hydrogen bond strength, 261–263, 266–268
multiple hydrogen bond analysis, 269–270
overview, 250–251
Tryptophan synthase, substrate channeling
alternative substrate studies, 121
aminoacrylate and α reaction activation, 121–122, 128
hydrophobic tunnel
conformational changes, 126–127
crystal structure, 116, 126–127
mutation studies, 122–124
mutation studies
β Cys-170, 122–124
Glu-109, 122
Lys-87, 122
rapid quench analysis, 117–118
solid-state nuclear magnetic resonance studies
advantages over solution-state studies, 128
aminoacrylate intermediate, 128–132
serine binding site, 132
stopped-flow analysis, 117–118
subunits
communication modeling
tests of model, 121–124
transient kinetic experiments, 118–121
reactions catalyzed, 116
X-ray crystallography
available structures, 124–125
flexible loop region of α subunit, 125–126
hydrophobic tunnel, 116, 126–127
Tubulin polymerization

minimal mechanism, 93
self-assembly models
ATP role
bidirectional polymerization with hydrolysis at time of monomer addition, 99–101
delayed hydrolysis after monomer addition, 101–102
difficulty of study, 110–111
overview, 94
timing control of assembly and disassembly, 102–103, 110
tubulin · GTP cap, 105–106
unidirectional polymerization with hydrolysis at time of monomer addition, 97–99
condensation equilibrium model, 94–96
elongation without nucleotide hydrolysis, 96–97
exchange-limited polymerization zone model
depolymerization zone, 109
nucleation, 108–109
oligoproline sequences, 107
rationale for model, 106–107
rates of elongation, 106–107
Turnover number, upper limit, 3

X

X-ray crystallography
$G_{i\alpha 1}$ with ligands, GTP hydrolysis mechanism
crystallization conditions, 74–75
data collection, 74, 76
data sets, 73–74
domains, 76
GDP complex, 86–88
GDP · P_i ternary complex trapping, 85–86, 91
ground state enzyme–substrate complex, 78, 80–81, 90
hydrogen bonding interactions, 76–78
intermediates, forcing within crystals, 88–90
regulators of G-protein signaling, mechanism of action, 92
transition state as modeled by GDP · $AlF_4^- \cdot Mg^{2+}$, 81–85, 90–91
organic solvent trapping of intermediates
chymotrypsin, see Chymotrypsin
cross-linking of crystals, 204
lysozyme, 205
side chain rearrangements, 204
stability of crystals, 205–206
substrate addition, 206–207
subtilisin, 205
transfer technique, 205
water structure changes on protein surface, 204
substrate–product equilibrium in intermediate trapping, 202–203
tryptophan synthase, substrate channeling studies
available structures, 124–125
flexible loop region of α subunit, 125–126
hydrophobic tunnel, 116, 126–127
ultrafast data collection of enzyme intermediates, 201

ISBN 0-12-182209-5

9 780121 822095   90038

UCSF LIBRARY MATERIALS MUST BE RETURNED TO:

THE UCSF LIBRARY

530 Parnassus Ave.
University of California, San Francisco
This book is due on the last date stamped below.
Patrons with overdue items are subject to penalties.
Please refer to the Borrower's Policy for details.
Items may be renewed within five days <u>prior to</u> the due date.
For telephone renewals -- call (415) 476-2335
Self renewals -- at any UCSF Library Catalog terminal in the Library, or
renew by accessing the UCSF Library Catalog via the Library's web site:
http://www.library.ucsf.edu
All items are subject to recall after 7 days.

28 DAY LOAN

28 DAY

JUN 2 8 2000
RETURNED

JUN - 8 2000

28 DAY

NOV 1 6 2000

28 DAY

DEC PAY 2000

JAN 1 2 2001

28 DAY

MAR 0 9 2001

28 DAY

FEB 0 9 2001

RETURNED

MAR 1 2 2001

28 DAY

MAY 1 4 2001

RETURNED

MAY 0 1 2001